45th Northeast Bioengineering Conference (NEBEC 2019)

21st Century Opportunities in Biomedical Engineering: Impacting Healthcare from the Clinic to Industry

New Brunswick, New Jersey, USA
20 – 22 March 2019

ISBN: 978-1-7138-1407-8

Printed from e-media with permission by:

Curran Associates, Inc.
57 Morehouse Lane
Red Hook, NY 12571

Some format issues inherent in the e-media version may also appear in this print version.

Printed with permission by Curran Associates, Inc. (2020)

Northeast Bioengineering Conference 2019
c/o Ronke Olabisi
3408 Eng Hall / UC Irvine
Irvine, CA, 92697

ronke.olabisi@uci.edu

Additional copies of this publication are available from:

Curran Associates, Inc.
57 Morehouse Lane
Red Hook, NY 12571 USA
Phone: 845-758-0400
Fax: 845-758-2633
Email: curran@proceedings.com
Web: www.proceedings.com

45TH ANNUAL **NORTHEAST BIOENGINEERING CONFERENCE**

21st Century Opportunities in Biomedical Engineering:

Impacting Healthcare
from the Clinic to Industry

March 20–22, 2019

Hosted by
Rutgers University
Department of Biomedical Engineering
at the
Livingston Student Center

Table of Contents

NEBEC 2019 **Organizing Committees** **4**

2019 NEBEC **Sponsors** **5**

2019 NEBEC **Keynote Speakers** **6**

KiBum Lee 6

Mark Pierce 6

Shridar Ganesan, MD, PhD 6

2019 NEBEC SPEAKERS **7**

Industrial Speakers 7

Other Speakers 7

Wednesday, March 20, 2019 . . . . **8**

Schedule 8

Abstracts 10

Posters 12

Thursday, March 21, 2019 . . . . **71**

Schedule 71

Abstracts 74

Posters 77

Friday, March 22, 2019 . . . . **147**

Schedule 147

Abstracts 149

Posters 150

NEBEC 2019 **Organizing Committees**

Conference Co-Chairs (Rutgers)
Joseph Freeman | Ronke Olabisi

Proposal PI (Rutgers)
Jeffrey Zahn

Industry and Sponsors
Kristen Labazzo | Biju Parekkadan | Susan Englehart

Judging Committee
Joseph Freeman | Jeffrey Zahn

Program Session Planning
Jay Sy | Adam Gormley | Ilker Hachihaliloglu | Noshir Langrana

Career Exploration
Martin Yarmush | Janet Alder

Publicity
Joseph Freeman | Lawrence Stromberg | Jay Sy

Finance
Robin Yarborough

Local Arrangements
Joseph Freeman | Laleh Najafizadeh

Social Program
Fernando Rebolledo Uscanga | Bioengineering Student Society (BESS) President

Abstract Review Coordination
Ronke Olabisi | Maribel Vasquez | Mark Pierce

Website Coordinator
Jay Sy

Smartphone App Developer
Ronke Olabisi

2019 NEBEC **Sponsors**

Silver

DELL

Bronze

HINJ
HealthCare Institute of New Jersey

CELLINK

NSF

Exhibit

SPRINGER NATURE

Recruitment

BROWN

RUTGERS

Drexel
UNIVERSITY

Sponsored Keynote

MYOSTRENS
TECHNOLOGY INC.

2019 NEBEC **Keynote Speakers**

KiBum Lee

Brain Health Institute

Dr. Lee's research develops and integrates nanotechnology and chemical biology to modulate signaling pathways in cancer and stem cells. His research focuses on identifying the various microenvironmental cues (e.g. soluble signals, cell-cell interactions, and insoluble/physical signals) affecting stem cell and cancer cell fate and thereafter utilizing these cues for the neuro-differentiation of stem cells and apoptosis of brain tumor cells. In addition, he develops novel nanomaterials for applications such as cancer therapy, molecular imaging and bio-sensing.

Mark Pierce

Global Health Institute

Dr. Pierce's broad interests lie in the development and application of engineering solutions to global health challenges. A particular focus is in the area of biomedical imaging technologies, ranging from new techniques in microscopy to point-of-care screening tools for identifying early-stage disease.

Shridar Ganesan, MD, PhD

Cancer Institute of New Jersey

Dr. Ganesan's research revolves around breast cancer biology and DNA repair. He is exploring how DNA repair defects in cancers can be exploited to develop novel effective treatments. He is also applying next-generation sequencing technology to identify specific genomic changes in cancers that can be therapeutically targeted. As a physician/scientist he runs a basic research laboratory and sees patients. Dr. Ganesan works collaboratively with experts across multi-disciplines and puts theory into practice as he strives to develop the next generation of targeted treatments for breast cancer.

2019 NEBEC **SPEAKERS**

Industrial Speakers

Steven Leist, PhD
SunP Biotech

Evi Chnari, PhD
Musculoskeletal Tissue Foundation (MTF) Biologics

C. Alex DePaula, PhD
Johnson & Johnson, DePuy Synthes

Other Speakers

Jinho Ki, PhD
Stevens Institute of Technology
Section: Regenerative Medicine I

Mehmet Kurt, PhD
Stevens Institute of Technology
Section: Brain Health/NeuroEngineering

Marcin Iwanicki, PhD
Stevens Institute of Technology
Section: Cancer Biology

Peter Galie, PhD
Rowan University
Section: Regenerative Medicine II

Carlotta Mummolo, PhD
NJIT
Section: Nutrition/Health or Device Development

Vincent Beachley, PhD
Rowan University
Section: Regenerative Medicine II

Sebastian Vega, PhD
Rowan University
Section: Biomaterials I

Jiwook Shim, PhD
Rowan University
Section: Molecular and Cellular Bioengineering

Raviraj Nataraj, PhD
Stevens Institute of Technology
Section: Rehabilitation

Alex Zhou
NJIT
Section: Rehabilitation

Wednesday, March 20, 2019

Schedule

9:00 am–5:00 pm	Registration
9:00 am–10:00 am	Breakfast
10:00 am–10:40 am	Opening Remarks & Keynote Speaker from Global Health Institute —*LIVINGSTON HALL*
10:40 am–12:00 pm	Oral Session on Global Health—*LIVINGSTON HALL*
12:00 pm–2:00 pm	Lunch/Vendors and Sponsors/Poster Session
2:00 pm–3:20 pm	**Parallel Platform Presentations: Session I** (12 minute talks with 3 minutes for questions)

	Session 1A Biomaterials I —*LIVINGSTON HALL*	**Session 1B Molecular and Cellular Bioengineering** —*201*	**Session 1C Medical Instrumentation / Global Health** —*202*
2:00–2:20 pm	1A-1:Sebastian Vega Rowan University	1B-1:Jiwook (Stephen) Shim Rowan University	1C-1:Development of Mobile Phone Based Transcutaneous Bilirubinometer
2:20–2:35 pm	1A-2:Controlled Cell Migration within Cell Selective Hydrogels	1B-2:Shear-Induced Extensional Response Behaviors of Tethered von Willebrand Factor	1C-2:Liver Classification Based on SVM for Ultrasound Images
2:35–2:50 pm	1A-3:A Method of Making Biocompatible Antio3idant Ceria Nanoparticles Loaded in Multi-layer Polymer Films on Device	1B-3:Calibration and Measurement of FRET Efficiency in a Vinculin Tension Probe	1C-3:Rapid, Label-Free Genetic Detection of Enteropathogens in Stool without Genetic Isolation or Amplification
2:50–3:05 pm	1A-4:Free Radical Responsive Biomaterials to Target Therapeutic Immobilization	1B-4:Endothelial cell organelle positioning exhibits an intrinsic left-right bias	1C-4:A Single Cell Pathogen Identification and Antimicrobial Susceptibility Testing System for Rapid Diagnosis of Infectious Diseases
3:05–3:20 pm	1A-5:Engineering the Osteochondral Interface: 3D-Printed Scaffolds with Tunable Peptide Organization	1B-5:In vitro human lymphocytes activation and inter-cellular activities from the co-culture of dermal fibroblasts	1C-5:Ultraviolet Germicidal Irradiation of Clinical Equipment for the Prevention of Hospital Acquired Infections

3:20 pm–3:40 pm Break/Vendors and Sponsors

3:40 pm–5:00 pm **Parallel Platform Presentations: Session II**
(12 minute talks with 3 minutes for questions)

	Session 2A Biomaterials II *—LIVINGSTON HALL*	**Session 2B Tissue Engineering / Regenerative Medicine I**—*201*	**Session 2C Biomedical Imaging I** *—202*
3:40–4:00 pm	2A-1:Bioprinting of Complex 3D Vascular Networks within Cell-Laden Hydrogels	2B-1:Jinho Ki Stevens Institute of Technology	2C-1:Ilker Hacihaliloglu Rutgers University
4:00–4:15 pm	2A-2:3D Printing with Peptide Polymer Conjugates to Control Spatial Peptide Concentration	2B-2:Selective Replacement of Damaged Airway Epithelium for Functional Recovery of Donor Lungs	2C-2:A Phantom Based Study of Short-Wave Infrared Emitting Nanocomposites for Fluorescence Guided Surgery
4:15–4:30 pm	2A-3:Design of Porous PEEK Topologies Using Fused Filament Fabrication	2B-3:Self-assembling peptide induces axon infiltration following spinal cord injury	2C-3:Evaluation of Performance of Fitness Functions in Adaptive Cuckoo Search for Differentiation of Indirect Immunofluorescence Images
4:30–4:45 pm	2A-4:Synthesis and Characterization of Self-Assembling Hyaluronic Acid-Isoleucine Biomimetic Hydrogels	2B-4:Effects of Combinational Treatment on Restoring Bone Morphology after Contused Spinal Cord Injury	2C-4:Disrupted Brain Network Topology in Motoric Cognitive Risk Syndrome: A Resting-state Functional Magnetic Resonance Imaging Study
4:45–5:00 pm	2A-5:A Polymerization Friendly Dibenzoazacyclooctyne Monomer for Two-Step Bioconjugation	2B-5:Dynamic Hydrogel Platforms for in vitro Cardiac Models	2C-5:Reduced Occipital Hemodynamic Response During Visual Attention Processing in Young Adults with Attention-Deficit Hyperactivity Disorder–a Functional Near-Infrared Spectroscopy Study

Abstracts

1A-2 **Controlled Cell Migration within Cell Selective Hydrogels** 14
E. Pashuck, Stephanie Fung, Neil Joshi, Joachim Kohn
New Jersey Center for Biomaterials at Rutgers University

1A-3 **A Method of Making Biocompatible Antioxidant Ceria Nanoparticles Loaded in Multi-layer Polymer Films on Device** 15
Yue Hu, Vivian Ling
Brown University

1A-4 **Free Radical Responsive Biomaterials to Target Therapeutic Immobilization** 16
Emily DiMartini, Christopher Lowe, Keana Mirmajlesi, Adam Gormley, David Shreiber
Rutgers, The State University of New Jersey

1A-5 **Engineering the Osteochondral Interface: 3D-Printed Scaffolds with Tunable Peptide Organization** 17
Paula Camacho [1], Hafiz Busari [1], Peter Schwarzenberg [1], Anne Behre [1], William De Long [2], Hannah Dailey [1], Lesley Chow [1]
[1] Lehigh University, [2] St. Luke's University Health Network

1B-2 **Shear-Induced Extensional Response Behaviors of Tethered von Willebrand Factor** 18
Yi Wang, Michael Morabito, Ciara Kavanagh, Avani Pisapati, Xiaohui F. Zhang, Edmund Webb III, Alparslan Oztekin, Xuanhong Cheng
Lehigh University

1B-3 **Calibration and Measurement of FRET Efficiency in a Vinculin Tension Probe** 19
Jeremy Stein [1], Daniel Sumetsky [1], Nicolas Emanuely [2], Mihir Patel [1], Marina Cararo Lopes [1], Bonnie Firestein [1], Nada Boustany [1]
[1] Rutgers, The State University of New Jersey, [2] Institut d'Optique Graduate School

1B-4 **Endothelial cell organelle positioning exhibits an intrinsic left-right bias** 20
Jie Fan, Haokang Zhang, Tasnif Rahman, Diana Stanton, Leo Wan
Rensselaer Polytechnic Institute

1B-5 **In vitro human lymphocytes activation and inter-cellular activities from the co-culture of dermal fibroblasts** 21
Yu Ting Lin , Kao Kun Lin [1], ZhenWang Jie [1,2]
[1] Hong Kong Life Sciences and Technologies Group, [2] Rensselaer Polytechnic Institute

1C-1 **Development of Mobile Phone Based Transcutaneous Bilirubinometer** 22
Brandon Harrison
Temple University

1C-2 **Liver Classification Based on SVM for Ultrasound Images** 23
Hui Che [1], John Nosher [2], Ilker Hacihaliloglu [1,2]
[1] Rutgers University, [2] Rutgers University, Rutgers Robert Wood Johnson Medical School

1C-3 **Rapid, Label-Free Genetic Detection of Enteropathogens in Stool without Genetic Isolation or Amplification** 24
Song Han [1], Mehmet Soylu [1], Ceyhun Kirimli [1], Wei Wu [2], Bhaswati Sen [3], Suresh Joshi [3], Christopher Emery [4], Giang Au [1], Xiaomin Niu [1], Richard Hamilton [5], Kyle Krevolin [6], Wei-Heng Shih [2], Wan Shih [1]
[1] School of Biomedical Engineering, Science, and Health Systems, Drexel University, [2] Department of Materials Science and Engineering, Drexel University, [3] Department of Microbiology and Immunology, Drexel University, [4] Department of Pathology, Drexel University, [5] Department of Emergency Medicine, Drexel University, [6] Microbiology & SIVM Laboratories, Hahnemann University Hospital

1C-4 **A Single Cell Pathogen Identification and Antimicrobial Susceptibility Testing System for Rapid Diagnosis of Infectious Diseases** . . . 25
Hui Li, Jian Gao, Pak Wong
The Pennsylvania State University

1C-5 **Ultraviolet Germicidal Irradiation of Clinical Equipment for the Prevention of Hospital Acquired Infections** . . . 26
Kendra Michaud, Andrew Strong, Sam Charpentier
University of Rhode Island

2A-1 **Bioprinting of Complex 3D Vascular Networks within Cell-Laden Hydrogels** . . . 28
Shen Ji, Emily Almeida, Murat Guvendiren
New Jersey Institute of Technology

2A-2 **3D Printing with Peptide Polymer Conjugates to Control Spatial Peptide Concentration** . . . 29
Hafiz Busari, Kelly Seims, Paula Camacho, Lesley Chow
Lehigh University

2A-3 **Design of Porous PEEK Topologies Using Fused Filament Fabrication** . . . 30
William Hartley, Kenny Cho, Anthony Law, Wing Ni Lee, Colin Burlingham
Drexel University

2A-4 **Synthesis and Characterization of Self-Assembling Hyaluronic Acid-Isoleucine Biomimetic Hydrogels** . . . 31
Sydney Yang, Hafiz Busari, Lesley Chow
Lehigh University

2A-5 **A Polymerization Friendly Dibenzoazacyclooctyne Monomer for Two-Step Bioconjugation** . . . 32
Shashank kosuri
Rutgers University

2B-2 **Selective Replacement of Damaged Airway Epithelium for Functional Recovery of Donor Lungs** 33
Jinho Kim
Stevens Institute of Technology

2B-3 **Self-assembling peptide induces axon infiltration following spinal cord injury** . . . 34
Kiet Tran
Rowan University

2B-4 **Effects of Combinational Treatment on Restoring Bone Morphology after Contused Spinal Cord Injury** . . . 35
Michael Feeley, Shawn McGinley, Anita Singh
Widener University

2B-5 **Dynamic Hydrogel Platforms for in vitro Cardiac Models** . . . 36
Andrew House, Murat Guvendiren
New Jersey Institute of Technology

2C-2 **A Phantom Based Study of Short-Wave Infrared Emitting Nanocomposites for Fluorescence Guided Surgery** . . . 37
Marissa Berger, Jay Shah, Vidya Ganapathy, Prabhas Moghe, Mark Pierce
Rutgers, The State University of New Jersey

2C-3 **Evaluation of Performance of Fitness Functions in Adaptive Cuckoo Search for Differentiation of Indirect Immunofluorescence Images** . . . 38
Kanchana Devanathan, Nagarajan Ganapathy, Ramakrishnan Swaminathan
Indian Institute of Technology Madras

2C-4 **Disrupted Brain Network Topology in Motoric Cognitive Risk Syndrome: A Resting-state Functional Magnetic Resonance Imaging Study** . . . 39
Yuyang Luo, Xiaobo Li
New Jersey Institute of Technology

2C-5 **Reduced Occipital Hemodynamic Response During Visual Attention Processing in Young Adults with Attention-Deficit Hyperactivity Disorder – a Functional Near-Infrared Spectroscopy Study** 40
Ziyan Wu, Ruichao Wang, Xiaobo Li
New Jersey Institute of Technology

Posters

P1-1 **Clearance of drugs from the brain: the potential effect of Acetazolamide and Verapamil as CSF modulating drugs.** . . . 41
Mariagemiliana Dessi, Sonia Yevick, Caroline Wood, Jay Sy
Rutgers, The State University of New Jersey

P1-2 **Optimizing NIR Quantum Dot Molecular Probe for Assessing Breast Cancer Tumor Margins** . . 42
Alexandra Jednorski [1], Wan Shih [1], Wei-Heng Shih [2]
[1] Drexel University - School of Biomedical Engineering, Science and Health Systems, [2] Drexel University - Department of Materials Science and Engineering

P1-3 **Analysis of Post-Injury Morphological Changes in Neuronal Mitochondria using Label-Free Imaging** . . . 44
Corey Ditmars, Nada Boustany, Rene Schloss, Erin Kelly, Mohammad Naser
Rutgers University

P1-4 **Study of Stability and Functionalization of Aqueous CdPbS Near-infrared Quantum Dots** . . . 45
Ozgun Acar [1], Wan Shih [2], Wei-Heng Shih [1]
[1] Drexel University - Materials Science and Engineering, [2] Drexel University - School of Biomedical Engineering, Science and Health Systems

P1-7 **Cortical Activation during Breath Hold using Functional Near-Infrared Spectroscopy** 46
Keerthana Deepti Karunakaran [1,2], Bharat Biswal [2]
[1] New Jersey Institute of Technology, [2] Rutgers Graduate School of Biomedical Sciences

P1-8 **Gadolinium Oxide Nanoplates as MRI Contrast Agents** . . . 48
Jake Villanova, Edward Esposito, Vicki Colvin
Brown University

P1-9 **Multifractal analysis of uterine EMG signals to differentiate Term and Preterm conditions using Hurst exponent** . . . 49
Punitha Namadurai, Ramakrishnan Swaminathan
Indian Institute of Technology Madras

P1-10 **Polymerized Hemoglobin for Enhanced Oxygen Transport in a Mini-bioartificial liver** 50
Nuozhou Chen [1], Donald Belcher [2], Josh Leipheimer [1], Andre Palmer [2], Francois Berthiaume [1]
[1] Rutgers, The State University of New Jersey, [2] The Ohio State University

P1-11 **Simplified Implementation of Optimized Whitening of the Electromyogram Signal** 51
He Wang [1], Kiriaki Rajotte [1], Haopeng Wang [1], Chenyun Dai [2], Ziling Zhu [1], Moinuddin Bhuiyan [1], Edward Clancy [1]
[1] Worcester Polytechnic Institute, [2] Fudan University

P1-12 **Compliant Multilayer Intracortical Microelectrodes to Improve Recording Potential** 52
Xin Liu
Rutgers University

P1-13 **Ultraviolet Germicidal Irradiation of Clinical Equipment for the Prevention of Hospital Acquired Infections** . . . 54
Sam Charpentier, Andrew Strong, Kendra Michaud
University of Rhode Island

P1-14 **A Versatile Neural Modulation Device for Vestibular Evoked Myogenic Potentials (VEMP)** . . . 57
Courtney Semkewyc, Christopher Kidchob[1], William Craelius, PhD[1], Ashley Wackym, MD[2]
[1] Rutgers University New Brunswick, [2] Robert Wood Johnson University Hospital

P1-15 **In Miniature Non-Continuous Oxygen Concentrator** . . . 58
Devin Belmar, Caitlin Cambio, Samantha Ferreras
Stevens Institute of Technology

P1-18 **Arduino-Based Multi-Point Smart Pressure Sensor for Improved Seating Posture** . . . 59
Andrew Hwang
Newark Academy

P1-19 **Response of endogenous neural stem/progenitor cells to traumatic brain injury** . . . 60
Jeremy Anderson, Misaal Patel, Rebecca Risman, Li Cai
Rutgers, The State University of New Jersey

P1-20 **The Effect of Electroporation Buffer Composition on Cell Viability and Electro-transfection Efficiency** . . . 61
Joseph Sherba, Stephen Hogquist, Jerry Shan, Hao Lin, David Shreiber, Jeffrey Zahn
Rutgers, The State University of New Jersey

P1-21 **Nkx6.1 Gene Therapy Induces Neurogenesis after Spinal Cord Injury** . . . 62
Misaal Patel, Jeremy Anderson, Shunyao Lei, Rebecca Risman, Li Cai
Rutgers University

P1-22 **Effect of PKC Substrate-Fascin on Cell Chirality** . . . 63
Haokang Zhang, Jie Fan, Leo Wan
Rensselaer Polytechnic Institute

P1-23 **Real-Time Transfer of Lentiviral Particles by Producer Cells using an Engineering Coculture System** . . . 64
Lauren Timmins
Rutgers University

P1-24 **Nanoporous Alumina Membrane Electroporation for Optimized Transfection and DNA Delivery to Cells** . . . 66
Stephen Hogquist
Rutgers University

P1-25 **Cerebral spinal fluid modulators for enhanced drug delivery in the brain** . . . 67
Caroline Wood, Mariagemiliana Dessi, Jay Sy
Rutgers University

P1-26 **Using a Novel Protein Energetics Model and Microfluidics to Improve the State of Cancer Screening** . . . 68
Zachary Fritz, Rene Schloss, Anil Shrirao, Lawrence Williams, Martin Yarmush
Rutgers University

P1-27 **Minimizing Infection and Revision Surgeries through Nitric Oxide Releasing Total Knee Replacement Prosthesis** . . . 69
Ashley Widing, Kagya Amoako
University of New Haven

P1-28 **Evaluation of Alginate-Encapsulated Mesenchymal Stromal Cells for Osteoarthritis Treatment** 70
Ileana Marrero - Berrios, Sarah Salter, Rishabh Hirday, Rene Schloss, Martin Yarmush
Rutgers, the State University of New Jersey

Controlled Cell Migration within Cell Selective Hydrogels

E. Thomas Pashuck, Stephanie L. Fung, Neil S. Joshi, Joachim Kohn
New Jersey Center for Biomaterials, Rutgers University

Statement of Purpose: All tissues in the body consist of multiple cells types that work together to perform biological functions. However, the majority of matrices used in biomaterials are optimized for only a single cell type. To more fully recapitulate the *in vivo* environment, biomaterials will need to be developed that can respond differently to the various cell types that exist within a tissue. This is especially true for injury sites, where the local inflammatory response and immune cell infiltration plays an important role in regenerative processes. There are many differences cell types, but enzymes are an especially useful tool for material design since they perform a chemical function. Proteases have been used extensively in biomaterials, but almost exclusively for degrading matrices for cell migration. In this work, we utilize proteases to create reactive groups which will form cross-links after cleavage from specific cell-secreted proteases.1 We used peptide sequences which generate an N-terminal cysteine after enzymatic cleavage (Figure 1A). These cysteines, but not cysteines within the peptide sequences, will react with a cyanobenzothiazole (CBT) moiety to form a new crosslink (Figure 1B). By creating materials in which crosslinks form from proteases secreted by specific cells we can use proteolytic activity to prevent cells from migrating in these hydrogels.

Methods: Hydrogels were fabricated using copper free click chemistry between dibenzocyclooctyne (DBCO), functionalized 8-arm poly(ethylene glycol) (PEG), and azido-peptides. All peptides were synthesized using solid phase peptide synthesis and purified with high pressure liquid chromatography (HPLC). Migration was quantified by staining cells with a live cell tracker dye, seeded into the different hydrogels and cultured for 24 hours before live cell confocal microscopy for 18 hours and quantified using the TrackMate plugin for ImageJ.

Results: In this work, we utilized a ligation reaction between N-terminal cysteine and a cyanobenzothiazole (CBT) moiety to form new crosslinks in hydrogels. We first validated the ligation reaction by mixing our thioester peptide with free cysteine in PBS, monitored the reaction by HPLC and found that the reaction went nearly to completion in a few minutes. We then used a trypsin substrate peptide which generates an N-terminal cysteine after trypsin cleavage and saw that mixing the trypsin peptide with the thioester did not induce a ligation reaction, but the further addition of trypsin did, indicating that the protease induced the ligation. We then formed a hydrogel using 8-arm PEG in which half of the PEG arms were crosslinked with peptides to induce gelation, and the other half were functionalized with either the thioester peptide or the protease substrate peptide. We found that THP-1 cells, a macrophage-like cell line, had decreased migration within hydrogels which had a control peptide versus those with an MMP substrate peptide. (Figure 1C).

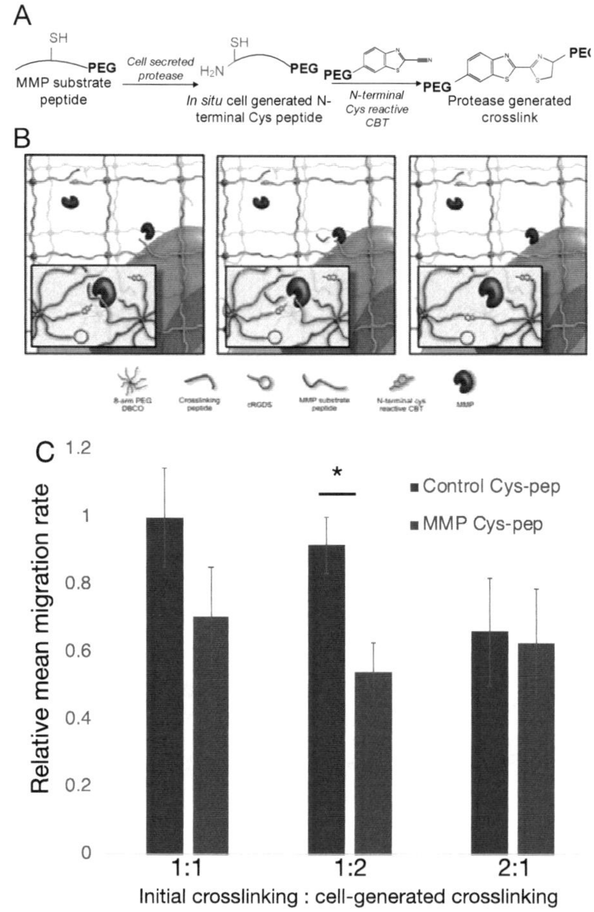

Figure 1. A) Protease induced crosslinking. MMP substrate peptides with cysteines at specific locations are cleave by proteases and generate N-terminal cysteines, these undergo a ligation reaction and form new crosslinks. B) Incorporation of protease induced crosslinking into a hydrogel. C) MMP-induced crosslinking slows migration of macrophages.

Conclusions: Utilizing ligation reactions which are specific for N-terminal amino acids is a promising method to generate new crosslinks in hydrogel matrices upon protease mediated peptide cleavage. These peptide sequences can be synthesized to have specificity for particular enzymes, and hydrogels can be fabricated which inhibit the migration of specific cell types. The use of cell-responsive materials to control the migration of individual cell types within a healing tissue is a promising method for improving regenerative efficacy.

References: Rosales, AM. Anseth, Nat. Rev. Mater. 2016, 1, 15012.

A Method of Making Biocompatible Antioxidant Ceria Nanoparticles Loaded in Multi-layer Polymer Films on Device

Yue Hu[1], Elaina Atherton[2], Vivian Ling[2], David Borton[2], and Vicki Colvin[1,2]
1.Department of Chemistry, Brown University.
2. School of Engineering, Brown University.

Introduction:

Reactive oxygen species (ROS) generated in vivo from severe immune responses are generally linked to cancer. Several kinds of antioxidant nanoparticles have been developed by chemists and biologists to reduce the ROS levels in vivo.[1] Cerium oxide (ceria) nanoparticles exhibit recyclable antioxidant properties due to facile electron transfer between Cerium (III) and Cerium (IV). Recent studies reveal that ceria nanoparticles with diameters of 5 nm are non-toxic and efficient for quenching ROS in vivo.[1,2] Here, we developed a highly reproducible method to generate stable, biocompatible ceria nanoparticles which can be used in vivo to control the ROS levels induced by immune responses.

Methods:

Ceria nanoparticles were synthesized using a hydrothermal decomposition method. Briefly, 1mmol Cerium (III) Nitrate Hexahydrate ($Ce(NO_3)_3 \cdot 6H_2O$; Aldrich, St. Louis, MO) was mixed with 3 mmol Oleylamine (Aldrich, St. Louis, MO) in 5.02 mL 1-Octadecene (Aldrich, St. Louis, MO). The mixture was heated at 80 °C for 30 minutes, then the temperature was increased to 100 °C and the mixture was kept at this temperature for 1 hour to remove low boiling point impurities. Finally, the mixture was heated to 260 °C and stirred for 2 hours until the particle growth was completed. Nanoparticles were washed by methanol and acetone at least 5 times. Oleylamine coated ceria nanoparticles with diameters of 5 nm were generated by this method.

Biocompatible surfactant Nitro-dopa-PEG was synthesized in the lab. Briefly, 1g mPEG-AA (Monofunctional PEG Carboxyl Methyl Acid, MW 5K) (Creative PEGWorks, Durham, NC) was dissolved in 5 mL DMF (Aldrich, St. Louis, MO) at 37 °C. After adding 25 mg COMU (Aldrich, St. Louis, MO), the mixture was cooled in an ice bath and became a pale-yellow solid. Then 0.2 mL 4-methylmorpholine (Aldrich, St. Louis, MO) and 50 mg 6-Nitrodopamine (premade in the lab through the nitrification of dopamine) was added into the mixture and stirred for at least 12 hours at room temperature. The final product was washed by cold acetone and purified by dialysis method.

Biocompatible ceria nanoparticles were generated through phase transfer by ligand exchange method. Briefly, 5 mg ceria nanoparticles were dispersed in 2 mL chloroform and mixed with 10 mg Nitro-dopa-PEG. Then 3 mL DI water was added into the solution and the mixture was stirred overnight until the chloroform evaporated. The particle solution was purified by using a centrifugal filter (Cole-Parmar, Vernon Hills, IL) and a syringe filter (Aldrich, St. Louis, MO).

Modified biocompatible ceria nanoparticles could be easily loaded in multi-layer polymers films for antioxidant applications. For example, Glass slides (Electron Microscopy Sciences, Hatfield, PA) were washed with Methanol 3x and Water 3x, dried and then attached to the dipper machine. Polymer solutions of 2mg/mL poly-L-lysine (Aldrich, St. Louis, MO) and 2mg/mL Polyglutamic acid (Aldrich, St. Louis, MO) were made fresh with PBS. Ceria was added to the PLL solution for a final concentration of 140ppm. Glass slides were dipped in PLL, PBS, PGA, and PBS to make 1 bilayer. Each film was prepared with 10, 15, 20, 25, 30, 35, or 40 bilayers. Films were washed in water briefly and allowed to dry before characterization by profilometer or release in PBS. The release was measured by placing the film in 1.5 mL PBS at 37°C and collected daily.

Results: Water-soluble antioxidant ceria nanoparticles were generated and could be saved and restored as a dry powder after freeze drying. Ceria nanoparticles made from this procedure exhibit highly active antioxidant properties which is demonstrated by degradation of Rhodamine after being treated with hydrogen peroxide and measured by using UV-Vis. Treatment of primary cortical cells with Ceria by MTT assay suggests that these nanoparticles are biocompatible and could be easily loaded in multi-layer films.

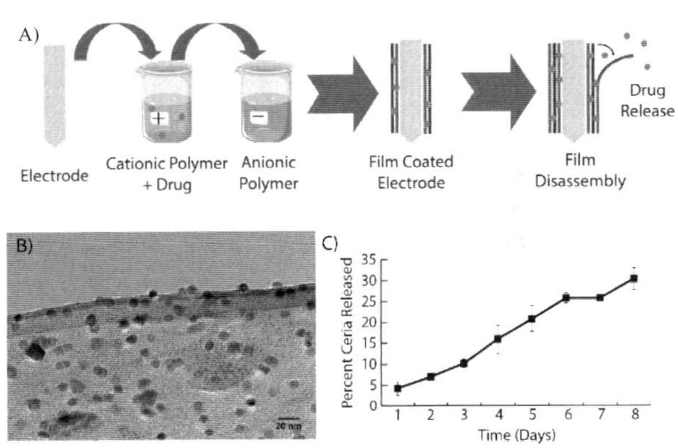

Figure 1. A) Ceria nanoparticles loaded in multi-layer films on the surface of an electrode; B) TEM image of ceria nanoparticles loaded in multi-layer films (10 layers), and the scale bar is 20 nm; C) Kinetic curve of ceria drug releasing from multi-layer films (10 layers).

Conclusions: We have developed a highly reproducible method to generate biocompatible antioxidant ceria nanoparticles. Ceria nanoparticles coated with Nitro-dopa-PEG could be loaded in multi-layer polymer films and used for antioxidant study.

References:

1. Lee, S. S. et al. Chem. Mater. 2012; 24: 424–432.
2. Lee, S. S. et al. ACS Nano 2013; 7: 9693–9703.

1A-4

Free Radical Responsive Biomaterials to Target Therapeutic Immobilization

Emily DiMartini, Christopher Lowe, Keana Mirmajlesi, Adam Gormley, David Shreiber.
Rutgers, The State University of New Jersey, Piscataway, NJ, USA.

Introduction: Targeted drug delivery promises to minimize off-target effects through precise delivery of therapeutics to target tissues. This approach generally requires that tissue-specific ligands elevated in disease are identified and leveraged for selective delivery, but this is costly. We propose a platform biomaterial system that leverages free radicals – highly reactive species native to a variety of different disease states – as a broad-based target for therapeutic delivery.[1] An elevated concentration of free radicals is characteristic of a wide range of disease and injury states including cancer, neurodegenerative diseases, and inflammatory diseases.[2,3] Free radicals are a promising target molecule because of their ability to initiate rapid polymerization.[4] We have hypothesized that free radicals native to various disease states can be used to initiate the crosslinking of acrylated PEG molecules. We propose that this *in situ* crosslinking will locally generate a polymer network and sustain the biomaterial in the diseased tissue. Here, a therapeutic factor can be coupled to the platform polymer, achieving targeted immobilization that is amenable to various diseases and conditions. Previous work in our lab has demonstrated the reactivity of acrylated PEG polymers with free radicals native to injury sites though reactive oxygen species (ROS) and reactive nitrogen species (RNS) assays. Further, this reaction protects neurons from free radical-induced damage. Building on previous results, we sought to characterize the crosslinking kinetics and its implications in immobilizing our polymer networks within tissue mimics.

Methods: *Size Exclusion Chromatography:* Gel permeation chromatography (GPC) was used to quantify the molecular weight of unreacted and cross-linked acrylated PEG. Horseradish peroxidase (HRP), hydrogen peroxide, and acetylacetone were mixed to generate reactive oxygen species (ROS) in vitro, which were reacted with acrylated PEG polymers overnight. Samples were passed through a PBS column, and peak height was used to quantify polymer conversion from a single molecule to a cross-linked species. *Diffusivity Model:* A collagen gel was used as a tissue-mimic to examine immobilization of our crosslinked conditions. Fluorescently-doped, acrylated PEG was allowed to diffuse through a collagen hydrogel overnight. Radicals were introduced with HRP and hydrogen peroxide, and gels were subsequently washed to mimic tissue clearance. Residual fluorescence in the gel and the supernatant was measured to demonstrate immobilization.

Results: As shown in Figure 1, the height of the peak at 33 min elution time, which represents unreacted 2,000 Dalton PEG diacrylate (2K PEGDA), decreases with increasing concentrations of HRP. Conversion rate was highest for 2K PEGDA (~60%) compared to 6,000 Dalton and 10,000 Dalton PEGDA (~30% conversion), indicating that the lower MW PEGDA crosslinks more efficiently. However, in the collagen gel immobilization assay, 10K PEGDA was the only species that showed an increase in residual fluorescence following washes. This indicates that while the 2K PEGDA crosslinks 2 to 3 times as frequently compared to higher molecular weight analogs, the small size requires many more coupling events to achieve immobilization. Together this suggests that a mix of small, rapidly crosslinking PEGs with larger, less diffuse molecules offers greater immobilization potential for therapeutic delivery.

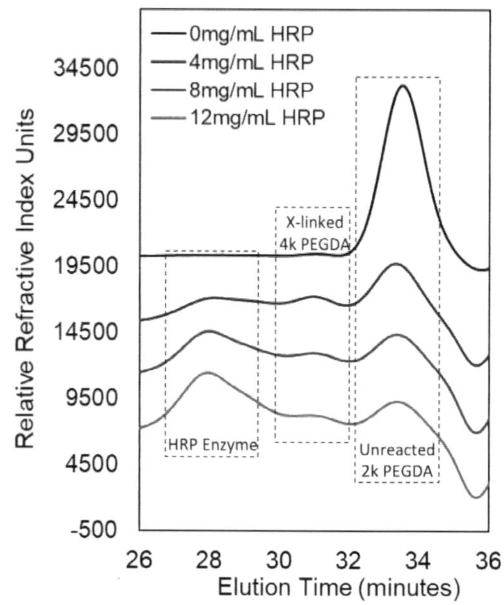

Figure 1. *GPC results for 2K PEGDA showing peaks indicative of different molecular weights as a function of sample elution time. In GPC, larger MW species pass through the column faster and have shorter elution times. At 0mg/mL HRP, a showing a single molecular weight peak (MW: 1779) is apparent, which that is reduced with the introduction of ROS. Percent conversion to doublet species was quantified by the reduction in peak height.*

Conclusions: These results indicate the strong potential of acrylate-acrylate crosslinking to form higher molecular weight polymers with reactive species generated with physiologically relevant methods. Further, these coupled PEGDA polymers have the ability to become immobilized within a tissue for sustained presentation of a coupled therapeutic. Future work will focus on identification of the optimal polymer configurations and incorporation of a therapeutic group to assess sustained presentation capabilities.

References: 1. Pham-Huy LA. et al. Int. J. Biomed. Sci. 2008;4;89-96.
2. Reuter S et al. Free Rad. Biol. Med. 2010;49;1603-1616.
3. Florence TM. Aust. N. Z. J. Ophthalmol. 2018;23;3-7.
4. Jean-François L. J. Polym. Sci. 2008;46,3459-3470.

Acknowledgements: This work is supported by the Rutgers Biotechnology Training Program (NIH Grant Number 5T32GM008339-20), as well as a pilot research grant (CBIR16PIL015), and a graduate fellowship (CBIR14FEL004) generously provided by the New Jersey Commission on Brain Injury Research (NJCBIR).

Engineering the Osteochondral Interface: 3D-Printed Scaffolds with Tunable Peptide Organization

Paula Camacho[1], Hafiz Busari[2], Peter Schwarzenberg[3], Anne Behre[1], William G. De Long, Jr[4], Hannah L. Dailey[3], Lesley W. Chow[1,2]
[1]Department of Bioengineering, [2]Department of Materials Science and Engineering,
[3]Department of Mechanical Engineering and Mechanics, Lehigh University, Bethlehem, PA, USA
[4]Department of Orthopedic Surgery, St. Luke's University Health Network, Bethlehem, PA, USA

Introduction: Functional regeneration of the osteochondral (OC) interface remains a challenge because the transition from bone to cartilage presents gradients in biochemical and physical properties from the nano-to macroscale.[1] Regeneration of this complex tissue requires a biomaterials-based approach to engineer a micro-environment that guides cells to restore the entire OC interface. Recent advances in direct-write 3D printing allow the fabrication of 3D geometries at room temperature in a freeform fashion.[2] Expanding on this technique, our lab has developed a new strategy to 3D print peptide-polymer conjugates to create biodegradable polymer-based scaffolds spatially functionalized with bioactive peptides. This platform enables fabrication of hierarchically organized scaffolds with controllable biochemical distribution and structure to mimic the extracellular matrix. During a single fabrication step, scaffolds can be printed to present one or more peptides in discrete locations while simultaneously varying fiber spacings to control scaffold architecture in a continuous construct. For osteochondral scaffolds, we synthesized peptide-poly(caprolactone) (PCL) conjugates[3] with hyaluronic acid (HA)-binding (HAbind-PCL) or mineralizing peptides (E3-PCL) designed to promote human mesenchymal stem cell (hMSC) chondrogenesis[4-6] or osteogenesis[7,8], respectively. The goal of this work was to investigate how multi-peptide organization influences early hMSC behavior.

Methods: Peptides were synthesized using standard Fmoc solid phase peptide synthesis techniques before covalently attaching to PCL using methods previously described.[3] The resulting peptide-PCL conjugates included bioactive sequences RYPISRPRKR (HAbind-PCL) and EEE (E3-PCL). They were co-dissolved with unmodified PCL in hexafluoro-2-propanol (HFIP) to produce printable "inks", which were 3D printed using a customized solvent-cast 3D printer to create four scaffold groups: PCL-only; E3-PCL; HAbind-PCL; and HAbind/E3-PCL with peptides presented in opposing regions. For HAbind/E3-PCL, two printer heads with inks containing HAbind-PCL or E3-PCL were used to control multi-peptide organization. Scaffolds were fluorescently labeled using fluorescein-HA and/or amino-Cy3 to specifically tag HAbind and E3 peptides, respectively, and visualize their spatial distribution on the fiber surface. Bone marrow-derived hMSCs (P3) were seeded at 200,000 cells per scaffold of each group. Cells were cultured in growth media without added differentiation factors for up to 28 days. Samples were evaluated for chondrogenesis and osteogenesis using qPCR and biochemical assays for glycosaminoglycan (GAG) content and alkaline phosphatase (ALP) activity. Histological staining with Alcian Blue (for GAGs) and Alizarin Red (for calcium) were used to assess overall spatial tissue formation.

Results: All peptide-PCL conjugates were successfully synthesized and used to produce bioactive inks for 3D printing. Rapid solvent evaporation after printing resulted in solid PCL fibers displaying peptides on the surface. Fluorescent labeling illustrated how this functionalization strategy produced peptide-modified scaffolds with distinct spatial organization. Gene expression data showed that the dual-peptide HAbind/E3-PCL scaffolds significantly enhanced chondrogenesis, showing upregulation of SOX9, and osteogenesis, with upregulation of RUNX2. Biochemical analyses showed a significant increase in GAG accumulation and higher ALP in the HAbind/E3-PCL scaffolds compared to PCL and single-peptide scaffolds. Histology corresponded with trends seen in biochemical assays and demonstrated complete cellular infiltration and matrix deposition throughout the scaffolds (Figure 1).

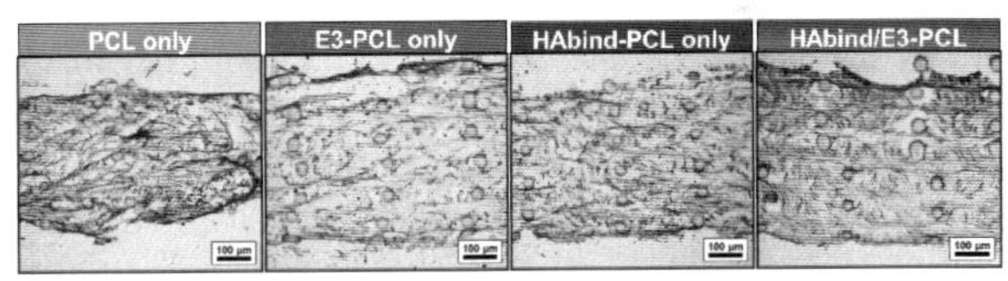

Figure 1. Histological sections staining for sulfated glycosaminoglycans (sGAG; Alcian Blue) and cell nuclei (Nuclear Fast Red) after 28 days of culture.

Conclusions: Solvent-based 3D printing with peptide-PCL conjugate inks allowed for single-step peptide functionalization of scaffolds without the need for post-processing. This work demonstrates a tunable platform to control multi-peptide organization within a continuous 3D-printed construct. Bioactive peptides designed for OC tissue engineering influenced chondrogenic and osteogenic differentiation of hMSCs, and the presence of both peptides in a single construct created a synergistic effect that enhanced cellular responses. Notably, these significant results occurred without the addition of exogenous GAGs or differentiation factors during culture. Ongoing work includes optimizing peptide concentration and porosity to evaluate their individual and combined effect on hMSC behavior. Biochemical and biophysical cues can therefore be optimized independently to synergistically guide local cell behavior and spatial tissue formation.

References: [1] Di Luca A et al. *Birth Defects Res. Part C.* 2015;105:34-52. [2] Guo SZ et al. *Small.* 2013;9:4118-4122. [3] Chow LW et al. *Adv Healthc Mater* 2014;3:1381-1386. [4] Goetinck P et al. *J Cell Bio.* 1987;105:2403-2408. [5] Puetzer JL et al. ORS Meeting 2016, poster 2226. [6] Parmar PA et al. *Adv Healthc Mater* 2016;5:1656-1666. [7] Polini A et al. *Biomater Sci.* 2014;2:1779-1786. [8] Karaman O et al. *J Tissue Eng Regen Med.* 2016;10:E132-E146.

Shear-Induced Extensional Response Behaviors of Tethered von Willebrand Factor

Yi Wang, Michael Morabito, Ciara Kavanagh, Avani Pisapati, Xiaohui F. Zhang, Edmund Webb III, Alparslan Oztekin, and Xuanhong Cheng.
Lehigh University, Bethlehem, PA.

Introduction: Functionality in many biological systems is strongly dependent upon changes in their biomechanical environments. In particular, flow plays an important role in the regulation of biomechanical state and behavior for a number of biomolecules and their assemblies. Von Willebrand Factor (vWF) is a plasma protein that initiates platelet plug formation in response to vascular damage and its functions are shear flow-regulated. Previous authors have used single-molecule experiments and simulations to examine the flow-induced dynamics and functionality of tethered and free-flowing vWF, which helped elucidate molecular-level behaviors crucial to vWF biology and physics. However, many domain-level details of the underlying flow-induced biomechanical responses of vWF remain unknown. By combining simulations with experiments, our work demonstrates how one could gain in depth understanding of flow-induced conformational changes and activation of biopolymers on single molecular and intra-monomer length scales.

Methods: Experimental tests were performed using a confocal fluorescence microscope. vWF multimer from human plasma was purchased and labeled by a sulfo-NHS-LC-Biotinylation kit and an Alexa 488 5-SDP Ester kit. The fluorophore and biotin tagged vWF was then applied for flow tests in a PDMS microfluidic device that generates both high shear and extensional flow. The microfluidic channel was pre-incubated with BSA-biotin at room temperature for 2 hours, followed by incubation with a casein based blocking solution, streptavidin, vWF and free biotin for 30, 10, 5 and 15 minutes respectively.

Computational works were based on a coarse-grained vWF monomer model and Brownian Dynamics (BD) simulations. The vWF model represents a single monomer by a system of two beads connected by a finitely extensible nonlinear elastic (FENE) spring as the A2 domain. BD simulations began with a 20-chain ensemble of non-interacting vWF multimers and continued until an ensemble average quasi-steady elongation length was achieved, which depended on the shear rate.

Results: As shown in Figure 1, we captured reversible unraveling behavior of tethered green-fluorescent single vWF molecules under a microscope at increasing shear rates and the results are in qualitative agreement with previous reported works from literature. Extension of each molecule increases as the shear rate goes up. Most chains begin with globular conformations that are 1.0 to 1.5 μm in diameter and reach maximum lengths between 2.0 to 4.0 μm at the shear rate 20,000 s^{-1}. The threshold shear rate associated with the onset of extension was found to be 400 s^{-1}. A large variation of the unraveling behavior is also observed. Comparably, computational results exhibit similar elongational behavior and variation when tuning molecular and binding parameters. Simulation results show that extensional response is larger for longer multimers, retarded as tethering location along the multimer contour tends toward the chain center and affected even larger by the locations of multiple binding sites. Additionally, we experimentally observed non-homogenous distribution of fluorescent intensity along many unraveled vWF contours indicating that the local monomer density may vary in the flow direction. The simulation snapshots (Figure 1 insets) captured similar features at different flow and binding conditions.

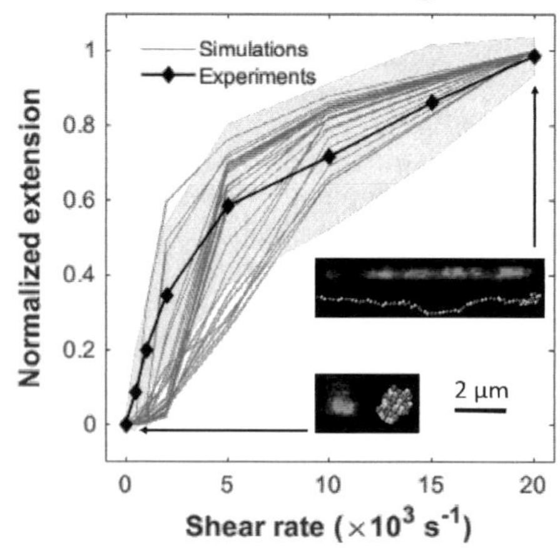

Figure 1. Experimental and simulation results of vWF unraveling at increasing shear rates. The shaded region indicates the behavior variation among 70 molecules observed experimentally. Simulations results are from varied parameters including molecular weight, tethered locations and multiple tethering points. Images demonstrate coiled and unraveled molecules captured both in experiments and simulations.

Conclusions: We performed single-molecule flow experiments for the shear-induced unraveling and elongation of tethered vWF multimers and BD simulations to help explain details of our experimental observations. We found good agreements between experimental results and computational predictions for the threshold shear rate of extension, existence of non-homogenous fluorescence distributions along unraveled multimer contours, and large variations in extensional response behaviors. BD simulations revealed the strong influence of varying chain length, tethering point location, and number of tethering locations on the underlying unraveling response. We showed that global conformational changes of tethered vWF can be accurately captured using a relatively simple mechanical model. This work demonstrates the power of working experiment and simulation in tandem for understanding flow-induced changes in biomechanical state and global conformations of macromolecules.

Acknowledgements: This work was supported by National Science Foundation grant DMS-1463234 and utilized the Extreme Science and Engineering Discovery Environment (XSEDE), which is supported by National Science Foundation Grant No. ACI-1548562.

Calibration and Measurement of FRET Efficiency in a Vinculin Tension Probe

J. L Stein, D. Sumetsky, N. Emanuely, M. Patel, M. M Cararo Lopes, B. L Firestein and N. N Boustany
Rutgers, The State University of New Jersey

Introduction: Fluorescence Resonance Energy Transfer (FRET) allows for the quantification of distances through the use of two fluorophores, a donor and an acceptor. When the two fluorophores are within a certain distance, typically 1-10nm, a nonradiative energy transfer occurs.[1] This dependence allows for us to quantify nanoscale distances between the fluorophores. When the two proteins are connected via an elastic linker with a known relationship between tension force and FRET efficiency, the linker's extension measured by FRET will report on the microscopic tension force across the linker. The quantification parameter that we are using is FRET Efficiency, defined as the proportion of the donor molecules that have transferred excitation state energy to the acceptor molecules, and decreases as the separation distance increases, according to a sixth power relationship. [2] Our construct is composed of two fluorescent proteins, mTFP and mVenus, both of which are GFP derived proteins[3], connected by an elastic linker. Our tension probe, named VinTS, is optimized to insert itself into the Focal Adhesion Protein Vinculin after amino acid 883.[4] This allow for us to determine tension in the Focal Adhesions within immortalized baby mouse kidney (iBMK) cells and bovine aortic endothelial cells (BAEC).[5] An additional construct, TSMod, will also be tested in the neuronal cells. This soluble construct is the same as the sensor molecule in the VinTS construct but without being inserted into the vinculin protein. [4]

Methods: We used control FRET constructs, with known FRET efficiency, that we expressed in iBMK cells, BAEC, and isolated cortical neurons. The calibration constructs consisted of either our donor, mTFP1, our acceptor, mVenus, or both connected by a long or short linker. The long linker protein was TRAF (TNF receptor associated factor) while the short linker was the amino acid sequence $(GGSGGS)_2$. These constructs were transfected into the selected cell types using lipofectamine LTX according to a previously developed protocol. All of the constructs were tested in the aforementioned cell lines. The methods involved imaging using a 3-channel "sensitized emission" FRET methodology, with the results analyzed in a lab-developed MATLAB Script. The collected images were also corrected for background. The calibrated system was then utilized to measure the FRET efficiency in the VinTS tension probe expressed in the BAEC and iBMK cells. For this, the focal adhesions were computationally isolated so that the FRET efficiency could be measured in solely those areas. The constructs were then transfected into isolated Cortical Neurons and imaged similarly to the control cells. After imaging the FRET control constructs in neurons, our goal will be to measure the FRET efficiency of VinTS in neurons and to compare the VinTS probe to the TSMod mentioned above.

Results: The investigation so far, has allowed for calibration within the BAEC and iBMK cell types. The calculated bleedthrough values were 0.65 and 0.47 for TFP and mVenus, respectively. In addition, we obtained a G Factor (as defined in Ref. 1 and shown in Fig. 1 as the negative slope of the line) of approximately 2.1, across both control cell lines and in neurons. This was done using a linear fit between the averages of the For the TRAF construct, we obtained an Efficiency of 0.03-0.06 in iBMK cells and 0.07-0.09 in the BAEC samples. For $(GGSGGS)_2$, we obtained an Efficiency of 0.47-0.49 in both the iBMK and BAEC samples. The VinTS construct also appears to produce expected values within the target range for each cell type. In neurons, we obtained TRAF values in the 0.02-0.06 range. For the $(GGSGGS)_2$ construct in neurons, we obtained values in the range of 0.22-0.24.

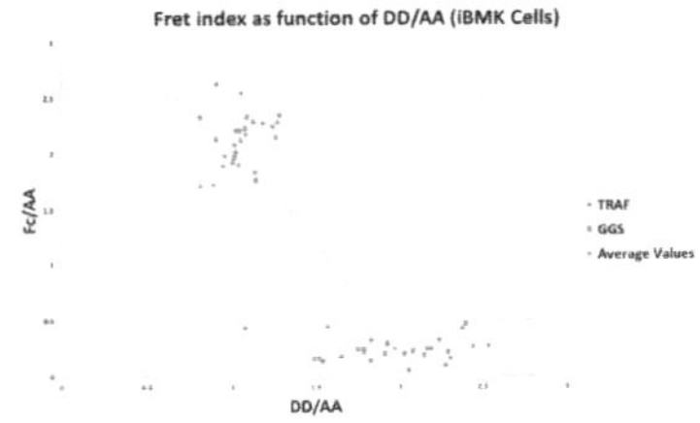

Figure 1. FRET Index as a Function of DD/AA for iBMK

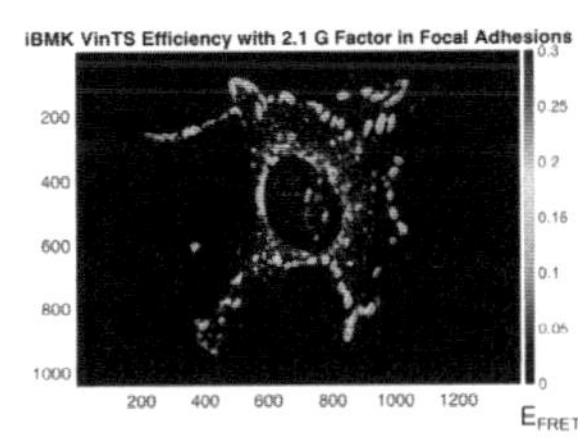

Figure 2. Image of VinTS Expressed in an iBMK cell after image processing in MATLAB

Conclusions: The obtained values for both of the control constructs were consistent with literature. For TRAF, the expected values should be less than 0.11.[6] For $(GGSGGS)_2$, the Efficiency should be approximately 0.5. The expected TRAF values are consistent with results across all cell types. The expected $(GGSGGS)_2$ values are consistent with the iBMK and BAEC samples, but not the neuronal ones. Further testing within neurons to better understand this is currently ongoing. The calibration values allow for us to continue testing the VinTS construct to obtain reproducible and consistent values. The calibration also establishes a baseline for future study in all three cell types. This will include investigations into how the mechanical and geometrical properties of various engineered substrates affect neuronal growth and development.

References:

1. Chen, H. et al. Biophysical journal. 2006;91(5):L39-L41.
2. Sun, Y. et al. Cytometry Part A. 2013;83(9):780-793.
3. Koushik, S. V. et al. Biophysical journal. 2006; 91(12): L99-L101.
4. Grashoff, C. et al. Nature. 2010; 466(7303), 263.
5. LaCroix, A. S. et al. Methods in cell Biology. 2015; 125:161-186.
6. Day, R. N. et. al. Journal of biomedical optics. 2008; 13(3): 031203-1-031203-9.

Acknowledgements: We wish to thank Dr. B. Hoffman and the his lab at Duke for the DNA constructs and pertinent advice throughout the project. This project is supported by NSF grant award #1825433 to N. N Boustany.

1B-4

Endothelial cell organelle positioning exhibits an intrinsic left-right bias

Jie Fan[1,2], Haokang Zhang[1], Tasnif Rahman[1], Diana Stanton[1], Leo Q. Wan[1,2,3,4,*]
1 Department of Biomedical Engineering, Rensselaer Polytechnic Institute, Troy, NY 12180, USA.
2 Center for Biotechnology & Interdisciplinary Studies, Rensselaer Polytechnic Institute, Troy, NY 12180, USA.
3 Department of Biological Sciences, Rensselaer Polytechnic Institute, Troy, NY 12180, USA.
4 Center for Modeling, Simulation and Imaging in Medicine, Rensselaer Polytechnic Institute, Troy, NY 12180, USA.
* Corresponding author. Email: wanq@rpi.edu

Introduction: The maintenance of blood vessel integrity requires the proper polarization of the cells, including the left-right polarity that is influenced by cell chirality (or handedness, left-right asymmetry) of endothelial cells[1]. Great efforts have been taken to measure cell chirality in vitro using engineering tools at both the single cell level and multi-cellular level, and demonstrated that cell chirality is an intrinsic property of the cell shown as the left-right biased polarization[2], cell migration[3-5], directional rotation of actin structure[6], nucleus[6], cytoplasmic flow[7], or cell itself in whole[8], and multicellular morphogenesis[9]. However, the inability to use these engineering tools in in vivo settings greatly limits the understanding of the role of cell chirality in many developmental and physiological processes. Therefore, it is crucial to develop an intracellular marker that determines cellular left-right asymmetry without any engineering setup. We have recently reported that endothelial cells exhibit a rightward bias in positioning of the cellular centroid relative to the nucleus-centrosome axis[1]. To investigate the relationship between this organelle bias and the multicellular chirality, here we further performed the analysis of cell organelle positioning on the ring-shaped micropatterns.

Methods: The human umbilical vascular endothelial cells (hUVECs) were patterned and cultured on the micro ring-shape as described previously[1,9]. Micropatterned cells were fixed, permeabilized, blocked, and incubated with ZO-1 and pericentrin antibodies followed by appropriate secondary antibodies. Finally, samples were mounted in Fluoromount-G with DAPI and imaged using a fluorescence microscope (BZ-X700, Keyence). Image analysis for cell organelle chirality was performed with ImageJ[1].

Results: As expected, the hUVECs showed a clockwise alignment with cell-cell junction shown by ZO-1 staining (Figure 1A). After segmentation along cell borders, the centroid of each cell was calculated using ImageJ (Figure 1B). Similarly, cell nuclei were segmented based DAPI staining and nuclear centroids were calculated acoordingly (Figure 1C). The centrosome stained with the pericentrin antibody shows a concentrated single dot located next to the nucleus of each cell. The combination of these three dots, nuclear centroid, centrosome, and cell centroid (Figure 1D) allows easy characterization of the left-right biases of the 2D topological cell shape by judging the left-right positioning of cell centroid relative to the direction of the nucleus-centrosome axis (Figure 1E). Finally, the biases of each cell could be color-coded, showing the overall distribution of left-, right- or non-biased (neutral) cells on the micropattern. According to this cell organelle-based analysis, significantly more hUVECs exhibited a right bias than a left bias, which was consistent with the chirality calculated using the micropatterning assay[1,9]. The mechanism of this right-biased positioning of cell centroid remains unclear, but it might result from the biased cell migration[4,5]. It should be also noted that a portion of cells would be characterized as neutral or left-biased with this proposed approach, indicating the substantial noise of this method for cell chirality determination, as a result of determining cell chirality at the single cell level.

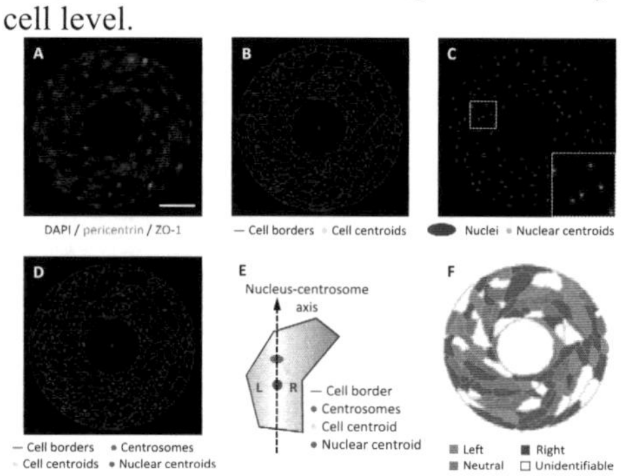

Figure 1. Left-right biases analyzed from the fluorescence images. **(A)** Immunofluorescence of hUVECs on a ring-shaped micropattern (Scale bar: 100 μm). **(B)** Cell borders (red) and cell centroids (yellow). **(C)** Cell nuclei (blue) and nuclear centroids (cyan). **(D)** Merged image for cell bias analysis. **(E)** A schematic of determination of the left (L) or right (R) cell bias. **(F)** Color-coded cell biases.

Conclusions: The centroid of endothelial cells is preferably polarized towards the right of the nucleus-centrosome axis on the clockwise oriented pattern, suggesting the consistency of right-biased organelle chirality of individual cell and clockwise chirality of multiple cells on a ring-shaped micropattern. This finding implies an intrinsic chirality of organelle positioning inside of individual cells and provides a new approach for studying cell chirality in situ/in vivo without the limitations of engineering tools such as microcontact printing.

References:

1. Fan J. et al. Science advances. 2018;4(10):eaat2111.
2. Chen T.H. et al. Circ Res. 2012;110(4):551-9.
3. Xu J, et al. PNAS. 2007;104(22):9296-300.
4. Tamada A. et al. Nat Commun. 2017;8(1):2194.
5. Sullivan K.G. et al. bioRxiv. 2018.
6. Yamanaka H. et al. Genes to cells. 2015;20(1):29-35.
7. Liu W. et al. ACS Nano. 2016;10(8):7409-17.
8. Chin A.S. et al. PNAS. 2018;115(48): 12188-12193.
9. Wan L.Q. et al. PNAS. 2011;108(30):12295-300.

Acknowledgements: This work was supported by the National Institutes of Health (OD/NICHD DP2HD083961). L.Q.W. is a Pew Scholar in Biomedical Sciences (PEW 00026185), supported by the Pew Charitable Trusts.

In vitro human lymphocytes activation and intercellular activities from the co-culture of dermal fibroblasts

Yu Ting Lin, JieZhen Wang, Kao Kun Lin
Rensselaer Polytechnic Institute, Hong Kong Life Science and Technologies.

Introduction:

The interaction of how immune cells activate and respond to a heterogeneous cell type has always been a quest for the optimizations of immunotherapeutic products. The heterogonous nature of human dermal fibroblast (HDF) could change the intercellular activities of T lymphocytes. HDF plays an important role in studying cell physiology, cell adhesion integrin, and gene activation Preliminary images of growth, patterns, and cell state can be determined by taking images from the individual in vitro cell culture. To determine the cell to cell interaction and to quantify the affects from a different cell type, the secretion of cytokines IL-2 and IL-6 can be used to record those data to estimate the activation and intercellular activities of a particular cell type in regards to time and cell morphologies.

Methods:

10 ml of peripheral blood samples are collected from healthy donor and HDF cell lines are retrieved from Kunming Cell Bank of the Committee of Typical Collection of the Academy of Science (KCB 200537). Isolation of Human lymphocytes and incubation of the cell culture are at 37°C and 5% CO_2 overnight, and images of the test groups under the microscope are taken daily. The lymphocytes will be seeded with HDF ($8.4*10^4$, 94.9%) at the ratio of 1:1 for three wells in a six well plate for three days. Two individual wells with only fibroblasts and lymphocytes group will be labelled as negative controls. Supernatant samples are collected and assayed by sandwich ELISA for levels of IL-2, IL-6, and cell death conditions. ELISA kits are from Nanjing SenBeiJia Biological Technology.

Results:

T lymphocytes attack HDF when they are being co cultured as shown in Figure 3. Prior to co culture, the quantity of lymphocytes is far less than before. The results can be seen at day 2 and day 3 when all of the fibroblasts are gone. From the image, it is very likely that the fibroblasts trigger the activation of T lymphocytes to attack HDF

Co-Culture of fibroblast enhances the T lymphocytes IL-2 secretion. From Figure 1, the levels of IL-2 for all the test groups increased all three days of the co culture. The value is also higher than that of lymphocyte control or fibroblast control. The number of cells in three different days is also recorded in regards to IL-2 level. The more stimulated T lymphocytes are present the higher the secretion level of IL-2. The data correlate with a high R^2 and an exponential function denote trend of the growth for the test groups. (1)

The inflammation effects of the T lymphocytes can be seen from the overall higher level of IL-6 in relative to HDF over the three days in Figure 2. Overall increase suggests the possibility of pro-inflammatory responses being elicited from the fibroblasts co culture (2).Overall increase in the number of lymphocytes could suggest a possibility of growth activation from culturing with a heterogeneous cell type.

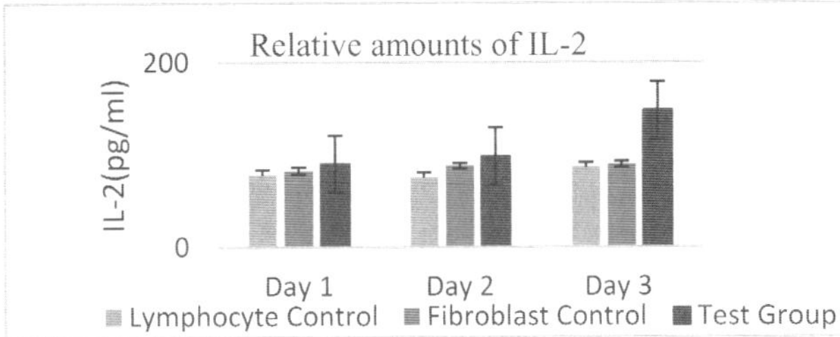

Figure 1.Relative levels of IL-2

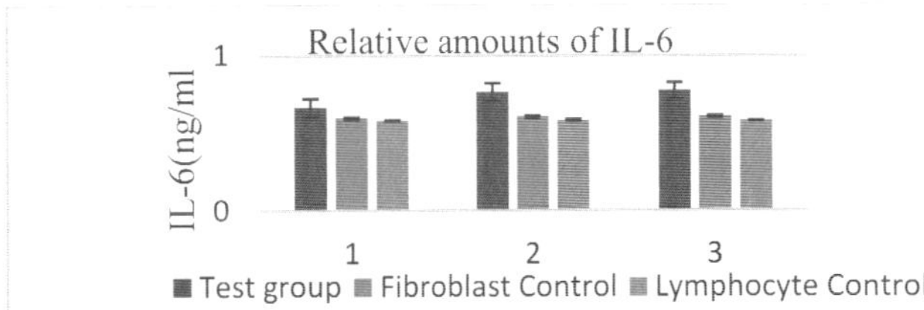

Figure 2. Relative level of IL-6

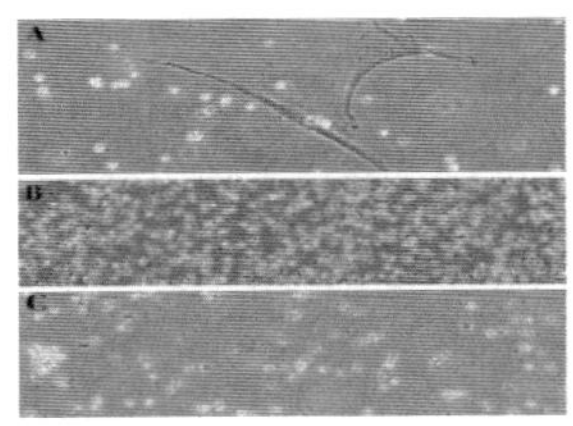

Figure 3. Images of the cell culture under the microscope at 200X magnification from day 1, 2, and 3 are denoted by A, B, and C respectively

Conclusions:

From the photo, fibroblasts are completely gone in day 2 and day 3.The aggregation of T lymphocytes may be a possibility of the expansion and activation of T lymphocytes culturing with HDF. This is supported by comparing the data of the levels of IL-2 in both the test and control groups. A significant increase of the rise of the IL-2 level can be illustrated by comparing with those of the control upon co culture. A pro inflammation response can be deducted when the IL-6 remained increasing as opposed to the control. IL-6 may contribute to the apoptosis of HDF from the disappearance of visible HDF and aggregations of T lymphocytes (3). More experiments at the molecular level, for instance, PCR and western blot, could be conducted to explain the mechanisms and patterns of the killing of the cytokines in which T lymphocytes secrete to promote the apoptosis of HDF.

References:

1. Rubinchik, E., & -Schaffer, F. L. (1996). Interleukin-2 inhibits 3T3 fibroblast proliferation. *Life Sciences*
2. Tanaka, T., Narazaki, T. (2014). IL-6 in inflammation, immunity, and disease. *Cold Spring Harbor perspectives in biology*, *6*(10)
3. O 0

Acknowledgements: Acknowledgements to funding from Hong Kong Life Science Technologies for reagent kits and research equipment

Development of Mobile Phone Based Transcutaneous Bilirubinometer

Brandon Harrison[1], Alexander Dumont[1], Jerry So[1], Shreyas Chandragiri[1], Amy Nwaba[2],Sarah Nwaba[2] Zack McCormick[3], Hendrik Weitkamp[2] Nuraddeen Lawal[4], Zubaida Ladan Farouk[4], Abdussalam Muhammedand[4], Chetan Patil [1]

1. Temple University, 2. Vanderbilt University Medical Department of Neonatology, 3.Trailblazing technologies 4.Department of Paediatrics, Bayero University Kano, Kano 7002 31, Nigeria

Introduction:

In the low and middle-income countries (LMIC), factors that increase the risks of neurodevelopmental impairment, such as sepsis, malnutrition and certain genetic conditions are common. Extreme or prolonged neonatal jaundice (hyperbilirubinemia) oftentimes results in permanent neurological impairment and even death. Treatment of neonatal jaundice can be simple; however, identifying infants with jaundice through visual screening is unreliable. In the United States, objective screening of infants for hyperbilirubinemia prior to hospital discharge is universal. Often, screening relies on transcutaneous bilirubinometry (TcB), a reliable optical technical based on diffuse reflectance. In LMIC's, TcB devices are not common; however, it is increasingly common for these first line care providers to carry mobile phones equipped with a camera and flash. We will report initial feasibility of TcB utilizing the built-in camera and flash of a mobile phone, a Monte Carlo model driven design of a snap-on optical assembly, testing and characterization using optical phantoms, and the results from ongoing clinical studies in newborns which compare mobile-phone based measurements of TcB with corresponding serum bilirubin levels.

Methods:

This study was conducted at Vanderbilt University and, Bayero University Hospitals, where a nurse tested our device by collecting images of a neonate's skin. A protocol was set in place for nurses to use our device, including identifying infants that were eligible for the study, obtaining consent and coordinating serum bilirubin blood sampling. Identification of the human subject included healthy infants at 35 weeks gestational age or greater, infants 24 hours of age and mothers staying in the hospital for longer than 24 hours. Use of our device was coordinated with serum bilirubin draws at the hospital. Comparing our data alongside well-established techniques used in hospitals will validate the use of our device for low-income settings. When collecting diffuse reflectance of the neonatal skin, it was necessary that we acquired images from high rates of blood perfusion in the skin locations. Skin locations with high perfusion rate offer more circulation of bilirubin and hemoglobin. The locations of the forehead, sternum and interscapulum were chosen due to their high blood perfusion rates. Multiple images of each location were taken to account for human error and determine variance in measurements.

Results:

To date 29 infants have been enrolled in the study and preliminary results indicate an $R^2 = .89$ and a relative standard error of 1.57 mg/dL. These promising results will lead to a discussion of feasibility and limitations for mobile-phone based TcB in LMIC settings.

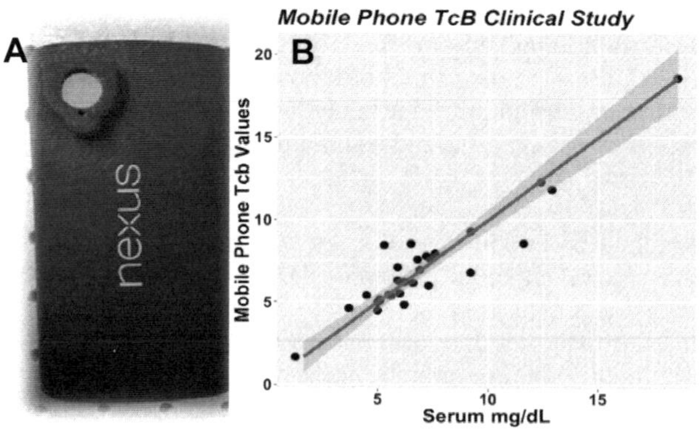

Figure 1a. A physical representation of our device that goes on the back of the cell phones casing. **Figure 1b.** Is a correlation map of Mobile phone TcB Values vs Serum Bilirubin levels.

Conclusions:

In this paper, we have presented our mobile phone device app design, reference standard, image analysis and preliminary clinical results. Our preliminary clinical results compare six images taken from the sternum of the neonates and correlates these images with serum bilirubin concentrations. Current results indicated a positive correlation, an R^2 value of .89 and a relative standard error of 1.57 mg/dL. With these promising preliminary results, we will continue to collect images from neonates from both facilities and monitor the correlation and error of our device.

Acknowledgements:

Bill & Melinda Gates, Bond-Hill Scholarship, Duco Jansen (Vanderbilt University), and Nuraddeen Lawal (Aminu Kano Teaching Hospital)

Liver Classification Based on SVM for Ultrasound Images

Hui Che[1], John Nosher[2], Ilker Hacihaliloglu[1,2]
[1]Department of Biomedical Engineering, Rutgers University, USA.
[2] Department of Radiology, Rutgers Robert Wood Johnson Medical School, USA

Introduction: The incidence of morbidity and mortality of liver disease is increasing yearly on a global scale. In order to increase the chances for survival, accurate detection of liver disease earlier is crucial to patients. Ultrasound (US), as a kind of imaging test, offers great improvements in liver disease diagnosis and therapy due to its real-time, non-ionizing imaging capabilities. However, the quality of US images is often influenced by many factors and the assessment from a doctor is subjective. There is an ongoing need of computer-aided diagnosis for liver states and classification. These years many efforts have been done to improve the accuracy of diagnosis. Adaptive filters, anisotropic diffusion and wavelets are effect methods to achieve enhancement of images after they have been captured. Machine-learning techniques such as support vector machine (SVM) have been applied for disease detection and classification, such as grading liver fibrosis, classification of focal liver lesions and liver disease prediction. In this work, we propose a novel method for more reliable and accurate classification of normal or abnormal liver. The method includes three stages: image enhancement based on the extraction of image phase features for improving representation of the liver tissue, feature extraction with Bag of Features (BOF) algorithm and classification with SVM.

Methods: The sample B-mode US liver images were acquired from Robert Wood Johnson University Hospital after obtaining institutional ethics board approval. The present work is conducted on 962 images, obtained from 56 subjects, out of which 674 are used for training (445 normal and 229 abnormal) and the remaining 288 (190 normal and 98 abnormal) are used for testing. In this study, local-phase liver tissue enhancement is used for accurate and robust feature detection, because of more contribution to the visual appearance than magnitude information [1]. The process includes Fourier transform, alpha-scale filtering, inverse Fourier transform, transmission map estimation and image regularization. Alpha-scale filter is a band pass filter which provides improved edge detection results. The US signal transmission map helps pertinent liver tissue structure recovered from the US images (Fig.1). Abnormal liver presents radial symmetric tissue structures which are not appeared in healthy liver images. Therefore fast radial symmetry transform (FRST) is used for detecting points of interest after image enhancement which is a variant on the circular Hough transform (Fig.1). As to feature extraction, bag of features (BOF) is an adopted technique to model image structure in a robust way [2]. The BOF framework leads that vocabulary can be used to explain each unique images as a histogram and implement clustering in feature extraction. In classification, a binary classifier using BOF descriptor is trained with support vector machine (SVM), to construct a hyperplane that separates the two classes. In order to investigate the improvements in liver tissue classification, by using the processed US data, we conduct several validation studies. In validation study one we only use the B-mode US data for training and testing. In our second validation we use B-mode US data together with the fast radial symmetry features for training and testing. Finally, during study three, we use B-mode US data with the enhanced local phase tissue features for training and testing.

Results: Figure 1 shows processing results and Table 1 lists the accuracy value with test set. Table 1 displays the obtained quantitative results. We can see that by combining the extracted liver tissue features the classification accuracy improves. Best results was obtained when local phase tissue signatures were combined with the B-mode US data. Two tailed t test showed statistically significant improvement ($p<0.05$).

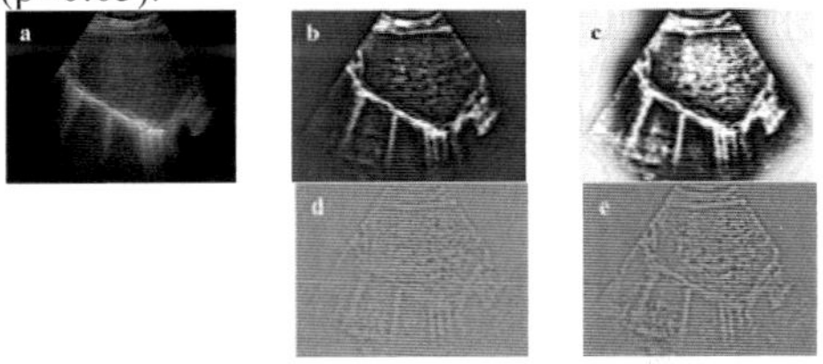

Figure 1. (a) B-mode in vivo US image of diseased liver. (b) - (c) Local phase tissue signatures. (d) FRST image S1; (e) FRST image S4.

Table 1. The classification accuracy

	Accuracy
Study1: B-mode only	75%
Study 2: B-mode + Fast Radial Symmetry Features	81%
Study 3: B-mode + Local Phase Features	86%

Conclusions: In this study, we propose a novel approach for more accurate liver classification. The image processing method highlights the tissue features of the US liver images. Using SVM with features extracted by BOF to do classification is very effective. The classification accuracies with processed images are improved compared to the B-mode US data. Especially generated images with the information from original and processed images lead best classification result. That proves the image enhancement is very useful. We expect this method will be applied in identification of liver diseases when the data set contains more disease details. Out future work will also include the training of novel classification algorithms based on deep learning.

References:

1. Hacihaliloglu I. International journal of computer assisted radiology and surgery.2017, 12(6), 951-960.
2. Nowak E. et al.. InEuropean conference on computer vision 2006, 490-503.
3. Abdar M. et al. Expert Systems with Application, 2017, 67(C):239-251.
4. Shiji T P. et al. Procedia Computer Science, 2017, 115: 518-525.

Rapid, Label-Free Genetic Detection of Enteropathogens in Stool without Genetic Isolation or Amplification

Song Han[1], Mehmet C. Soylu[1], Ceyhun E. Kirimli[1], Wei Wu[2], Bhaswati Sen[3], Suresh G. Joshi[3], Christopher L. Emery[4], Giang Au[1], Xiaomin Niu[1], Richard Hamilton[5], Kyle Krevolin[6], Wei-Heng Shih[2], and Wan Y. Shih[1]

1. School of Biomedical Engineering, Science, and Health Systems, Drexel University, Philadelphia, PA 19104
2. Department of Materials Science and Engineering, Drexel University, PA 19104
3. Department of Microbiology and Immunology, Drexel University, Philadelphia, PA 19102
4. Department of Pathology, Drexel University, Philadelphia, PA 19102
5. Department of Emergency Medicine, Drexel University, Philadelphia, PA 19102
6. Microbiology & SIVM Laboratories, Hahnemann University Hospital, Philadelphia, PA 19102

Introduction: In underdeveloped countries where resources are limited, diarrheal infections (DIs) are a leading cause of illness and a major healthcare challenge, with 4.6 million episodes and 3 million deaths each year worldwide. In developed countries, DIs such as that caused by *Clostridium difficile* (CD) are a serious healthcare-associated infection. In the US alone in 2011, CD caused more than 500,000 infections and 29,000 deaths within 30 days of the initial diagnosis. The cost to treat CD infection (CDI) was over $3B a year in the US alone. On the other hand, current genetic detection methods require gene isolation, gene amplification and detection with a fluorescent-tagged probe. They typically require sophisticated equipment and expensive fluorescent probes, rendering them not widely available for rapid acute infection diagnoses at the point of care to ensure timely treatment of the diseases.
Piezoelectric plate sensor (PEPS) consisting of a lead magnesium niobate-lead titanate $(Pb(Mg_{1/3}Nb_{2/3})O_3)_{0.65}$-$(PbTiO_3)_{0.35}$ freestanding film 8 µm in thickness is a unique new sensor platform developed in Shih and Shih laboratories. For genetic detection, probe DNA complementary to a target DNA (tDNA) is immobilized on the PEPS surface. Binding of tDNA to the probe on the PEPS surface shifts the PEPS resonance frequency. What is unique about PEPS is that it can self-enhance its detection resonance frequency shift by 1000-fold through binding induced crystalline orientation switching. As a result, it can detect DNA with PCR sensitivity but without DNA isolation or amplification.

Methods: Taking advantage of the unprecedented sensitivity of PEPS, in this study, we further investigated *in situ* bacterial genetic signature detection directly from stool by employed a unique continuous flow system consisting of two temperature zones—a higher-temperature reservoir kept at 95 °C, where bacteria were loaded, to completely lyse the bacteria and de-hybridize the DNA and a low-temperature detection cell at the detection temperature where detection of the tDNA took place. A PEPS coated with the probe complementary to the tDNA was located at the center of the detection cell to detect the tDNA for 30 min. A low-cost, portable electrical impedance analyzer was used to measure the f due to the binding of the tDNA to the probe on the PEPS surface in real time.

Results: Using *Escherichia coli* O157:H7 (EC) as a model organism spiked in stool, genetic signature of EC was detected directly from stool at 150 CFU/ml, which was further validated by (1) *in situ* fluorescent reporter microspheres (FRMs)—fluorescent microspheres coated with reporter DNAs complementary to tDNA but different from the probe—following the stool detection, (2) subsequent microscopic visualization of the captured FRMs on the PEPS surface, (3) PCR verification of the captured tDNA on the PEPS surface, and (4) correct EC detection in blind spiked stool samples.
The same method was applied to detect *Clostridium difficile* (CD) infection (CDI) in 40 blind patient stools with 95% sensitivity (95%) and specificity (95%) against the gold standard, toxin culture (TC) testing, the same as commercial qPCR (Cepheid).

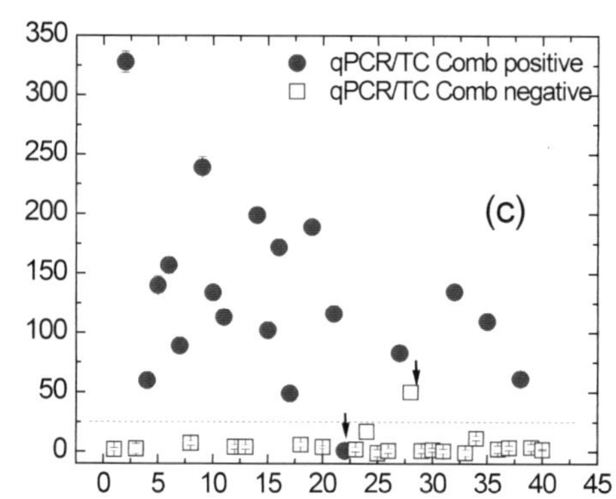

Figure1. - f of the 40 blinded patient stool samples.

Conclusion: Rapid, label-free bacterial genetic signature detection directly from stool without DNA isolation or amplification was demonstrated using PEPS with a temperature-controlled flow system both with EC as a model bacteria and in 40 clinical CDI patient stools with 95% sensitivity and 95% specificity. This indicates the potential of using PEPS as a rapid, low-cost, point of care (POC) detection modality to detect genes directly from patient samples for a wide range of applications.

Acknowledgment: This work was supported in part by the National Institute of Health Grants No. 1R41AI112224 and 1R41AI120445 and Pennsylvania Department of Health.

A Single Cell Pathogen Identification and Antimicrobial Susceptibility Testing System for Rapid Diagnosis of Infectious Diseases

Hui Li[1], Jian Gao[1], and Pak Kin Wong[1,2,3].
[1]Department of Biomedical Engineering, [2]Department of Mechanical Engineering, [3]Department of Surgery, The Pennsylvania State University, University Park, PA, USA

Introduction: Bacterial infection is a leading cause of morbidity and mortality. Current diagnostic methods typically require 3-5 days. Rapid microbiological analysis techniques are essential to properly manage infectious diseases and combat multidrug resistant pathogens. We developed an integrated system for rapid pathogen identification and antimicrobial susceptibility testing (AST) at the single cell level. A nanotube assisted microwave electroporation (NAME) technique delivers molecular biosensors into viable bacteria for multiplex single cell pathogen identification. We demonstrate the feasibility of NAME for identifying clinical isolates of bloodborne and uropathogenic pathogens and detecting bacterial pathogens directly from patient's samples in 30 minutes. Furthermore, the identifed bacteria were captured in microchannels for rapid AST. Using this approach, the time for microbiological analysis reduces from days to as few as one hour, which will have a significant impact on the clinical management of bacterial infections.

Methods: Molecular probe consisted of two DNA strands. The donor strands were complementary to the target RNA of interest. The fluorophores were labeled at the 3' end of the donor strands. The quencher strands were labeled with a dark quencher at the 5' end of the quencher strands immediately adjacent to the fluorophores on the donor strands. To transform the double-stranded probe into bacteria, 100 µl of bacterial sample was washed twice with phosphate buffer. The pellet was resuspended into 100 µl solution containing multiwall carbon nanotubes. Ten microliter of molecular probe was added and incubated at room temperature for 10 minutes. The sample was incubated in ice for 5 minutes and then loaded to a microwave oven for 10 seconds. Multiwall carbon nanotubes enhance the local electric field by ~90 times on the CNT tips and facilitate the electroporation with a minimal effect on the cell viability. The electroporated bacteria was incubated at room temperature for 30 minutes and washed 3 times with PBS for imaging. To conduct AST, the bacteria were loaded onto the microfluidic device by capillary force. The microfluidic device was mounted onto a heated stage for microscopic observation.

Results: We developed a microwave electroporation approach for transforming molecular probes into viable bacteria for pathogen identification and combined with a micrfluidic AST technique (**Figure 1A-C**). To demonstrate the feasibility of multiplex single cell pathogen identification using NAME, uropathogenic *E. coli* and *P. aeruginosa* were transformed with corresponding probes for pathogen identification (**Figure 1D**). High sensitivity and specificity were achieved for monomicrobial and polymicrobial scenarios. The identified bacteria were captured for AST by monitoring bacterial growth at the single cell level (**Figure 1E-F**). The resistant profiles were obtained in as few as 30 min, which is comparable to the doubling time of the pathogen.

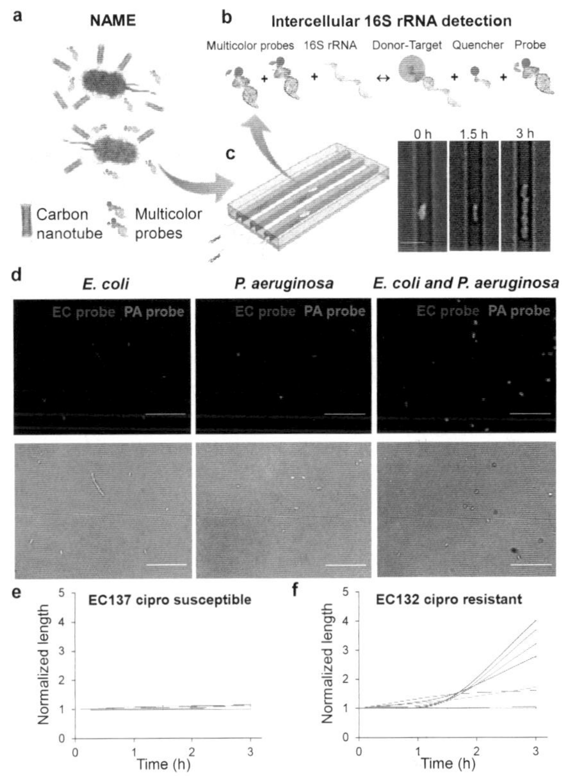

Figure 1. **An integrated system for single cell pathogen identification and AST.** (**A**) Schematic of NAME for pathogen identification. (**B**) Mechanism of molecular probing for 16s rRNA. (**C**) The viable cells were captured in microchannels for AST. (**D**) Multiplex detection of *E. coli* and *P. aeruginosa* by NAME. Scale bars, 25 µm. (**E-F**) AST profiles for two uropathogenic clinical isolates.

Conclusions: In this study, we develop an integrated system for pathogen identification and AST. The identification was achieved using NAME technique that delivers molecular biosensors for detecting intracellular 16S rRNA. AST was performed by monitoring the growth of the bacteria at the single cell level.

References:

1. J. Gao, et al. ***Nanomedicine: NBM***, in press.
2. H. Li, Y. Lu and P.K. Wong. ***Lab Chip***, 2018,18, 3086-3089.
3. Y. Lu, et al. ***Ana. Chem.*** 85, 3971-3976, 2013

Acknowledgements: This work is supported by NIH NIAID (R01AI117032) and Penn State Convergence of Materials and Life Sciences grant.

Ultraviolet Germicidal Irradiation of Clinical Equipment for the Prevention of Hospital Acquired Infections

Samuel Charpentier, Kendra Michaud, Andrew Strong, Jordan Anderson, Ying Sun, Ph.D.
University of Rhode Island, Kingston, RI 02881~*Correspondence:* Jordan.Anderson@va.gov

Introduction: According to the Center for Disease Control and Prevention, at any given time in the United States 4-6% of all patients in a hospital are fighting at least one infection that they acquired since entering the hospital. The spread of *Clostridium difficile* and Methicillin-resistant *Staphylococcus aureus* within the healthcare environment puts patients at risk and raises costs of healthcare across the board. We can begin to lower this number by enabling more effective methods of cleaning equipment that patients come into contact with on a regular basis. An inexpensive and efficient method of disinfecting medical equipment is through the use of UV light. This method has already been proven effective on instruments and microbiology lab equipment and has been expanded recently to include full patient room disinfection systems. The room disinfection method, however, requires an empty room to perform, and does not get a 360 degree clean of all the equipment, leaving infected equipment that will soon be used by patients. We will create a UV cabinet to be used to disinfect non-critical medical equipment commonly used in hospitals, without removing the ability to safely enter the room. This will solve the problem that infected medical devices and equipment cause due to the compromised immune system of the patients who come in contact with the equipment.

Methods: The method of disinfection proposed is germicidal ultraviolet light delivered by fluorescent lamps in a sealed cabinet. Short-wavelength UV-C (100-280 nm), although the most damaging type of UV radiation, has the strongest germicidal effect. The UV-C tube used in this project is the Philips model G4T5, emitting the desired wavelength of UV light: 254 nm. Many healthcare acquired infections (HAIs) are due to microorganisms that are antibiotic-resistant. Germicidal irradiation, using UV-C, effectively disables the metabolic functions of microorganisms, by creating thymine dimers in their genetic material. These areas of criss-crossed nucleic acids make it impossible for the bacterium to replicate or transcribe its DNA, effectively neutralizing or killing it. By placing the equipment inside the cabinet, closing the door, and activating the timer, you begin the disinfection process. The enclosure has a locking mechanism that is operated by a heavy duty dual latch system which ensures the prevention of UV radiation bleed through. With safety features, such as mechanical and magnetic reed switches, in-series, the cabinet will deactivate the lights when either connection is not detected. Ensuring the UV-C will not omit if the door is opened even slightly during the disinfection process. Along with the onboard safety features, the cabinet has a user interface LCD affixed with start and stop buttons. It displays a countdown timer for the amount of time left in the disinfection and readings of the temperature within the cabinet throughout the process. Once the user is ready, pressing start sends a signal to an Arduino Uno that controls a Light Relay, which turns the light circuit on. The Uno is also sending information to the temperature sensors inside the cabinet. The main components of the cabinet include the 36”x17”x18” aluminum enclosure, the UV light circuitry, and the external serviceable component box. The circuit was designed to accommodate two UV lights, although the number of UV lights within the enclosure could be increased if desired. To determine the effectiveness of the germicidal disinfection of the cabinet, a UV meter (SPER Scientific, UVC Light Meter 850010, SPER Scientific, Scottsdale, AZ) was used. A study was conducted to monitor the UV intensity inside the enclosure continuously for a duration of 200 s. The study was repeated having the sensor face different directions within the box, due to its unidirectional property.

Results: After experimental data was collected, where the UV intensity was measured at different points and sides within the cabinet, a graphical representation of the UV dosage was formed. Figure 1 was calculated by gathering the UV intensity reading every 10 s for 200 s. After gathering these readings they the intensities were multiplied by 10 s and added to the previous value, to get dosage. This gave us a graphical representation for the necessary exposure time to reach the UV dosage that would be needed to disinfect the targeted C. diff bacteria. The lowest dosage was from the left side of the box, due to this position not being directly in front of a UV bulb. After 150 s the left side will deliver the minimum UV dosage for the disinfection of C. diff bacteria at 6,510 µWs/cm2, this is shown on Figure 1 as the disinfection threshold, labeled as the dashed black line. In order to ensure complete disinfection, the exposure time is increased to 180 seconds.

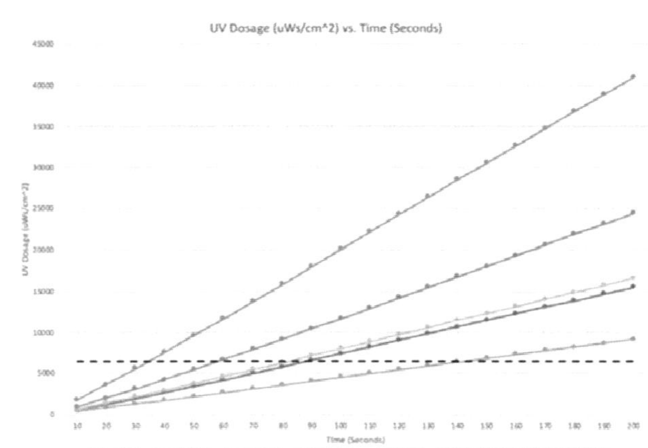

Figure 1. Graphical data representing the UV Dosage from the UV-C lights within the Cabinet. Data was gathered using a unidirectional UV intensity sensor called SPER Scientific, UVC Light Meter 850010.

Conclusion: A prototype germicidal device was developed in this study, which showed a cost-effective UV-C radiation enclosure for disinfecting portable hospital equipment. The device was able to deliver 2,000-10,000 μWs/cm2 of UV-C exposure in under four minutes on all sides of the cabinet. This would greatly reduce the time necessary for sanitation of certain items that require regular cleaning. This device is designed to have the appropriate amount of space to be used with the hospital setting. Making it a strong asset in a hospital environment.

Acknowledgment: This project was funded by a research grant from the University of Rhode Island 2017-18 Undergraduate Research and Innovation Awards. Providence VA Medical Center is thanked for the project direction and guidance.

References: Riley, Kyle, et al. *Germicidal Ultraviolet Light Enclosure*. URI, 2017, pp. 1–2

Bioprinting of Complex 3D Vascular Networks within Cell-Laden Hydrogels

Shen Ji,[1] Emily Almeida,[1] and Murat Guvendiren[1,2]
[1]Otto H. York Department of Chemical and Materials Engineering, New Jersey Institute of Technology
[2]Department of Biomedical Engineering, New Jersey Institute of Technology

Statement of Purpose: Creating microchannels with user-defined and tunable shape and size within cell-laden 3D hydrogels is a major challenge.[1] This limits our ability to fabricate tissue models with built in vasculature and human-scale tissues and organs. In this work, we report a novel bioprinting approach to fabricate microchannels within cell-laden hydrogels. We believe that our approach could potentially be used to fabricate hydrogel devices with complex microchannels to create vascularized tissue models.

Methods: *Bioprinting:* Matrix inks were prepared from 9% (w/v) methacrylated alginate (MeAlg) or 15% (w/v) methacrylated hyaluronic acid (MeHA, 90%) in PBS containing 0.05% (w/v) photoinitiator, LAP (lithium phenyl-2,4,6-trimethylbenzoylphosphinate). Pluronic (F-127, 40% (w/v) in DI water) was used as the sacrificial hydrogel ink. After printing was completed, the sacrificial hydrogel was washed away to form channels.

Cell culture studies: Human mesenchymal stem cells (hMSCs, passage 4, Lonza) were used for viability studies, within scaffolds with and without channels. Scaffolds were cultured for 4 days, and live-dead staining was done at day 1 and day 4 time points. For live-dead staining, cells were stained with calcein-AM (live) and ethidium homodimer (dead) for 30 min. Human umbilical vein endothelial cells (HUVECs, passage 4-6, Lonza) were used for cell attachment studies on the channel surfaces. Cell suspension (3×10^7 cells/ml) was injected into the channels using a micropipet (0.1-10 µL tip). Scaffolds were then incubated up to 9 days.

Results: We developed a novel approach to create complex microchannels within cell-laden hydrogels. Photocurable matrix hydrogels (MeHA or MeAlg) were printed layer-by-layer. Each printed layer was briefly (5 s for MeHA and 15 s for MeAlg) exposed to light to induce partial crosslinking to create self-supporting layers. When the desired thickness was reached, the top layer was not exposed to light and sacrificial hydrogel was directly printed within this viscous uncrosslinked layer. The layer was then exposed to light, and additional layers were added as needed. The hydrogel is fully crosslinked after printing was completed After fully crosslinking the system (240 s for MeAlg, and 90 s for MeHA), sacrificial hydrogel is washed away to form channels. We were able to create channels from 150-1200 mm by controlling the print pressure and speed (using 0.1 and 0.16 mm diameter needles) (Figure 1A and 1B). No significant change in size was observed (for both MeHA and MeAlg) when (sacrificial) strut and channel (after washing the sacrificial strut) size was compared. A representative strut is shown in Figure 1C. HUVECs attached, proliferated and formed confluent layers on channels (Figure 1D). Cell (hMSC) viability was 88-90% within MeAlg and MeHA hydrogels at day 1 (with or without channels) (Figure 1E and 1F). For hydrogels without channels, viability was ~86% for MeAlg and 81% for MeHA at day 4, whereas remained same for constructs with channels.

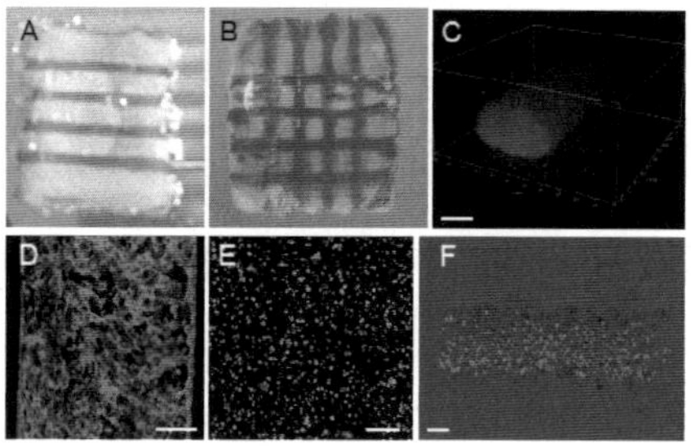

Figure 1. (A) MeAlg hydrogel with perfusable channels (600-µm in diameter). (B) MeAlg hydrogel (10x10 mm) with interconnected channels in two different planes (y-direction). Channels were injected with red food dye (A-B). 3D designs of hydrogels (grey) with sacrificial struts (blue). (C) Confocal scanning image of the sacrificial hydrogel (doped with a red fluorescent dye) printed within matrix hydrogel. (D) Confocal scanning image of HUVECs (F-actin: green, and cell nuclei: blue) forming a layer within the channel. (E-F) Confocal scanning images of hMSCs (stained for live (green) and dead (red) stain) within MeAlg hydrogels (top scan (E) and cross-section (F). Scale bars are 200 microns.

Conclusions: We successfully showed a novel approach to create microchannels, with user-defined and tunable shape and size, within photocurable, cell-laden hydrogels. We showed that endothelial cells (HUVECs) attached and formed confluent monolayers when seeded within channels. We also incorporated stem cells (hMSCs) into the ink formulation prior to printing to investigate the effect of our approach on stem cell viability. Stem cells remained viable within the 3D printed constructs with or without embedded channels. This approach could potentially be used to fabricate hydrogel devices with complex microchannels for fabricating vascularized tissues and studying cell behaviors on diverse channel surfaces.

Conflict of Interest: New Jersey Institute of Technology has a financial interest in the bioprinting approach reported in this manuscript, and has already filed a provisional patent application for the technology described in this paper (U.S. Application No. 62/715,869).

Acknowledgements: NJIT New Faculty Startup Fund (M.G.), NSF Award# 1714882 (M.G.), Faculty Seed Grant from CEMB, NSF CMMI: 15-48571.

References:

1. Liaw, C. Y.; Guvendiren, M., Current and emerging applications of 3D printing in medicine. *Biofabrication* **2017,** *9* (2), 024102.

3D Printing with Peptide-Polymer Conjugates to Control Spatial Peptide Concentration

Hafiz Busari[1], Kelly Seims[1], Paula Camacho[2], Lesley W. Chow[1,2]
[1]Department of Materials Science and Engineering, Lehigh University, Bethlehem, PA, USA
[2]Department of Bioengineering, Lehigh University, Bethlehem, PA, USA

Introduction: Functional regeneration of biological tissues remains a challenge because native tissues contain gradients in biochemical, mechanical, and structural properties critical for tissue function. Biomaterials for tissue engineering must therefore be designed to provide spatially organized biochemical and physical cues to guide functional tissue formation.[1] Here, we introduce a novel platform that combines multi-material 3D printing with peptide-functionalized polymer-based materials to produce scaffolds with controllable peptide organization. This strategy allows us to independently and simultaneously tune the concentration and location of multiple peptides within a single construct. In this work, we synthesized peptide-polymer conjugates that include amino acid sequences to promote cell adhesion (RGDSK(biotin)-PCL) or prevent cell adhesion[2] (RGESK(azide)-PCL). Biotin and azide groups were included for specifically labeling each peptide on 3D-printed scaffold surfaces. This study provides a proof-of-concept to demonstrate how the combination and spatial organization of biochemical cues can be used to tune cell behavior within a continuous scaffold.

Methods: Poly(caprolactone) (PCL) was modified with peptides to synthesize peptide-PCL conjugates using methods previously described.[3] Briefly, peptides were synthesized using a CEM Liberty Blue automated microwave peptide synthesizer before purifying and covalently coupling to PCL.[3] Peptide-functionalized scaffolds were 3D printed using a customized solvent-cast 3D printer with inks containing the conjugate at 0, 1, 5, and 20 mg/mL concentrations dissolved with unmodified PCL in a volatile solvent hexafluoro-2-propanol (HFIP). Two printer heads, each containing an ink with a different peptide-PCL conjugate, were used to independently control peptide deposition within a single scaffold. The presence and location of RGDSK(biotin)-PCL and RGESK(biotin)-PCL peptides were detected by labeling with streptavidin-tagged markers (colloidal gold nanoparticles or fluorescein isothiocyanate (FITC)) or dibenzocyclooctyne-cyanine3 (DBCO-Cy3), respectively. The 3D-printed scaffolds were imaged with scanning electron microscopy (SEM) and confocal fluorescence microscopy to observe scaffold structure and peptide spatial concentration. NIH-3T3 mouse fibroblasts were seeded on PCL, single-peptide, and dual-peptide scaffolds at 75,000 cells per scaffold. Samples were fixed and stained after 24 hours with rhodamine-phalloidin and Hoechst to label actin filaments and nuclei, respectively. The number of adhered cells and cell morphology were characterized using fluorescence confocal microscopy.

Results: Solvent-based 3D printing with peptide-PCL conjugates allowed for single-step fabrication of peptide functionalized scaffolds without the need for post-processing. SEM of PCL only or PCL with RGDSK(biotin)-PCL scaffolds labeled with streptavidin-gold nanoparticles confirmed the presence of peptides. Further, increasing RGDSK(biotin)-PCL conjugate concentrations corresponded with statistically significant increases in the number of gold nanoparticles on the surface, indicating ability to fine-tune peptide density. RGDSK(biotin)-PCL and RGESK(azide)-PCL scaffolds labeled with streptavidin-FITC and DBCO-Cy3, respectively, showed how these local changes in peptide density translated to a global increase in fluorescence in the entire scaffold and statistically significant differences in fluorescence intensity. Scaffolds were also 3D printed to localize peptides to alternating fibers or layers to demonstrate how peptide functionalization can be controlled within the same construct (Figure 1). NIH-3T3 mouse fibroblasts seeded on these alternating scaffolds showed preferential adhesion and increased cell spreading on RGDS(biotin)-PCL fibers compared to RGES(azide)-PCL fibers. This illustrates our ability to tune local cell behavior within a continuous biomaterial.

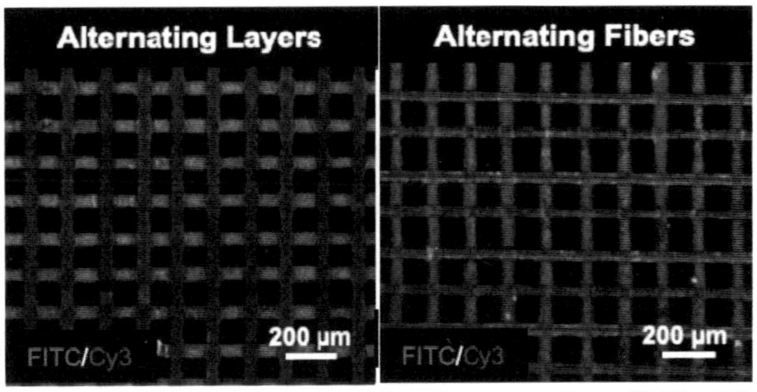

Figure 1. Fluorescence overlay images of scaffolds 3D printed with 20 mg/mL RGDS(biotin)-PCL and 20 mg/mL RGES(azide)-PCL in alternating layers or fibers. Peptides were labeled with streptavidin-FITC and DBCO-Cy3 to visualize peptide organization in user-defined locations within a continuous, 3D-printed scaffold.

Conclusions: This work introduces a versatile platform to control multi-peptide concentration and organization within a continuous, 3D-printed scaffold to guide tissue formation. Spatial control of biochemical cues can be exploited to fine-tune local cell behavior. Future studies are focused on utilizing these biorthogonal chemistries to modify the biochemical cues over time for temporal control of cellular behavior.

References: [1] Chow & Fischer, *Exp Biol Med* 241(10): 1025-1032, 2016.; [2] Hersel et al., *Biomaterials* 24(24), 4385-4415 [3] Chow et al., *Adv Healthc Mater* 3(9): 1381-1386, 2014.

Design of Porous Polyetheretherketon Topologies Using Fused Filament Fabrication

William Hartley UG, Kenny Cho UG, Anthony Law UG, Wing Ni Lee UG.
Drexel University.

Introduction: Aseptic loosening of non-porous implants utilized for bone reconstruction applications remains one of the most common implant pathologies experienced by recipients of said implants, and accounts for a large percentage of total hip arthroplasties revisions and total knee arthroplasties (51.9% and 21.9%, respectively).[1,2] Aseptic loosening can also result in inflammation, pain, and imaging complications. Conversely, the use of porous implants can increase bone in-growth and reduce the likelihood for recipients to experience aseptic loosening. Although conventional metallic porous orthopedic implants significantly reduce post-implantation pathologist, they exhibit much higher elastic moduli compared to that of human bone, and are therefore still prone to stress shielding as well as aseptic loosening. The design proposed by the group utilizes Polyetheretherketone (PEEK), and a novel tripply periodic minimal surface (TPMS) design to reduce aseptic loosening, minimize stress shielding in orthopedic implants, and promote osseointegration. The design will be manufactured via the 3D printing technique known as Fused Filament Fabrication (FFF).

Methods: After determining the initial medical need, the team conducted a systematice literature review to explore conventional solutions, novel solutions, physiological constraints, potential material candidates, and potential surface geometries. Using keywords such as orthopedic implant, biomaterial, porous biomaterial, PEEK, porous PEEK, 3D printing, additive manufacturing, surface topography, and TPMS, 87 papers were initially catologued. Each paper was read and assesed for relevance to the medical need. A total of 69 papers were used in the final version of the review. This review served as a guide to designing the proposed solution.

To create a porous topology (or surface archetecture) mechanically and physically similar to that of trabecular bone, PEEK, a thermoplastic polymer that exhibits an elastic modulus similar to that of human bone, was utilized. In adition to choosing a material, careful consideration was given to determining which structure geometry would adress the medical need. Ultimately a TPMS gyroid structure was chosen. TPMS's are mathematically derived, smooth infinite surfaces that free form self-intersections and are periodic in the X, Y, and Z directions. By integrating two surfaces together, a solid surface can be created with controllable strut width and unit cell density. The gyroid structure was chosen due to its printability, high porosity, pore interconnectivity, and surface area to volume ratio. To manufacture the designed structure PEEK was drawn into filament which was then extruded by the Apium 220 (ANIWAA, New York, NY). This printer utilizes FFF techniques; it extrudes filament from a heated nozzle with 3 axis of movement onto a build plate with 3 axis of movement in horizontal stacks. After manufacturing samples, porosity, permeability, pore size, surface area, printability, and mechanical properties were experimentally measured to ensure resemblance of the material to trabecular bone. The quantitative requirements and justification for the design are as follows: porosity >60% (increased cell ingrowth potential), permeability coeficient 200 μm^2 - 1000 μm^2 and Reynold's number 1 – 10 (reported ranges for trabecular bone), pore size 400 – 700 μm (optimal range based on testing data), surface area 101 – 162 cm^2 (optimal range based on testing data), and mechanical properties, namely young's modulus 0.02 – 2.0 GPa, compressive yield stress 2 – 17 MPa, and ultimate compressive strength 0.22 – 10.44 Mpa (reported ranges for trabecular bone). [3] To verify porosity, pore size, and surface area, MicroCT tests were performed on a Skyscan 1172 (Micro Photonics Inc., Allentown PA) with a source voltage of 44 kV, source current of 222 μA, and an image pixel size of 7.93 μm. Scans were subsequently reconstructed and processed using Analyze 12.0 software (AnalyzeDirect, Inc., Overland Park, KS), and the relevent metrics were calculated. Permeability coefficient and Reynold's number were verified using falling head test method, which experimentally calculates flow type and mass transfer through a porous structure. Mechanical properties of the design were tested via unconfined uniaxial compression test (strain rate: 0.001 s^{-1}) on an MTS Criterion loadframe (MTS, Eden, MN).

Results: The printing of gyroid structures using FFF techniques was verified. MicroCT images scanned by Widener University were scanned at inappropriate current and voltage levels, rendering the first round of testing obselete. The permeability testing setup has been verified via [4]. The average permeability coefficient taken from n=3 samples tested n=3 times was 557 $\mu m^2 \pm$ 24.7. The average young's modulus of the design was 238 ± 11.9 MPa and the average compressive yield stress was 14.2 ± 0.12 Mpa.

Conclusions: The manufacturing technique used to print the proposed design (FFF) was verified and will not be changed moving forward. Additional MicroCT tests will need to be conducted in the future to establish statistical significance. Both permeability and mechanical requirements were met in version 1 of the design. In the future, the group will evaluate further design solutions and options to further the robustness of the design.

References:

1. American Joint Replacement Registry Annual Report 2016
2. AAOSNow. 2018
3. D. Apostu. Et al. Journal of International Medical Research. 2017;46(6):2104–2119
4. Bobbert et al. Acta Biomaterialia. 2017;53: 572-584

Acknowledgements: The team would like to acknoloedge Dr. Kurtz, Dr. Frohbergh, Dr. Bhatt, Dr. Balasubramanian, Exponent Inc., Implant Research Center, and Widener University.

Synthesis and Characterization of Self-Assembling Hyaluronic Acid-Isoleucine Biomimetic Hydrogels

Sydney Yang[a], Hafiz Busari[b], Lesley W. Chow[a,b]
[a]Department of Bioengineering, Lehigh University, Bethlehem, PA 18015 USA
[b]Department of Material Science and Engineering, Lehigh University, Bethlehem, PA 18015 USA

Introduction: The properties and composition of the extracellular matrix (ECM) are crucial to tissue function and regeneration. Specifically, the ECM is significantly affected by structure-function relationships found in its major components, such as proteoglycans. Proteoglycans are primarily made up of glycosaminoglycans (GAGs), which are linear polysaccharides diverse in type, number, and length and involved in many cell-ECM and cell-cell interactions[1]. This diversity introduces complex yet specific biochemical and mechanical properties that are difficult to recreate with synthetic systems. Hydrogels composed of GAGs introduce the potential to mimic the biochemical and biophysical properties of the native ECM[2]. In this study, hyaluronic acid (HA) was used as the hydrogel backbone onto which β-sheet-forming peptides were attached at varying degrees of modifications. The HA-peptide molecules peptides self-assemble via non-covalent cross-linking to enable dynamic hydrogel formation.

Methods: The β-sheet-forming peptide DDIIIIDDY (I4) was synthesized using standard Fmoc solid phase peptide synthesis on an automated microwave peptide and purified with high performance liquid chromatography (HPLC). Electrospray ionization (ESI) was used to confirm the expected mass. The N-terminal amine on the purified peptides were modified with dibenzocyclooctyne-acid (DBCO-COOH) via EDC-NHS to form DBCO-I_4. The carboxylic acids along the HA backbone was reacted with amine-PEG3-azide at either 8% or 20% modification via EDC-NHS chemistry to form HA-azide. The DBCO-peptides were conjugated to HA-azide through copper-free strain-promoted azide-alkyne cycloaddition[3]. Each synthesis step was verified using ^{1}H nuclear magnetic resonance (NMR) and Fourier-transformed infrared (FTIR) spectroscopy. HA-I_4 molecules were self-assembled in phosphate buffer saline (PBS) at 10, 15, and 20% w/v, and the resulting hydrogels were characterized using rheology via a parallel plate rheometer. The HA-I_4 sample was gelled on the rheometer for 30 minutes before applying 1000% strain for 10 seconds. After shear breakage disrupted the β-sheet peptide crosslinks, the sample was monitored for an additional 30 minutes to verify hydrogel re-assembly.

Results: I_4 peptides were successfully synthesized and purified. NMR confirmed the peptides were modified with DBCO. FTIR showed HA-azide was successfully synthesized at low (8%) and high (20%) degrees of modification and that DBCO- I_4 was conjugated to HA-azide (Figure 1). Specifically, the peak at 2200 cm^{-1} indicating the azide group was present in HA-azide but absent or reduced in the HA and HA-I_4 peaks, respectively. This verified that the DBCO-azide reaction was successful. Rheology demonstrated the hydrogel formed and reassembled after shear breakage, confirming self-assembly of HA-I_4 molecules into a hydrogel with viscoelastic properties. Before strain, the storage modulus reached equilibrium at 60 kPa and 75 Pa for the 20% and 8% HA-I_4 modifications respectively. After shear breakage, the the elastic storage modulus was greater than the value for the loss modulus for both the 20% and 8% modified HA-I_4. Both modifications demonstrated reassembly of the hydrogel over time after shear breakage. The 20% modified HA-I_4 formed a hydrogel with a larger elastic storage and loss moduli, at 50 kPa and 15 kPa respectively, compared to the 8% modified HA-I_4 hydrogel, at 80 Pa and 70 Pa respectively.

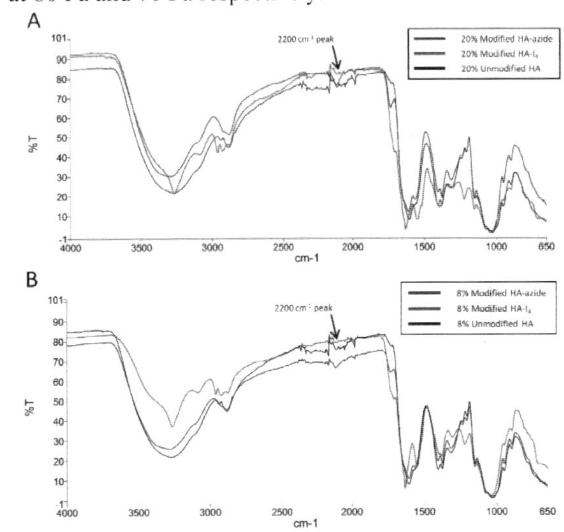

Figure 1. FTIR of unmodified HA, HA-azide, and HA-I_4 at (A) high (20%) or (B) low (8%) degree of modification (arrow indicates absence/presence of azide at 2200 cm^{-1}).

Conclusions: HA-I_4 molecules self-assembled into hydrogels at high and low degrees of peptide modification. As expected, the 20% modified HA-I_4 showed larger elastic storage and loss values than that for the 8% modified HA-I_4. The greater density of β-sheet peptides increased the degree of crosslinking between HA chains, resulting in higher stiffnesses. This illustrates how we can tune the mechanical properties of the hydrogel to influence cell behavior. Ongoing experiments include fully characterizing hydrogel mechanical properties at additional degrees of modification to observe the effect of peptide modification on gelation time and stiffness. Future research will investigate the effect of hydrogel stiffness on encapsulated cell behavior. This information will be utilized to optimize hydrogel properties to more accurately imitate the native ECM, providing a versatile platform for tissue and organ repair.

References: [1] Christian+ *J Cell Sci.* 2010; 123:4195-4200; [2] Drury+ *Biomaterials.* 2003; 24(24):4337-51; [3] Fu S. et al. *Carb Poly.* 2017; 169:332-340.

Acknowledgements: SY acknowledges the Grants for Experimental Learning in Health (GELH) grant funded through the P.C. Rossin College of Engineering at Lehigh University. HAB and LWC acknowledge start-up funds awarded to LWC from Lehigh University.

A Polymerization Friendly Dibenzoazacyclooctyne Monomer for Two-Step Bioconjugation

Shashank Kosuri, Adam J. Gormley.
Department of Biomedical Engineering, Rutgers University, Piscataway, New Jersey

Statement of Purpose:
Bioorthogonal click chemistry has revolutionized the field of bioconjugate chemistry. Strain promoted alkyne-azide cycloaddition (SPAAC) is particularly attractive since SPAAC does not require the presence toxic catalysts for bioconjugation.[1,2] For this reason, a variety of dibenzoazacyclooctynes (DIBAC) and their derivatives have been developed for efficient bioconjugation to azides[3]. Unfortunately, the intolerance of these functional groups to radical polymerization reactions necessitates multi-step post-polymerization modifications. To overcome this challenge, here we report the synthesis of a novel dibenzocyclooctyne monomer masked with a cyclopropenone (DIBACm), that can be directly copolymerized into the polymer backbone. Subsequent addition of the clickable azide with UV deprotection of the cyclopropenone results in rapid and efficient bioconjugation via SPAAC.

Methods:
DIBACm is synthesized via a five step process and is characterized by UV VIS spectroscopy, mass spectrometry and NMR. Copolymerization proceeds in a 96 well plate along with dimethyl acrylamide (DMA monomer) by PET-RAFT in DMSO for incorporation into the polymer backbone[4]. After polymerization, crude polymer is purified using Sephadex G25 resin to remove any unreacted monomer and characterized by size exclusion chromatography with multi-angle light scattering (SEC-MALS). The purified polymer is then irradiated under UV light (365 nm) for 15 mins to unmask the cyclopropenone group and generate the azide-reactive alkyne. PEG-N_3 (polyethylene glycol-azide) is then added to the polymer (1:1 azide:alkyne eqiv) to enable SPAAC. This reaction proceeds in less than 1 hr but is generally left overnight. Molecular weight shifts and conjugation are confirmed by SEC-MALS.

Results:
Cyclopropenone-masked DIBACm does not react with azides under ambient conditions but very efficiently produce reactive dibenzoazacyclooctynes with UV deprotection. After synthesis, the presence of the cyclopropenone group is confirmed by UV-VIS spectrophotometry (Figure 1). The UV spectra of DIBACm in DMSO (Fig 1, red) shows two close-lying intense bands (»$_{max}$ = 331 and 347 nm). UV deprotection results in the loss of cyclopropenone group and formation of acetylenes (Fig 1, blue). After deprotection, sodium azide addition results in complete loss of the spectrum indicating the formation of triazole, which has no absorbance above 340 nm (Fig 1, yellow).

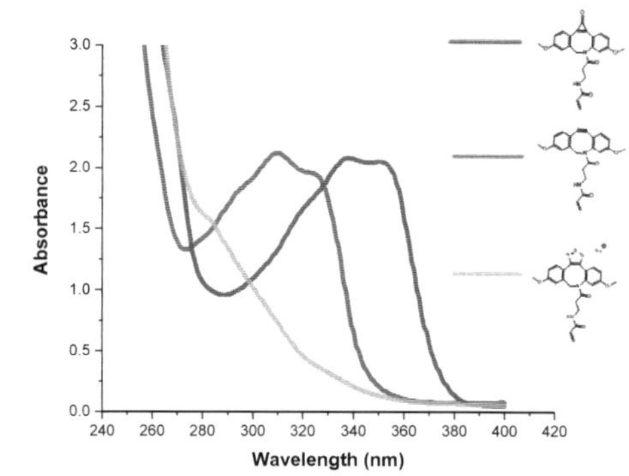

Figure 1: Spectrum of **DIBACm** (red), its deprotection (blue) and after SPAAC (yellow).

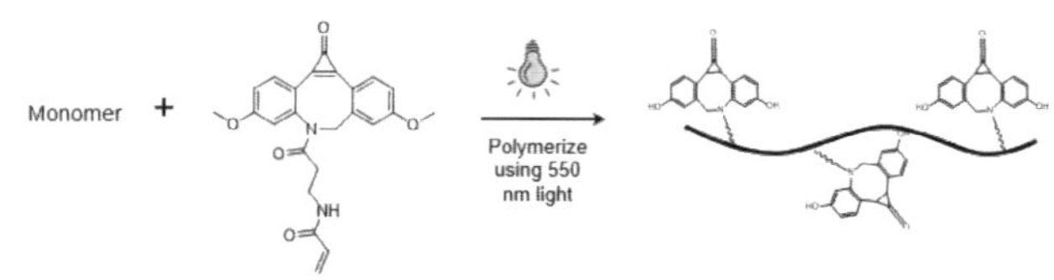

Step 1: Copolymerization with **DIBACm**.

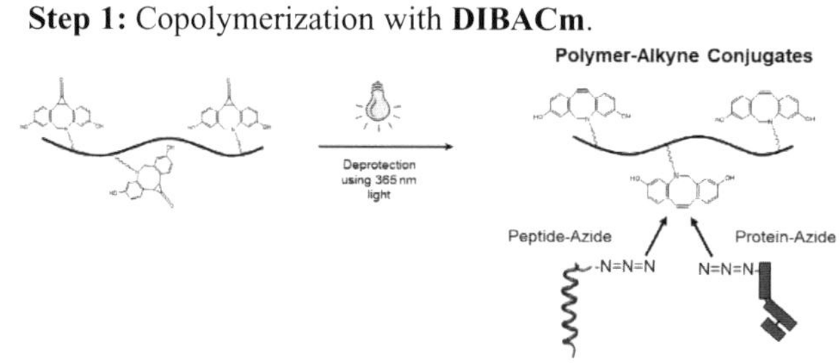

Step 2: UV deprotection and SPAAC bioconjugation.

We have tested the ability of our DIBACm to be copolymerized into the polymer backbone (0-10 mol%) and are able to achieve good reaction control up to 5 mol%. Molecular weight shifts after PEG-azide addition are observed (data not shown) confirming SPAAC.

Conclusions: Moving forward, this monomer may be a valuable tool for bioconjugation due to its versatility as post-polymerization modifiers with few synthetic steps.

References:
1. Pickens, C. J. Bioconjugate chemistry, 29(3), 686-701.
2. Poloukhtine, A. A. Journal of the American Chemical Society, 131(43), 15769-15776.
3. Debets, M. F., Chemical communications, 46(1), 97-99.
4. Gormley, A. J., Angewandte Chemie International Edition, 57(6), 1557-1562.

Selective Replacement of Damaged Airway Epithelium for Functional Recovery of Donor Lungs

Jinho Kim[1], Zidong Zhang[2], Yufei Wang[2], John O'Neill[3], Brandon Guenthart[4], Ahmed Hozain[4], Yulia Tipograf[5], Matthew Bacchetta[5], and Gordana Vunjak-Novakovic[3]
Departments of [1]Biomedical Engineering and [2]Electrical Engineering, Stevens Institute of Technology, NJ, USA
Departments of [3]Biomedical Engineering and [4]Surgery, Columbia University, NY, USA
[5]Department of Thoracic Surgery, Vanderbilt University, TN, USA

Introduction: Lung transplant is the only clinical option available for patients with end-stage lung disease. This procedure, however, is often delayed due to the severe shortage of viable donor lungs [1]. As a result, nearly 25% of the patients do not receive suitable lungs for more than a year. While waiting, 11% of the patients die or are taken off the transplant list due to deteriorating illness. While timely organ transplantation is one of the major contributors to favorable treatment outcomes, more than 80% of donor lungs are rejected for transplant due to edema, contusion, or diminished lung function [2]. Notably, lung damage is mostly located in lung epithelium. Aiming to increase the number of transplantable lungs, we established a cell-based lung regeneration approach where epithelium in selected regions (e.g., most severely damaged regions) of the donor lung is replaced with healthy epithelium, thereby achieving recovery of lung functions without compromising the delicate lung tissues.

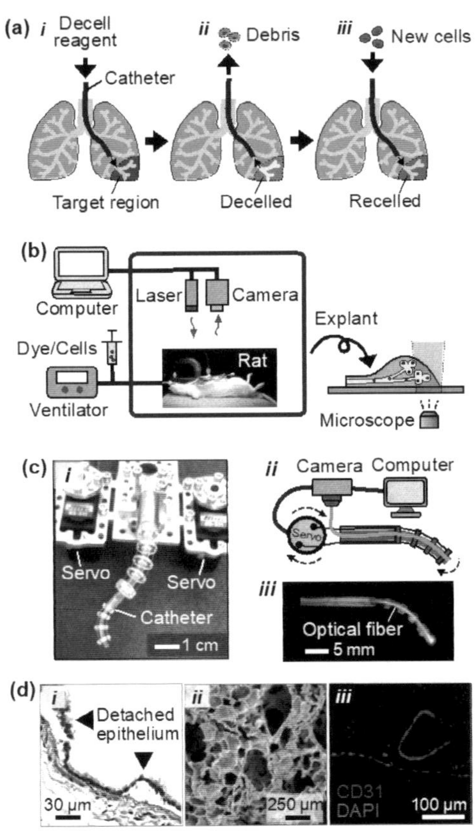

Figure 1. Selective replacement of epithelium in lungs: (**a**) Overview of epithelium replacement in selected lung regions. (**b**) Deep-tissue imaging setup. (**c**) Image-guided device for cell removal and delivery. (**d**) Removal of airway epithelium with preservation of underlying tissues.

Methods: For removal of the airway epithelium in a selected region of the lung, a visually guided steerable catheter (or video bronchoscope) is placed near a target area. A small volume of reagents for cell removal (e.g., hundreds microliters of detergents or enzymes) is then instilled into the airways through the catheter creating a thin liquid layer (thickness: several micrometers) of decellularization solution onto the epithelium [3] (**Fig. 1a,** ***i***). Following incubation, the epithelial cells detached are removed from the lung by repeated washing of the airways with saline solutions (**Fig. 1a,** ***ii***). The denuded airways through the decellularization procedure are then reseeded with new epithelium by infusion of a liquid bolus carrying the cells, which will generate a cell layer on the airway surface (**Fig. 1a,** ***iii***). When carefully controlled, the airway epithelial cells can be gently removed and replaced without damaging underlying airway tissues.

Results: To monitor delivery of liquid (i.e., decellularization solution and cell-carrying solution) into selected regions of the lungs, we constructed a deep-tissue imaging system that allows visual inspection of fluorescently labeled reagents and cells administered to the lungs without processing the lung tissues (**Fig. 1b**). In addition, we developed a visually guided steerable delivery device that allows minimally invasive instillation of decellularization solution and new cells into predefined lung regions. The translation movement of this device is manipulated by computer-controlled servomotors that enable gentle insertion of the device into a desired location for removal and replacement of the airway epithelium (**Fig. 1c**). Using this approach, we were able to detach the epithelium layer from the airway while preserving the underlying lung tissues such as blood vessels. Also, we demonstrated seeding of cells, labeled with red quantum dots (Qdot), onto the airway surfaces via liquid instillation (**Fig. 1d**).

Conclusions: We present here a cell-based approach for donor lung regeneration through targeted cell replacement enabled by minimally invasive imaged-guided techniques for cell removal, delivery, and monitoring. We envision that this innovative approach could expand the pool of donor lungs that are more acceptable for transplantation.

References:

1. J. Kim et al., P. Natl. Acad. Sci. USA, 2015;112(37), 11530-11535
2. J.D. O'Neill et al., Nat. Biomed. Eng., 2017;0037
3. B.A. Guenthart et al., J. Heart Lung Transplant., 2019;38(2):215-224

Acknowledgements: New Faculty Startup Fund from Stevens Institute of Technology, Research Grant from New Jersey Health Foundation, NIH grants (HL120046, EB002520, and L134760), and Mikati Foundation.

Self-assembling peptide induces axon infiltration following spinal cord injury

Kiet A. Tran, Paul P. Partyka, Peter A. Galie
Rowan University

Introduction: Spinal cord injury (SCI) often results in neuronal death, which leads to permanent paralysis and sensational loss below the injury area [1]–[3], due to the lack of innate axon regeneration in the CNS [1]. The complex pathophysiology behind SCI presents challenges, including demyelination, glial scarring, apoptosis of neurons, and cyst formation [4] Recent work in our laboratory has demonstrated that aligning vasculature within a collagen-based scaffold can direct the orientation of regenerating axons following spinal cord injury [5]. However, incorporation of the collagen-based scaffold into the host tissue remains a challenge. Self-assembling peptide nanofiber scaffolds (SAPNS) are found to be biocompatible and promote axon regeneration [6]. Thus, the goal of this study is to evaluate the effect of injecting a RADA-16 peptide hydrogel with endothelial cells at the site of a contusion spinal cord injury on axonal infiltration.

Materials: Eight Sprague Dawley female rats were exposed to a C4-C5 spinal cord compression with a force of 150,000 dynes, which is considered a moderate contusion level. One week later, 10-μL peptide-cell injections were administered into the cavity that formed in the injury site. GFP-labeled human cerebral microvascular endothelial cells (HCMEC/D3) were cultured on gelatin-coated polystyrene petri dishes and GFP-labeled human brain vascular pericytes (hBVP) were cultured in regular cell culture dishes. The cells were suspended in 20% sucrose at cellular densities of either 5M/mL or 2.4M/mL, with a ratio of 5:1 hCMEC/D3 to hBVP. The cell solution was combined with self-assembling peptide to yield final concentrations of 10% sucrose and 1 mg/mL peptide. Two rats received 10μL of acellular peptide to serve as negative controls, three received 10μL peptide containing cells at a density of 2.4M/mL and three received 10μL of the higher cellular Matrigel was applied outside of the injection site to keep the injected peptide within the dura. Animals were sacrificed after two weeks.

Results: Cellular peptide transplants displayed a high level of permissivity to Tuj-positive host axons compared to the acellular controls. The high cellular density condition appeared to have the highest amount of Tuj-positive host axons within the transplant compared to the other conditions. As suspected, the acellular peptide condition showed low Tuj-positive host axon infiltration at the scaffold-host interface, with even fewer axons at the center of the cavity. 5-HT positive axons were also present in the animals treated with cell-seeded scaffolds. The endothelial cells and pericytes organized into multicellular structures, and lumens in the vessels were observed in the low-density condition. Neurovascular interaction

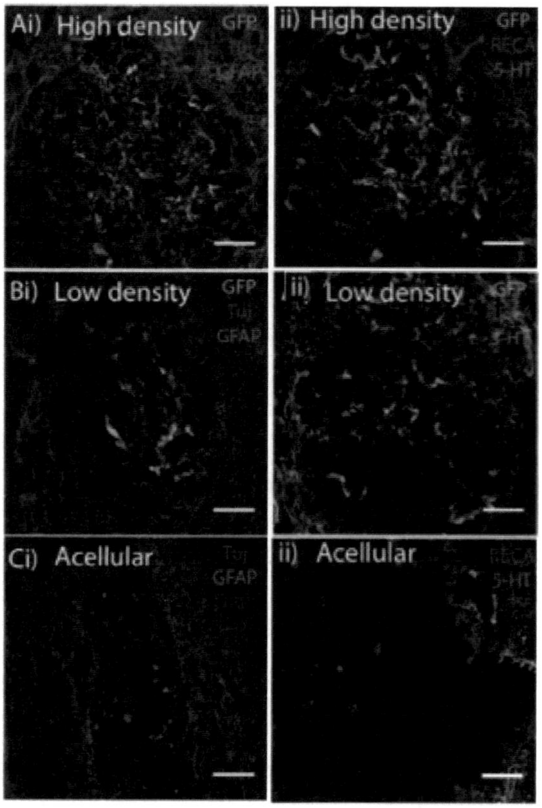

Figure 1. Host axon (Tuj) and 5-HT infiltration within the cell-seeded peptide for A) high density, B) low density, C) acellular conditions

was observed between infiltrating Tuj-positive axons and transplanted GFP-positive cells, similar to our findings in collagen-based scaffolds. Immunostaining for ZO-1 indicated that vessels with lumens formed tight junctions, consistent with vasculature within the central nervous system. Additionally, staining for ED-1 indicated that the cell-seeded scaffolds mitigated an immune response in the area surrounding the injury.

Conclusion: SAPNS are an effective biomaterial for spinal cord injury because they incorporate well into the host tissue and support growth of vasculature following injury. Our results indicate strong cell survival and axon infiltration into the cavity area. Although the results demonstrate that axons do infiltrate regenerate, further experiments are being conducted to determine behavioral changes in response to the treatment.

References:
1. Moradi F. et al. J Neurosci. 2012;90:2335-2348.
2. Guo J. et al. Biol Med. 2007;3:311-321
3. Liu Y. et al. Acta Biomater. 2013;9:8075-8088
4. Thuret S. et al. Nat Rev Neurosci. 2006;7:628-643
5. Partyka P. et al. Nature Sci Rep. 2019; in press
6. Sang L. et al. Biol Med. 2015:11:611-620

Effects of Combinational Treatment on Restoring Bone Morphology after Contused Spinal Cord Injury

Michael Feeley, Shawn McGinley, Anita Singh
Department of Biomedical Engineering, Widener University, Chester, PA, USA

Introduction:
Spinal cord injury (SCI) refers to permanent damage to any part of the spinal cord as well as the resulting changes in strength, sensations, and other functions that are located below the injury site. As of 2017, SCI is seen in about 54 people per million residents of the United States with approximately 17,500 new cases seen each year. The leading four causes of SCI are vehicular crashes (38.4%), falls (30.5%), violent acts (13.5%), and sports injuries (8.9%) (1). Treatment is available in the form of passive and active training rehabilitation techniques, but effectiveness is hard to evaluate in humans (2). This study evaluates the physical and biomechanical changes in fore limb and hind limb bones in adult rats that have undergone incomplete spinal cord injury and have been trained through (active) Body Weight Supported Treadmill Training (BWSTT) or (passive) bike training.

Materials and Methods:
Details of animal groups are provided in Table 1. Each group had n=8. Forelimb and hind limb bones (obtained from another study) were harvested and stored in a freezer until used. Bones were scanned using a SkyScan 1172 Micro-CT X-Ray machine and the voltage and current were set to the correct parameters for each bone type prior to scanning. NRecon was used to reconstruct the bone scans and CTan was used to select the regions of interest in the metaphyseal and diaphyseal areas of the bones. CTan was also used to analyze the scan and acquire data. Following completion of micro-CT scans, the bones underwent three-point bending mechanical testing. By using the three-point bending, the ultimate load was found for data analysis (Figure 1).

Table 1: Study Details

	Week 0	Week 1	Week 2	....	Week 8	Week 9	Week 10
Group 1	Normal		→		Harvest		
Group 2			→		Harvest		
Group 3	Contusion	BWSTT		→		Harvest	
Group 4		Transplant		→		Harvest	
Group 5		Transplant BWSTT			→		Harvest

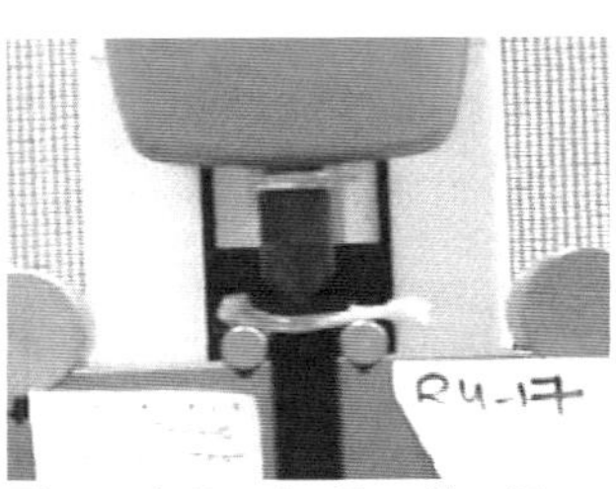

Figure 1: 3-point Bending Test

Results and Discussion:
From this study, SCI and BWSTT showed a significantly higher cross-sectional thickness and cortical BV/TV compared to normal bones. It was seen that cortical perimeters and medullary volume were both higher in normal animals when compared to SCI and BWSTT (Figure 2). The three-point bending indicated an increase in the mechanical properties after SCI and a return to normal after receiving BWSTT (Figure 3). Changes in trabecular and cortical parameters also suggest that a combination of BWSTT and a neurotrophin releasing scaffold shows even more improvement than BWSTT by itself.

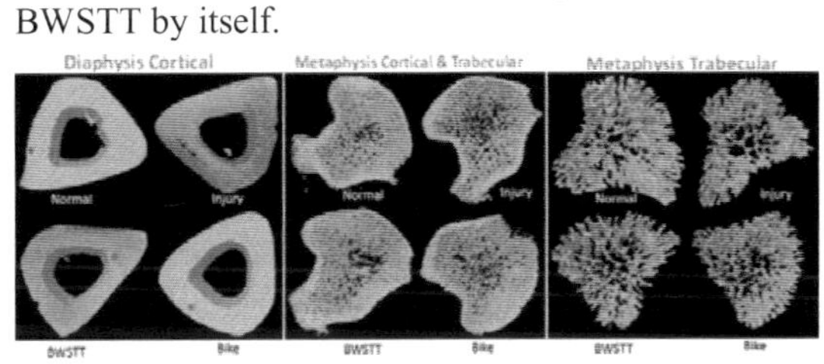

Figure 2: Micro CT data

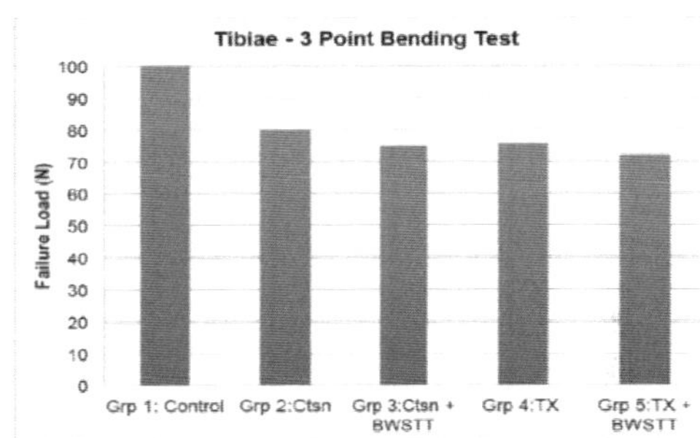

Figure 3: 3-point Bending test data

Conclusion:
SCI has very extensive effects on subject forelimb and hind limb bones, and BWSTT greatly assists rehabilitation. Physically and mechanically, BWSTT leads to properties much closer to normal rats then that of SCI rats. This in combination with a neurotrophin releasing scaffold shows great promise for recovery to normal after experiencing SCI.

Acknowledgement:
This work was supported by New Jersey State Commission Grant #CSCR14ERG001

References: [1]Spinal Cord Injury Facts and Figures at a Glance, NSCISC. [2]Corcoran et al., *Western Journal of Medicine*. 1991;154(5):536-538.

Dynamic Hydrogel Platforms for in vitro Cardiac Models

Andrew House[1] and Murat Guvendiren[1,2]
[1]Otto H. York Department of Chemical and Materials Engineering, New Jersey Institute of Technology
[2]Department of Biomedical Engineering, New Jersey Institute of Technology

Statement of Purpose: Heart Disease is currently the leading cause of death in the United States, and is most often caused by myocardial infarction (MI)[1]. There is a desire to create novel therapies and treatments. However, due to a lack of fundamental understanding of the heart's inflammatory response to infarction, the development of new therapies is both time consuming and costly. In this work, we report a novel approach to fabricating a dynamic hydrogel cardiac platform which better mimics the five major structural components of the heart (Figure 1). We believe that our approach could potentially replace both *in vivo* animal models and static *in vitro* models.

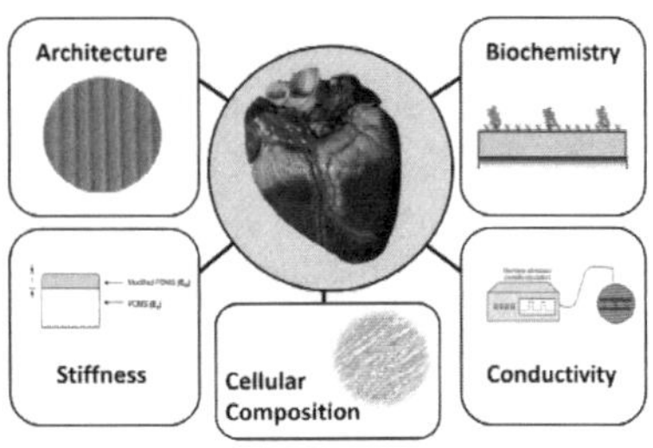

Figure 1. The 5 major structural components of the heart. The heart has a distinct cellular composition, biochemistry, architecture, matrix stiffness, and conductivity. These five components must be considered when designing new cardiac tissue platforms. Images are reproduced from [2-5].

Methods: *PDMS wrinkle formation:* Polydimethylsiloxane (PDMS) is prepared by mixing a base and curing agent at a ratio of 10:1 and placing it under vacuum to remove any air bubbles. A 1 mm sheet of cured PDMS is then cut into rectangles (10x20 mm) and stretched to 20% of its original strain. The stretched PDMS rectangles then undergo UV/ozone treatment for 3 hours (Figure 2A).

Hydrogel synthesis: Hydrogels were fabricated using Michael-type addition polymerization. Crosslinking occurs with the introduction of 30% (w/v) dithiothreitol (DTT) to a 3% (w/v) methacrylated alginate (MeAlg) solution in PBS buffer containing 0.2M triethanolamine (at pH 10) and 0.05% (w/v) I2959. For confocal imaging, 2μg/mL rhodamine was added to the hydrogel solution. Afterwards, 4μL of MeAlg is pipetted on the surface of the stretched PDMS and incubated for 1 hour at room temperature (Figure 2B-C).

Cell culture studies: Human mesenchymal stem cells (hMSCs, Lonza) were used for viability studies, within platforms with and without channels (Figure 2D). After 7 days of incubation, cells were stained with Dapi and AlexaFluor 488 Phalloidin.

Results: We developed a novel approach to create a dynamic cardiac platform with complex architecture that better mimics the heart. Upon release, the PDMS undergoes uniform buckling at the surface, creating wrinkle patterns which closely resemble the striations seen in natural cardiac tissue. We use a 2-step crosslinking to dynamically control the stiffness of the hydrogel. Initial crosslinking is undergone using DTT and a secondary crosslinking takes place under exposure to UV light using a photoinitiator (I2959). Cells attached, proliferated and formed striation patterns (Figure 2E). These cells can be observed resting within the wrinkle patterns in Figure 2F.

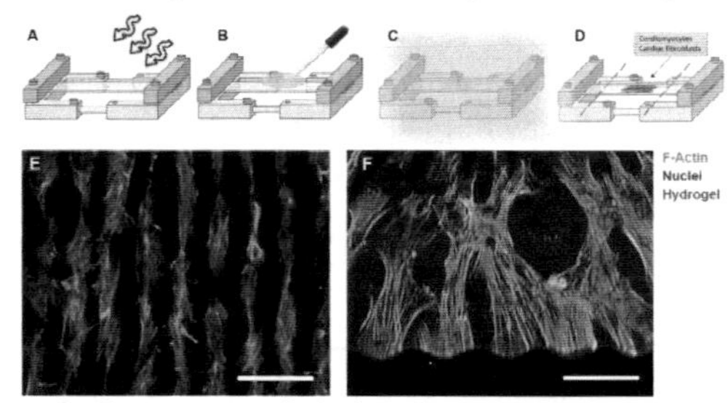

Figure 2. Our proposed design for dynamic cardiac tissue model. (A) PDMS strip is stretched to 20% original strain and exposed to UV/ozone for 3 hours. (B) The surface is functionalized with TMS overnight. (C) A 4μL drop of methacrylate hydrogel is placed on top of the PDMS strip and covered with a glass coverslip. It is then crosslinked to a specified stiffness. (D) Cells are seeded onto the hydrogel. (E) Fluorescent image of the hMSCs seeded onto a wrinkled MeAlg hydrogel. (F) Confocal scanning image of hMSCs (F-actin: green, cell nuclei: blue, hydrogel: red) seeded onto the wrinkles. Scale bars for (E-F) are 200 and 100 microns respectively.

Conclusions: We successfully showed a novel approach to create a dynamic hydrogel platform, with tunable stiffness and architecture. We showed that hMSCs attached and formed striation patterns when seeded on these surfaces. This approach could potentially be used to fabricate hydrogel devices that can mimic the dynamic changes in the extracellular matrix post MI and develop cardiac drug screening models.

Conflict of Interest: New Jersey Institute of Technology has a financial interest in the approach explained in this abstract.

Acknowledgements: This research has been funded in part by a grant from New Jersey Health Foundation.

References:

1. Benjamin EJ, et al., Circulation, 2018.
2. Takizawa, P., *Histology@Yale.* medcell.med.yale.edu/histology/muscle_lab/acute_myocardial_infarction.php.
3. Engler, A.J., et al., J Cell Biol, 2004. **166**(6): p. 877-87.
4. Guvendiren, M., et al., ACS Biomater Sci Eng, 2016. **2**(10): p. 1679-1693.
5. Bowden, N., et al., Nature, 1998. **393**: p. 146.

A Phantom Based Study of Short-Wave Infrared Emitting Nanocomposites for Fluorescence Guided Surgery

Marissa Berger, Jay Shah , Vidya Ganapathy, Prabhas V. Moghe, Mark C. Pierce
Rutgers, The State University of New Jersey

Introduction: Accuracy of solid tumor resection is often hindered by poor margin visibility. Fluorescence-guided surgery (FGS) is an imaging technique that aims to enhance the contrast between fluorescently-labeled cancerous cells and background tissues, potentially improving tumor discernibility. Clinical FGS systems operate in the near infrared (NIR) wavelength range (700-1000 nm) using FDA-approved indocyanine green (ICG) dye for applications including angiography and perfusion diagnostics. Recently, the short-wave infrared (SWIR) range (1000-1700 nm) has been proposed for deeper tissue imaging due to its reduced scattering over NIR light. To this end, SWIR-emitting Erbium-doped rare earth albumin nanocomposites (ReANCs) have previously been developed[1]. In this study, an FGS system capable of illuminating and imaging these nanocomposites has been developed. A phantom study compares ReANC-guided, ICG-guided, and visual inspection-guided resection of simulated tumors with the purpose of evaluating imaging depth and resolution in a biological context.

Methods: The SWIR-detecting FGS system illuminates specimens with a 980 nm fiber-coupled laser that is collimated and scanned across the whole specimen. Emissions are filtered with a 1497-1579 nm bandpass filter and detected by an InGaAs camera (Sensors Unlimited) with an exposure time of 32.98 ms and a frame rate of 33.3 fps. LabVIEW-based acquisition software stores the maximum pixel intensities detected in the fluorescent image for overlay in real time on a white light image providing anatomical context. For validation, tissue-simulating gelatin phantoms containing tumor-simulating agarose inclusions were imaged. Inclusions of known size and shape were prepared by heating a mixture of Type I agarose (Sigma Aldrich), 1x TBS (BioRad), and the contrast agent of interest and pouring it into custom 3D-printed molds. Inclusions were embedded in a phantom composed of 225 g bloom bovine gelatin (Sigma), 1x TBS, nigrosin (Sigma Aldrich), and Intralipid 20% (Sigma Aldrich). Inclusions were placed at known depths by fabricating the gelatin phantom in layers of known thickness. Inclusion depths and sizes were chosen to be biologically relevant based on imaging data from inclusions placed under chicken breast slices of known thickness. After assessment of scattering and resolution as functions of inclusion size, depth, shape, and contrast agent, the phantoms will be resected and assessed for margin accuracy.

Results: Initial imaging of 5 mm diameter ReANC-containing inclusions placed 2.5 mm and 5 mm deep into a gelatin phantom demonstrated a reduction in emission intensity and signal-to-background ratio (SBR) with increased depths. Identical inclusions placed under 1 mm, 2 mm, 2.5 mm, 3 mm, 4 mm, 5 mm, 6 mm, and 8 mm thick slices of chicken breast also reflected a decrease in signal intensity and SBR with increased thickness. SBR became negligible at 1.1 for 5 mm thickness. Intensities and SBRs of gelatin and chicken cases were similar in value.

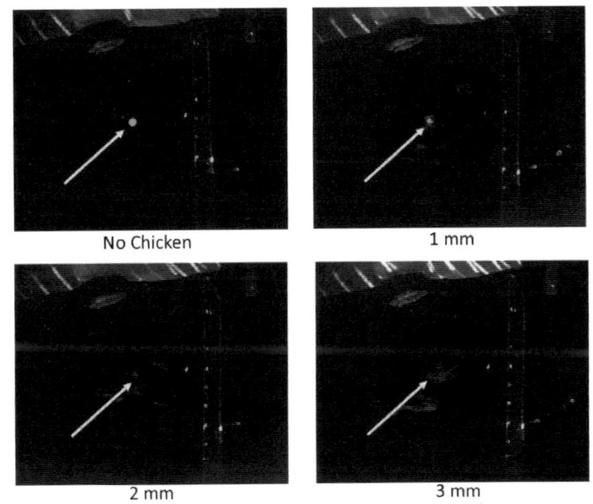

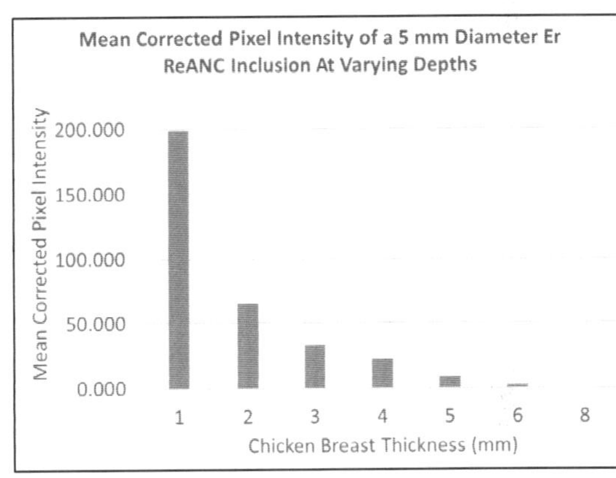

Figure 1. Background-corrected pixel intensity of ReANC-containing inclusion at varying depth

Conclusions: This study demonstrates that the proposed SWIR-based FGS imager can detect SWIR-emitting Er-doped ReANC particles. Analysis of chicken breast and gelatin phantom images confirm that phantom composition is biologically relevant. This is evident in the similar pixel intensities and trends seen between the two. To remain biologically relevant, phantoms must be designed with a limiting inclusion diameter of 5 mm and limiting depth of 5 mm based on the loss of signal in chicken breast experiments by 6 mm thickness. Future work will compare image scattering, resolution, and depth limitations between ReANCs and ICG. Resection accuracy against ICG-guided resection and gold standard visual inspection and palpation techniques will be assessed.

References:

1. Naczynski, D. J. *et al.*, Nat. Commun. 2013;4:1-10.

Evaluation of Performance of Fitness Functions in Adaptive Cuckoo Search for Differentiation of Indirect Immunofluorescence Images

Kanchana. D, Nagarajan. G and Ramakrishnan. S.
Department of Applied Mechanics, Indian Institute of Technology Madras, Chennai, India - 600036.

Introduction: Indirect Immunofluorescence (IIF) imaging is the benchmark laboratory test for screening of autoimmune diseases. Bag of Keypoint Features (BoKF) model has been used in automated analysis of IIF images. In BoKF, keypoint features are extracted using Speeded-up Robust Feature (SURF) descriptor. BoKF generates high dimensional features which require large memory space. To reduce dimensionality, several feature selection methods have been reported. Adaptive Cuckoo Search (ACS) algorithm is one of the most popular method for optimal feature selection. Fitness function plays a vital role in feature selection by obtaining the optimal values that provides better classification performance. In this study, three different fitness functions namely Generalized Rastrigin's function (GR), Six-Hump Camel-Back (SHC) function and Sphere model (SM) function are evaluated to determine optimal features for IIF image classification.

Methods: For this study, 404 IIF images belonging to nucleolar and centromere pattern are obtained from publicly available ICPR 2016 dataset. The images are pre-processed using edge aware local contrast manipulation method. The contrast enhanced images are subjected to BoKF model and SURF keypoints are extracted. These SURF keypoints with similar feature descriptors are categorized into 'k' clusters. The clustered data is then mapped to form a visual vocabulary of size 'k'. Here, k is initialized as 500. Further, ACS optimized feature selection is implemented on the visual vocabulary. ACS is a metaheuristic algorithm with adaptive step size derived from fitness function. GR, SHC and SM function are used as fitness functions. Further, the selected features are applied to k-nearest neighbor (kNN) classifier to discriminate nucleolar and centromere pattern.

Results: Fig.1 shows the representative contrast enhanced IIF images. The nucleolar pattern is observed to exhibit dotted fluorescence inside nucleus whereas centromere pattern has more granular structures.

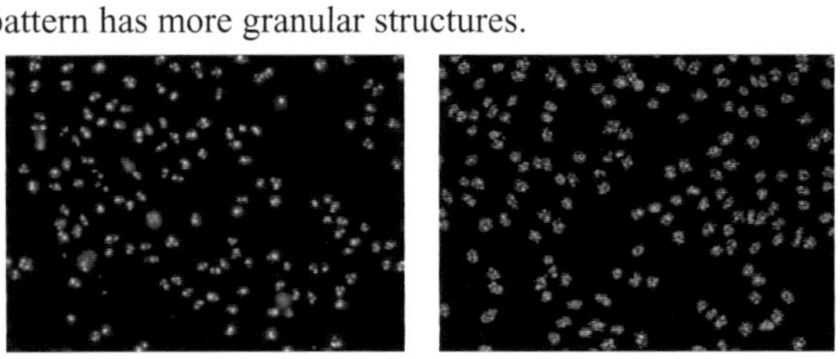

(a) (b)

Figure 1. Representative pre-processed IIF images of (a) nucleolar and (b) centromere pattern

Fig. 2 depicts the association of ACS optimal features with the classification performance. Among the fitness functions, SM has selected 25 features and achieved maximum accuracy of 81.6%. This might be due to one local minima obtained by SM. SHC provides an accuracy of 79.7% with 28 significant features. This may be because of two global minima in the search space of SHC. Fig. 3 represents the ACS ranking of the optimal features using SM. The red circles highlight the significant features and the blue dots indicate the rank assigned to each feature of the image by ACS. It is observed that SM is able to identify the most significant features.

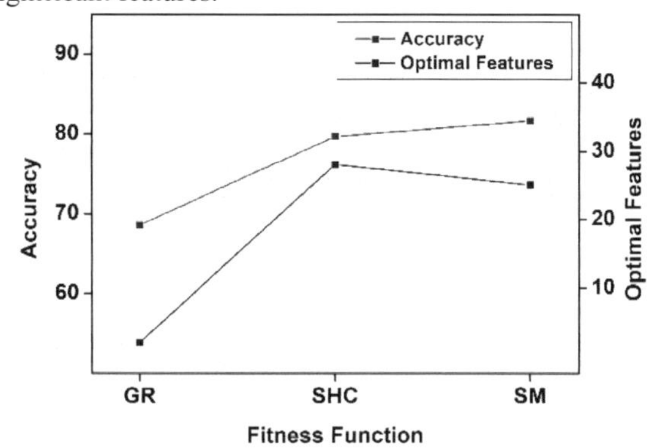

Figure 2. Fitness function performance in ACS

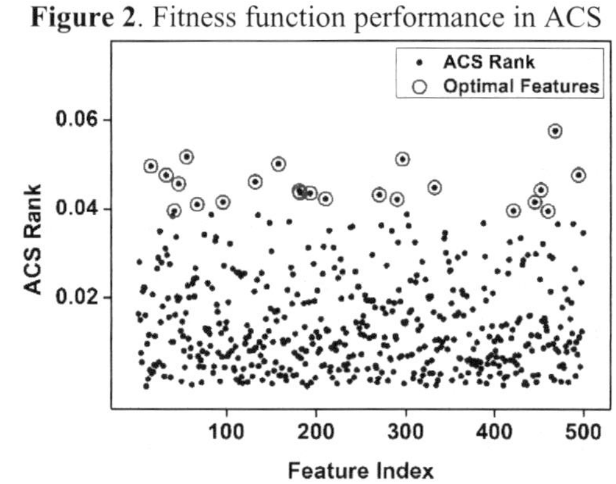

Figure 3. Optimal Feature Selection using SM in ACS

Table. 1 provides the performance evaluation of kNN classifier with precision, recall and F-measure. It is found that precision is better for centromere pattern and recall is higher for nucleolar pattern.

Table 1. Classifier Performance Metrics

Pattern	Precision (%)	Recall (%)	F-measure (%)
Nucleolar	80.29	83.50	81.86
Centromere	83.16	79.90	81.49

Conclusions: The authors have proved that ACS optimized feature selection is able to discriminate the IIF images. In this work, SM in ACS has achieved the best F-measure of 81.8% and 81.4% for nucleolar and centromere patterns respectively. Hence, SM could be used as fitness function in ACS to select the significant features for classification of IIF images.

References:

1. Zhao J, IEEE T Image Process, 2016; 25(3):1084-94.
2. Naik M K, Appl Soft Comput, 2016; 38:661-675.
3. El Aziz M, Neural Comp Appl, 2018; 29(4):925-934.
4. Kanchana D, IEIE T Smart Process Comput, 2019 (accepted).

Disrupted Brain Network Topology in Motoric Cognitive Risk Syndrome: A Resting-state Functional Magnetic Resonance Imaging Study

Yuyang Luo[1], Xiaobo Li[1,2]
[1]Department of Biomedical Engineering, New Jersey Institute of Technology, Newark, NJ
[2]Department of Electrical and Computer Engineering, New Jersey Institute of Technology, Newark, NJ

Introduction: The motoric cognitive risk (MCR) syndrome is a predementia syndrome that is characterized by slow gait and cognitive complaint with prevalence of 9.7% in older adults [1]. Very limited structural magnetic resonance imaging studies have observed smaller gray matter volume in premotor, supplementary motor, prefrontal cortices, insular in MCR patients as compared with normal controls (NCs) [2, 3]. No functional study has been conducted to identify functional alterations in MCR patients. This study collected the resting state functional magnetic resonance imaging data to assess the topological properties in both MCR and NCs. We hypothesized that relative to NCs, older adults with MCR will show deficient functional networks, including the fronto-parietal network and cingulo-opercular network(maintained top-down control of executive function [4]), as well as default mode network(observed disrupted in Alzheimer's disease [5]), that may result in the cognitive and gait abnormalities in this cohort.

Methods: A total of 13 MRC patients and 20 group-matched NCs were involved in this study. For each subject, the collected resting-state functional magnetic resonance imaging data were first preprocessed using Data Processing Assistant for Resting-State fMRI (DPARSF) and SPM12, including slice-timing correction, realignment, coregistration, segmentation, normalization, and linear detrending using bandpass filter (0.01-0.1 Hz) to remove low-frequency drift and high-frequency physiological noise. GRETNA software was then utilized to construct the whole-brain functional network. Dosenbach atlas was employed to segment the whole brain into 160 cortical and subcortical regions-of-interest (ROIs) using spheres with radius of 5 mm. For each subject, the average time series of each ROI were obtained by averaging the blood-oxygen-level dependent time series over all voxels within each ROI. Then Pearson correlation coefficients between the regional average time series of all possible pairs of nodes were calculated to generate a 160×160 correlation matrix for each subject. The correlation matrix was then converted into a binary matrix according to a predefined threshold from 0.1 to 0.3. For brain networks at each sparsity threshold, both global and regional network measures were calculated, including clustering coefficient, characteristic path length, normalized clustering coefficient, normalized characteristic path length, small-worldness, global and local efficiency, as well as regional measures including degree (D), betweenness centrality (BC) and efficiency. The area under the curve (AUC) for each network metric was calculated for each subject. Network hubs were determined by using measures of D and BC. For group-level analyses, two-sample t-test was performed on the AUC of each network metric. Bonferroni correction was applied to control multiple comparisons at significance level of 0.05.

Results: Relative to NCs, adults with MCR had significantly reduced D, BC, and nodal efficiency in left parietal lobe, as well as significantly decreased efficiency in right middle insular, and significantly increased BC in left anterior insular (**Table 1**). Insufficient functioning in left parietal lobe, bilateral occipital and right cerebellar were observed in adults with MCR; while these regions were important to maintain the resting state functional networks in NCs [6] (**Figure 1**).

Table 1. Significant between group differences in nodal topological properties in adults with MCR and NCs.

	Measures	Node	Sub-Network	*p*-value
NC > MCR	Degree	L. IPL	Fronto-parietal	0.0021
NC > MCR	Degree	L. AG	Cingulo-opercular	<0.001
NC > MCR	BC	L. AG	Cingulo-opercular	<0.001
NC > MCR	E_nodal	R. mIns	Cingulo-opercular	0.0044
NC > MCR	E_nodal	L. IPL	Fronto-parietal	<0.001
NC > MCR	E_nodal	L. AG	Cingulo-opercular	0.0011
NC < MCR	BC	L. aIns	Cingulo-opercular	<0.001

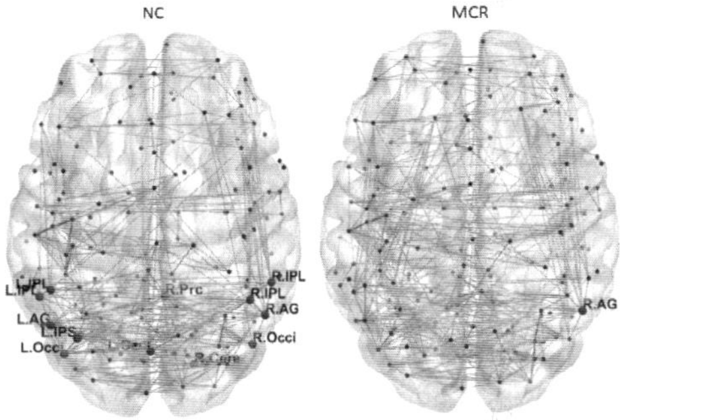

Figure 1. Group-specific patterns of the networks in NC and adults with MCR

Conclusions: The topological property anomalies in the left parietal, bilateral occipital and insular as well as right cerebellar regions may impact the function of fronto-parietal, cingulo-opercular, cerebellum, and occipital circuits and thus result in symptoms of slow gait and cognitive deficits of MCR syndrome.

References:

1. Verghese, J., et al. Neurology. 2014;83(8):718-726.
2. Beauchet, O., et al. J Gerontol A Biol Sci Med Sci. 2016;71(8):1081-1088.
3. Blumen, H.M., et al. J Gerontol A Biol Sci Med Sci. 2018.
4. Sadaghiani, S. and M. D'Esposito. Cereb Cortex. 2015;25(9):2763-2773.
5. Lustig, C., et al. Proc Natl Acad Sci U S A. 2013;100(24):14504-14509.
6. Mastrandrea, R., et al. Sci Rep. 2017;7(1):4888

Acknowledgements: This work was partially supported by NIH (R03MH109791) and NJCBIR (CBIR17PIL012).

2C-5

Reduced Occipital Hemodynamic Response During Visual Attention Processing in Young Adults with Attention-Deficit Hyperactivity Disorder – a Functional Near-Infrared Spectroscopy Study

Ziyan Wu, Ruichao Wang, Xiaobo Li
New Jersey Institute of Technology, Newark, NJ, USA

Introduction: Attention-Deficit Hyperactivity Disorder (ADHD)is a major public neurodevelopmental disorder in children and adolescents with high pervasiveness into adulthood [1]. Functional near-infrared spectroscopy (fNIRS) has emerged as an alternative neuroimaging technique which enables non-invasive clarification of brain functions in psychiatric disorders. It offers better temporal resolution and less sensitivity of head motion relative to typical neuroimaging modalities (i.e. functional magnetic resonance imaging). Previous fNIRS studies in adults with ADHD demonstrated reduced prefrontal hemodynamic responses [1]. However, there are few of them focused on occipital lobe which also played an important role in visual attention processing. In this study, we utilized fNIRS technique to examine characteristics of both frontal and occipital hemodynamic responses during a visual sustained attention task (VSAT) (designed to challenge human's sustained attention capacity) [2] in adults with ADHD and group-matched normal controls (NCs).

Methods: Sixteen young adults (18-27 years old) with ADHD and 18 group-matched NCs participated in the study. Hemodynamic responses during the VSAT were recorded continuously using a continuous-wave, multichannel fNIRS system (TechEn, Inc., Milford, MA) utilizing two wavelengths (~690 and 830 nm) at a sampling of 50 Hz. NIR light was delivered via fiber optic cables to a customized rubber head cap. A total of 8 source and 16 detector optodes were placed, creating a 16-channel array (each channel represented by a source-detector combination separated by ~3cm). The optodes were placed using a neural navigator to cover 6 regions of interest (ROIs) including bilateral middle frontal gyri (MFG), bilateral calcarine gyri (CG), and bilateral inferior occipital cortices (IOC). Individual level data analyses were processed using customized MATLAB program. Changes of fNIRS optical intensity data were first converted to hemoglobin concentration changes based on the modified Beer-Lambert law [3]. A bandpass filter between 0.01Hz and 0.15Hz was then applied to remove baseline fluctuations and other physiological noise. Regional cortical activation magnitude in each ROI was generated based on the general linear model (GLM) analysis [4]. Pairwise functional connectivities (FC) among 6 ROIs were constructed by calculating the Pearson correlation coefficient of the average time series of each pair. Group comparisons of regional cortical activation magnitude and pairwise FC measures between two diagnostic groups were carried out using unpaired two-sample t test. Multiple comparisons were corrected using the false discovery rates at ± = 0.05.

Results: Relative to controls, adults with ADHD had significantly decreased cortical activation in left CG ($t = -2.598$; $p = 0.014$) and reduced activation in right CG which approached the borderline of significance ($t = -1.959$; $p = 0.059$) (Figure 1A). Pairwise FC between left CG and right IOC was significantly increased in ADHD group (marked in blue) relative to controls ($t = 3.934$; $p < 0.001$). In addition, ADHD group showed reduced FC between the left MFG and left IOC (marked in yellow) which approached the borderline of significance ($t = -2.024$; $p = 0.051$) relative to controls (Figure 1B and C).

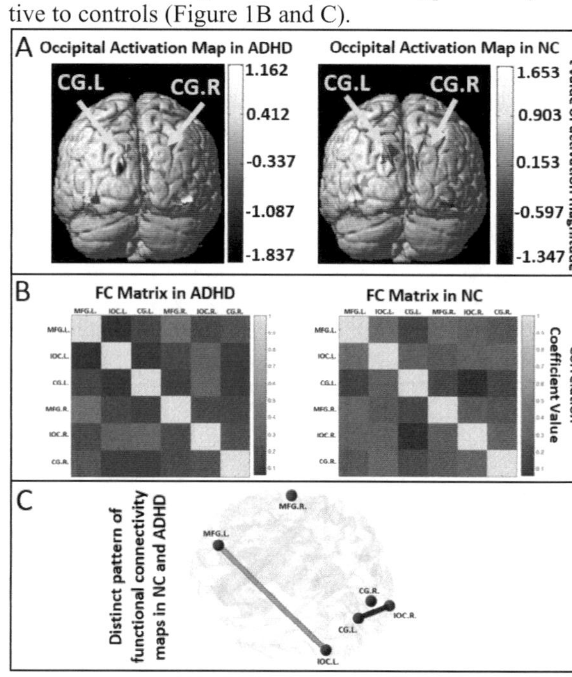

Figure 1 (A) Brain activation in occipital lobe; (B) FC matrices among 6 ROIs in ADHD group and NCs; (B) Distinct patterns of FC maps in ADHD group and NCs.

Conclusions: Abnormally decreased activation magnitude in bilateral occipital cortices, hyper-communication among occipital cortices and fronto-occipital dysconnectivity, responding to visual sustained attention processing in the ADHD group may suggest important involvement of these brain regions for the attention-related impairments in this disorder. Future work can focus on understanding the whole brain neurobiological mechanisms of ADHD in adults in a larger study sample, by integrating broad clinical and neurocognitive assessments.

References:

1. Ueda, S., et al. Psychiatry Clin Neurosci, 2018. 72(6): 380-390.
2. Wu, Z., et al. CNS Neurosci Ther, 2018; 1-10.
3. Baker, W.B., et al. Biomed Opt Express, 2014. 5(11):4053-4075.
4. Santosa, H., et al. Algorithms 2018, 11, 73;

Acknowledgements: This work was supported by the National Institutes of Health(R03MH109791) and New Jersey Institute of Technology Start-up Award to Dr. Xiaobo Li.

Clearance of drugs from the brain: the potential effect of Acetazolamide and Verapamil as CSF modulating drugs.

M Dessi', SS Yevick, C Wood, JC Sy
Biomedical Engineering, Rutgers The State University of New Jersey

Introduction: The ability to deliver drugs directly to the brain, circumventing the blood brain barriers holds great potential in the treatment of brain pathologies. Cerebrospinal Fluid (CSF) has accepted role as fluid cushion and responsible of the transport of nutrients, metabolic by-products, and other compounds vital for brain health [1]. Increasing evidence suggests that in absence of a true lymphatic system in the parenchyma, CSF and brain interstitial fluid could be an alternative pathway for the clearance of solutes from the brain. [2]. Pharmacological strategies that can up or down-regulate CSF production through the brain are under intense investigation to advance brain treatments, including Alzheimer's disease, TBI, cancer and hydrocephalus. In this work we tested the hypothesis that drug pharmacokinetic in the brain is affected by the efflux of CSF. Using Near Infrared (NIR) live brain imaging technique, we investigate the impact of two potential CSF-modulatory agents on the clearance kinetics of small molecule tracers. Acetazolamide (ACZ, carbonic anhydrase inhibitor) and Verapamil (VRPL, calcium channel inhibitor) have been chosen as inhibitor and accelerator of CSF-production. Their use via intravenous injection is FDA approved [3], however their local effect on CSF production has not been fully investigated yet, nor their in vitro and in vivo toxicity.

Methods: In vitro toxicity studies were conducted on mixed primary cells (glia and neurons) isolated through enzymatic dissociation from E-18 embryonic rat brains. Cells were exposed for 24 hours to drugs at concentrations of 0.1, 1, 10, 50, 100, 500 and 1000 μM and cell viability was measured by metabolic activity (MTS). Craniotomies were performed on adult Fisher F344 rats fixed in a stereotaxic frame. ACZ or VER in combination with a NIR tracer (Genhance750, Perkin Elmer) were perfused into the third ventricle at a rate of 1/μl/min for 10/min. The in vivo distribution and intensity of the tracer in the brain was evaluated using an MsFxPRO system. Animals were imaged at 0, 2, 6, 10 and 24hr after the injection.

Results: Both ACZ and VRPL do not reduce viability of primary brain cells *in vitro* within the tested range of concentrations (Fig.1).

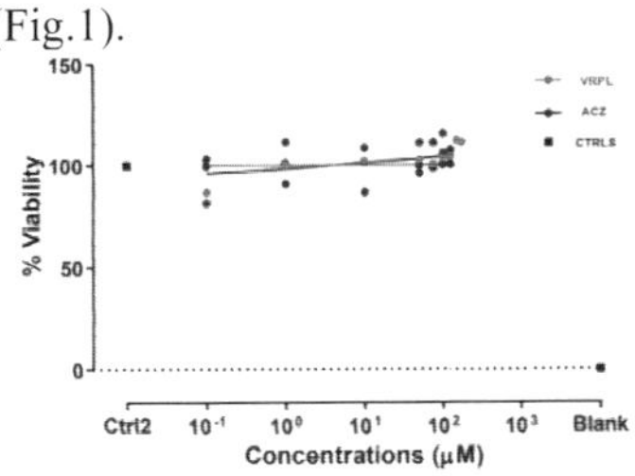

Figure 1. ACZ and VRPL do not reduce viability of primary brain cells. Data are fluctuating around 100% up to concentrations for all the tested concentrations. n = 3; Mean ± SEM.

When locally injected in the brain, the drugs prove to notably impact the CSF production. NIR images of rats treated with ACZ show enhanced fluorescence intensity in the first 10hr after the injection compared to the control. The pharmacokinetics of both tracer clearance fit a zero-order kinetic. Quantitative analysis show that the bolus injection of a single dose of ACZ results in about 30% more tracer retrained in the first 24 hours. On contrary, fluorescence analysis of rats treated with VRPL highlights a faster tracer clearance from the brain, which takes place in 6 hours and with a residual tracer percentage of 26% compared to the injection time, indicating a lower retention time of the tracer in the brain (Fig.2).

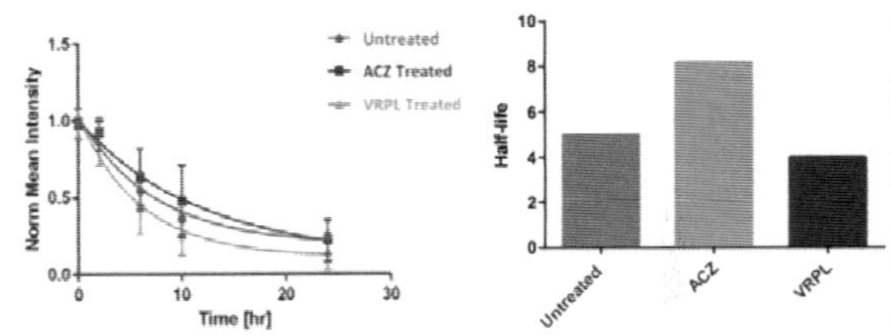

Figure 2. ACZ and VRP affect the clearance of the tracer in the brain All the samples show a clearance kinetic of zero order. Half-life of the fluorescence tracer is 5hr, in presence of ACZ the drug remains confined into the brain with half-life of 8.5 hr. On contrary, the half-life of the tracer is reduced to 4 hrs. n = 12; Mean ± SEM

Conclusions: This study unveils the powerful use of NIR brain imaging technique as conservative tool for non- invasive imaging and the key role of CSF on drug exposure time, distribution and clearance from the brain. ACZ and VRPL proved to be two CSF-modulatory agents, impacting the clearance of small molecule from the brain. These findings might propel their potential use in the pharmacological treatment of neural pathologies where the CSF production could be a prominent feature.

References:

1. Johanson CE et al. Cerebrospinal Fluid Research 2008, 5:10.
2. Plog BA and Nedergaard M. Annu Rev Pathol. 2018 Jan 24; 13: 379–394.
3. Yonai Yet al. J Cereb Blood Flow Metab. 2010; 30(3): 516–521.

Acknowledgements: This work has been supported by the grant NIH/NIBIB R00 EB016690.

Optimizing NIR Quantum Dot Molecular Probe for Assessing Breast Cancer Tumor Margins

Alexandra Jednorski[1], Wan Y Shih[1], Wei-Heng Shih[2]
[1] School of Biomedical Engineering, Science and Health Systems, Drexel University, Philadelphia, PA
[2] Department of Materials Science and Engineering, Drexel University, Philadelphia, PA

Introduction: Breast conserving surgery (BCS) or lumpectomy is the preferred method of treatment for 60-75% of breast cancer patients in early stages. Of these patients, about 30-60% must undergo at least one additional surgery as the cancer tumor margins were found to be positive by routine post-operative histological examination.[1] Re-excision is not only traumatic, painful, and costly, but is also associated with a higher recurrence rate. Developing an intra-operative method for assessing breast cancer tumor margins could drastically reduce the incidence of tumor re-excisions.

Near-infrared (NIR) aqueous quantum dots (AQDs) conjugated to anti-Tn antigen antibodies (Tn Ab) are capable of imaging breast cancer cells on tumor margins with low background from surrounding tissue. The Tn antigen is specific to carcinoma; therefore, the AQD-Ab complex is suited to stain cancer cells with high specificity and sensitivity. The goal of this study is to create AQDs that are stable at neutral pH and able to bind to the crosslinker efficiently. This would allow optimal conjugation of AQD to Ab and subsequently optimal staining of cancer cells in the future.

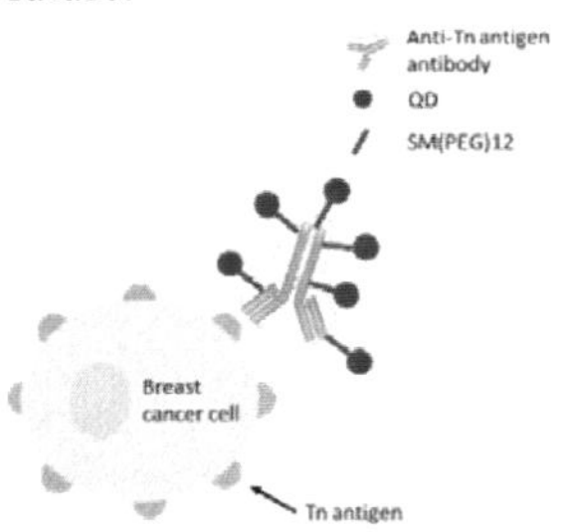

Figure 1. Overview of CdPbS AQD-Ab complex and aimed target of Tn antigen on cell surface.

Methods: First, cadmium lead sulfide NIR AQDs are synthesized at pH 12 using 3-mercaptoprionic acid (MPA) as a capping molecule. Then, (3-mercaptopropyl)trimethoxysilane (MPS) is added to replace MPA capping molecules after a series of filtering steps to remove the excess MPA. This creates a silica shell to protect the AQDs as well as a functional thiol group for conjugation. Different amounts of MPS were explored to understand if the stability of the AQDs is affected by MPS concentration. The MPS-replaced AQDs are stored at pH 11.5.

To start forming the AQD-Ab conjugate, the AQDs are desalted to reduce pH to neutral and the crosslinker, $SM(PEG)_{12}$ (ThermoFisher, Waltham, MA) in PBS, is added.

Different ratios of $SM(PEG)_{12}$:AQD were examined as there were aggregation issues with the addition of too much $SM(PEG)_{12}$. Because colloidal stability may be affected by ion concentration, AQDs incubated with $SM(PEG)_{12}$ in DI water were compared to those incubated with $SM(PEG)_{12}$ in PBS.

Stability of the suspensions can be assessed by the physical characteristics of the suspension (visually noticing aggregation) as well as changes in the scattering and photoluminescence (PL). These properties are measured using a fluorometer (Horiba, Irvine, CA).

If cell staining is performed, Ab is added after crosslinker incubation. Unused NHS esters are quenched with hydroxylamine. AQDs unbound to Ab are removed through 300 kDa filtering at 10k rpm for 5 minutes, 3 times. The conjugate is incubated for 1 hour with cells blocked with 1% BSA in TBST (1X TBS+0.1% Tween20) for 1 hr. Images analyzed using ImageJ.

Results: A systematic $SM(PEG)_{12}$:AQD study was performed using AQDs with 0.72 mM MPS addition and 0.60 mM MPS addition. In the ratio of 300 $SM(PEG)_{12}$ per QD there was much less aggregation in the sample with 0.72 mM MPS compared to that with 0.6 mM MPS.

Figure 2 shows a subsequent study examining $SM(PEG)_{12}$ addition, but also solvent affect (DI water or PBS). MPS AQDs in DI water showed higher PL and lower scattering (more stability) compared to MPS AQDs in PBS. Scattering increased slightly more when $SM(PEG)_{12}$ was added to samples in DI water as opposed to PBS (less stable), but this was only studied in the 300 crosslinker per AQD ratio and other ratios have less scattering with comparable PL. The 600 $SM(PEG)_{12}$ per AQD had noticeable aggregation and this is backed by low PL and high scattering data.

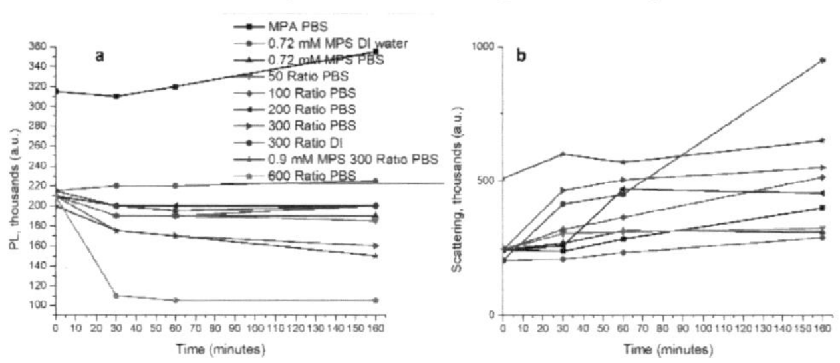

Figure 2. a) Photoluminescence (PL) data for AQD samples with varying $SM(PEG)_{12}$ ratios, including one sample with 0.9 mM MPS, samples in DI water instead of PBS, and an unreplaced MPA sample in PBS **b)** Scattering data for the same samples with the "600 Ratio" excluded as data was off the graph

Conclusions: The addition of 0.72 mM MPS creates more stable AQDs at neutral pH with less aggregation. There is still aggregation at higher $SM(PEG)_{12}$ ratios, though 50 to 300 should be sufficient to create stable Ab-AQD conjugate. The 0.90 mM MPS samples showed less stability, but only the 300 ratio in PBS was tested at this time. Future studies would include finding using more ratios of $SM(PEG)_{12}$ with the 0.90 mM MPS. Also, because MPS AQDs in DI water showed greater stability, more $SM(PEG)_{12}$ ratios should be tested using DI water or DI water with slight base addition to bring pH closer to 7-7.4. Studies on cancer cell staining can also begin to examine the optimal $SM(PEG)_{12}$ ratios (as 50 to 300 ratios appear similar in stability) and, next, ratios of AQDs per Ab, with the goal of having as many AQDs per one Ab as possible.

References:

1. McCahill LE, Single RM, Aiello Bowles EJ, et al. Variability in Reexcision Following Breast Conservation Surgery. *JAMA*. 2012;307(5):467–475. doi:10.1001/jama.2012.43

Acknowledgements: This work is supported by NCI STTR grant No. 1R41CA221595. One author (AJ) would like to acknowledge Ozgun Acar for his helpful discussions.

Analysis of Post-Injury Morphological Changes in Neurons using Label-Free Imaging

Corey Ditmars, Erin Kelly, Mohammed Naser, Rene S. Schloss Ph.D., Bonnie L. Firestein, Ph.D., Nada N. Boustany PhD,
Rutgers University, Biomedical Engineering Department.

Introduction: Mitochondria, key energy-providing subcellular organelles, are crucial to maintaining the functionality of cells within the body. Specifically, within the nervous system, mitochondrial dysfunction has been associated with various neurodegenerative diseases and conditions, such as Traumatic Brain Injuries (TBIs), prompting researchers to investigate therapies that will protect the mitochondria within neurons (Lin & Beal). We investigate label-free optical imaging methods based on Fourier filtering and dark-field microscopy to quantify subcellular morphological changes in neurons, in response to TBIs, modeled by chemical stimulation through the use of glutamate. Glutamate causes excitotoxicity leading to mitochondrial rounding and fission (Rintoul, et al.). We are utilizing our data to guide the development of label-free imaging of mitochondria within the axons in an in vitro microfluidic model of mechanical injury. Our ultimate goal is to identify various chemicals that may be useful in reversing/limiting mitochondrial damage associated with neuronal cell death.

Methods: Glutamate, at a molar concentration of 120uM, was used in this experiment to model the effect of a traumatic brain injury on the mitochondria of cortical neuronal cells. The cortical neuronal cells used were harvested from embryonic E18 rat cortices and were plated at cells/ density. The studies performed in this experiment took place on day-in-vitro (DIV) 10, 11, 12 and 15 because these time points allow for neuronal cells to grow and make connections to one another while remaining healthy. DIV 0 corresponds to the day of cell harvest and plating. Some samples are transfected with M-Cherry on DIV 6 to allow for fluorescent imaging.

The samples were analyzed under a Zeiss microscope and images were captured using filtered dark field or fluorescence. The filtered dark-field method consists of sequentially acquiring a stack of 48 filtered images in response to Fourier filters implemented on a liquid crystal device in the optical setup. The filters correspond to different spatial periods and orientations as described previously (Naser, et al.) For each region of interest (ROI) that was studied, a baseline image was first acquired. The sample was then treated with the 120 M glutamate (or with a buffered saline as control) and then images were taken every 60 seconds following treatment. The filtered dark-field images were analyzed using an algorithm that utilizes k-means clustering to bin the pixels into four separate classes on the basis of intensity. The lowest intensity pixels are binned together and corresponded to the image background. We subtract this bin from the others to eliminate the background and leave the remaining three bins for analysis. The analysis yields the filter period, S_{max} which maximizes the signal at each pixel within the ROI. Thus, a pixel histogram of these S_{max}-encoded images results in the distribution of S_{max} values within the ROI. The mode of this histogram is then extracted and plotted over time.

Results: As these images continue to be analyzed, we found that a difference exists between experimental and control samples when plotting the mode of the S_{max} distribution in the analyzed images. So far, 5 control ROI's and 8 experimental ROI's were analyzed. As seen in Figures 1 and 2, there is less variation in the mode of the S_{max} image histograms for the control samples compared to treated samples. Figure 3 shows the S_{max} textured image of a given ROI for the experimental condition, while figure 4 shows the corresponding histogram for the S_{max} image. If the mode remains the same, then the binning of the pixels has not dramatically changed and the bin with the highest number of pixels has remained the same.

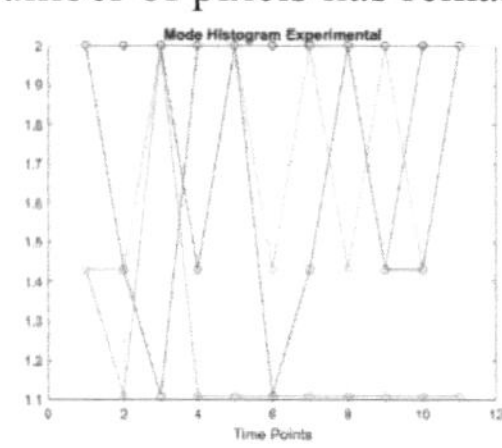

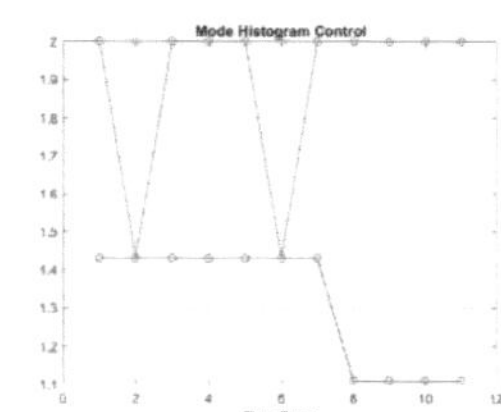

Figure 1. Experimental Only **Figure 2.** Control Only

Figure 3. Example of ROI with classified pixels.

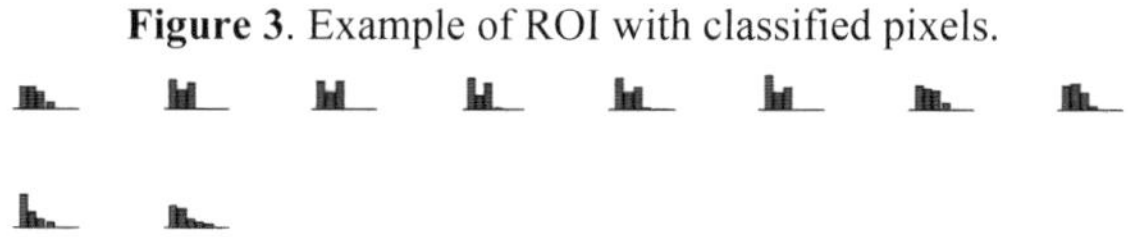

Figure 4. Examples of S_{max} histograms for the ROI in Fig.3

Conclusions: As seen in the above images, there is more change associated with experimental samples as compared with control samples, leading to the conclusion that there is a change in the binning of the pixels due to subcellular morphological changes in response to glutamate treatment. Additionally, these images were analyzed using the filtered images which are not reliant on fluorescence. This leads to the conclusion that label-free imaging is possible using k-means classification as a source of analysis.

References:

1. M.T. Lin, and M.F. Beal. "Mitochondrial Dysfunction and Oxidative Stress in Neurodegenerative Diseases." *Nature*, vol. 443, no. 7113, 2006, pp. 787–795.
2. G.L. Rintoul, et al. "Glutamate Decreases Mitochondrial Size and Movement in Primary Forebrain Neurons." *The Journal of Neuroscience*, vol. 23, no. 21, 2003, pp. 7881–7888.
3. M. Naser, R. Schloss, P. Berjaud, and N.N. Boustany. "Label-Free Dynamic Segmentation and Morphological Analysis of Subcellular Optical Scatterers", *Journal of Biomecial Optics*, 23(9), (2018).

Acknowledgements: This work was partially supported by Grant CBIR14PIL005 from the New Jersey Commission on Brain Injury Research and NSF grant CBET1512170.

Study of Stability and Functionalization of Aqueous CdPbS Near-infrared Quantum Dots

Ozgun Acar[1], Wan Y Shih[2], Wei-Heng Shih[1]
[1] Department of Materials Science and Engineering, Drexel University, Philadelphia, PA
[2] School of Biomedical Engineering, Science and Health Systems, Drexel University, Philadelphia, PA

Introduction: Quantum dots (QDs) are nanocrystalline semiconductors. Some of them are photoluminescent which makes them potential candidates for applications in the area of biomedical imaging [1]. There has been some efforts towards synthesizing aqueous QDs (AQDs), however, it is complicated in terms of precursors which are soluble in water [2]. Shih et al. developed a direct one-step aqueous synthesis of QDs with emission wavelengths between 400-700 nm [3].

For bioimaging applications, near-infrared (NIR) QDs can reduce tissue autofluorescence interference [4]. CdPbS emits in NIR range and is a potential candidate as molecular probes [5].

CdPbS AQDs were typically made with 3-Mercaptopropionic acid (MPA) as the capping molecules at high pH, which became unstable when pH was reduced to neutral. Furthermore, thiol group functionalization is preferred for bioconjugation. Earlier studies have used (3-Mercaptopropyl)trimethyloxysilane (MPS) (Ozgun, please spell out MPS) to replace MPA. However, replacing MPA with MPS was not a reliable process. Another approach is to lengthen the capping molecule to make the AQDs more stable while also facilitating conjugation. Cysteamine has amine end that can react with the carboxyl of the MPA to form a peptide bond to lengthen the capping molecule while the thiol end of the cysteamine can facilitate bioconjugation. The objective of the study is to create bright, stable, aqueous NIR suspensions that can be conjugated with biomolecules for bioimaging.

Methods: Precursors were prepared for the synthesis of quantum dots. For this, lead nitrate $Pb(NO_3)_2$ from Fisher Scientific, Cadmium nitrate tetrahydrate ($Cd(NO_3)_2.4H_2O$, 99% metal basis) from Alfa Aesar and, Sodium sulfide nonahydrate ($Na_2S.9H_2O$) from Sigma-Aldrich. Desired molarity for precursors were 0.08 M in DI water.

CdPbS quantum dots were prepared by 3-Mercaptopropionic acid (MPA), 99% from Alfa Aesar capping molecule and EDC, NHS and cysteamine were used for surface functionalization. Tetramethylammonium hydroxide, 1.0 M from Alfa Aesar (TMAH) was used for pH adjustment. After synthesizing, MPA-capped CdPbS QDs were desalted and mixed with EDC and NHS. Right after that, different amounts of cysteamine was added to the suspension and mixed for 45 minutes. Some of the samples were desalted again whereas rest of them were not. After desalting, pH of the suspension becomes neutral. The photoluminescence (PL) spectra were measured using a fluorometer (QuantaMaster 40, PTI, Birmingham, NJ).

Results: Three different samples were prepared with different cysteamine concentrations. First, 0.15 mM of both EDC and NHS were added to desalted MPA-capped QDs. They were left for mixing for 5-10 minutes. After that, cysteamine was added in different concentrations. For 0.23 mM ratio, it was added in one round, for 0.31 mM and 0.42 mM, cysteamine was added in two rounds. Each round was stirred for 20-25 minutes. Some of the suspensions were further desalted again and their stability was monitored. Moreover, desalted MPA-capped QDs which do not include any EDC, NHS or cysteamine were also monitored as control.

In terms of color and overall QD concentration, pH 12 sample was darker and more concentrated. Considering these, it can be said that except 0.41 mM sample, all samples have reasonable photoluminescence (see Figure 1). 0.41 mM sample aggregated after 3 days. Aggregation was also seen from previous studies with higher cysteamine or EDC, NHS concentrations. MPA pH 12 sample slowly degraded over a week but there was no visible aggregation. Desalted sample degraded relatively faster. The 0.23 mM sample had the highest photoluminescence among the thiol functionalized samples. However, the photoluminescence started to decrease after 4 days. Only the 0.31 ml sample has stable PL up to 7 days and is expected to be the most stable one.

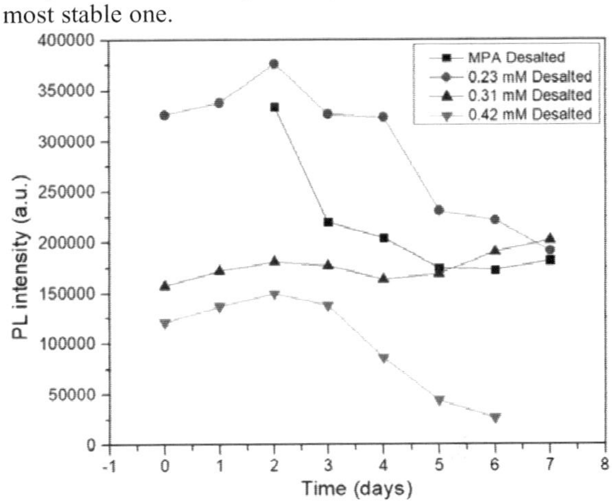

Figure 1. Photoluminescence stabilities of as-synthesized MPA desalted pH 7 MPA-capped sample, 0.23mM, 0.31mM, and 0.42mM cysteamine functionalized samples. Some fluctuations are due to the instrument.

Conclusions: It is shown that adding an appropriate amount of cysteamine (0.31 ml) can create stable suspensions of CdPbS QD at pH 7. This is very encouraging. Further work on conjugation of these stable QD suspensions with biomolecules will be performed to examine whether the thiol functionalization is working.

References: 1. A. P. Alivisatos et al., *Annu. Rev. Biomed. Eng.*, vol. 7, no. 1, pp. 55–76, 2005.
2. D. Gerion *et al.*, *J. Phys. Chem. B*, vol. 105, no. 37, pp. 8861–8871, 2001.
3. W.-H. Shih et al., 2006.
4. M. Yang *et al.*, *Proc Natl Acad Sci U S A*, vol. 97, no. 3, pp. 1206–1211, 2000.
5. G. H. T. et al., *Nanotechnology*, vol. 23, no. 27, 2012.

Acknowledgements: This work is supported by the national Cancer Institute (NCI) STTR grant No. 1R41CA221595. We would like to thank Alexandra Jednorski for her valuable discussions.

P1-7

Cortical Activation during Breath Hold using Functional Near-Infrared Spectroscopy

Keerthana Deepti Karunakaran , Bharat B Biswal
New Jersey Institute of Technology, Department of Biomedical Engineering, University Heights, Newark, NJ 07102 [a]

Introduction: Breath hold (BH) is a straightforward and inexpensive vasoactive stimulus that induces systemic changes in the cerebral blood oxygenation levels in as little as 3 sec [1, 2]. It induces vasodilation by increasing the partial pressure of carbon dioxide in the cerebrovascular system, making it a convenient technique to measure cerebrovascular reactivity. Cerebrovascular reactivity from BH is routinely implemented in fMRI studies to determine differences in the neurovascular coupling of different regions in an individual [3]as well as between individuals during health[4], and disease[5]. Although, it is unclear how effective BH technique is in identifying non-neuronal hemodynamic activity using functional near-infrared spectroscopy (fNIRS). fNIRS is a cost-effective, flexible and portable alternative to fMRI and has made brain hemodynamic measurement feasible in infants and children[6]. Hence, the objective of this study is to evaluate the net hemodynamic response to 15 second breath hold at two time points and quantify the test re-test reliability of breath holding paradigm measured using fNIRS.

Methods: Ten right-handed healthy participants (3 females, 23.3 ±3.4 years) with no history of brain trauma, neurological conditions, addiction or psychological conditions were scanned using Tech En FNIRS system at 690 and 830 nm at NJIT, Newark. The participants performed 15 seconds of breath holding alternated with 30 seconds of normal breathing for a total duration of 330 seconds. The 10 participants visited for a second session/run at least 1 week after the first run. The same paradigm was repeated. Hemoglobin concentration changes were recorded from the bilateral sensorimotor cortex and prefrontal cortex using 25 channels (Figure 1c). Data analysis was performed using in-house scripts in MATLAB. Outcome measures include the block-averaged hemodynamic response and the maximum oxy-hemoglobin increase after 15 second breath holding. Intraclass correlation coefficient was calculated for maximum oxy-hemoglobin increase to compute the test-retest reliability of cortical activity measured using fNIRS during breath holding.

Preliminary Results: The block averaged hemodynamic response to BH from the 10 participants in Run 1 and Run 2 is shown in Figure 1a. In general, the response curves look very similar between the two runs. Further, the HbO concentration changes in response to a 15 second breath hold task peak at slightly different time periods in the different sources. Also, apparent differences in the stimulus induced overshoot and post-stimulus undershoot is observed between sources. For e.g., the post-stimulus undershoot is greater in source C and F in Run1 and shows a similar trend in Run 2 for source F. Intra-class correlation coefficient computed for the maximum HbO increase in the 30 seconds after BH onset show an ICC of at least 0.5 in 17 out of the 25 channels (Figure 1b).

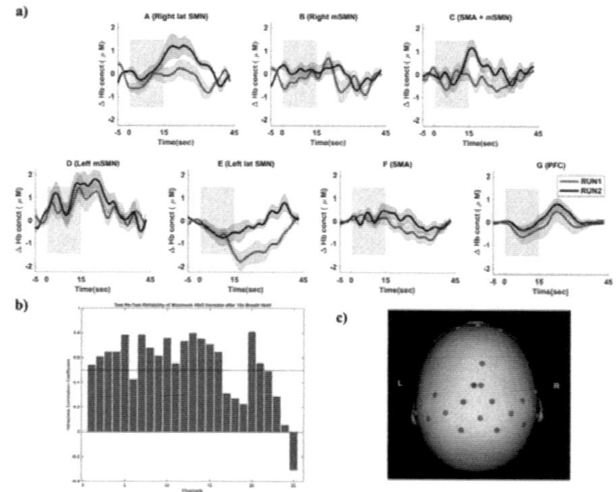

Figure 1. a) Mean HbO response to 15 seconds breath hold from 7 sources (A-F) during Run1 and Run 2. Shaded area indicates the standard error of mean and shaded block indicate the task period. b) Intraclass correlation coefficient measures for the maximum HbO increase after stimulus obtained for Run 1 and Run 2. The line represents an ICC of 0.5. c) Optode placement in one subject where pink denotes the sources A-F, blue denotes the detectors and purples denotes the short separation detector.

Discussion and Conclusion: Although preliminary in nature, the findings show that the hemodynamic response curves display distinct temporal characteristics depending on the channel location measured -likely reflecting the underlying macrovasculature. On that note, the large post-stimulus undershoot in the medial sources could result from the delayed vascular compliance from a large draining vessel such as sagittal sinus. Furthermore, the hemodynamic response to 15 second breath holding measured from 25 channels during two sessions shows a fair reliability (>0.5) in majority of the channels. Additionally, modelling any possible delay in hemodynamic response to BH from the different regions may result in a more accurate measurement of vascular dynamics. Further work is required to understand the different temporal and spatial characteristics of BH hemodynamic response using fNIRS and achieve high test re-test reliability.

Acknowledgment: This study was funded by a pre-doctoral fellowship from NJCSCR (CSCR15FEL002).

References:

1. Abbott, D.F., et al., *Brief breath holding may confound functional magnetic resonance imaging studies.* Hum Brain Mapp, 2005. **24**(4): p. 284-90.

2. Kastrup, A., et al., *Cerebral blood flow-related signal changes during breath-holding.* AJNR Am J Neuroradiol, 1999. **20**(7): p. 1233-8.
3. Kastrup, A., et al., *Assessment of cerebrovascular reactivity with functional magnetic resonance imaging: comparison of CO(2) and breath holding.* Magn Reson Imaging, 2001. **19**(1): p. 13-20.
4. Kastrup, A., et al., *Functional magnetic resonance imaging of regional cerebral blood oxygenation changes during breath holding.* Stroke, 1998. **29**(12): p. 2641-5.
5. Shiino, A., et al., *Estimation of cerebral perfusion reserve by blood oxygenation level-dependent imaging: comparison with single-photon emission computed tomography.* J Cereb Blood Flow Metab, 2003. **23**(1): p. 121-35.
6. Villringer, A. and U. Dirnagl, *Coupling of brain activity and cerebral blood flow: basis of functional neuroimaging.* Cerebrovasc Brain Metab Rev, 1995. **7**(3): p. 240-76.

Gadolinium Oxide Nanoplates as MRI Contrast Agents

Jake Villanova[1], Edward Esposito[1], and Vicki L. Colvin[1,2]
1. Department of Chemistry, Brown University.
2. Center for Biomedical Engineering, School of Engineering, Brown University.

Introduction: Magnetic resonance imaging (MRI) can provide useful diagnostic information about cancer for therapy optimization. MRI procedures often require contrast agents (CAs) to enhance their images.[1,2] The most common, clinically-used CAs, gadolinium-based chelates (GBCs), though capable of inducing efficient contrast enhancement, are limited by toxicity issues, non-specific biodistribution, and short blood half-lives. Gd_2O_3 nanoparticles (GONs) are also capable of contrast enhancement similar to, or better than, GBCs and do not have the same limitations due to their nanoscale size and properties.[3,4] In particular, Gd_2O_3 nanoplates (GONPs) have boasted contrast efficiencies up to 10 times greater than commercially available GBCs.[5] This work is concerned with a reproducible approach for synthesizing GONPs and transferring them into water for future use in cancer diagnostic applications.

Methods: Briefly, 1.8 g gadolinium nitrate ($Gd(NO_3)_3 \cdot 6H_2O$; 99.99 %; Aldrich) was dissolved in 2.5 mL oleic acid (90 %; Aldrich) and 12.7 mL 1-octadecene (90 %; Aldrich) and heated to 110 °C under inert atmospheric conditions for 1 hr. 2 mL oleylamine (70 %; Aldrich) and 7.6 mL 1-octadecene were added to the reaction vessel before heating to 290 °C for 8 hr until the solution is opaque brown. TEM and XRD were used to characterize size, morphology, and crystal structure of hydrophobic GONPs.

To synthesize PAMPS-LA, 0.135 mL lauryl acrylate (LA, 90 %; Aldrich), 0.5175 g 2-acrylamido-2-methyl-1-propanesulfonic acid (99 %; Aldrich), and 3.75 mg 2,2'-azobis(2-methylpropionitrile) photoinitiator (AMPS, 98 %; Aldrich) were added to 3 mL DMF (Certified ACS; Fisher) in a 20 mL glass scintillation vial and stirred for 20 min. The reaction mixture was then placed in a UV reactor (253 nm) and stirred for 30 min to initiate polymerization. SE-HPLC and NMR were used to confirm polymerization and measure polymer size.

To disperse GONPs in water the following encapsulation procedure was used. Diluted 2 mL of highly concentrated GONPs with 2 mL DEE (Certified ACS; Fisher) in a 20 mL glass scintillation vial and stirred for 5 min. Separately, 2 mL of PAMPS-LA were dissolved in 2 mL DMF and stirred for 5 min. The two solutions were mixed and stirred for 12 hr to allow encapsulation to occur. 10 mL of DI water was added to the mixture and stirred for 12 hours. ICP-AES and DLS were used to measure transfer efficiency.

Results: According to TEM images, synthesized Gd_2O_3 nanoplates (GONPs) – which have been shown elsewhere to have exceptional T1 relaxivity – have small faces (d = 12.0 nm), thin edges (1.5 nm), and have uniform size distribution in both dimensions (Figure 1A).[5] XRD patterns indicate GONP structure is consistent with monoclinic (C2/m) and mostly cubic ($Ia\bar{3}$) bulk Gd_2O_3 (Figure 1B). NMR spectra and Se-HPLC data confirm the synthesis of approximately 80 kDa poly(2-acrylamido-2-methyl-1-propanesulfonic acid-lauryl acrylate) (PAMPS-LA) as an ideal

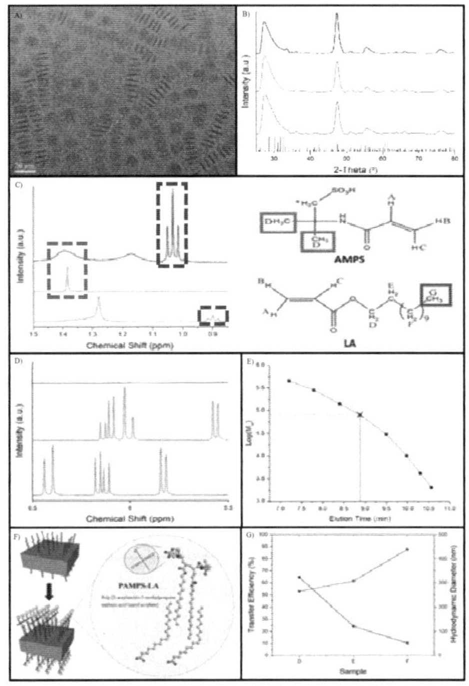

Figure 1. A) TEM images of GONPs. B) XRD patterns for three GONP and JCPDS cards for cubic (red) and monoclinic (blue) Gd2O3. C-D) NMR spectra for LA, AMPS, and as-synthesized PAMPS-LA (bottom to top) confirming polymerization by showing C) the presence of each monomer and D) absence of vinylic hydrogens in the sample. E) SE-HPLC elution times for as-synthesized PAMPS-LA (8.88 min, M_n = 80.9 kDa) measured against a sulfonated-polystyrene curve. F) Diagram depicting encapsulation-mediated phase transfer of GONPs with PAMPS-LA. G) Phase transfer efficiency and hydrodynamic diameter of three similar GONP samples (D, E, and F) collected using ICP-AES and DLS, respectively

encapsulating agent for GONPs (Figure 1C-F). Combined ICP-AES and DLS results indicate that transfer efficiency increases with decreasing polymer coating hydrodynamic diameter (HD) (Figure 1G).

Conclusions: A reproducible approach for GONP synthesis and encapsulation-mediated phase transfer was achieved. Smaller PAMPS-LA coatings result in more efficient transfer because more polymer is available to coat and transfer other GONPs. Reproducible production of water dispersible GONPs is essential for their future use in cancer diagnostics.

References:

1. Wei et al. *PNAS*, **2017**, 114 (9): 2325–2330.
2. Faucher et al. *ACS Applied Materials & Interfaces*, **2012**, 4 (9): 4506-4515.
3. Taheri, N. Ph.D. Thesis, Rice University, **2016**.
4. Park et al. *ACS Nano*, **2009**, 3 (11): 3663-3669.
5. Cho et al. *Nanoscale*, **2014**, 6 (22): 13637–13645.

Multifractal analysis of uterine EMG signals to differentiate Term and Preterm conditions using Hurst exponent

Punitha N. and Ramakrishnan S.
Department of Applied Mechanics, Indian Institute of Technology Madras, Chennai, India - 600036.

Introduction: Premature birth, referred as the delivery before 37th week of gestation, is a major problem during pregnancy. Neonates born preterm may suffer from neuro-impairment and have higher threat of mortality. Diagnosis of labor depends on the assessment of uterine contractions. The existing devices to monitor uterine contractions are subjective and not efficient. Uterine Electromyography (uEMG) is a non-invasive technique that records the electrical activity of the uterus and provides data on the nature of uterine contractions. Hence, analysis of uEMG signals is essential as it give good insights of the pregnancy and labor processes that help in preterm diagnosis. In this study, an attempt has been made to differentiate term (gestation age > 37 weeks) and preterm conditions using uEMG signals and Multifractal Detrended Fluctuation Analysis (MFDFA) based Hurst exponent.

Methods: For this work, 143 term and 19 preterm delivery records from the Term-Preterm ElectroHysteroGram (TPEHG) database are considered. These signals are recorded during second trimester at a sampling rate of 20 Hz. These signals are subjected to MFDFA method and the generalized Hurst exponent is computed. From the exponent, four features namely, maximum (HE_{MAX}), minimum (HE_{MIN}) and zero order exponents (HE_0) and degree of multifractality (DOM) are extracted. Synthetic Minority Over-sampling Technique (SMOTE) is applied on the preterm features to address the class imbalance issue. The normality of the data is verified using Q-Q (Quantile-Quantile) plot and features are statistically analyzed using t-test.

Results: The representative uEMG signals recorded before 26 weeks of gestation are shown in Figure 1. It is seen that the amplitude of the signal in preterm group is comparatively higher than that in term condition which may be due to increase in the strength of uterine muscle contraction towards labor.

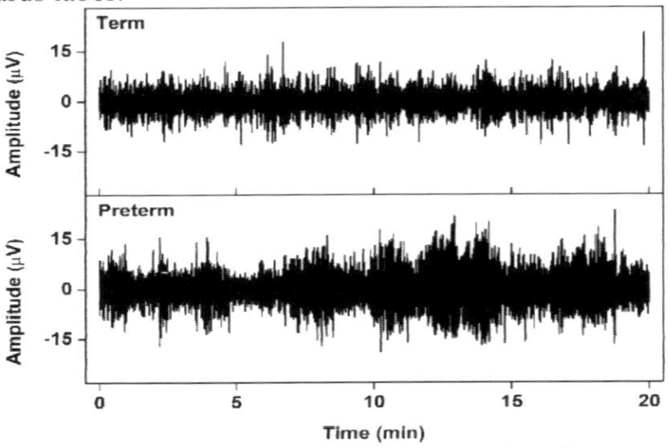

Figure 1. Representative uEMG signals

The variation of generalized Hurst exponent $h(q)$ for increasing orders of the fluctuation is presented in Figure. 2. It is seen that $h(q)$ in both term and preterm conditions vary nonlinearly with increasing orders indicating that uEMG signals are multifractal in nature. Also, $h(q)$ values for preterm are higher than that of term groups which can be due to the existence of more fluctuations in the amplitude of the signal.

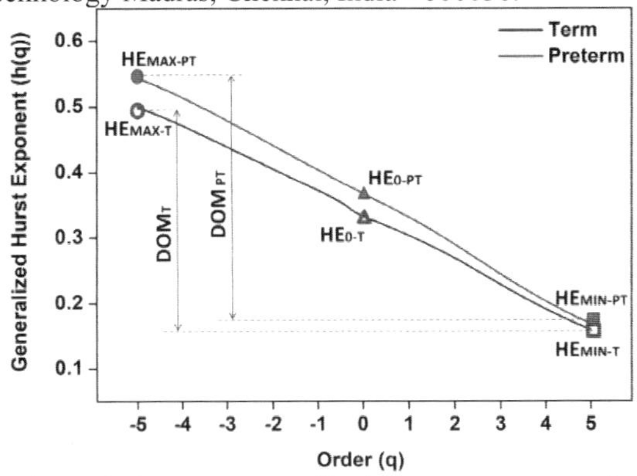

Figure 2. Average generalized Hurst exponent variations in term and preterm conditions

The statistical measures of the extracted features are shown in Table 1. Three features namely, HE_{MAX}, HE_0 and DOM are statistically significant in differentiating the signals in term and preterm groups.

The positive and negative orders of fluctuations relate to large and small fluctuations in the signal respectively. The increase in the average value of HE_{MAX} and HE_0 in preterm group can be due to the increase in smaller fluctuations which influence the low frequency components in the signals. A higher value of DOM is a representation of stronger multifractal characteristics. Increase in DOM in preterm group indicates that the uEMG signals are shifting to higher dimension as the labor approaches. HE_0 is not found to be significant in differentiating the signals.

Table 1. Statistical Measure of the extracted features

Hurst exponent features	Mean (Standard Deviation)	
	Term	Preterm
HE_{MAX}*	0.49 (0.15)	0.55 (0.06)
HE_0*	0.33 (0.05)	0.37 (0.03)
HE_{MIN}	0.16 (0.08)	0.17 (0.07)
DOM*	0.34 (0.15)	0.37 (0.09)

*Statistically significant ($p < 0.05$)

Conclusions: The multifractal characteristics of uEMG signals is proven from the variations of generalized Hurst exponent. It is seen that three features namely, HE_{MAX}, HE_0 and DOM are able to distinguish signals in term and preterm groups. Rise in multifractality in the preterm condition can be due to the synchronization of action potentials generated from the uterine cells towards labor. Hence, it appears that the multifractal Hurst exponent features can help in predicting the term or preterm delivery of pregnant women.

References:

1. Fele-Žorž G. et al. Med. Biol. Eng. Comput. 2008;46(9):911-922.
2. Kantelhardt JW. et al. Physica A Stat. Mech. Appl, 2002;316:87-114.
3. Punitha N. et al. Conf Proc IEEE Eng Med Biol Soc. 2018;1-4.
4. Punitha N. et al. Proc Inst Mech Eng H. 2019.

Polymerized Hemoglobin for Enhanced Oxygen Transport in a Mini-bioartificial liver

Nuozhou Chen[1], Donald Andrew Belcher[2], Josh Leipheimer[1], Andre Palmer[2], Francois Berthiaume[1]
[1]Biomedical Engineering Department, Rutgers, The State University of New Jersey
[2]Department of Chemical and Biomolecular Engineering, The Ohio State University

Introduction: The only clinically efficient treatment for end-stage liver disease (ESLD) is liver transplantation. Approximately 15% of patients with chronic liver failure and 30% of patients with acute liver failure die while on the donor waiting list. Thus, there is a need for a liver-supporting therapy that could serve as bridge to transplantation. A liver assist device, which would be connected to the patient as an extracorporeal support system, was previously developed by various commercial ventures (e.g. ELAD® from Vital Therapies). Such a system typically consists of a hollow-fiber based bioreactor containing liver cells that mimics native liver function; however, clinical efficacy studies have not been very conclusive. This may be due, in part, because these systems did not emulate the metabolic zonation that exists in actual liver. This zonation involves a functional gradient along with an oxygen gradient between the periportal inlet and the perivenous outlet. We hypothesize that by recreating an *in vivo*-like oxygen gradient in a hollow fiber-based bioreactor, metabolic zonation will be recapitulated with improved device performance. To explore the operational conditions in a cost and time effective manor, we propose a mini-bioreactor which utilizes few hollow fibers of the same type that would be used in the large-scale bioreactor, and culture media containing polymerized hemoglobin (PolyHb), an oxygen carrier, to provide gas and nutrient transport. With this design, we hypothesize that we will be able to maintain a liver cell line (HepG2/C3A) under an in vivo-like oxygen gradient, and that will, in turn, improve metabolic performance.

Methods: HepG2/C3A cells were purchased from the American Type Culture Collection (ATCC) and maintained as instructed. This immortalized human liver carcinoma cell line, unlike adult hepatocytes, can be easily propagated thus providing a reliable cell supply for bioartificial liver systems. This cell line also expresses several of the normal liver functions. Polymerized hemoglobin (PolyHb) and unmodified hemoglobin (Hb) were purified and synthesized as described previously [1]. PolyHb exists in a relaxed (R-) and a tense (T-) state, each with a different oxygen affinity, which makes it possible to control the oxygen gradient in the device [2]. HepG2/C3A monolayers were incubated with PolyHb or Hb to evaluate their impact on cellular function. A tolerable concentration of PolyHb and Hb was determined first based on viability measurements, and then several liver-specific functions of HepG2/C3A cells were evaluated at that concentration. Albumin production was evaluated using a human albumin ELISA. Modified Jung's method was used to measure the ammonia removal activity. Several cytochrome P450 (CYP)s involved in phase I xenobiotic metabolic activity were also evaluated. Prototyping of the mini-bioartificial liver was done by laser engraving on a poly (methyl methacrylate) (PMMA) sheet. Two hollow fibers were embedded into the laser engraved channel to mimic a unit space of the large-scale bioreactor. The surface profiling was done using 3-dimentional Extended Depth of Field (EDF) tomography. An oxygen sensitive layer of polystyrene/ platinum (II) octaethylporphyrin ketone (PtOEPK) was used as an oxygen sensor [3].

Results and Discussion: HepG2/C3A could tolerate up to 10mg/ml PolyHb (not shown). The 'leaked' fraction of PolyHb through the hollow fiber wall into the cell culture chamber is expected to be lower. Under 10mg/ml of PolyHb, albumin production was stable, while it was transiently stimulated by Hb (Figure 1). Urea production was better preserved in the R-state than in the T-state PolyHb or with free Hb. PolyHb groups exhibited elevated CYP450 1A1 activity and similar CYP450 1A2 activity compared to control, suggesting that their use during bioreactor culture prior to clinical use may be beneficial.

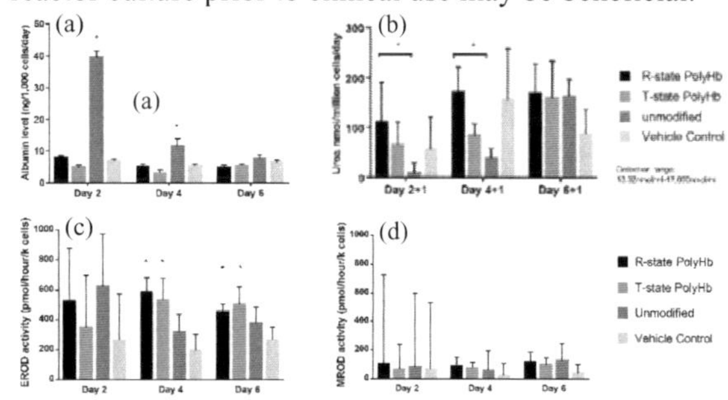

Figure 1. Panels show HepG2/C3A functions for up to 6 days of exposure to 10mg/ml PolyHbs and free Hb. (a) shows albumin production level, (b), (c), (d) are detoxification functions, namely (b) urea production, (c) cytochrome P450 1A1 (EROD) activity and (d) cytochrome P450 1A2 (MROD) activity.

The mini-bioreactor was successfully fabricated (Figure 2) and cell viability could be maintained in the device. The general performance under the oxygen gradient of the bioreactor remains to be examined.

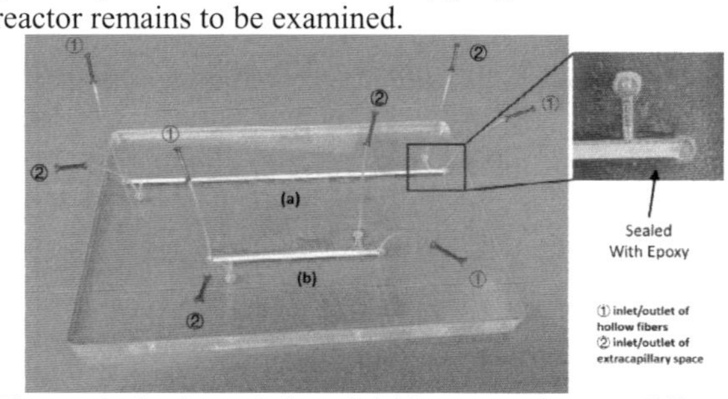

Figure 2. Setting of the mini-bioreactor in two different lengths (a) 10cm and (b) 5cm.

References:

1. Zhang, N. et al. 2011 Tissue Engineering Part A, 17(7-8), 927-940.
2. Belcher, D. A. et al. 2017. PloS one, 12(10), e0185988.
3. Sinkala, E., & Eddington, D. T. 2010. Lab on a chip, 10(23), 3291-3295.

Acknowledgements: This project is supported by a National Institutes of Health Grant (R01 EB02192601A1).

Simplified Implementation of Optimized Whitening of the Electromyogram Signal

He Wang[1], Kiriaki J. Rajotte[1], Haopeng Wang[1], Chenyun Dai[2], Ziling Zhu[1], Moinuddin Bhuiyan[1], Edward A. Clancy[1]
[1] Worcester Polytechnic Institute (WPI), USA; [2]Fudan University, China

Introduction: The surface electromyogram (EMG) signal is well modeled as an amplitude modulated, correlated random process. The amplitude modulation, defined as the time-varying standard deviation (EMGÃ) of the signal, is used in various applications as a measure of muscle effort, e.g., EMG-force models, prosthesis control, clinical biomechanics and ergonomics assessment. EMGÃ can be estimated by rectifying the EMG and then lowpass filtering (cutoff ~1 Hz). However, it has long been known that the correlated nature of EMG reduces the statistical efficiency of the EMGÃ estimate, producing a large variance.

To combat this problem, a whitening filter can be used prior to the rectifier. Whitening removes signal correlation—while preserving signal standard deviation—producing a substantially improved EMGÃ. The advantages of whitening filters have been known since at least 1974 [3]—yet, few researchers use them. A key limitation to widespread use is that most whiteners are "calibrated" to each subject, making them cumbersome to implement.

Since EMG whitening filters have low gain at low frequencies and higher gain at high frequencies, Potvin [4] implemented simple whitening via a fixed, *low-order*, FIR, *highpass* filter that was *not* calibrated to individual subjects. This approach was not compared to the established technique of subject-specific whitening filters.

Our work reported herein describes development of a simplified whitening technique that relies only on EMG magnitude normalization (a measure that is already common). We compare this technique to state-of the art subject-specific whitening.

Experimental Methods: Pre-existing data from 64 subjects [5] were used and did not require human studies supervision per the WPI IRB. Four electrodes over the biceps and four over the triceps muscles were acquired during three trials of 30-s duration, constant-posture, force-varying elbow contractions in which subjects followed a target displaying a 1 Hz bandlimited, uniform and random process, spanning 50% maximum voluntary contraction (MVC) flexion to 50% MVC extension. Using our existing subject-specific technique to form whitening filters for each electrode (calibrated from additional 5-s rest recordings and constant-effort 50% MVC trials, and limited to 600 Hz in frequency [6,7]), we related EMGÃ to force. This EMGÃ-force model used each of the eight EMGÃ values as inputs, a 15th-order dynamic FIR model per EMGÃ, additionally included the squared value of each EMGÃ at the 15 time lags (to model the EMG-force non-linearity), and was trained from two trials using least squares. The average ± std. dev. test error on the distinct third trials was 4.84±1.98% flexion MVC (%MVC$_F$). This error served as our "baseline" performance.

Analysis Methods and Results: Our whitening filters (Fig. 1) are comprised of a fixed whitening filter followed by an adaptive noise canceller (with variance preservation). The first stage is a fixed linear filter whose magnitude response is the inverse of the square root of the power spectral density (PSD) of the noise-free EMG signal (estimated by subtracting the 0% MVC PSD from the 50% MVC PSD). This filter has low gain at low frequencies and higher gain at high frequencies—the opposite of the spectral content of EMG. The second stage cancels high frequency noise, above the dominant frequency of EMG. This filter is a time-varying lowpass filter, with a cut-off frequency that is lower at lower effort levels. The time adaptation is set via a first-pass unwhitened EMGÃ estimate. The gain of this stage preserves the overall power of the noise-free signal, so that the full whitening process does not alter EMGÃ.

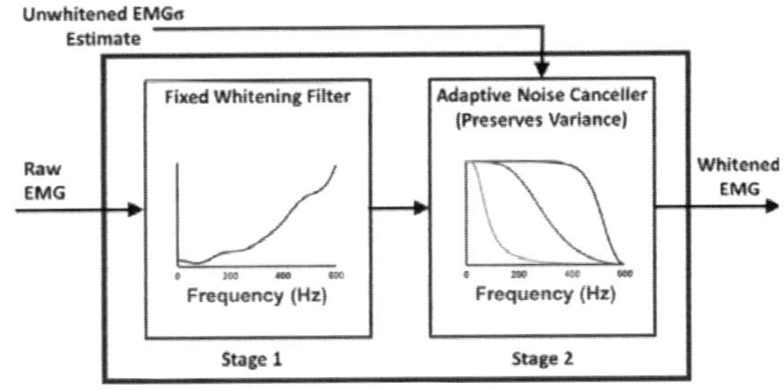

Fig. 1. Two-stage adaptive whitening filter [6].

We contrasted subject-specific whitening filter calibration to "universal" calibration. Each EMG was gain normalized, to account for gain variations between channels. Thereafter, the 0% MVC PSDs and (separately) the 50% MVC PSDs were ensemble-averaged across the 512 calibration recordings (64 subjects x 8 electrodes/subject). The one, ensemble-averaged 0% MVC and the one, ensemble-averaged 50% MVC were then used to form a single "universal" two-stage whitening filter. This filter was then similarly evaluated on the EMG-force data, producing an average ± std. dev. test error of 4.80±2.03 %MVC$_F$—the same as that of subject-specific whiteners.

Conclusions: Our work, combined that of Potvin [4], suggest that the PSD of EMG is sufficiently consistent subject-to-subject that subject-specific calibration of PSDs for EMG whitening may not be necessary (for noise cancellation). Only a gain normalization may be needed per channel. Note that PSD shapes are known to vary with inter-electrode distance [1] and might vary muscle-to-muscle. Also, this set of dynamic contractions may not be particularly sensitive to the magnitude of the noise power, since few of the active-trial contractions were near 0% MVC. (Noise is most impactful at low contraction levels.)

References:

1. Hogan N et al. IEEE TBME. 1980;27:396–410.
2. Clancy EA et al. IEEE TBME. 1995;42:203–211.
3. Kaiser E et al. In "Control of Upper-Extremity Prosthetics & Ortho.," Charles C. Thomas, 1974:54–57.
4. Potvin et al. J Electromyo Kinesiol. 2004;14:389–399.
5. Dai et al. IEEE TNSRE. 2017;25:1529–1538.
6. Clancy et al. IEEE TBME. 2000;47:709–719.
7. Dasog et al. IEEE TNSRE. 2014;22:664-670.

Compliant Multilayer Intracortical Microelectrodes to Improve Recording Potential

Xin Liu[1], Elena Bibineyshvili[2], Fernando Rebolledo[1], Sunshine Littlecreek[1],
David J. Margolis[2], Hilton M. Kaplan[3], Joachim Kohn[3], David I. Shreiber[1], Jeffrey D. Zahn[1]
[1]Department of Biomedical Engineering, [2]Department of Cell Biology and Neuroscience
[3]New Jersey Center for Biomaterials, Rutgers University, Piscataway, NJ

Introduction: Brain–computer interfaces (BCIs), which have emerged as a disruptive technology to interface with central nervous system (CNS) enable users to control external devices with neural activities recorded from the brain [1]. By recording movement-related neural signals, and then decoding them to control signals to drive external prosthetic devices, BCIs could help patients with paralysis to restore some movement functions. Extracting neural activities of high quality from the brain is crucial in the functioning of BCIs [2]. Of all the neural signal recording approaches, implanting intracortical neural probes to record local field potentials has been identified as the optimum one, due to its high signal noise ratio (SNR) and high spatial resolution. However, this method is invasive and suffers from chronic instability. The tissue inflammation around the implantation site tends to induce an immune cascade and subsequent gliosis, which will eventually isolate the implanted probe from nearby neurons, leading to degradation of the recorded signal quality. The hypothesis that smaller and more compliant intracortical probes produce less damage to brain and thus mitigate the tissue response have been the impetus for researchers to explore soft polymers as neuroprobe substrate materials, and attempts to scale down probe size [3]. However, conventional methods of patterning electrodes on a single layer limit the number of recording sites particularly for narrower probe widths. We present a novel strategy of scaling up the number of recording sites without proportionally increasing the size of the probe by patterning recording electrodes vertically on multiple Parylene C support layers. Each layer will consist of recording traces insulated by a thin polymer layer electrically isolating the traces from their neighbors.

Methods: Fabrication of the multilayer probe involves repeated photolithographic definition of titanium/platinum (Ti/Pt) electrodes and chemical vapor deposition (CVD) of Parylene C. Reactive Ion Etching (RIE) was adopted to define the final probe geometry and open recording sites. The functionality of electrodes was characterized by performing an impedance frequency sweeping within a 1X PBS solution.

Results: A series of multilayer Parylene C-based probes have been successfully fabricated with 16 recording channels supported by four different layers (4 electrodes on each layer) along a shank with of 80 µm. Recording windows are 30 µm in diameter, and connection traces are 3-5 µm in width with 5-8 µm spacing (Fig.1). Electrodes were stimulated at 50 mV with a frequency sweeping from 100 Hz to 10 kHz and the resulting impedance values at 1 kHz were comparable to other probes reported in literature.

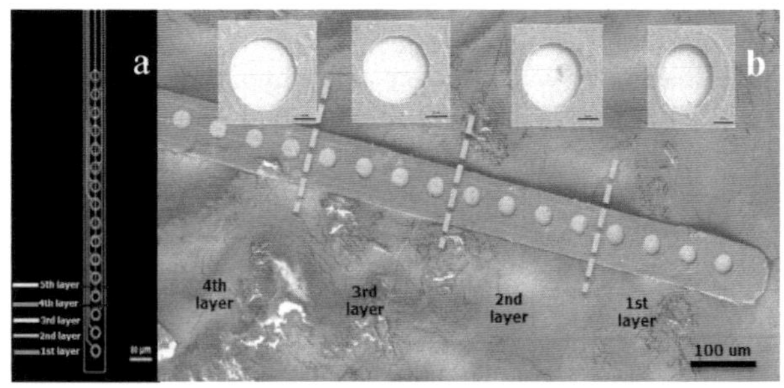

Figure 1. Four layer electrode neural probe with shank containing 16 electrodes (a) Schematic of the mask design with a different color for each electrode layer illustrating of how the 16 electrodes are patterned. (b) SEM image showing the open 16 recording sites. Each layer of electrodes is delineated by the dashed line. The insets show a zoomed-in image of a recording window from each layer.

Conclusions: A multilayer approach was adopted on Parylene probe fabrication, allowing for scaling up the number of recording channels without proportionally increasing the invasiveness of the probe. The SEM image and electrical characterization confirms no structure defect and good functionality of the probe. Future work involves temporarily strengthening the probe to mitigate buckling during brain insertion and conducting acute and chronic recordings.

References:

1. Brown, E.N., et al., *A Statistical Paradigm for Neural Spike Train Decoding Applied to Position Prediction from Ensemble Firing Patterns of Rat Hippocampal Place Cells.* The Journal of Neuroscience, 1998. **18**(18): p. 7411.

2. Schwartz, A.B., et al., *Brain-controlled interfaces: movement restoration with neural prosthetics.* (0896-6273 (Print)).

3. Weltman, A., J. Yoo, and E. Meng, *Flexible, Penetrating Brain Probes Enabled by Advances in Polymer Microfabrication.* Micromachines, 2016. **7**(10): p. 180.

Acknowledgements: This work is supported by New Jersey Commission on Spinal Cord Research Award #CSCR12IRG001 and Award #CSC16IRG007

P1-13

Ultraviolet Germicidal Irradiation of Clinical Equipment for the Prevention of Hospital Acquired Infections

Kendra Michaud, Samuel Charpentier, Andrew Strong, Jordan Anderson, Ying Sun, Ph.D.
University of Rhode Island, Kingston, RI 02881~*Correspondence:* Jordan.Anderson@va.gov

Significance: According to the Center for Disease Control and Prevention, at any given time in the United States, 4-6% of all patients in a hospital are fighting at least one infection that they acquired since entering the hospital [2]. The spread of *Clostridium difficile* and Methicillin-resistant *Staphylococcus aureus* within the healthcare environment puts patients at risk and raises the costs of medical care across the board. We can begin to lower the number of hospital-acquired infections (HAIs) by enabling more effective methods of cleaning clinical equipment that patients often come into contact with, effectively limiting the spread of harmful microbes in the healthcare environment. An inexpensive and efficient method of disinfecting medical equipment is the use of ultraviolet (UV) light. This method has already been proven effective on instruments and microbiology lab equipment and has been expanded recently to include full room disinfection systems[2]. The room disinfection method, however, requires an empty room to perform on and does not get a 360-degree clean of all the equipment, leaving infected equipment that will soon be used. In the design proposed in this paper, we created a UV cabinet, as a way to disinfect non-critical medical equipment commonly used in hospitals, without removing the ability to safely enter the room during cleaning. This will eliminate the complications that infected non-critical medical equipment can cause due to the compromised immune system of those who come in contact with the insufficiently disinfected surfaces.

Design Description: The framework of the cabinet is a 36"x17"x18" aluminum enclosure, the inside of which is wired with fluorescent UV light circuitry cabled to an external serviceable component box, shown in Figure 1. The external serviceable box houses two ballasts and starters, a 5 volt to 120 volt relay and an Arduino Uno which acts as our control system. The circuit was designed to accommodate two UV lights, although the number of UV lights within the enclosure could be increased if desired. By placing the clinical equipment inside the cabinet, closing the door, and activating the timer, the disinfection cycle begins.

The enclosure has a locking mechanism that is operated by a heavy-duty dual latch system making certain to prevent UV-C radiation bleed through. A series of safety switches, a mechanical pressure switch and a magnetic Reed switch, work to alert the cabinet's control system when the door is not properly fastened, completely disabling the lights. This ensures the harmful UV-C light will not emit if the door is opened even slightly during the disinfection process. Along with the onboard safety features, the cabinet has an LCD affixed with start and stop buttons that act as a user interface. It displays a countdown timer for the amount of time left in the disinfection cycle and readings of the temperature within the cabinet throughout the process.

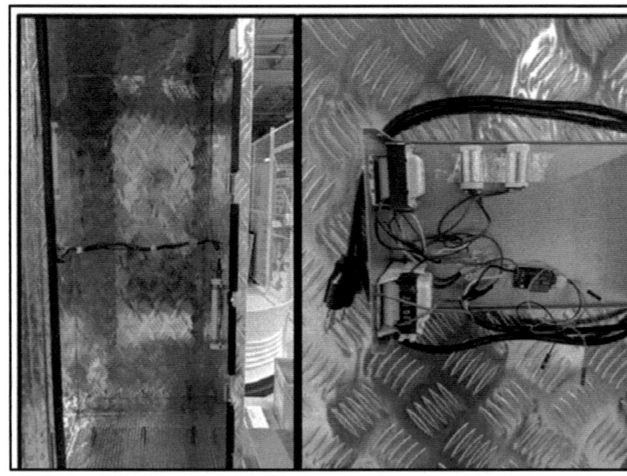

Figure 1. Left: Inside of the cabinet with exposed wires to depict circuitry. Right: External serviceable component box inside.

Once the user is ready, pressing 'Start' sends a signal to the Arduino Uno. The Uno controls a magnetic relay, in response to a signal from the start button the relay activates the light circuit. The Arduino control system also receives and displays information from the temperature sensors mounted inside the cabinet.

Design Criteria: The use of our product within a hospital environment required a clear design consideration. It was essential to create a device capable of effectively disinfecting stored medical equipment within a large chamber that is safe to keep in a hospital setting. Aluminum was chosen as the cabinet's enclosure because of its reflectivity, increasing the efficiency of the installed UV lights. It was also chosen because it's practically weightless compared to its size, making transferring the cabinet effortless. All exposed circuitry had to be contained to make it safe for commercial use. In Figure 1 left, the wires are covered with reflective, heat resistant aluminum tape in order to prevent wire damage. Additionally, Figure 1 shows the inside of the exterior serviceable box, that keeps all the components secured within. Figure 2 details the required circuitry, broken into 2 subsections; the light circuit and the user-interface/microcontroller. The Uno is used as the primary control system for the UV cabinet. It is connected to two external push buttons mounted on perfboard next to the LCD display, which read as the inputs start and stop. The LCD prompts the user to press the 'Start' button to activate the circuit and will start a 180-second timer. If the 'Stop' button is ever pressed while the timer is on, the lights will be turned off and the timer will be reset to 180 seconds, where it will wait for the user to press the 'Start' button once again. If the timer is counting down and the door is opened, the safety switches mounted on the door will register, opening the circuit and as a result deactivating the UV lights.

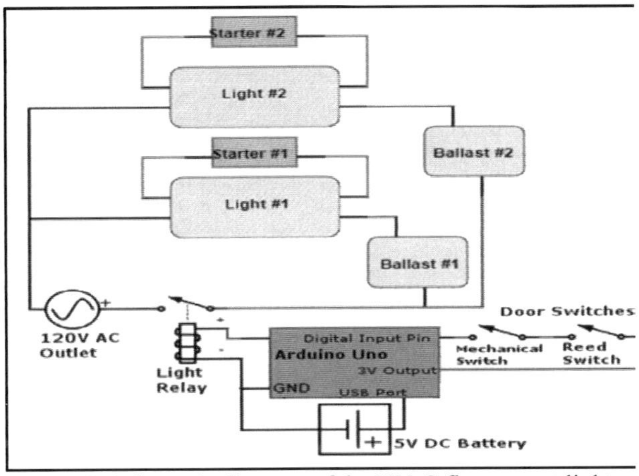

Figure 2. Schematic diagram of the UV-C fluorescent light circuit installed in the enclosure.

The Uno reads a signal from these safety switches at all times. Before the UV lights are able to be activated with the 'Start' button, the Uno confirms that the door on the cabinet is closed. These safety measures were very important to our design. Along with the safety considerations, other criteria was crucial. The method of disinfection proposed is germicidal ultraviolet light delivered by fluorescent lamps in a sealed cabinet. Short-wavelength UV-C (100-280 nm), although the most damaging type of UV radiation, has the strongest germicidal effect. The UV-C tube used in this project is the Philips model G4T5, emitting the desired wavelength of UV light: 254 nm. Many HAIs are caused by microorganisms that are antibiotic-resistant. Germicidal irradiation using UV-C light effectively disables the metabolic functions of microorganisms by creating thymine dimers within in their genetic material. These areas of crisscrossed nucleic acids make it impossible for the bacterium to replicate or transcribe its DNA, effectively preventing binary fission (replication) or killing the microorganism outright. To determine the effectiveness of the germicidal disinfection of the cabinet, a UV meter (SPER Scientific, UVC Light Meter 850010, SPER Scientific, Scottsdale, AZ) was used. A study was conducted to monitor the UV intensity inside the enclosure continuously for a duration of 200 s. The study was repeated having the sensor face different directions within the cabinet, due to the sensor's unidirectional limitations.

Results: After conducting an experiment where the UV intensity was measured at different points and sides within the cabinet, a graphical representation of the UV dosage was formed. Figure 3 was calculated by gathering the UV intensity reading every 10 s for 200 s. Using the equation intensity multiplied by time, the UV dosage was calculated. After gathering these readings, the intensities were multiplied by 10 seconds and added to the previous value. This gave us a theoretical graphical representation of the necessary exposure time for the cabinet to reach a UV dosage that would be needed to neutralize the C. diff bacteria that we are targeting. This is only theoretical based dosage data but can be further tested using microbes to observe growth after UV exposure.

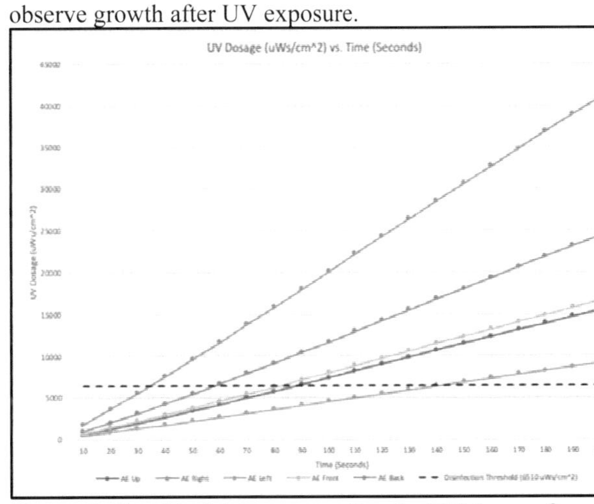

Figure 3. Graphical data representing the UV dosage from the UV-C lights within the cabinet. Data was gathered using a unidirectional UV intensity sensor called SPER Scientific, UVC Light Meter 850010.

Figure 3 shows the UV dosage for 5 sides within the cabinet; the study needed to be conducted in this manner due to the unidirectional nature of the UV intensity sensor. The experimental layout within the cabinet has a UV light located to the right of the sensor and another UV light about a foot above and to the left. The sensor was placed in the center of the cabinet. While observing the data collected and shown graphically in Figure 3, it can be seen that the right side of the UV cabinet has the greatest UV dosage over 200 seconds. While the lowest UV Dosage was from the left side of the cabinet, due to this side being not directly in front of a UV bulb. After 150 s, the left side will deliver the minimum UV dosage for the disinfection of C. diff bacteria at 6,510 µWs/cm2, this is shown in Figure 3 as the "disinfection threshold," labeled as the dashed black line. In order to ensure complete disinfection, the exposure time is increased to 180 s [3].

Future Implications: A prototype germicidal device was developed in this study, which showed a cost-effective UV-C radiation enclosure for disinfecting portable hospital equipment. The device was able to deliver 2,000-10,000 µWs/cm2 of UV-C exposure in under four minutes on all sides of the cabinet. This would greatly reduce the time necessary for sanitization of certain items that require regular cleaning. This device is designed to have the appropriate amount of space to be used with the hospital setting. For the future of this project, the implementation of more UV light fixtures would be highly advised. An increase in UV-C light output would allow for an increase in UV intensity at every point within the UV cabinet, in turn decreasing the exposure time an object needs. While increasing the number of UV tubes within the cabinet theoretically should decrease the required time of exposure, a study should be conducted using bacteria collected from common clinical instruments. Nutrient-rich agar plates should be inoculated with the collected bacteria, placed in the UV cabinet and exposed for different intervals of time, then be set to incubate for 24 hours. The resulting bacteria growth will be compared to a 'control' plate that was not treated in the UV cabinet. This will show the effectiveness of the UV cabinet's germicidal irradiation at varying time intervals, allowing clinicians to determine an appropriate exposure time for their equipment. Using data gathered from this kind of experiment would further increase the integrity of the cabinet in verifying its experimental ability to disinfect surfaces. The cabinet is a cost effective and manufacturable device, that fits the current need for a UV disinfection device for medical equipment in a hospital. The future of the device is to expand upon current testing with clinical testing in a hospital setting.

References:

1. Riley, Kyle, et al. *Germicidal Ultraviolet Light Enclosure*. URI, 2017, pp. 1–2, *Germicidal Ultraviolet Light Enclosure*.
2. Health Quality Ontario. "Portable Ultraviolet Light Surface-Disinfecting Devices for Prevention of Hospital-Acquired Infections: A Health Technology Assessment" *Ontario health technology assessment series* vol. 18,1 1-73. 7 Feb. 2018
3. "Ultraviolet light disinfection data sheet." www.clordisys.com/pdfs/misc/UV.

Acknowledgment:

This project was funded by a research grant from the University of Rhode Island 2017-18 Undergraduate Research and Innovation Awards. We would also like to give a warm thanks to Jordan Anderson of the Providence VA Medical Center for the project direction and guidance provided throughout the development of this prototyped device.

A Versatile Neural Modulation Device for Vestibular Evoked Myogenic Potentials (VEMP)

Abstract Disclosure Statement

I have no other conflicts of interest to disclose.

Introduction

Vestibular-based balance disorders affect an estimated 35% of the American population in the ages ranging 40 years and above. Current methods for testing for vestibular disorders tend to be uncomfortable for patients and suffers from low specificity between components of the vestibular organ. Audio signals, at high volumes, can stimulate the vestibular reflexes, however, this is uncomfortable for patients. Taps to the forehead with a physical hammer is more comfortable and effective at eliciting the reflex. This project details the design and fabrication of a tapping device that for triggering the vestibular reflex. The device will offer the ability to alter the pulse rate, pulse magnitude, pulse with, and allow for a bursting mode. The bursting mode would allow for a cycle of rapid strikes by the device followed by a relaxation period.

Methods and Materials

Our current device is a software-driven stimulus box that utilizes both internal and external circuit boards. We have developed a custom LabVIEW program and GUI to run the device. The internals of the device includes an internal circuit board for device control, a voice coil motor that actuates a physical striker, an internal Force Sensitive Resistor (FSR) sensor used to obtain the signal data, and a switching circuit. Initially, an external Arduino board considered in order to run our custom LabVIEW program however it was ineffective. Instead, this device currently utilizes two custom circuit boards, one for patient isolation and the other for current control for Pulse Width Modulation (PWM). Our current iteration of the LabVIEW deck requires these two external circuit boards in order to function.

Results

Constant or low frequencies tend to cause discomfort while higher or random frequencies do not appear to cause patient discomfort.

Conclusion

Our device was effective in nerve stimulation but future testing will be done in order to further confirm results.

In Miniature non-continuous oxygen concentrator

Catilin Cambio, Samantha Ferreras, Devin Belmar
Steven's Institute of Technology

Introduction: One of the many techniques to filter and concentrate air is in a process known as rapid pressure swing adsorption (rapid psa). Air is placed under high pressure and pumped through a molecular sieve. The sieve then removes nitrogen and carbon dioxide for a higher percentage of oxygen. This technique is used in many oxygen concentrators which allow them to be light weight portable. This study was performed to test that rapid pressure swing adsorption could be used at a smaller scale for a non-continuous system.

Methods: The testing apparatus consisted of a swing pressure piston pump that would pump air at a pressure of 5 Bar into the molecular sieve beds at a rate of 5L/min. The sieve beds were based on the smallest dimension found in literature, which were two 10 cm by 4 cm cylinders in parallel[1]. As seen in Figure 1, one chamber will be pressurized to produce purified oxygen while the other chamber is depressurized to release waste. The pressure differential required to cycle between each bed can be observed in the (A) and (B) diagrams in Figure 1. Each cylinder contained 46.65 g of 13X molecular sieve, as it is optimal for filtering out Nitrogen[2].

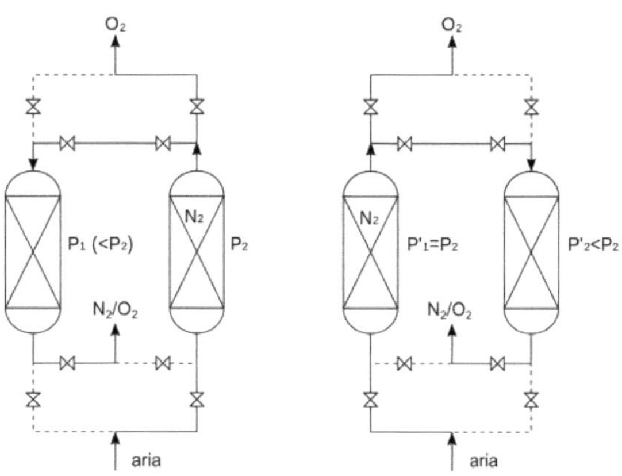

(A) (B)

Figure 1. Rapid Pressure Swing Adsorption Diagram[1]

The oxygen was then pumped into a holding tank at the outlet, where a percent oxygen sensor would measure the difference between ambient air and the purified product. Ideally, the sensor should measure an increase from 21% oxygen to about 75%. Also, to ensure that the apparatus is air tight, a barometer was included in the design. Any leakage would cause errors as the data is collected. A process flow diagram of the apparatus can be seen below in Figure 2.

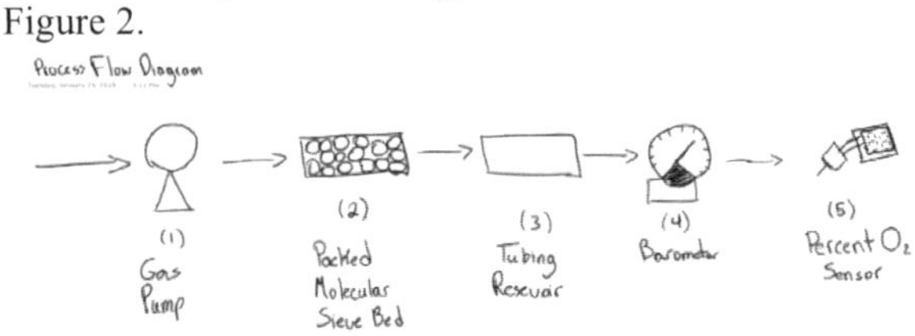

Figure 2. Process Flow Diagram for Testing Apparatus

To test the rapid psa process, the apparatus was first pressurized to 5 Bar using the air pump. The oxygen percentage was then recorded along with the time it took to fill the apparatus using different sensors. After the chamber was filled and the data was collected, it was crucial to fully release the pressure in the system to allow it to reset. The process was repeated 10 times to ensure consistency in data. The results were then compared to the theoretical calculations.

To calculate theoretical results for percent oxygen, the following equation was used:

$$p_{O2}=V_{O2} / (V_{N2}-V_S)+V_{O2}$$

Equation 1. Percent Oxygen

The following equation was used to calculate the cycle time

Cycle time = total volume pumped into sieve / flow rate

Equation 2. Cycle Time

Results: The testing apparatus showed an increase in percent oxygen from ambient air, 21%, to 75%. The apparatus also proved that 5 Bar was the most effective pressure for rapid psa to take place.

Conclusions: The theoretical results that were calculated using equation 1 and 2 stated that the molecular sieves with dimensions of 10 cm by 4 cm could produce a gas mixture with 82% oxygen when subjected to a pressure of 5 Bar at a rate of 21.4 mL/s. The cycle time was also calculated to be 11.6 seconds. The test apparatus was able to prove that rapid pressure swing adsorption works on a small scale. To improve this test, one may want to consider using different sized molecular sieve beds to further test how small the beds could get.

References:

1. Chai, Siew Wah. Ind. Eng. Chem. Res. 2011; 50:8703-8710
2. Sarker, Ariful Sarker. Energy Procedia. 2017; 114(2017): 2450-2459

Acknowledgements: The team would like to thank Dr. Vikki Hazelwood, Dr. Becky Tucci, Dr. Valerie DeAngelo, and Dr. Ritter for their help with this project

Arduino-Based Multi-Point Pressure Sensor for Improved Sitting Posture

Andrew Hwang
Newark Academy

Introduction:

A major factor that contributes to the development of back pain is poor sitting posture. Poor sitting posture can be defined as an imbalanced distribution of pressure while sitting. The optimal sitting posture is one that distributes pressure evenly, which prevents extra stress from harming the back. Poor sitting postures can damage the back by contorting the spinal cord and applying excessive pressure to one area of the back. The combination of the Arduino Uno control board, pressure sensors, and a sitting cushion allow for the collection of pressure values over a period of time while a person sits. The ability to obtain numeric pressure values related to an individual's sitting posture allows for the mathematical comparison between a desirable posture and that of an individual. The posture data of any given individual could also be compared with different controlled postures to determine where the individual applies the most pressure and how best to correct poor posture.

Methods:

A memory foam seat cushion (Novashion, CA) was divided into 4 quadrants by separating the cushion horizontally and vertically. A single DF9-40 Resistance-Type Pressure Sensor (Adafruit.com, NY, NY) was placed in the center of each quadrant, and a fifth sensor was placed in the middle of the seat back in the approximate location that the tailbone would rest. An individual must sit on the seat for at least one hour to collect a sufficient amount of data. The sensors were all connected to an Arduino Uno control board (Mouser Electronics, Mansfield, TX), with specifically designed software. The data was compiled into graphs and analyzed. The Smart Cushion was used in a two-part trial. In part one, I sat on the Smart Cushion using my normal posture (position 1) for one hour. After the hour concluded, I recorded my back pain on a scale from 1-10. In part two, I sat using a balanced, ideal posture (position 2) for one hour and recorded my back pain at the end of the hour.

Results:

On average, Position 1 applied more pressure to the left side. Sensors 1 and 2 read higher values on average in comparison with the values reported by sensors 3, 4, and 5. This posture could result in back pain on the left side of the body due to the unbalanced distribution of pressure and could be corrected by sitting up straight, with both feet on the ground, which would create a balanced posture. According to the data shown in Figure 2, in Position 2, the individual leaned towards the back right corner of the cushion. This posture could be corrected in the same manner as Position 1.

Conclusions:

The smart cushion was able to detect how different sitting postures distribute weight across a cushion, which could enable the user to correct poor posture and potentially reduce back pain with regular use.

Figure 1: Change in Pressure for Position 1

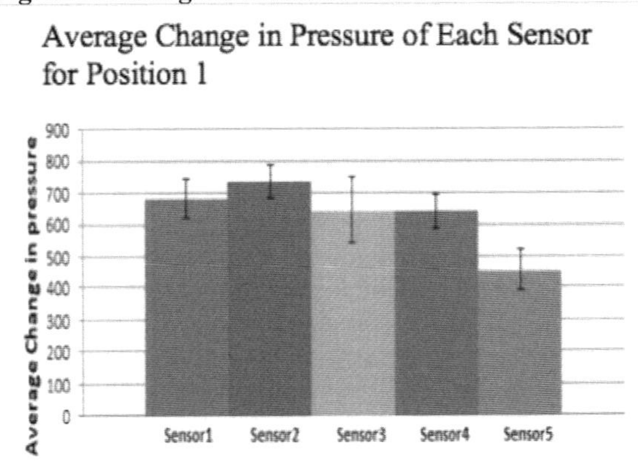

Figure 2: Change in Pressure for Position 2

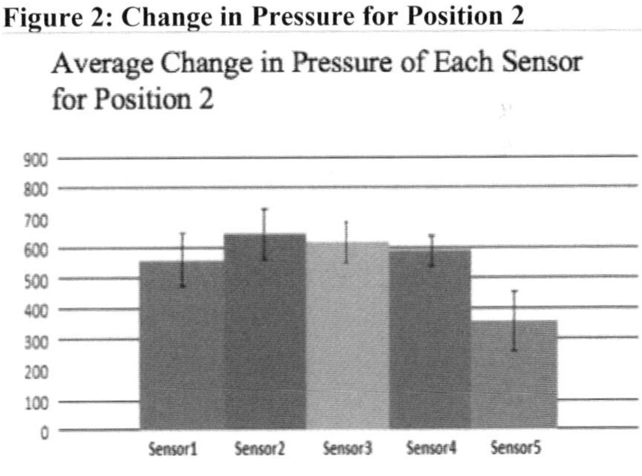

References:

[1] Mary Bond, The New Rules of Posture, 2006; 54-72
[2] Applied Ergonomics, 37(5), 619-628. doi:10.1016/j.apergo.2005.09.002.
[3] David M. Antle, Linda L. Miller. "Sitting disease – working postures and alternative workstations", EWI Works, Innovative Ergonomic Solutions, 2014; 1-22 [4] Dr. Heidi Roberts, Katie Roberts,, The Sitting Disease, 2017; 25-40

Response of endogenous neural stem/progenitor cells to traumatic brain injury

Jeremy Anderson, Misaal Patel, Rebecca Risman, Quinn Wade, and Dr. Li Cai
Department of Biomedical Engineering, Rutgers University, NJ

Introduction: Traumatic brain injury (TBI) causes temporary or permanent loss of memory, motor and cognitive function, and even death. TBI is responsible for approximately 2.2 million emergency room visits and 56,000 deaths annually (CDC, 2017). Cells die from the impact immediately after injury (primary injury), and the primary injury is followed by bleeding and inflammation (secondary injury) which leads to additional cell death. Currently, there are no treatments to recover cells lost from the TBI, mostly due to the lack of knowledge regarding cellular and molecular response to injury. The adult brain harbors endogenous neural stem/progenitor cells (NSPCs) that respond to injury and integrate into neural networks in the subgranular zone (SGZ) of the dentate gyrus (DG) in the hippocampus and the subventricular zone (SVZ) of the neocortex [1]. Although the response of adult NSPCs is insufficient to recover significant damage, an increased understanding of endogenous NSPC injury response will help development of regenerative medicine for TBI. The Notch1 gene is highly involved in CNS development, neurogenesis, injury response, and production of progenitor cells and neuroblasts in the SVZ [2]. However, the role Notch1 plays in injury response and lineage adoption is not understood. Using the Notch1CR2-GFP transgenic mouse model where a subset of Notch1 activated NSPCs are tagged with GFP will identify an unstudied population of NSPCs and begin to define their role in injury response. This research characterizes the role of injury-induced Notch1 signaling NSPC activation and differentiation.

Methods: Using a closed head injury (CHI) model in Notch1CR2-GFP transgenic mice [3], where the subset of Notch1-activated NSPCs express GFP reporter, CHI was performed on 8-12 week old mice. At the designated time point, animals were euthanized and perfusion with 4% PFA was performed. Brain tissue was preserved in 4% PFA followed by 30% sucrose before sectioning. After sectioning, immunohistochemistry was performed to identify expression of different cell markers. This stained brain tissue was analyzed under fluorescence microscopy 2 to 14 days post injury (dpi) to characterize the Notch1 injury-activated NSPCs. Cell counting was performed for e 3 cryosections on male (n=3) and female (n=3) mice at each time point. Statistics performed on each group with error bars represented as Mean ± SEM using students t-test.

Results: At 2 dpi, GFP+ cells were found to have an increased co-labeling with NSPC marker Nestin in the SVZ/DG of injured transgenic mouse brains (2.58±0.53% vs. 0.36±0.12% in sham; $p<0.05$). These injury-activated GFP+ cells were also with increased proliferation marker Ki67 (4.06±0.93% vs. 0.45±0.21% in sham; $p<0.05$). At 14 dpi, GFP+ cells in the injured brain had an increased co-labeling with astrocyte marker S100b (0.25±0.06% vs. 0.02±0.01% in sham, $p<0.05$), GABAergic marker GABA (5.82±2.47% vs. 0.37±0.14% in sham, $p<0.0001$); and glutamatergic marker vGlut2 (1.95±0.65% vs. 0.21±0.05% in sham, $p<0.0001$). However, there is no significant difference in co-labeling with oligodendrocyte marker NG2 (CHI: 0.17±0.04% vs. sham: 0.03±0.02%).

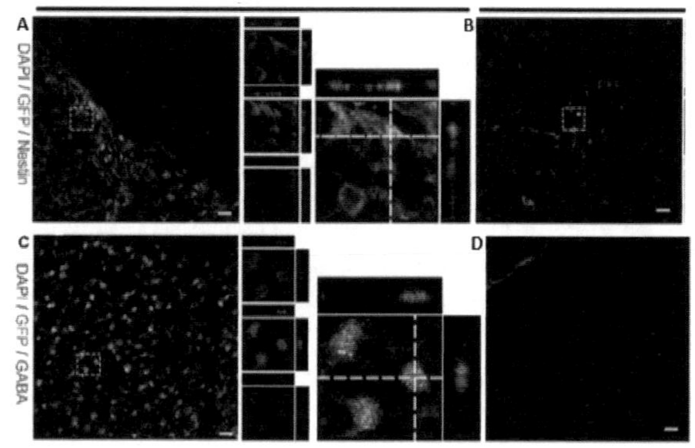

Figure 1. Injury activates NSPCs to proliferate and differentiate. Increased GFP+ NSPCs (Nestin+) were present in the injured mouse brain tissue (A) compared to sham (B) in the SVZ at 2 dpi. Increased GFP+ GABAergic neurons (GABA+) were present in the injured brain tissue (C) compared to sham tissue (D) in the cerebral cortex at 14 dpi.

Conclusions: These results demonstrate that 1) injury-induced NSPCs proliferate and have the potential to generate all the cell types, e.g., neurons, astrocytes and oligodendrocytes, after injury; 2) Notch1CR2-GFP+ cells preferentially differentiate into GABAergic interneurons; 3) Notch1CR2-GFP mouse is a useful tool that greatly facilitate cellular and molecular investigation of TBI-activated NSPCs. Further cellular and molecular characterizations of Notch1 subpopulations of NSPCs after brain injury will provide a better understanding of the NSPC response to injury. In addition, our findings support the potential of endogenous NSPCs as a cell source for TBI therapeutics. Future work includes performing single-cell RNA-sequencing to identify injury-induced transcriptome changes and elucidate the mechanisms of NSPC activation post-injury. Understanding how NSCs respond to injury and develop into various lineages will help the development of regenerative medicine for TBI.

References:

1. Ming, G.L. Neuron. 2011; 70: 687-702.
2. Patel, K. Brain Res. 2016; 1640: 104-113.
3. Tzatzalos, E. Dev Biol. 2012; 372: 217-228.

Acknowledgements: This study is supported by the New Jersey Commission on Spinal Cord Research, the Busch Biomedical Research Award, and the Rutgers Biotechnology Training Program.

The Effect of Electroporation Buffer Composition on Cell Viability and Electro-transfection Efficiency

Joseph J. Sherba[1], Stephen Hogquist[1], Jerry W. Shan[2], Hao Lin[2], David I. Shreiber[1], Jeffrey D. Zahn[1].
[1]Rutgers, The State University of New Jersey, Department of Biomedical Engineering
[2]Rutgers, The State University of New Jersey, Department of Mechanical and Aerospace Engineering

Introduction: Electroporation (EP), is a technique used to introduce foreign molecules, such as plasmid DNA vectors, into the intracellular space of biological cells using an external electric field. Although this technique has been in use for a few decades, there is still an incomplete understanding of the effects that different molecular solutes in the cell suspension buffers has on the overall transfection outcomes. With gene therapy beginning to be adopted in clinical settings and EP as one of the techniques used for DNA transfection, a better understanding of how different buffer compositions effect EP outcomes will aid in the rational development and optimization of DNA transfection protocols. Thus, the purpose of this study is to investigate how EP buffer composition affects cell viability and electro-transfection efficiency (eTE) following EP.

Methods: Sterile EP buffers were prepared with matching physiological pH (7.4) and osmolality (~300 mOsm). Either sucrose or trehalose was used to balance osmolality. Final buffer conductivity was adjusted to 500 µS/cm or 2000 µS/cm, using $MgCl_2$, KCl, $MgSO_4$, or a mixture of $MgCl_2$ and KCl. 3T3 fibroblast cells were harvested and washed in EP test buffer prior to resuspension at a concentration of 3x106 cells/mL in a 2 mm gap EP cuvette. This resuspension contained the Lonza pMAX GFP vector at a final concentration of 20 µg/mL. A single square wave pulse was applied to each cell population. The prescribed electric field and duration ranged from 1.2 kV/cm to 4.8 kV/cm and 0.063 ms to 1.0 ms, respectfully. Electric field / pulse duration combinations were chosen to either keep the total electrical energy dissipation (ÃE^{2}t) or charge movement (ÃEt) constant (Ã—conductivity, E—electric field, t—pulse duration) during the pulse. Following EP, cells were re-plated and incubated for 24 hours then imaged under phase contrast and fluorescence microscopy to determine the resulting cell viability and TE, respectfully. Cell viability was determined as the total number of cells imaged in the experimental condition normalized to the total number of cells imaged under a 'no-pulse' control. TE was defined as the ratio of total number of GFP (+) cells to total number of cells per condition. In a separate experiment, lidocaine, an ATPase ion channel inhibitor, was added to both $MgCl_2$ and KCl buffers (500 µS/cm) to a final concentration of 10 mM. Electric pulses of different energies were applied to cells in suspension and cell viability was assessed 24 hours following pulse application.

Results: Figure 1 is a color map representing the EP outcome score, the product of cell viability and efficiency percentages for each condition. Different pulse applications, E-Fields, are plotted here as symbols whose colors represent EP buffer composition and its conductivity (µS/cm). The results show three distinct regions. Some conditions, denoted as section 1, resulted in good cell viability, but poor eTE. This was observed primarily in conditions with lower electrical energy dissipation pulse applications or buffers that contained Mg^{2+} at higher pulse energy dissipation. The second region in the lower right displayed the opposite: high eTE with low cell viability. This region is primarily of EP buffers that lack Mg^{2+} as the cation used to balance conductivity entirely or have the highest electrical energy dissipation pulse applications. The third region resulted in a balance of both good cell viability and eTE. This region is composed of EP buffers that had the lowest concentrations of Mg^{2+} tested or buffers without Mg^{2+} but with moderate electrical energy dissipation pulse applications. From these results it appears that an optimal concentration of Mg^{2+} is required to preserve cell viability while still allowing for high eTE. Mg^{2+} is known to be a cofactor for many biological enzymes, in particular DNases and ATPase membrane ion channels. High concentrations of Mg^{2+} may result in lower TE due to higher activity of DNase enzymes. However, it is believed that removing Mg^{2+} altogether does not allow the cell to re-establish physiological equilibrium, resulting in higher amounts of cell death. To test this hypothesis, lidocaine, an ATPase inhibitor, was added to the cell suspension during EP. The results (not shown here) indicate that even with Mg^{2+} present, a significant decrease in cell viability is observed at higher energy dissipation pulse applications, suggesting that ATPase activity plays a role in the recovery of cells following EP. Ultimately, the EP buffer containing both K^{+} and Mg^{2+} is found under all pulse conditions tested to result in both high viability and eTE, with the highest EP outcome score of 52. We believe this is because the lower overall Mg^{2+} concentration promotes cell recovery while limiting DNase activity.

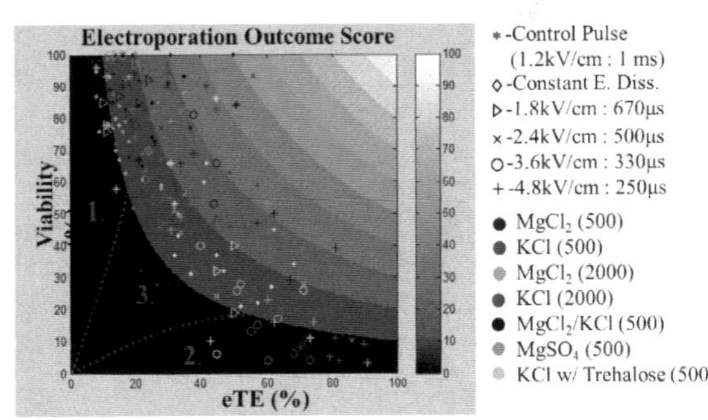

Figure 1. Electroporation outcome score.

Conclusions: The composition of EP buffer appears to play a major role in terms of cell viability and TE following EP. Mg^{2+} is crucial for preserving cell viability, but at the same time detrimental for transfection success. Future work will investigate both different buffer compositions and cell types, with emphasis on the mechanisms by which Mg^{2+} plays a role in the transfection process, with hopes of further enhancing overall EP outcome scores.

Acknowledgements: NSF DBI IDBR 1353918.

Nkx6.1 Gene Therapy Induces Neurogenesis after Spinal Cord Injury

Misaal Patel, Jeremy Anderson, Shunyao Lei, Rebecca Risman, and Li Cai
Department of Biomedical Engineering, Rutgers University, New Brunswick, NJ

Introduction: Spinal cord injury (SCI) results in permanent loss of function leading to paralysis [1]. Unfortunately, there are no therapeutics that promote complete repair and regeneration after SCI. Identification of neural stem cells (NSCs) and their role in injured spinal cords have provided promising opportunities for spinal cord regeneration. Adult NSCs mostly generate astrocytes and oligodendrocytes [2, 3]. However, the potential for adult NSCs to generate neurons after SCI has not been clearly understood. Our previous studies have established that transcription factor Nkx6.1 regulates Notch1 signaling during spinal cord development [3]. Nkx6.1 may represent a therapeutic target for treatment of SCI patients.

Methods: In this study, we used the lentivirus-mediated gene expression to transduce Nkx6.1 in the injured mouse spinal cord. *In vivo* hemi-section SCI was performed on 8-12 weeks old mice followed by injection of lentivirus encoding Nkx6.1 with a reporter RFP. For control, we inject lentivirus carrying RFP reporter. Injured animals are harvested at 3, 14, and 56 days post injury (DPI) to determine cellular and molecular changes surrounding the injection and injury sites using immunohistochemistry. RNA isolation and quantitative PCR is performed to validate the immunohistochemistry results.

Results: We found that overexpression of Nkx6.1 leads to an increase in i) cell proliferation (control: 29.93%, Nkx6.1: 46.95%); ii) the number of Nestin+/RFP+ cells (control: 62.34%, Nkx6.1: 80%) at 3 DPI compared to control, indicating the activation of NSCs at early stages post SCI. To examine the effect of the overexpression Nkx6.1 on the fate of virally transduced NSCs, we examined spinal cord tissues at 14 DPI with neuronal progenitor marker (DCX), astrocyte marker (GFAP), and oligodendrocyte progenitor marker (PDGFRa). Our results indicate that overexpression of Nkx6.1 induces NSPC differentiation towards neuronal over oligodendrocyte lineage, with no significant change in astrocyte lineage. To further determine the ability of Nkx6.1 to generate mature neurons and specific type of interneurons, we examined spinal cord tissues at 56 DPI after viral injection. Immunohistochemical analysis was performed using mature neuronal marker (NeuN), glutamatergic neuronal marker (vGlut2), GABAergic neuronal marker (GABA), and cholinergic neuronal marker (ChAT). Quantification of the NeuN+/RFP+ cells, we found a significant increase in Nkx6.1 (10.37%) group as compared to control (2.68%). Furthermore, there was no significant difference in the percent of glutamatergic and GABAergic interneurons between control and Nkx6.1 treatment. However, Nkx6.1 treatment significantly increases ChAT+/RFP+ cells compared to control (control: 32.08%, and Nkx6.1: 53.00%) (Figure 1).

Conclusions: Nkx6.1 expression in the injured spinal cord induces cell proliferation and NSC activation, and promotes NSCs differentiate into mature neurons, specifically into cholinergic interneurons. Collectively, our findings suggest that Nkx6.1 is a potential therapeutic agent for neuroregeneration after SCI.

References:

1. Silver J. Nat Rev Neurosci 2004, 5(2):146-156.
2. Sabelstrom H. Exp Neurol 2014, 260:44-49.
3. Li Y. Sci Rep 2016, 6:38665.

Acknowledgements: This work is supported by the State of New Jersey Commission on Spinal Cord Research 15IRG006 (L.C.), U.S. Department of Education GAANN Precision and Personalized Medicine Pre-Doctoral Training Fellowship (M.P.), and NIH Biotechnology Pre-Doctoral Training Fellowship T32GM008339 (M.P.).

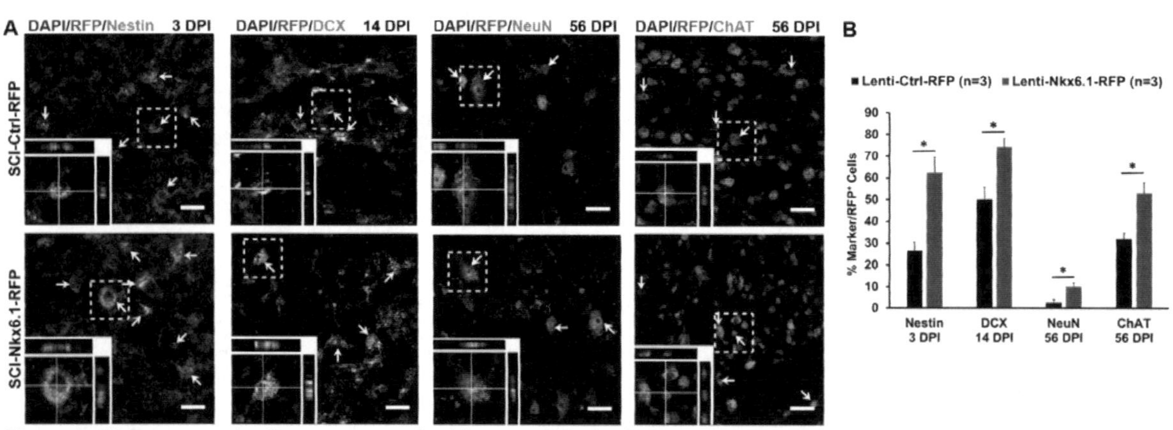

Figure 1. (A) Representative fluorescence confocal images of Nestin+/RFP+ cells at 3 DPI, DCX+/RFP+ cells at 14 DPI, and NeuN+/RFP+ and ChAT+/RFP+ cells at 56 DPI. Scale bar = 20µm. (B) Quantification of RFP+ cells co-labeled with neuronal markers. Mean±SEM; n = 3. * = $p<0.05$

Effect of PKC Substrate-Fascin on Cell Chirality

Haokang Zhang[1], Jie Fan[1,2], Leo Q. Wan[1,2,3,4,*]
1 Department of Biomedical Engineering, Rensselaer Polytechnic Institute, Troy, NY 12180, USA.
2 Center for Biotechnology & Interdisciplinary Studies, Rensselaer Polytechnic Institute, Troy, NY 12180, USA.
3 Department of Biological Sciences, Rensselaer Polytechnic Institute, Troy, NY 12180, USA.
4 Center for Modeling, Simulation and Imaging in Medicine, Rensselaer Polytechnic Institute, Troy, NY 12180, USA.
* Corresponding author. Email: wanq@rpi.edu

Introduction: Fascin is an actin binding protein which creates tight, paralleled bundling of actin filaments, plays important roles in the formation of cell motility structures such as filopodia or invadopodia, and is highly related to the cell invasion in metastatic carcinomas.[1] As one of the mechanisms that regulate fascin function, protein kinase C (PKC) activation has been reported to dissociate fascin from actin bundles through phosphorylation of S39 amino acid.[2] The left-right asymmetry of cells, also known as cell chirality, is an intrinsic cell property that demonstrates great importance during development and other physiological processes, the establishment of which requires functional actin cytoskeleton.[3] Our lab has previously reported the chirality reversal of several cell types after PKC activation when cultured on defined ring-shape micropatterns.[4] In this study, we aim to investigate the possible involvement of fascin as a PKC substrate in the PKC activation-mediated chirality reversal, with the purpose of gaining insights of the regulation of cell chirality.

Methods: The cell chirality analysis was used as previously described. Briefly, ring shaped micropatterns was printed on a glass slide with fibronectin (EMD Millipore) using the standard microcontact printing technique. The mouse fibroblast (3T3) cells (ATCC) were seeded onto the micropatterns at a density of 200,000 cells/mL and allowed for adhering for 15 minutes before rinsing with PBS to remove unattached cells. The cells were cultured in fully supplemented DMEM combined with G2 (Xcessbio), a small molecular inhibitor that specifically inhibits the actin-bundling function of fascin, at various concentrations for 24 hours, then fixed by 4% paraformaldehyde (Sigma-Aldrich) upon confluence. Phase-contrast images of each ring pattern was then taken using a Zeiss AxioVision microscope (Carl Zeiss). The chirality of cells was characterized by tracking the intensity gradient and circular statistics through a custom-developed MATLAB (Mathworks) program, and the chirality of cells on each ring pattern was designated as clockwise (CW), counter clockwise (CCW) and non-chiral (NC). Data from 3 repeats was analyzed via Tukey HSD using JMP12 software (SAS).

Results: After culturing with G2 for 24 hours, the cells became confluent on the ring-shaped micropatterns in all groups. After imaging and chirality analysis, the control group demonstrated a significant clockwise bias, which was consistent with our previous findings of 3T3 cell chirality[3]. Similarly, no significant change was observed in 10 μM group. However, when the G2 concentration was increased to 25 μM, the significance of difference between the percentages of CW rings and CCW rings was abrogated. And when G2 concentration was further raised to 50 μM, the chirality of the 3T3 cells was clearly reversed from CW to CCW, together with a slight increase of non-chiral rings. Higher concentrations (> 50 μM) resulted in cell death.

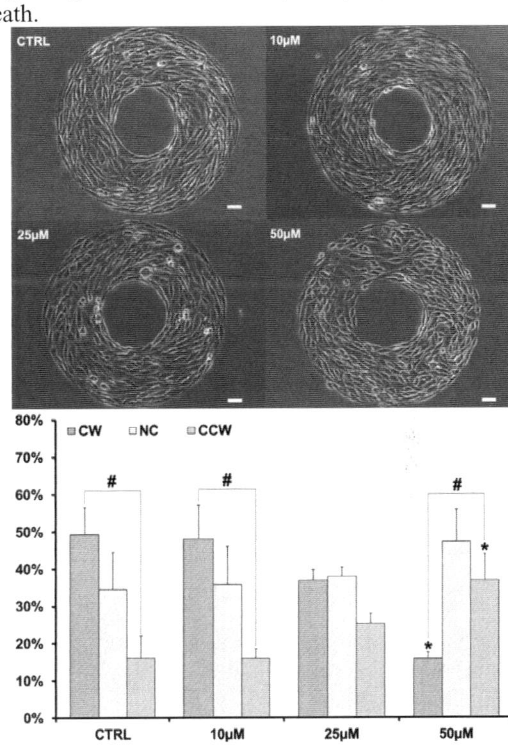

Figure 1. Ring pattern images and chirality analysis results withG2 treatment *(* p<0.05 vs. consecutive concentrations, # p<0.05 vs. different chirality. Scale bar: 50μm)*

Conclusions: Based on our results, the inhibition of fascin using the small molecular drug G2 successfully reversed the chirality of 3T3 cells, indicating fascin's potential involvement in PKC activation-mediated chirality reversal, and once again confirmed the importance of a functional actin cytoskeleton in regards to cell chirality. Further validation such as genetic knockdown will be needed in order to testify this result. Future study is also necessary to examine the interaction between facin and PKC substrate and the implication of cell chirality in fascin regulated cancer metastasis.

References:

1. Jayo A. et al. Devcel. 2016;38:371-383
2. Larsson C. Celsig. 2006;18:276-284
3. Wan L.Q. et al. PNAS. 2011;108(30):12295-12300
4. Fan J. et al. Sci. Adv. 2018;4:eaat2111

Acknowledgement: This work was supported by the National Institutes of Health (OD/NICHD DP2HD083961). L.Q.W. is a Pew Scholar in Biomedical Sciences (PEW 00026185), supported by the Pew Charitable Trusts.

Real-time Transfer of Lentiviral Particles by Producer Cells using an Engineered Coculture System

Timmins, Lauren M., Parekkadan, Biju
Rutgers University Department of Biomedical Engineering

Introduction: Lentiviruses are derived from human immunodeficiency virus (HIV) which allows them to be quite effective gene delivery systems for stable production of genetically engineered human cells. However, prior to using lentiviruses to deliver genetic materials to cells of interest, the normal course of production of these viruses involves a lengthy collection, purification, preservation, and quantification process. In this report, we demonstrate the ability for producer HEK293T cells to simultaneously produce lentiviral particles and transduce (i.e., infect) target cells through a membrane-based coculture system in a continuous, real-time mode which negates the need for a separate viral collection and quantification process. The coculture system was evaluated for major design features such as variations in HEK293T seeding density, -target cell type densities, -as well as membrane porosities to identify key relationships between lentiviral particle production rate and infection kinetics for adherent and suspension cell types. As a proof-of-concept for the creation of an engineered cell immunotherapy, we successfully describe the ability to engineer human T cells isolated from PBMCs under the control of this coculture system in under six days with a GFP construct. These studies suggest the capability to combine and more closely automate the transfection/transduction process in order to facilitate well-timed and cost-effective transduction of target human cell types. Our experiments provide novel insight into the forthcoming transition into improved manufacturing systems for viral production and subsequent cell engineering.

Materials and Methods:

Vector Production: Generation of lentiviral vectors was accomplished using Human 293T kidney fibroblast cell line (HEK293T) seeded at varying densities in transwell inserts depending on experimental conditions. Transfection with packaging plasmid psPAX2, envelope plasmid pMD2.G and transfer vector with GFP insert took place in DMEM/F12 media in the presence of Polyethylenimine (PEI) and transduction with target cell types took place in the presence of 8 ug/mL polybrene. Transduction efficiency in all groups was measured using fluorescence-activated cell sorting (FACS) analysis.

Coculture System: The engineered system is based on transwell inserts of varying porosities from 0.4, 1, 3 and 8 μm.

PBMC Culture and T Cell Isolation: PBMCs were isolated from whole blood from patient PW. T Cell isolation from PBMC population was carried out using Miltenyi Biotec Pan T Cell Isolation Kit (MACS Miltenyi Biotec) following the provided protocol. Briefly, manual magnetic labeling was carried out using Pan T Cell Biotin-Antibody Cocktail per 10^7 cells added. After labeling, manual cell separation was done using LS column placed in magnetic field of MACS separator. Flow through containing unlabeled, enriched T cells was collected.

Results: In order to assess the effectiveness of the system to transduce a target human cell type, isolated T cells were seeded at $1x10^6$ cells per well in a 24 well format, while HEK293T cells were seeded in corresponding inserts on day 0 to reach 75% or 90% confluency on day of transfection (day 1). T cell groups were stimulated with IL-2 one day prior to transfection. Flow cytometry analysis and ZEISS images are shown below. All groups were run in triplicate and viral particles directly cultured with T cells were used as control.

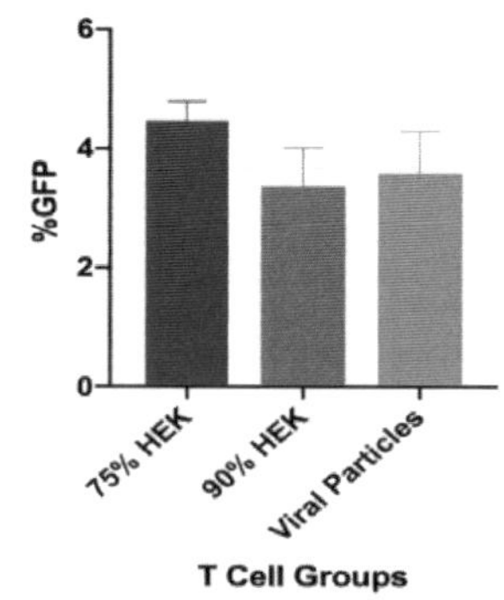

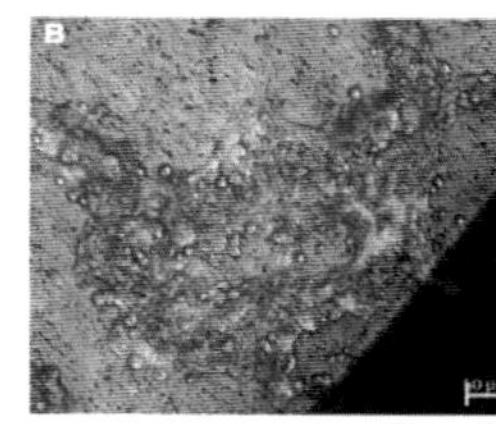

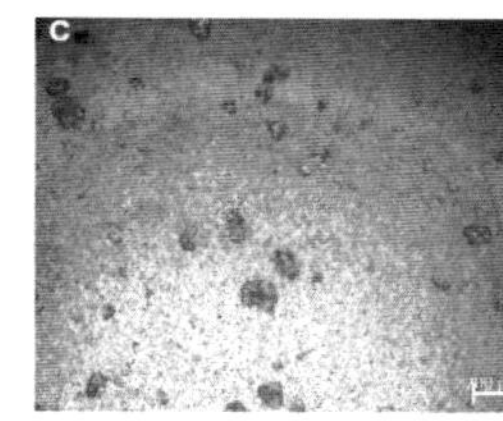

Figure 1. Transducing target human T cells. (**A**) Transduction efficiency of T cells in coculture system. (**B**) GFP positive HEK293T cells in insert confirming successful transfection and particle production. (**C**) GFP positive human T cells in the well of transwell confirming successful transduction.

Conclusions: Our study is an initial investigation into the potential to have a single step continuous system for engineering target cell types through combining the standard processes of transfection using HEK293T cells and transduction. First, our study focused on the potential for cells to be transduced in such a continuous system, and secondly, we assessed the various parameters of the system that can be changed in order to determine the impact on transduction efficiency in order to improve the degree to which cells can be transduced.

In summary, we developed a novel platform to combine the key transfection and transduction processes for viral production and cell engineering in order to improve the timeline and quality of virus used to engineer key human cell types. We assessed a variety of cell types, both adherent and suspension, to confirm the feasibility of the system and tested initial parameters to prove the potential for system optimization. The data collected herein suggest that an optimal range for particle output and subsequent infection can be determined for both adherent and suspension cell types in a transwell based system to match demands of a cell therapy product. Furthermore, combination of the two key processes of transfection and transduction can address many of the current issues associated with timelier and cost-effective cell manufacturing methods using lentiviral vectors. This report can help future development of closed, semi to fully automated systems for manufacturing of cell therapy products with the intent of minimizing cost, reagent use and time.

Acknowledgements: The work was supported in part by the Shriners Hospital for Children (B.P.), National Institutes of Health Grants R01EB012521 (B.P.).

Nanoporous Alumina Membrane Electroporation for Optimized Transfection and DNA Delivery to Cells

Stephen Hogquist[1], Joseph Sherba[1], Jerry W. Shan[2], Hao Lin[2], David I. Shreiber[1], Jeffrey D. Zahn[1]
[1]Department of Biomedical Engineering, Rutgers University, Piscataway, NJ
[2]Department of Mechanical and Aerospace Engineering, Rutgers University, Piscataway, NJ

Introduction: Electroporation is the technique of applying an electric field to a cell to achieve temporary permeabilization and allow the passage of drugs, chemicals, or DNA through a cell membrane. Electroporation experiments are a function of various parameters such as the applied electric field, cell type used, and the contents and electrical parameters of the buffer solution through which a field is applied. This experiment aims to demonstrate a technique by which a lower voltage can be used to electroporate cells when compared to conventional experiments with the goal of increasing cell viability and transfection efficiency with plasmid DNA. This is achieved utilizing a biocompatible, 60μm thick nanoporous alumina membrane substrate with 100nm pores at 45% porosity. Computational modeling predicts applying an electric field through this nanoporous substrate in solution results in electric field amplification near nanopores to achieve electroporation while maintaining a lower field strength elsewhere in the area surrounding a cell. While existing electroporation experiments work with cell suspensions in cuvettes to achieve electroporation, this technique presents a method to achieve electroporation of adherent cells. Preliminary results suggest that DNA plasmid can be electrophoretically driven[1,2] through these nanopores and enter a cell with an electroporation pulse. The results of this modeling and experiment describe a technique to further investigate the application of nanoporous substrates in electroporation experiments.

Methods: Nanoporous alumina membranes with 100nm pores (Sterlitech, Kent, WA) are soaked in 50 μg/mL of fibronectin (Sigma, St. Louis, MO) for 4 hours to promote cellular adhesion and allowed to dry. 3T3 Mouse Fibroblast cells are cultured onto these coated nanoporous alumina membranes and incubated for 24 hours at 37°C and 5% CO_2. After culturing, cells attached to the alumina membrane are transferred to a Franz Cell diffusion chamber device (PermeGear, Hellertown, PA) which serves as the experimental housing. Using the Franz Cell device, electrodes are positioned immediately above and below the alumina membrane to supply voltage to the system, and the cells are immersed in a HEPES based-buffer with 500 μS/cm conductivity containing fructose and balanced with magnesium and potassium ions, optimized from previous study. The chamber of the system opposing the cells contains plasmid GFP at 20 μg/mL concentration in buffer solution. A DNA delivery pulse below the threshold for electroporation at 0.4 kV/cm is first applied to the system with varying pulse durations, 10ms, 100ms, and 500ms. This pulse is designed to transport the pDNA through the alumina membrane to be positioned adjacent to cells. Next, an electroporation pulse is applied to these cells with 2 kV/cm field strength for 500μs. This second pulse serves to permit cellular uptake and transfection with the cross- membrane transported pDNA. All pulses applied utilize a BTX 830 Square Wave Pulse generator system (BTX, Holliston, MA).

Computational modeling utilizing COMSOL Multiphysics software is used to determine the appropriate pulse strength conditions based on the predicted electric field intensity with the resistive, field-amplifying component of the alumina membrane taken into account.

Results: The results of these experiments show that, for pulse durations greater than 10ms, DNA delivery through the alumina membrane is successful. With a significantly increased pulse duration, a decrease in the transfection efficiency of cells is observed.

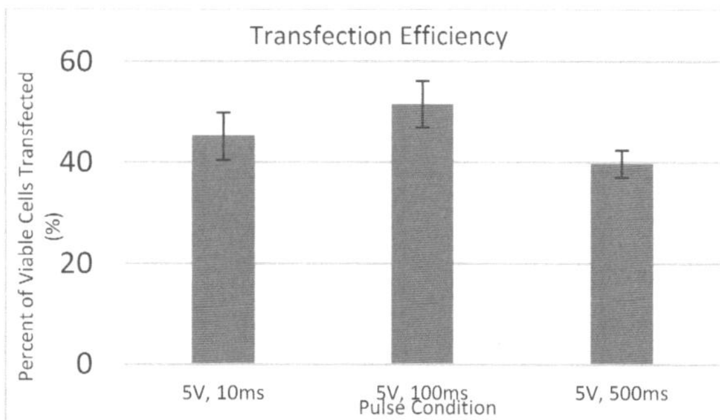

Figure 1. Transfection Efficiency of 3T3 Cells based on length of pDNA delivery pulse

Conclusions: This work demonstrates that plasmid DNA can be successfully transported through the nanoporous alumina membrane and delivered to cells with an electroporation pulse. While shorter pulse durations show that some pDNA is being delivered across the membrane, the decrease in efficiency at the 500ms time point indicates that raising the delivery pulse duration can have a negative effect on transfection efficiency potentially due to cell damage or the pDNA being driven away from cells. The theoretical prediction for DNA electrophoretic transport through the alumina membrane is 14ms, so further investigation is required to validate this prediction. Future studies utilizing this adherent cell electroporation technique will focus on studying the impact of shorter DNA electrophoretic transport pulse durations on transfection efficiency, optimizing electroporation pulse parameters such as pulse strength and pulse duration for this setup once an ideal DNA transport pulse has been determined, and will incorporate improved field strength modeling in and around the alumina membrane nanopores to better understand the electric field amplification.

References:

1. Sadik M. et al. BBA-Biomembranes. 2013;1828(4):1322-1328.
2. Chen P. et al. Nano Letters. 2014;4(11):2293-2298

Cerebral spinal fluid modulators for enhanced drug delivery in the brain

Caroline Wood, Mariagemiliana Dessi Ph.D., Jay C. Sy Ph.D.
Department of Biomedical Engineering, Rutgers, The State University of New Jersey, Piscataway, NJ

Introduction:

Drugs delivered to the brain are cleared rapidly by cerebral spinal fluid (CSF) efflux mechanisms, inhibiting their ability to achieve optimal distributions and exposure. The choroid plexus, a layer of epithelial cells found lining the lateral ventricles, is responsible for secreting most of the CSF found in the central nervous system. Various enzymes, ion channels, and transporters are involved in CSF production. We hypothesize that drug clearance rates in the brain can be altered through the modulation of CSF secretion at the choroid plexus. The goal of this project is to identify and evaluate potential CSF modulators and investigate their ability to increase drug efficacy in the brain. Preliminary studies have demonstrated *in vivo* that the carbonic anhydrase inhibitor, Acetazolamide, and calcium ion agonist, Verapamil, alter CSF dynamics when injected intraventricularly in rats. To further identify novel CSF modulators, an *in vitro* model of the choroid plexus will be fabricated to rapidly screen drugs and their formulations' effects on CSF secretion without animal sacrifice. Ultimately, the identified CSF modulators will be tested synergistically with sustained release chemotherapeutic formulations *in vivo* with the hope that CSF modulation will allow currently established therapies to significantly increase patient survival.

Methods:

To optimize the *in vitro* CSF production model, multiple cell lines and primary cell isolation methods are being compared, including: the Z310 epithelial cell line [1], primary cells from direct choroid plexus dissection of rat brains, and E-cadherin labeled magnetic isolated cell separation of primary epithelial cells from a mixed rat brain cell population. Cells are characterized for epithelial purity using flow cytometry, immunocytochemistry, and western blot with epithelial cell markers including e-cadherin and EPCAM. Transwell membranes are used to create a cell culture model of the choroid plexus barrier and monitor CSF secretion. Epithelial barrier integrity and tight junction formation is quantified using transepithelial resistance (TEER) values measured with an epithelial voltohmmeter and analyzed with immunocytochemistry. This model will be used to screen carbonic anhydrase inhibitors, Na+/K+/2Cl- pump inhibitors, Na+/K+ATPase inhibitors, and Na+ channel inhibitors that have been previously FDA approved for alternate applications.

To study the effect of using CSF modulators with previously FDA approved sustained release chemotherapeutics, such as the GLIADEL® wafer [2], polyanhydride polymers, including (Poly-[1,3] bis(*p*-carboxyphenoxy)propane-co-sebacic acid], have been synthesized using melt polycondensation. Products have been analyzed using Proton nuclear magnetic resonance (H-NMR) and size exclusion chromatography. Drugs will be formulated with chemotherapeutic polymers and CSF modulators, formed into wafers, and implanted into the brains of rat glioblastoma models to monitor the effects of CSF modulation on sustained release drug pharmacokinetics.

Results:

Flow cytometry, immunocytochemistry, and western blot analyses are currently being finalized, but studies thus far have indicated high epithelial purity of the various cell sources.

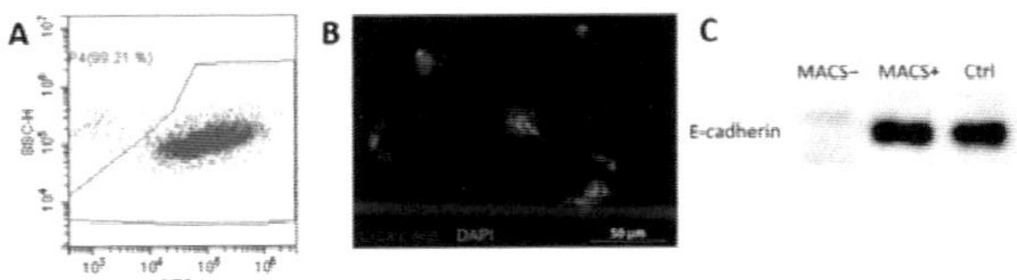

Figure 1: A. E-cadherin positive cells from choroid plexus isolation measured by flow cytometry n=8,000. B. Immunostaining of MACS isolated epithelial cells. C. Western blot of MACS isolated cells and MACS remaining cells

Immunocytochemistry of Transwell membranes has demonstrated the presence of tight-junction formation of various cell types. H-NMR analysis has indicated successful synthesis of the polyanhydride sustained release system. Current studies are being performed to fully characterize the optimized *in vitro* system and prepare polymers for future *in vivo* testing with identified CSF modulators.

Conclusions:

We have demonstrated the ability for two candidate CSF modulators to significantly change CSF production rates in *in vivo* rat models. An *in vitro* model is being developed to discover additional, more effective drugs and formulations capable of altering CSF secretion at the choroid plexus. The optimal cell choice for this model will be determined based on characterization studies, and their use with Transwell membranes allows for secretion analysis to be performed with potential modulators. These modulators will be studied *in vivo* with sustained release polymer systems of chemotherapeutic drugs to obtain more desirable drug distributions and exposure times in brains of rodents with glioblastoma. If successful, this study will highlight a novel approach in improving drug efficacy and patient survival.

References:

1. Zheng W, Zhao Q. Establishment and characterization of an immortalized Z310 choroidal epithelial cell line from murine choroid plexus. Brain Res. 2002;958(2):371-80.
2. Fleming A, Saltzman W. Pharmacokinetics of the Carmustine Implant. Clin Pharmacokinet 2002; 41 (6): 403-419.

Acknowledgements:
The Sy Lab
NIH/NIBIB R00 EB016690
Biotechnology Training Grant NIH T32 GM008339

P1-26

Using a Novel Protein Energetics Model and Microfluidics to Improve the State of Cancer Screening

Zachary R. Fritz[1], Rene S. Schloss[1], Anil B. Shrirao[1], Lawrence J. Williams[2], Martin L. Yarmush[1]
[1]: Department of Biomedical Engineering, Rutgers University, Piscataway, NJ
[2]: Department of Chemistry and Chemical Biology, Rutgers University, Piscataway, NJ

Introduction: Cancer is the second leading cause of death in the U.S., but detecting the disease before it reaches its metastatic stages can improve survival rates. Autoantibodies generated in response to tumor-associated antigens, such as mutated versions of the tumor suppressor protein p53, have the potential to be useful diagnostic biomarkers for cancer; they occur in a variety of malignancies, have long-term stability in the bloodstream, and can precede formal diagnosis by years. However, immunoassays for the detection of these autoantibodies suffer from low sensitivity, with one large study detecting p53 autoantibodies in only 26% of colorectal cancer patients[1]. *We hypothesize that the diversity inherent in cancer patients' immune responses and the choice of antigens incorporated into these immunoassays has hindered their diagnostic power.* We aim to develop a superior autoantibody immunoassay by using a new protein energetics model to design optimized peptide antigens that correspond to multiple epitopes on multiple molecules. These optimized antigens should better recapitulate the free energy profiles of their native (whole protein) counterpart sequences, enabling the detection of low to high affinity polyclonal autoantibodies and improving assay sensitivity. Finally, the assay will be carried out in a centrifugal force-based microfluidic platform that will have integrated detection and automation capabilities.

Methods: The model is a coarse-grained energetics model based on the scale invariant properties of proteins that takes into account amino acid identity and protein secondary structure. For a given protein or peptide it generates an "interaction factor" value for each amino acid residue that is related to free energy. To optimize the peptide antigens, flanking residues and terminal acetylations can be added in order to match the energy values of the isolated peptide antigen to its counterpart found within the whole native protein. We are starting with a single antigenic sequence in p53 and have designed 20 flanking sequence permutations to fully encompass all possible energetic profiles (peptides synthesized by Genscript). These antigens will be conjugated to 6.5 μm Luminex microbeads and can later be made in the form of branching dendrimer-like multi-antigen peptides (MAPs) to increase surface area and antigen presentation. The centrifugal microfluidic device assay platform will be made from alternating layers of laser-cut acrylic and pressure sensitive adhesive. Proof-of-concept experiments consist of immunoassay (Bio-Plex 200 system for detection, Bio-Rad) and surface plasmon resonance (Biacore T200, GE Healthcare) affinity measurements using peptide antigens predicted by the model to have superior, inferior, or more non-specific affinity for autoantibodies as compared to an unmodified control. These experiments will use the same monoclonal anti-p53 antibody standards (including an isotype negative control, all from Abcam) and pooled human sera as a matrix. To assess the optimized assay's diagnostic power and response to a typical polyclonal sample, a randomized case-control study will be conducted. This will use serum samples from colorectal cancer patients (from the Cancer Institute of New Jersey's biorepository) and control samples from patients without cancer taken from pre-admission testing specimens (Robert Wood Johnson University Hospital).

Results: One of the primary hypotheses of the model, *that peptides with higher interaction factors will have a higher affinity for antibodies*, has so far been borne out by preliminary immunoassay and surface plasmon resonance experiments. **Figure 1** shows this trend with a set of 4 peptides, with a lower dissociation constant (Kd) corresponding to a higher affinity for the monoclonal antibody.

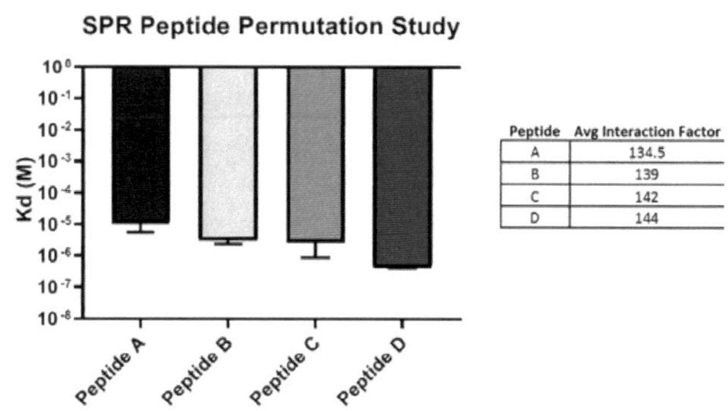

Figure 1.

Additionally, our model was able to successfully predict mutation-induced changes in free energy of folding (ΔΔG) in T4 lysozyme with a high degree of accuracy relative to experimental data (R^2=0.73).

Conclusions: The protein energetics model has thus far shown that it can be a powerful tool for designing optimized antigens with enhanced affinity for antibodies. Besides testing all of proof-of-concept peptides, our other work will focus on examining specificity effects (i.e. susceptibility of the peptides to bind to off-target species), designing a user-friendly microfluidic assay platform with point-of-care potential, and seeing if the model's predictions translate to improved clinical sensitivity with a case-control study.

References:

1. Pedersen, J.W.,et al. *British Journal of Cancer.* (2013). 108(1), 107-114

Acknowledgements: This work was partially funded through NIH grants T32GM008339, P41EB002503, R01EB020036, and NSF IDBR grant DBI1353918.

Minimizing Infection and Revision Surgeries through Nitric Oxide Releasing Total Knee Replacement Prosthesis

Ashley Widing[1], Kagya Amoako, Ph.D.[1]
[1]University of New Haven, West Haven, CT

Introduction: Replacing a weight-bearing joint is accompanied by long procedures, longer recovery times with intensive monitoring, and prolonged recovery times if the surgical site becomes infected. A study predicts that by 2030, there will be approximately 3,800,000 Total Knee Replacements (TKRs) with a risk of 1.2% of patients becoming infected. Without prophylactic antibacterial properties incorporated into knee replacements, revision surgery is usually the only corrective option after infection. It is hypothesized that the antibacterial properties of nitric oxide (NO) can be translated into addressing TKR-related infections via NO releasing Polydimethylsiloxane (PDMS) coatings. Observing sessile bacterial colonies and treating them with NO can help determine the appropriate flux levels needed to kill biofilm-protected bacteria. In addition, a prosthetic device coated with NO releasing polymer, will help determine the coatings' safety and efficacy.

Methods: To test the bactericidal properties of NO released from PDMS, customized petri dishes were 3D printed and *Pseudomonas aeruginosa* was grown in them to represent *in vitro* bacterial adhesion (Figure 1). The dish has a pour plate that seats flush to initially pour nutrient agar. This plate was then replaced with a six well dish; each well containing Phosphate-buffered saline (PBS) and two 1mm diameter NO-modified PDMS samples. This allows for selective colony treatment which are analyzed using a LIVE/DEAD staining kit. The NO samples were synthesized using NO-to-amine conjugation under hyperbaric conditions.

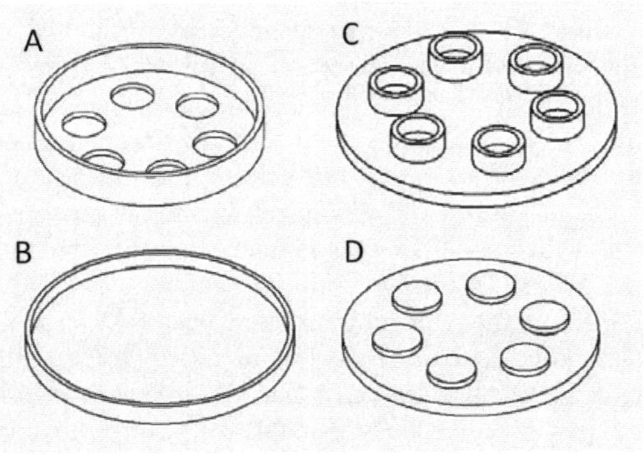

Figure 1: A: Base Plate, B: Top Plate, C: Six Well Dish, D: Pour Plate

Results: A customized petri dish was designed to determine the ability of NO releasing polymer to kill bacteria without disturbing the colonies. Bacterial adhesion and LIVE/DEAD staining assays have been developed and tested. Our lab has shown that NO from PDMS at a flux of 21.18 $\pm$ 5.31 $\times 10^{-10}$ mol/min/cm^2 inhibits bacterial growth. This flux level is being investigated for its ability to kill biofilm-protected bacteria. A 3D knee model was designed using Autodesk Inventor Professional 2018 and 3D printed using a CraftBot XL Printer and will be used for further *in vitro* studies.

Conclusion: Observing these bacteria colonies after introduction of nitric oxide, will determine the appropriate flux levels needed to kill bacteria. The overall goal of this study is to evaluate the effectiveness of a nitric oxide releasing biomaterial on biofilm protected bacteria. The biomaterial coating will then be applied to a prosthetic knee replacement device. Observing the devices' levels of bacterial adhesion will prove the efficacy of compound.

References:

1. Amoako KA. *deepbluelibumichedu*. 2011:1-190.

Evaluation of Alginate-Encapsulated Mesenchymal Stromal Cells for Osteoarthritis Treatment

Ileana Marrero – Berrios ME, Sarah E. Salter, Rishabh Hirday, Rene Schloss PhD, and Martin Yarmush MD PhD
Department of Biomedical Engineering, Rutgers, the State University of New Jersey

Introduction: Osteoarthritis (OA) is a chronic age-related disease characterized by the progressive destruction of articular cartilage, accompanied by changes in the whole joint. Although OA is considered a non-inflammatory joint disease, pro-inflammatory cytokines have been implicated as key mediators in the disease. There is currently no cure for OA. Existing treatments alleviate symptoms initially; however, they are not able to alter disease progression and disease development eventually proceeds. Therefore, there is a need to develop effective therapies that could alter OA progression and promote healing in osteoarthritic joints. One approach has been intra-articular administration of mesenchymal stem cells (MSC), which secrete anti-inflammatory and regenerative factors such as interleukin 1 receptor antagonist (IL-1RA), prostaglandin E_2 (PGE_2) and transforming growth factor (TGF)-^{2} 1 that could alter the underlying pathophysiology of OA. However, these cells are required in large numbers and are not long-lasting when freely administered. We have previously shown that encapsulation of MSC lengthens their survival, promotes their anti-inflammatory secretory function, and enhances TGF-^{2} 1 secretion without the addition of growth factors. We aim to ascertain whether encapsulated MSC (eMSCs) can provide sustained reduction of OA mediated joint inflammation and destruction and promote re-growth and healing. Such experiments could provide a powerful new therapy in an otherwise irreversible progressive disease.

Methods: Using an electrostatically assisted cell encapsulation unit (Nisco), human bone marrow-derived MSC were encapsulated in alginate and co-cultured with primary bovine chondrocytes or human macrophages in the presence or absence of inflammatory stimuli (interleukin (IL-) 1 and/or tumor necrosis factor (TNF)- ±) to simulate the inflammatory environment in OA. Media supernatants were collected to quantify secretion of cytokines by enzyme-linked immunosorbent assay (ELISA) or multiplex bead assays. In addition, chondrocyte cell pellets were collected to measure gene expression changes using an inflammatory cytokine qRT-PCR gene array (Qiagen).

Results: To investigate the effects of free or eMSCs on bovine chondrocytes, we performed immunoassays and RT-PCR gene arrays. IL-1 inflammatory stimulus induced enhanced secretion of anti-inflammatory cytokines such as IL-1RA in free or eMSCs when co-cultured with bovine chondrocytes (Figure 1A). In addition, treatment with free or eMSCs induced gene expression changes for pro-inflammatory cytokines in bovine chondrocytes (Figure 1B). IL-1 stimulus induced upregulation of pro-inflammatory cytokine genes and MSCs were able to attenuate expression of inflammatory genes, such as IL-8 and IL-1^{2} .

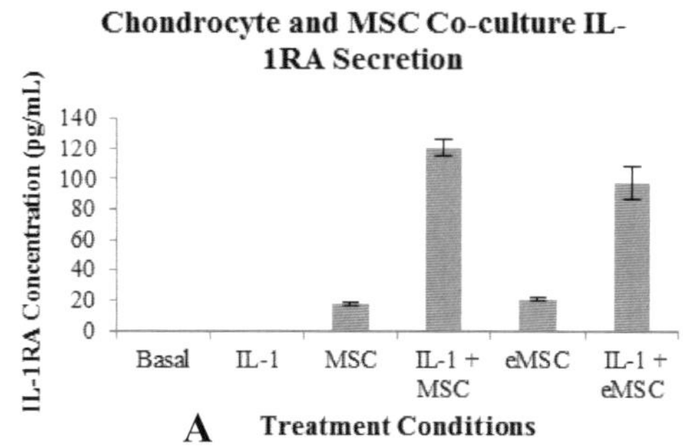

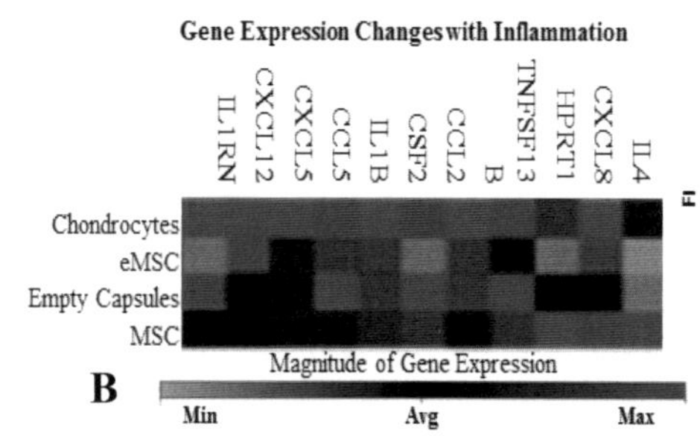

Figure 1: A) Co-culture with bovine chondrocytes and free or eMSCs. MSCs respond to IL-1 stimulus by producing IL-1RA, an anti-inflammatory cytokine that blocks IL-1- IL-1 receptor interaction. **B)** Bovine chondrocyte gene expression. MSC or eMSC co-culture down-regulated inflammatory gene expression after pro-inflammatory stimulus.

Conclusions: MSCs therapies could become a viable option for OA treatment by promoting tissue healing and regeneration. By modifying the MSC microenvironment it is possible to induce functional changes in the MSCs that may promote their anti-inflammatory and chondrogeneic function. eMSC secrete anti-inflammatory factors such as IL-1RA, that target OA inflammatory mediators. They are also capable of modulating chondrocyte inflammation by paracrine factor secretion. Further studies will evaluate the anti-inflammatory and chondrogenic effects of free or eMSC in co-cultures with both macrophages and chondrocytes and thereby investigate the responses of all the relevant OA promoting cell types. Our studies highlight the possibility of utilizing eMSC for OA treatment.

Acknowledgements: GAANN Fellowship in "Precision and Personalized Medicine" in Biomedical Engineering and Rutgers-UMDNJ NIH Biotechnology Training Program 2T32GM008339-21

Thursday, March 21, 2019

Schedule

8:00 am–5:00 pm Registration

9:00 am–10:00 am Breakfast

10:00 am–10:40 am Keynote Speaker from Brain Health Institute—*LIVINGSTON HALL*

10:40 am–12:00 pm **Parallel Platform Presentations: Session III**
(12 minute talks with 3 minutes for questions)

	Session 3A Biomechanics I *—COFFEE HOUSE*	**Session 3B Brain Health / NeuroEngineering** *—201*	**Session 3C Micro/Nanodevices** *—202*
10:40–11:00 am	3A-1:Multiscale Mechanics of Embryonic Tendon	3B-1:Mehmet Kurt Stevens Institute of Technology	3C-1:Forming Libraries of Magnetic Multicore Nanoparticles with Tunable Dimensions and their Biomedical Applications
11:00–11:15 am	3A-2:Impact of Contact Constraints on the Dynamics of Idealized Intracranial Saccular Aneurysms	3B-2:Jay Sy Rutgers University	3C-2:Effect of Composition on Nanomechanics of Dental Adhesives
11:15–11:30 am	3A-3:Upper Airway Area-Pressure Dynamic Analysis Tool for Studying Obstructive Sleep Apnea Syndrome	3B-3:None	3C-3:Biogenic metallic nanoparticles. A nanometric trojan horse approach
11:30–11:45 am	3A-4:Hemodynamic model of left atrial appendage thrombus risk in patients with atrial fibrillation	3B-4:An In Vitro Bioreactor for Testing Brain Implant Biocompatibility	3C-4:A Nanoscale Drug Delivery Platform with Controlled Drug Release Modulated by Aptamer Engineering
11:45 am–12:00 pm	3A-5:Mechanisms of Hand-Rung Force after a Ladder Climbing Perturbation	3B-5:Quantitative Assessment of Stress Levels with EEG and Heart Rate Variability	3C-5:Application of disturbed fluid flow in a three dimensional cerebral bifurcation model

12:00 pm–2:00 pm Lunch/Vendors and Sponsors/Poster Session

2:00 pm–4:00 pm **Parallel Platform Presentations: Session IV**
(12 minute talks with 3 minutes for questions)

	Session 4A Biomanufacturing *—COFFEE HOUSE*	**Session 4B Tissue Engineering / Regenerative Medicine II***—201*	**Session 4C Nutrition/Health or Device Development** *—202*
2:00 pm–2:40 pm	4A-1:Steven Leist, SunP Biotechnology	4B-1:Evi Chnari, MTF Biologics	4C-1:C. Alex DePaula, DePuy Synthes
2:40–3:00 pm	4A-2:High Throughput Identification of Synthetic Polymers with Globular and Protein-Like Features	4B-2:Peter Galie Rowan University	4C-2:Carlotta Mummolo NJIT
3:00–3:15 pm	4A-3:Solution Spun Protein-Based Polymer Fibers for Biomedical Applications	4B-3:Vincent Beachley Rowan University	4C-3:Release of Naloxone for Long-term Management of Opioid Addiction
3:15–3:30 pm	4A-4:Adaptation of Liquid Handling Robotics for High Throughput Customizable RAFT-Polymerization	4B-4:Alignment of Human Cardiomyocytes through Nano-Wrinkles on Shape Memory Polymers	4C-4:Significant Associations of Urinary Essential Elements and Autism Spectrum Disorder
3:30–3:45 pm	4A-5:High Throughput Screening of Random Heteropolymers that Stabilize Enzymes	4B-5:Platelet-Derived Growth Factor Fused with Elastin-Like Polypeptides for Pressure Ulcer Healing	4C-5:Developing an Inexpensive Powered Myoelectric Prosthetic Arm for Persons with Amputation
3:45–4:00 pm	4A-6:A Facile Method of Protein Crystallization using Gold Nanoparticles	4B-6:Biodegradable PCL-PGA-beta TCP Scaffolds for Bone Tissue Engineering	4C-6:Instrumented Platform for the Quantitative Assessment of Human Balance Control

4:00 pm–4:15 pm Break/Vendors and Sponsors

4:15 pm–5:30 pm **Parallel Platform Presentations: Session V**
(12 minute talks with 3 minutes for questions)

	Session 5A Biomechanics II *—COFFEE HOUSE*	**Session 5B Stem Cells / Gene editing***—201*	**Session 5C Biosystems Engineering** *—202*
4:15–4:30 pm	5A-1:Biomechanical Threshold of in vivo Neonatal Brachial Plexus after Stretch Injury	5B-1:Trophoblast Enrichment Using Surface Adhesion Properties	5C-1:A Modular Cell-Cell Adhesion Toolbox for Engineering of Novel Multicellular Systems
4:30–4:45 pm	5A-2:Protecting the Integrity of Surfactant-Stabilized, Oxygen Filled Microbubbles	5B-2:Wnt and Nodal Signaling Pathways Influence Chiral Bias in Human Embryonic Stem Cells	5C-2:Model Based Approach To Measuring Human Endothelial Arterial Function
4:45–5:00 pm	5A-3:Multimodal Foot-Ground Contact Interaction in Human Postural Stability	5B-3:Constitutively Active Ras Alters Cellular Chirality and Actin Structure of Mammary Epithelial Cells	5C-3:Maddox Components of Vergence
5:00–5:15 pm	5A-4:Effects of Walking Forces on Shoe Wear Rate	5B-4:The Effect of Zinc on the Osteogenic Differentiation of Mesenchymal Stem Cells	5C-4:The Impact of Vascular Ehlers-Danlos Syndrome Mutations on Collagen III Structure, Dynamics, and Function
5:15–5:30 pm	5A-5:Effects of diabetic therapeutic footwear and traditional athletic footwear on gait: a cross-sectional investigation	5B-5:Kinetics of MSC-Based Enzyme Therapy for Immunomodulation	5C-5:Effects of Orthoptic Treatment on Convergence Insufficiency

6:00 pm–9:00 pm Career Night-Tapas—*LIVINGSTON HALL*

Abstracts

3A-1 **Multiscale Mechanics of Embryonic Tendon** . . . 79
Benjamin Peterson
Pennsylvania State University

3A-2 **Impact of Contact Constraints on the Dynamics of Idealized Intracranial Saccular Aneurysms** . 80
Manjurul Alam, Padmanabhan Seshaiyer
George Mason University

3A-3 **Upper Airway Area-Pressure Dynamic Analysis Tool for Studying Obstructive Sleep Apnea Syndrome** . . . 81
Kok Ren Choy [1], Sanghun Sin [2], Yubing Tong [3], Jayaram Udupa [3], Can Wu [4], Mark Wagshul [2], Raanan Arens [2], Dirk Luchtenburg [1], David Wootton [1]
[1] The Coopert Union, [2] Albert Einstein College of Medicine, [3] University of Pennsylvania, [4] Philips Healthcare

3A-4 **Hemodynamic model of left atrial appendage thrombus risk in patients with atrial fibrillation** . 82
Soroosh Sanatkhani, Prahlad Menon, Sanjeev Shroff
University of Pittsburgh

3A-5 **Mechanisms of Hand-Rung Force after a Ladder Climbing Perturbation** . . . 83
Erika Pliner, Kurt Beschorner
University of Pittsburgh

3B-4 **An In Vitro Bioreactor for Testing Brain Implant Biocompatibility** . . . 84
Erika Davidoff, Jay Sy
Rutgers University

3B-5 **Quantitative Assessment of Stress Levels with EEG and Heart Rate Variability** . . . 85
Eyad Attar, Mehmet Kaya
Florida Institute of Technology

3C-1 **Forming Libraries of Magnetic Multicore Nanoparticles with Tunable Dimensions and their Biomedical Applications** . . . 86
Zhen Xiao, Qingbo Zhang, Vicki Colvin
Brown University

3C-2 **Effect of Composition on Nanomechanics of Dental Adhesives** . . . 87
Zinah Alabdali [1,2], Jennifer Lynch [1], Adrian Mann [1,3]
[1] Materials Science and Engineering Department, Rutgers University, [2] Materials Engineering Department, University of Technology, [3] Biomedical Engineering Department, Rutgers University

3C-3 **Biogenic metallic nanoparticles. A nanometric trojan horse approach** . . . 88
David Medina Cruz, Junjiang Chen, Thomas Webster
Northeastern University

3C-4 **A Nanoscale Drug Delivery Platform with Controlled Drug Release Modulated by Aptamer Engineering** . . . 89
Robert Mosley [1], Ricky Whitener [1,2], Matthew Talarico [1], Jacek Wower [2], Mark Byrne [1]
[1] Biomimetic & Biohybrid Materials, Drug Delivery Laboratories, Rowan University, [2] RNA Biochemistry Laboratories, Auburn University

3C-5 **Application of disturbed fluid flow in a three dimensional cerebral bifurcation model** . . . 90
Nesrine Bouhrira
Rowan University

4A-2 **High Throughput Identification of Synthetic Polymers with Globular and Protein-Like Features** 91
Rahul Upadhya, Supriya Atta, N. Sanjeeva Murthy, Shashank Kosuri, Matthew Tamasi, Adam Gormley
Rutgers University

4A-3 **Solution Spun Protein-Based Polymer Fibers for Biomedical Applications** 92
XIAO HU
Rowan University

4A-4 **Adaptation of Liquid Handling Robotics for High Throughput Customizable RAFT-Polymerization** 93
Matthew Tamasi, Shashank Kosuri, Rahul Upadhya, Adam gormley
Rutgers University

4A-5 **High Throughput Screening of Random Heteropolymers that Stabilize Enzymes** 94
Matthew Tamasi, Shashank Kosuri
Rutgers University

4A-6 **A Facile Method of Protein Crystallization using Gold Nanoparticles** 95
Hyewon Kim, Xiaoting Guo, Becka Padgett, Vicki Colvin
Brown University

4B-4 **Alignment of Human Cardiomyocytes through Nano-Wrinkles on Shape Memory Polymers** . . 96
Sarah Moore [1,2], Shiyang Sun [1,2], Chenyan Wang [1,2], Plansky Hoang [1,2], James Henderson [1,2], Zhen Ma [1,2]
[1] *Syracuse University,* [2] *Syracuse Biomaterials Institute*

4B-5 **Platelet-Derived Growth Factor Fused with Elastin-Like Polypeptides for Pressure Ulcer Healing** 97
Suneel Kumar [1], Mehma Chawla [1], Henry Hsia [2], Kyle Quinn [3], Rick Cohen [1], Martin Yarmush [1], Francois Berthiaume [1]
[1] *Rutgers, The State University of New Jersey,* [2] *Yale University,* [3] *University of Arkansas*

4B-6 **Biodegradable PCL-PGA-beta TCP Scaffolds for Bone Tissue Engineering** 98
Alok Kumar, Seyed Mir, Carlos Leon, Xiaojun Yu
Stevens Institute of Technology

4C-3 **Release of Naloxone for Long-term Management of Opioid Addiction** 99
Ruhi Naik [1], Kayla Mitchell [1], Zain Siddiqui [1], Vivek Kumar [1,2]
[1] *New Jersey Institute of Technology,* [2] *Rutgers School of Dental Medicine*

4C-4 **Significant Associations of Urinary Essential Elements and Autism Spectrum Disorder** 101
Fatir Qureshi [1], James Adams [2], Devon Coleman [2], Juergen Hahn [1]
[1] *Rensselaer Polytechnic Institute,* [2] *Arizona State University*

4C-5 **Developing an Inexpensive Powered Myoelectric Prosthetic Arm for Persons with Amputation** 102
Ryan Rattazzi [1], Emad Haque [1], Ricardo Whitaker [1], Nicole Baldassini [1], Sergei Adamovich [1], Ghaith Androwis [1,2]
[1] *New Jersey Institute of Technology,* [2] *Kessler Foundation*

4C-6 **Instrumented Platform for the Quantitative Assessment of Human Balance Control** 103
Karen Ayoub, Karina Dsouza, Swathi Pavuluri, Carlotta Mummolo
New Jersey Institute of Technology

5A-1 **Biomechanical Threshold of in vivo Neonatal Brachial Plexus after Stretch Injury** 104
Rachel Magee [1], Sriram Balasubramanian [2], Virginia Orozco [2], Maria Delivoria [2], Anita Singh [1]
[1] *Widener university,* [2] *Drexel University*

5A-2 **Protecting the Integrity of Surfactant-Stabilized, Oxygen Filled Microbubbles** 105
Brian Oeffinger [1], Purva Vaidya , Iman Ayaz [1], John Eisenbrey [2], Margaret Wheatley [1]
[1] *Drexel University,* [2] *Thomas Jefferson University*

5A-3 **Multimodal Foot-Ground Contact Interaction in Human Postural Stability** 106
Giulia Vicentini [1], Carlotta Mummolo [2]
[1] *New York University,* [2] *New Jersey Institute of Technology*

5A-4 **Effects of Walking Forces on Shoe Wear Rate** 107
Sarah Hemler, Jessica Sider, Kurt Beschorner
University of Pittsburgh

5A-5 **Effects of diabetic therapeutic footwear and traditional athletic footwear on gait: a cross-sectional investigation** 108
Kelly Poretti, James Peters, Anita Singh
Widener University

5B-1 **Trophoblast Enrichment Using Surface Adhesion Properties** 109
Sumaiya Sayeed, Christina Bailey-Hytholt, Anita Shukla, Anubhav Tripathi
Brown University

5B-2 **Wnt and Nodal Signaling Pathways Influence Chiral Bias in Human Embryonic Stem Cells** . . .110
Kathryn Worley, Leo Wan
Rensselaer Polytechnic Institute

5B-3 **Constitutively Active Ras Alters Cellular Chirality and Actin Structure of Mammary Epithelial Cells** 111
Amanda Chin, Kathryn Worley, Leo Wan
Rensselaer Polytechnic Institute

5B-4 **The Effect of Zinc on the Osteogenic Differentiation of Mesenchymal Stem Cells** 112
Jennifer Moy, Ateka Khader, Treena Livingston Arinzeh
New Jersey Institute of Techology

5B-5 **Kinetics of MSC-Based Enzyme Therapy for Immunomodulation** 113
Alexandra Burr, Biju Parekkadan
Rutgers University

5C-1 **A Modular Cell-Cell Adhesion Toolbox for Engineering of Novel Multicellular Systems** 114
Nathaniel Borders, Timothy Wannier, George Church
Harvard Medical School

5C-2 **Model Based Approach To Measuring Human Endothelial Arterial Function** 115
Gary drzewiecki[1], Kristin Distefano[1], Peter Lenehan[2]
[1] Rutgers University, [2] Everist Health

5C-3 **Maddox Components of Vergence** 117
Sebastian Fine[1], Elio Santos[1], Mitchell Scheiman[2], Tara Alvarez[1]
[1] The New Jersey Institute of Technology, [2] Salus University

5C-4 **The Impact of Vascular Ehlers-Danlos Syndrome Mutations on Collagen III Structure, Dynamics, and Function** 118
Cody Hoop, Sonal Ghalawat, Allysa Kemraj, Baifan Wang, Jie Zhu, Madison Godesky, Haley Warren, David Shreiber, Jean Baum
Rutgers, The State University of New Jersey

5C-5 **Effects of Orthoptic Treatment on Convergence Insufficiency** 119
Ayushi Sangoi[1], Elio Santos[1], Mitchell Scheiman[2], Tara Alvarez[1]
[1] New Jersey Institute of Technology, [2] Salus University

Posters

P2-1 **Controlled Synthesis and Solubility Characterization of Polymer-Peptide Conjugates for Biomedical Applications** . . . 120
Reem Eldabagh
William Paterson University of New Jersey

P2-2 **Bioinspired Vascularized Polymers for the Delivery of Bioactive Compounds at Surfaces** . . . 121
Caitlin Howell, Kayla Marquis, Benjamin Chasse
University of Maine

P2-3 **Synthesis and Characterization of Glycosaminoglycan Mimics for Cartilage Repair Applications** 122
Richard Vincent, Treena Arinzeh, Willis Hammond, George Collins
New Jersey Institute of Technology

P2-4 **Zein Protein Fibrous Matrices for Promoting Cell Adhesion and Osteogenic Differentiation** . . 123
Apurva Limaye, Jessica Cardenas Turner, Treena Arinzeh
New Jersey Institute of Technology

P2-5 **Stop Go Flow Encapsulation** . . . 124
Scott McKirgan
Rutgers University

P2-6 **The biomechanics of distal colon and rectal wall and its implication in visceral sensation and hypersensitivity** . . . 125
Saeed Siri [1], Franz Maier [2], Longtu Chen [1], Stephany Santos [2], David Pierce [3], Bin Feng [4]
[1] PhD Student of Biomedical Engineering Department, [2] PhD Student of Mechanical Engineering Department, [3] Assistant Professor of Mechanical Engineering Department, [4] Assistant Professor of Biomedical Engineering Department

P2-7 **Synthetic Tear Duct Drainage System** . . . 126
Alexis Rader, Kyle Fehn
Western New England University

P2-8 **Effect of Bifurcation Angle on Red Blood Cell Lingering and Partitioning** . . . 128
Andrew Pskowski, Jeffrey Zahn
Rutgers University

P2-9 **Injectable Solution of Epigallocatechin Gallate as a Potential Preventative Measure against Cartilage and Connective Tissue Damage** . . . 129
Mary Reiter, Adrian Mann, Joseph Freeman
Rutgers University

P2-10 **Effect of Counterface Surface Roughness on Tribological Rehydration of Articular Cartilage** . . 130
Meghan Kupratis [1], Margot Farnham [1], David Burris , Christopher Price [1]
[1] University of Delaware

P2-12 **Biophysical Model of Ebola Virus Interactions with TIM Proteins** . . . 131
Xinyu Cui, Luke Wang, Chuqian Xiong, Nicole Lapinski, Frank Zhang, Anand Jagota
Lehigh University

P2-13 **Developing Methods to Validate a Subject-Specific Magnetic Resonance Based Finite Element Model to Predict Strain in the Femur** . . . 132
Bryan Choate, Karen Troy
Worcester Polytechnic Institute

P2-14 **Bio-mechanical Characterization of TIM Protein mediated Ebola Virus Host Cell Adhesion** . . . 133
Matthew Dragovich, Chuqian Xiong, Nicole Fortoul, Anand Jagota, Wei Zhang, Xiaohui (Frank) Zhang
Lehigh University

P2-15 **Biomechanics of Arteries under Cyclical Pulsations in Hypertension and Drug Delivery** 135
Yueya Ge [1], Mehmet Kaya [2], Vignesh Balasubramanian [2], Peter Kerkhof [3], John Li [1]
[1] Rutgers University, [2] Florida Institute of Technology, [3] Vrije Universiteit Amsterdam

P2-16 **Pericellular Collagen VI Reorganization After Painful Temporomandibular Loading in the Rat** 136
Melissa Franklin [1], Megan Sperry [2], Evan Phillips [1], Eric Granquist [3], Beth Winkelstein [2], Michele Marcolongo [1]
[1] Drexel University- Materials Science and Engineering, [2] University of Pennsylvania- Bioengineering, [3] University of Pennsylvania- Oral & Maxillofacial Surgery

P2-18 **Glycosaminoglycan Mimetic Aligned Fibrous Scaffolds Promote Neurite Extension and Myelination** 137
Sharareh Hashemi [1], Patrice Maurel [2], Treena Arinzeh [1]
[1] New Jersey Institute of Technology, [2] Rutgers-The State University of New Jersey

P2-19 **Uric Acid Decreases Cell Death and Reactive Oxygen Species following Glutamate-induced Excitotoxicity in Organotypic Slice Culture** 138
Salman Khaliq, Bonnie Firestein
Rutgers University

P2-20 **The Effect of Phase-Specific Optogenetic Stimulation on Memory Recall in Mice** 139
Benjamin Lahner, Melanie Quick, Bahar Rahsepar, Jad Noueihed, John White
Boston University

P2-21 **Antioxidant Nanoparticle Films for Improving Deep Brain Recording** 140
Victoria Vafaee
Brown University

P2-22 **Microphysiological Modeling of Adipose Tissue for High-Throughput Applications** 141
Michael Struss
Temple University

P2-23 **Antimicrobial Susceptibility Testing using Piezo Electric Plate Sensors** 142
Pawan Rao [1], Dolores Conover [1], Wei-Heng Shih [2], Wan Shih [1]
[1] School of Biomedical Engineering, Science and Health Systems, Drexel University, [2] Department of Materials Science and Engineering, Drexel University

P2-24 **Highly sens detection of nucleic acids (DNA and RNA) using the "ESSENCE" biosensor electrochemical platform** 143
Yu-Hsuan Cheng, Zhenglong Li, Mahima Hariharan, Pedro Moura, Sagnik Basuray
New Jersey Institute of Technology

P2-25 **Detection of anti-Tn antibody in serum with 1000-fold better sensitivity than enzyme-linked immunosorbent assay (ELISA) using Piezoelectric Plate Sensor** 144
Song Han [1], Manoj Kumar Baskaran [1], Pawan Rao [1], Hassan Alotaibi [1], Alexandra Jednorski [1], Dolores Conover [1], Wan Shih [1], Wei-Heng Shih [2]
[1] Drexel University - School of Biomedical Engineering, Science, and Health Systems, [2] Drexel University - Department of Materials Science and Engineering

P2-26 **Optimizing Design of Electrochemical Bio-Sensor for Enhanced Sensitivity** 145
Zhenglong Li, Yu-Hsuan Cheng, Sagnik Basuray
New Jersey Institute of Technology

P2-27 **Discovery of Novel Photoreceptor Gene by Single-Cell RNA-seq Data Analysis** 146
Alexandria Pinto, Xin Ai, Li Cai
Rutgers University

Multiscale Mechanics of Embryonic Tendon

Benjamin E. Peterson (1), Spencer E. Szczesny, PhD (1,2).
(1) Department of Biomedical Engineering, Pennsylvania State University, University Park, PA.
(2) Department of Orthopedics and Rehabilitation, Pennsylvania State University, Hershey, PA

Introduction: During embryonic and neonatal development, the tendon tensile modulus and ultimate strength dramatically increase as the collagen fibrils grow in diameter and length.[1,2,3] However, how tenocytes carefully regulate this structural development and its corresponding mechanics is unknown. Tenocytes are sensitive to mechanical stimuli,[4,5] suggesting a mechanobiological feedback loop between tenocyte behavior and the changes in their mechanical microenvironment driving tendon development.[5,6] Previous studies examined macroscale mechanical properties of embryonic tendons, however this likely does not reflect he local microscale mechanics.[2,3,6] Studies investigating the multiscale mechanics of rat tail tendon fascicles showed that the microscale tissue strains are significantly less that those applied at the tissue level, which can be explained by relative sliding between collagen fibrils.[7,8] Identifying whether this occurs within embryonic tissue will provide critical knowledge about the local mechanical stimuli and the mechanobiological pathways active during tenogenesis. **The objective of this study is to measure the multiscale mechanics of embryonic tendons.** We hypothesize that the fibril strains will be less than the applied tissue strains, suggesting that the fibrils are discontinuous and exhibit interfibrillar sliding under tensile load.

Methods: Four digitorum longus tendons were isolated from the hindlimbs of white Leghorn day 20 (E20) chicken embryos. Samples were stained with DTAF at 10 ug/ml and secured in the grips of a tensile testing device mounted on a confocal microscope with a 5 mm reference length.

Prior to testing, the samples were rotated to measure the major and minor diameters, which were used to calculate the cross-sectional area. Samples were preconditioned to 1% strain for 5 cycles at 0.03 Hz. Three sets of photobleached lines (PBL) (4 lines, 125 μm apart) were bleached at the sample center and ± 2 mm from the center. Samples were then loaded to 20% grip-to-grip strain in 5% increments at 10%/min followed by a 15 min stress relaxation. At the end of the relaxation, z-stack images were acquired at all PBL locations and the stage positions were recorded.

Custom MATLAB scripts were used to track the photobleached lines and quantify the microscale fibril strains and interfibrillar sliding.[5] Macroscale strains were obtained by tracking the spatial displacement of the peripheral PBL sets relative to one another (PBL-to-PBL strain). The coefficient of variance for the ratio between the fibril and PBL-to-PBL tissue strains was calculated across all PBL locations for each sample. The applied load during the last 30 seconds of the relaxation was averaged to determine the quasi-static stress at each incremental strain. The macroscale tissue modulus was calculated using a linear regression fit to the quasi-static stress versus both the grip-to-grip and PBL-to-PBL strains.

Results: The fibril strains were significantly lower than the applied tissue strain ($p < 0.01$) but were not significantly different with increasing strain ($p = 0.43$) **(Fig. 1A)**. The interfibrillar sliding was significantly greater than zero ($p < 0.05$) and there was an increase in fibril sliding observed with increasing strain ($p < 0.01$) **(Fig. 1B)**.

Linear regressions of the quasi-static stress versus the PBL-to-PBL ($p < 0.01$) and grip-to-grip ($p < 0.02$) tissue strains indicated an average quasi-static modulus of 85.3 MPa and 31.7 MPa respectively **(Fig. 1C)**.

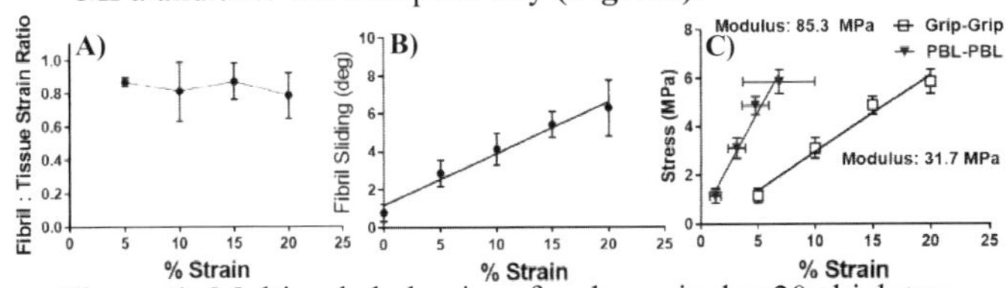

Figure 1: Multiscale behavior of embryonic day 20 chick tendon. (A) Fibril:tissue strain ratio and (B) interfibrillar sliding as a function of the applied grip-to-grip strain. (C)Linear regressions of the quasi-static tendon behavior as a function of both the grip-to-grip and optical (PBL-to-PBL) strains.

Conclusions: Consistent with our hypothesis, the microscale strains were less than the applied macroscale tissue strains **(Fig. 1A)**. Together with the non-zero interfibrillar sliding measurements **(Fig. 1B)**, suggests that the fibrils are discontinuous in E20 tissue and slide relative to each other with loading. Our data is consistent with previous testing of mature rat tail tendons displaying a discontinuous architecture[5]. However, in contrast to mature tendon, the fibril:tissue strain ratio was unaffected by increasing tissue strain. This suggests that a different mechanism governing interfibrillar sliding may exist in embryonic tendon; however, additional work is required.

There was a large discrepancy in moduli depending on whether the grip-to-grip or optical (PBL-to-PBL) strains were utilized **(Fig. 1C)**. This resulted in a nearly three-fold difference between our calculated modulus values. While our modulus values calculated from the grip-to-grip strains are comparable to previous tests of late embryonic chick tendons,[2] our optical strain data suggest that prior studies may underestimate the stiffness of embryonic tendon. Our findings suggest that cells are exposed to complex mechanical stimuli in developing tendons. Future work will be conducted across additional developmental time points to determine how changes in local tissue mechanics influence cell behavior and subsequent tendon development.

References:

1. Birk, D. et al. *Dev. Dyn.* 1995;**202**:229-243
2. McBride, D. *Int. J. Biol. Macromol.* 1988;**10**:194-200
3. Heather, A. et al. *Ann. Biomed. Eng.* 2011;**465**:1904-1913
4. Maeda, E. et al. *Biochem. Biophys. Res. Commum.* 2015: 465:281-286
5. Richardson, S. et al. *Mol. Cell. Biol.* 2007;**27**:6218-6228
6. Kalson N.S. et. al. *Matrix Biol.* 2010;**29**:678-689
7. Szczesny, S.E. et. al. *Acta Biomater.* 2014;**10**,2582-2590
8. Screen. H. R. et. al. *Proc. Inst. Mech. Eng. H.* 2014;**218**:109-119

Impact of Contact Constraints on the Dynamics of Idealized Intracranial Saccular Aneurysms

Manjurul Alam, Padmanabhan Seshaiyer.
George Mason University, Fairfax, VA 22030

Introduction: Intracranial brain aneurysms are common phenomena in cerebral vasculature and defined as the localized dilatations of the arterial wall that mostly happen at or near bifurcations in the circle of Willis [1]. One of the most apprehensive and serious complications of intracranial aneurysms is rupture, which results in a subarachnoid hemorrhage, a disastrous occurrence in the brain with a high mortality rate of 25% to 50%. Although most intracranial saccular aneurysms are asymptomatic prior to rupture, some unruptured lesions do present symptoms when pressed against adjacent structures such as nerve tissues. An important question to be investigated is whether the rupture potential of symptomatic lesions (lesions constrained with nerve tissues or other constraints) is different than the asymptomatic lesions (lesions without constraints). The main purpose of this study is to build on an earlier work and test the hypothesis that a sub-class of stiff contact constraints on an idealized axisymmetric intracranial saccular aneurysm can have a protective role on the aneurysm [2].

Methods: To capture the dynamics of an idealized axisymmetric intracranial saccular aneurysm (ISA) we employ finite element software ABAQUS CAE. The geometry considered allows for large deformation and in this work, we consider both geometric non-linearity and material non-linearity. The ISA wall is meshed using linear brick, hybrid, constant pressure, reduced integration, hourglass control (C3D8RH) material. Aneurysm wall is considered as a hyper-elastic anisotropic material and the constraint or obstacle is taken as an isotropic elastic material.

Results: The dependence of cauchy stresses along the dimensionless undeformed arc length for different types of constraints is illustrated in figure 1. The ratio of the arc length with includes $X \in [0,R]$, where $X=0$ corresponds to the fundus or pole of the lesion and $X=R$ corresponds to the neck of the lesion. From the graph, it can be seen that the stresses on the lesion behave differently for short and long constraints at the pole and near the neck, than that of the lesion with no constraint.

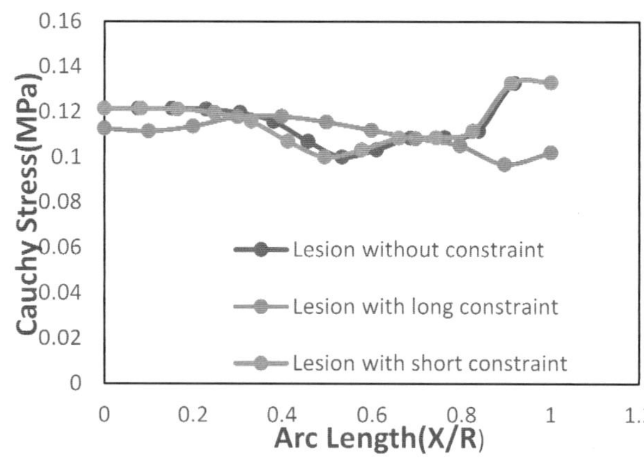

Figure 1. Cauchy Stresses (MPa) with the undeformed arc length(X/R) on the Axisymmetric ISA

Conclusions: The results suggest that lesions under short and long contact constraint generate different meridional and circumferential stresses compared to the lesions without constraint and hence the rupture potential of these lesions is lower compared to the ones that do not have constraint. However, since every ISA is biomechanically distinct and unique, it is unrealistic to address this statement for the lesion specific model. Although this work has provided some insights about a simple, axisymmetric idealized lesion under contact constraints, it is necessary to go further on this analysis by doing lesion specific model.

References:

1. Chalouhi N, Hoh BL, Hasan D. *Stroke.* 2013 Dec;44(12), 3613–22.
2. Seshaiyer P, Humphrey JD. *J Biomech.* 2001 May, 34(5), 607–12.

Upper Airway Area-Pressure Dynamic Analysis Tool for Studying Obstructive Sleep Apnea Syndrome

K. R. Choy MS[1], S. Sin MS[2], Y. Tong PhD[3], J.K. Udupa PhD[3], C. Wu PhD[4], M. Wagshul PhD[2],
R. Arens, MD[2], D.M. Luchtenburg PhD[1], D.M. Wootton, PhD[1]
[1]The Cooper Union, NYC, United States; [2]Albert Einstein College of Medicine, Bronx, NY, United States;
[3]University of Pennsylvania, Philadelphia, PA, United States; and [4]Philips Healthcare, Andover, MA, United States

Introduction: We recently used retrospectively gated dynamic MRI, critical airway closing pressure (P_{crit}), and image-based computational fluid dynamics (CFD) to study upper airway dynamics during wakeful tidal breathing in obese children (1). The slope of airway cross-sectional area vs. CFD airway pressure (called the "effective compliance") in the nasopharynx was negative in children with obstructive sleep apnea syndrome (OSAS) and significantly different from controls. Effective compliance was also inversely correlated with P_{crit}. For the present study we developed a new tool to investigate the pressure-area characteristics of the airway based on nasal resistance, real-time dynamic MR imaging, and synchronized flow rate and nasal pressure recordings for awake and sleeping subjects.

Methods: Real-time dynamic MRI of the upper airway was performed on a 3T scanner (Philips Medical Systems, Best, The Netherlands). Three hundred axial images (every 350ms) were acquired at retropalatal and retrolingual sites. Nasal-oral mask airflow and pressure and scanner synchronization signals were recorded at 500Hz using a BIOPAC MP150 data acquisition system coupled respectively to TSD-160B, TSD117 flow transducer (BIOPAC Systems, Inc. Camino Goleta, CA), and a custom-build pulse extender. Images were filtered using a Single Value Decomposition (SVD) image compression technique (2). The MR images were assembled as column vectors in a snapshot matrix A of dimension m by n, where m is the number of pixels of each image and n the number of images. The filtered images were obtained from the truncated SVD of A, $A_r = U_r S_r V_r^T$, where U_r and V_r are orthonormal matrices of size m by r and n by r respectively, $S = diag(\tilde{A}_1, \tilde{A}_2, ..., \tilde{A}_r)$, and r is the rank of the approximation. The airway was segmented from the filtered image using a k-means clustering algorithm (Matlab; Mathworks) in a region containing the airway. Average flow rate during each image was calculated, and airway pressure was estimated using measured nasal pressure and nasal pressure drop, calculated using flow rate and rhinomanometry Rhorer coefficients. An animated panel shows flow, MR images, area, and pressures to visualize pressure-area relationships.

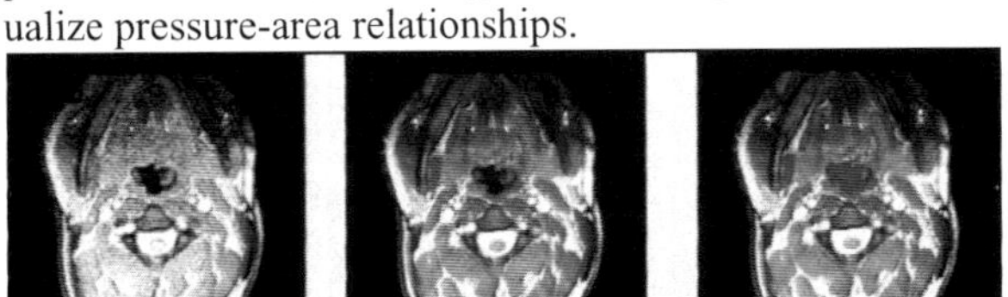

Figure 1: Original (left), filtered (middle), and segmented example airway image (right, airway area mask in red).

Results: Sample results from the retrolingual airway in two children during wakefulness and sleep are presented (OSAS subject ID 35, male, 17 years old & control subject ID 42, female, 18 years old). Original, filtered (r=6), and segmented retrolingual airway images of subject ID 35 are shown in Figure 1. The cross-sectional area and pressure relationships of both subjects during wakefulness and sleep are shown in Figure 2. Data during swallowing movements were manually excluded.

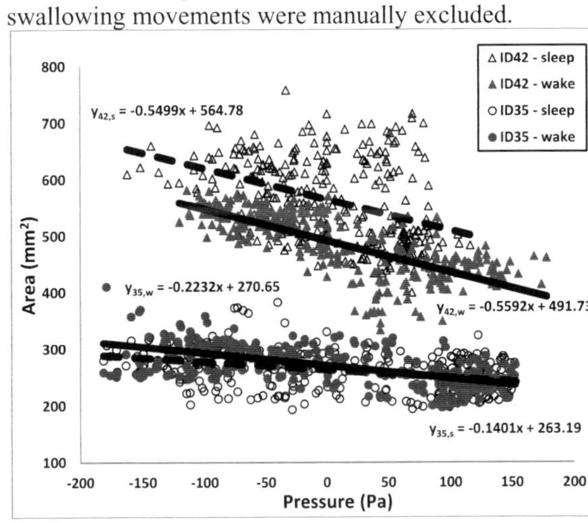

Figure 2: Pressure vs. cross-sectional area of subjects ID 35 &42 during wakefulness and sleep. Linear trend line slopes indicate effective compliance (in mm^2/Pa).

Although the pressure range is similar for both subjects, ID35 has approximately 50% smaller cross-sectional area than ID42 (figure 2). The average slope of pressure vs. area is negative in both subjects, meaning pressure and area are out of phase and suggesting phasic muscle activation to resist negative pressure; this result was common for other subjects in our previous study (1) and in this study (not shown). The slope magnitudes of ID42 are significantly higher than ID35 during both wakefulness and sleep, in contrast to our previous results (1). The slope magnitude is significantly reduced during sleep for the OSAS subject (ID35), and barely reduced in the control subject, suggesting a relatively higher loss of phasic airway muscle activation in OSAS during sleep.

Discussion and Conclusions: The SVD image compression technique has been incorporated as a novel and efficient image processing tool in this study. This new tool integrates noninvasive MRI, synchronized pressure and flow measurements, and rhinomanometry, enabling researchers to visualize the qualitative dynamics of the pressure-area relationship. This tool will be applied to statistical correlation of mechanical biomarkers such as effective compliance to clinical parameters in OSAS.

References: (1) Wootton et al., J. Applied Physiol. 121(4):925–931, 2016. (2) Aase et al., *Proc. 1999 IEEE ISCIS VLSI,* 4:13-16, IEEE Press, Piscataway NJ.

Acknowledgements: Grant NIH-1R01HL130468-A1.

Hemodynamic model of left atrial appendage thrombus risk in patients with atrial fibrillation

Soroosh Sanatkhani[1], Prahlad G. Menon[1], Sanjeev G. Shroff[1]
[1]Department of Bioengineering, University of Pittsburgh, Pittsburgh, PA, USA

Introduction: Atrial fibrillation (AF) currently afflicts more than 3 million people in the USA and this number is expected to rise to 7.56 million by 2050[1]. AF is the most common sustained arrhythmia, especially among elderly patients, costing the health care system ~$6 Billion each year[2]. AF patients have 3-5 folds higher risk of stroke. All thromboprophylaxis decisions require physician assessment of the patient's stroke risk, which is currently determined by clinical data included in the CHA_2DS_2-VASc score. However, we remain very limited in predicting who will have a stroke in the setting of AF. Any patients with CHA_2DS_2-VASc score above two is considered high risk for thromboembolic events and are considered for oral anticoagulation. The left atrial appendage (LAA) has been shown to be a culprit location for formation of thrombi in non-valvular AF (91%) and in valvular AF (50%)[3]. This suggests a possible correlation between the LAA shape and stroke risk suggesting that LAA geometry and its corresponding hemodynamics could be an important factor in risk assessment. However, left atrium (LA) hemodynamic does not have any value in the current risk assessment metrics. *Hypothesis:* A hemodynamic and a statistical model could fill the current void in risk stratification in AF patients.

Methods: We implemented our models to a cohort of 40 patients with symptomatic AF where six of them had prior stroke. Patients had CCT before catheter ablation. Contrast-enhanced CCT DICOM images were cropped and smoothed using a median filter with kernel matrix of 5 in every direction. Images were segmented in ParaView 5.2 (Kitware Inc., USA) using marching cubes method to create an iso-surface, replicating the LA surface. The extracted surface is including pulmonary veins (PV) (two right-PV and two left-PV), mitral valve and LAA. Extracted surfaces were smoothed out for CFD simulation by removing spikes and reducing noise filter (i.e. simplifying polygons) in Geomagic Studio 10 (Geomagic Inc., USA) and ANSYS SpaceClaim 18.2 (ANSYS Inc., USA). For each case, 300,000-500,000 tetrahedrons with five layers of prisms near walls were generated. Since atrium of patients with AF have a low contractility, the LA walls were assumed to be rigid with no slip condition in our model. Further, the mitral valve is assumed to be open throughout the simulations (simulative of atrial systole) with gauge pressure set to zero pressure and velocity gradient set to zero. Velocity profile is prescribed at inlet boundaries (aka pulmonary veins) based on a normal inflow rate waveform. The blood flow treated as an incompressible flow with blood assumed to be Newtonian with a density of Á= 1060 kg m^{-3} and dynamic viscosity of μ = 0.00371 Pa s. We solved a transient transport equation for a scalar using a solver developed from IcoFoam and ScalarTransportFoam solvers in OpenFOAM (The OpenFOAM Foundation, UK). Residence time distribution (RTD) curves were generated as an impulse response to neutrally buoyant scalar concentrations (simulative of blood cells) injected at four pulmonary veins entering the LA. The under the curve area of RTD curves in the tail region (44-75 cardiac cycles) is named RTD value and is used as metric for the probability of particles residing inside the LAA after a long time and subsequently a quantification for clot formation risk.

Results: A scatter plot of RTD values against the modified CHA_2DS_2-VASc score (to assess the ability of CHA_2DS_2-VASc in stroke prediction points corresponding to prior strokes were omitted.) is shown in Fig. 1 (large circles correspond to patients with prior stroke). A combination of these two indices (solid line) does a better job of identifying stroke patients; there are examples of stroke patients wherein one index predicts no stroke while the other predicts stroke (two labeled large circles). By combining CHA_2DS_2-VASc score and RTD value we could increase the specificity and sensitivity of the predicting stroke from 26% to 50% and 83% to 100%, respectively (length of the dotted lines in Fig. 1 represents the stroke risk predicted by combination of RTD and CHA_2DS_2-VASc). Please note that some of the false positives (small circles above the solid line) may become true positives over time, but only a longitudinal study can address this issue.

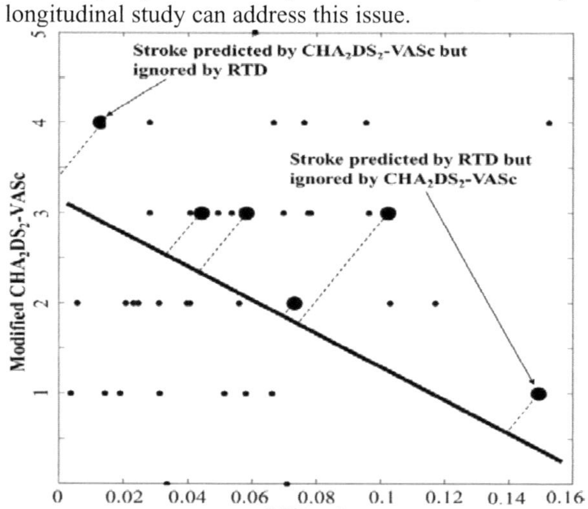

Figure 1. Stroke risk prediction improvement by adding RTD as one of the CHA_2DS_2-VASc factors.

Conclusions: The present study shows the importance of using patient-specific LA geometries along with patient-specific LAA. The RTD values may not be able to stratify the stroke risk independently, however, this preliminary study showed that a combination of RTD value and CHA_2DS_2-VASc score has a potential for providing a better stratification metric for the risk of stroke.

References:

1. Go AS et al. JAMA. 2001;(285):2370-2375.
2. Vainrib AF et al. Curr Cardiovasc Imaging Rep. 2017;(10):1-13.
3. Reddy VY et al. Circulation. 2013;(127):720-729.

Mechanisms of Hand-Rung Force after a Ladder Climbing Perturbation

Erika M Pliner and Kurt E. Beschorner
University of Pittsburgh

Introduction: Fatal fall injuries from a height have increased 26% since 2011 with ladders as the most common source [1]. To reduce ladder fall injuries, there is a need to understand mechanisms that arrest a ladder falling event.

After a ladder climbing perturbation (e.g. foot slip or misstep), the climber attempts to maintain and reestablish hand placement with the ladder handholds to arrest the falling event. The hand-rung forces generated in these events are similar to the maximum force limits achievable between a hand and handhold [2]. Yet, there is a lack of knowledge on mechanisms influencing applied hand-rung force.

Hand-rung force observed after a ladder climbing perturbation may be influenced by the climber's upper body strength or the severity of the falling event. Individuals with greater upper body strength may rely more on their upper body to arrest the fall, generating higher hand-rung forces. Alternatively, increased hand-rung force may be caused by a more severe falling event, requiring the hand to withstand higher hand-rung force due to greater downward momentum of the falling body. Understanding mechanisms of hand-rung force will assist in determining if the hand is at risk of decoupling. Therefore, the purpose of this study is to assess applied hand-rung force after a ladder climbing perturbation with individual upper body strength and severity of the falling event.

Methods: Thirty-one healthy adults (13 females; 24.5±5.4 years; 1.75±0.10; 74.2±12.1 kg) completed two testing sessions for this study. Approval was obtained by the University of Wisconsin-Milwaukee Institutional Review Board.

The first testing session consisted of an upper body strength test. Participants were asked to hold onto a rising rung until the rung broke free from their grasp. The peak force exerted onto the rung before the hand decoupled was recorded (referred to as breakaway strength). Participants were secured in a seated position for all trials and completed six trials per hand. Upper body strength was defined as the averaged breakaway strength across all trials and normalized to body weight.

The second testing session consisted of 30 climbing ascents and descents on a vertically fixed ladder. Participants were exposed to 6 random climbing perturbations, three per climbing direction. A perturbation below the foot was simulated by releasing the fourth ladder rung when 5% of the participant's body weight remained on the previous rung. Participants wore a safety harness with a load cell attached to measure the weight supported by the harness. Fall severity was defined as the peak harness force between perturbation onset and end of perturbation. Fall severity was normalized to body weight. Unidirectional load cells were attached to the 8th and 9th ladder rungs to measure the vertical and horizontal forces applied to the rungs. The resultant force was calculated for each rung. Applied hand-rung force was defined as the peak resultant hand-rung force by the combined hands between perturbation onset and the end of perturbation. Peak resultant hand-rung force was normalized to body weight.

A linear regression was performed for each climbing direction with hand-rung force as the dependent variable. Harness force and breakaway strength were entered into the models as predictor variables.

Results: The average (standard deviation) normalized peak resultant hand-rung force of the combined hands after an ascending and descending climbing perturbation was 1.08 (0.33) and 1.30 (0.24), respectively.

Higher peak hand-rung force was associated with increased harness force ($p<0.001$) and increased breakaway strength ($p=0.018$) after an ascending ladder climbing perturbation (Figure 1). Hand-rung force was not associated with harness force ($p=0.768$) or breakaway strength ($p=0.486$) after a descending ladder climbing perturbation. A participant's hand decoupled from the rung in four trials (2 ascent, 2 descent).

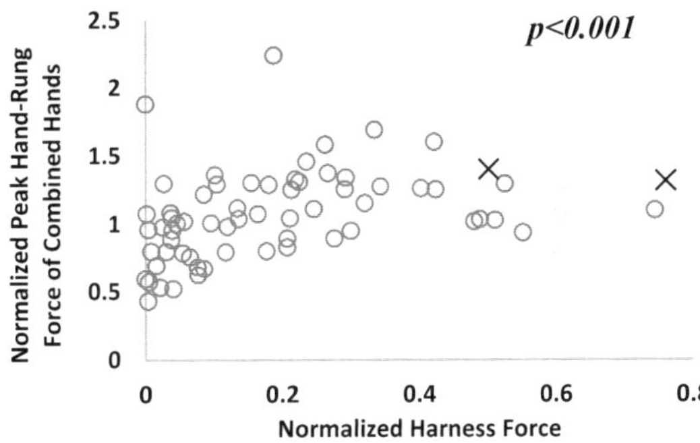

Figure 1. Hand-rung force of the combined hands after an ascending climbing perturbation across harness force. Crosses 'x' represent cases where the hand decoupled.

Conclusions: Applied hand-rung force after an ascending climbing perturbation appears to be dependent on the upper body strength of the climber and the severity of the falling event. However, participants are at risk of their hand decoupling in more severe falling events (supported by hand decoupling at high harness force magnitudes). There is a need to minimize the applied hand-rung forces after a ladder climbing perturbation to limit hand decoupling and prevent ladder fall injuries.

References:

1. BLS. *TheEconomicsDaily,* Washington, DC, 2018
2. Pliner et al. *Proc. of ASB'17,* Boulder, CO, 2017

An *In Vitro* Bioreactor for Testing Brain Implant Biocompatibility

Erika J. Davidoff, Jay C. Sy
Rutgers University Department of Biomedical Engineering

Introduction: Common treatments for epilepsy, tremor, Parkinson's disease, and other neurological disorders involve implanting electrodes into specific brain nuclei. However, these treatments' effectiveness is hampered by the brain's inflammatory foreign body response to the implant, including gliosis, in which astrocytes form dense glial scar tissue around the implant, reducing signal fidelity by insulating the electrode & displacing neurons away from recording sites. Gliosis is exacerbated by micromotion between brain tissue and implant due to respiration & blood circulation and by mechanical mismatch between the metallic implant (~100 GPa modulus) and brain tissue (~10 kPa). Hydrogel coatings for implanted electrodes may attenuate these factors as gels' elastic moduli more closely resemble brain tissue. The coatings are also able to absorb energy to cushion the tissue. Covalently linked hydrogels such as poly(ethylene glycol) diacrylate (PEGDA) are strong candidates due to availability and biocompatibility; however high strains can permanently sever their covalent bonds. In addition to these coatings, we are investigating non-covalently linked gel systems, and we have been developing a custom *in vitro* bioreactor to compare the effectiveness of each type of soft hydrogel coating.

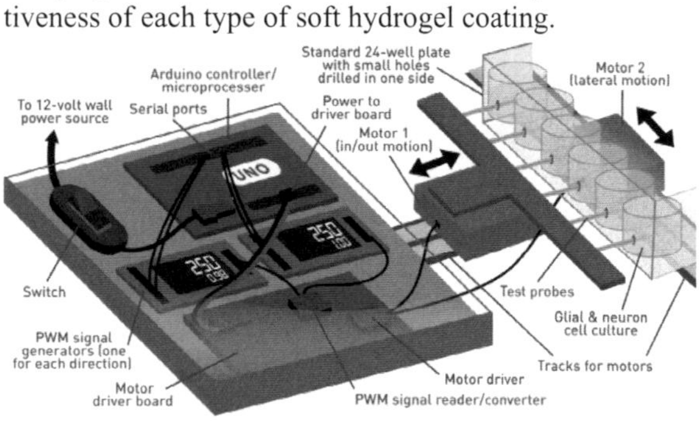

Figure 1: Overview of custom micromotion bioreactor

Methods: In order to test potential gel coatings, we have been developing an *in vitro* bioreactor (Fig 1) that simulates micromotion between implant and brain tissue. The bioreactor is comprised of two Arduino-controlled piezo linear actuators which are fed pulse-width modulation signals in order to control their speed and frequency. The motors drive a platform on which up to six test probes can be mounted. The probes are inserted into gels made up in the wells of a standard 24-well plate; these can be simply agarose gels in order to map the physical strain field around the implant, or can be complex three-dimensional cell culture gels with neurons, astrocytes, and microglia that can be stained to quantify the inflammatory factors activated in response to the probe. At present, we have been focusing on refining and characterizing the bioreactor.

To perform the *in vitro* experiments, covalently (e.g. PEGDA) and non-covalently cross-linked (e.g. ^{2}-cyclodextrin/adamantine guest-host interaction network and nanoparticle-linked polysaccharide network) hydrogels are synthesized and applied to 250µm silane-functionalized glass probes. The coated probes are inserted into our custom bioreactor which is loaded with either fluorescent bead-seeded agarose gel or with glial and neuron cells cultured in Matrigel and stained for GFAP (glial scarring), ED1/CD8 (activated macrophages), and NeuN (neuron cell bodies). The bioreactor simulates micromotion by moving the probes in the gel in 30µm increments and fluorescent images are captured at various time points (from minutes to days after probe insertion depending on the particular experiment). Particle image velocimetry analysis is performed on bead gels to quantitatively analyze the strain field around the probe. Histology analysis is performed on cell cultures to quantify foreign body response. Additionally, all coating hydrogels are analyzed rheologically.

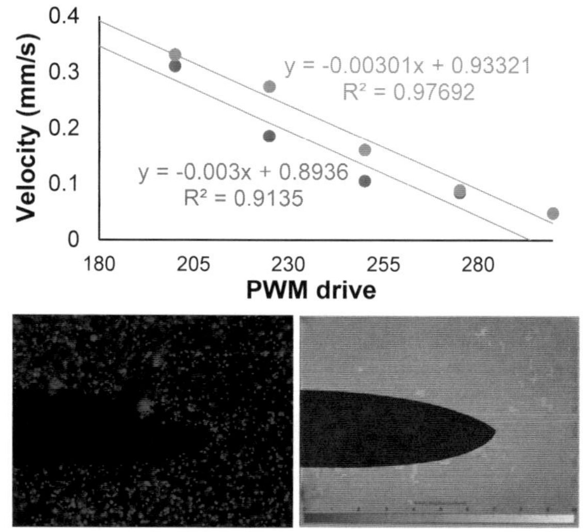

Figure 2: (a) Characterization of PWM frequency and probe velocity in 0.1% agarose (orange) and 0.6% agarose (blue). (b) Image of probe in fluorescent bead-seeded agarose gel. (c) PIV analysis of probes in agarose gel.

Results: Using commercially available PWM signal generator chips (each under $10) we are able to accurately and reliably simulate the 2 Hz, 30µm increments (requiring a velocity of 0.12 mm/sec) that have been found to represent micromotion in the brain[1] (Fig 2a). We have also begun quantitatively characterizing strain fields around uncoated and PEGDA-coated probes using PIV (Fig 2b,c).

Conclusions: Our inexpensive bioreactor proves a robust tool for quantitatively comparing hydrogel coatings for brain probes *in vitro*, reducing the need for more extensive *in vivo* testing. We are presently synthesizing non-covalently linked gel networks including guest-host interaction-based gels, polymer-nanoparticle gels, and peptide beta sheet networks, and will be analyzing their efficacy at reducing foreign body response using this bioreactor with the fluorescent bead agarose and 3D cell culture tissue mimics.

References: [1]Aaron G. & Jit M. *J Neural Eng* 2006, 3, 189

(Quantitative Assessment of Stress Levels with EEG and Heart Rate Variability)

Eyad Talal Attar, Dr. Mehmet Kaya.
Florida Institute of Technology.

Introduction: Stress is one of the major risk factors for many diseases such as hypertension, coronary artery disease, heart attack, and even sudden death. Stress has been also linked to decision making, performance, learning, etc. In many studies heart rate variability (HRV) is a relatively new method for evaluating stress. What makes HRV remarkable is the fact that it can reflect changes in stress while other physiological factors, like blood pressure, are still in normal ranges. HRV, in fact, is the only independent risk factor for sudden cardiac death. That's why HRV measurement is an increasingly popular parameter in health related fields as well as relaxation training therapy. EEG is a vital technique for studying human brain activities. EEG provides good evaluation data of changeability in mental status. The two main objectives are the benefits of using HRV in combination with EEG features for the detection of stress levels and how the findings can be applied to stress treatment and if these findings can be used to monitor the effect of possible treatments or therapies.

Methods: The EEG signals were measured by using DSI-24 dry electrode EEG headset. The DSI-24 consists of 21 electrodes positioned according to the 10-20 International System. EEG signals were recorded using the provided data acquisition software DSI-streamer. The EEG signal were recorded using 4 electrodes (F3, F4, Fp1, and Fp2) which are commonly used in stress measurements. The Stroop Color Word Test (SCWT) was used to induce stress. The Power Spectral Density (PSD) analysis was used for the EEG analysis. The PSD method shows the strength of the variations in energy as a function of frequency. The ECG signals were calculated using BioRadio150. This device is a wireless 12-channel lightweight programmable monitor for viewing and recording physiological signals, such as electrical activity from the heart. The Kubios software package was deployed to perform comprehensive HRV signal analysis. The participants had their EEG signals and HRV recorded for a total duration of 15 minutes: The test began with a 5 minutes 'No-stress session' (Relax) followed by a one minute rest period, and then a 5 minutes stress session (elaborate Stroop Color Word Test) followed with another 1 minutes rest period. Finally, we performed a 5 minutes relaxation/meditation session with the participants (watching and listening to wave beach video).

Results: The analysis of HRV in frequency domain showed a significant increase in the Low Frequency (LF) power under stress condition (21%). Also, there was a decrease in the Standard Deviation of NNs (SDNN) by 5% (Table 1). EEG frequency bands activities changed across the scalp in both conditions, while alpha activity was present to a high degree in no-stress condition and it disappeared (green) in electrode Fp1 showing the effect of stress (Figure 1). However, EEG showed a good relationship with the mental stress in terms of suppression of alpha waves. These results indicated a correlation between EEG and HRV in stress condition.

Frequency Domain			Time Domain		
Feature	No-Stress (ms^2)	Stress (ms^2)	Feature	No-Stress	Stress
VLF	819	808	NN50	210	201
LF	1457	1758	pNN50	59 %	58%
HF	2019	2037	RMSSD	92 ms	88 ms
LF/HF	0.7	0.9	SDNN	77 ms	81 ms

Table 1. HRV Analysis

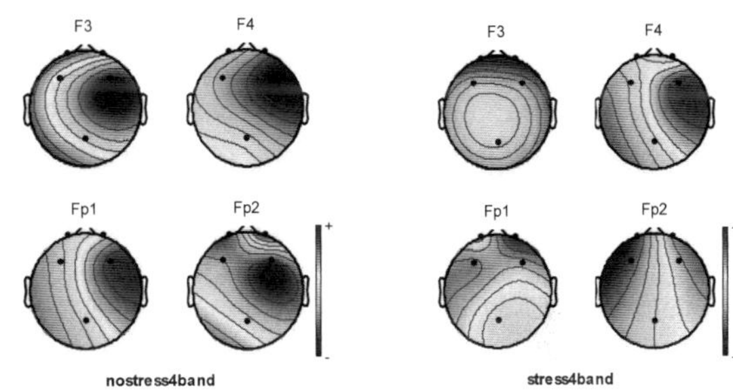

Delta	Theta	Alpha	Beta
(0-4) Hz	(4-8) Hz	(8-12) Hz	(12-16) Hz
Red	Yellow	Green	Blue

Figure 1.Topographic EEG Maps and Frequency Bands

Conclusions: Investigating the correlation between EEG and HRV under stress conditions is significant since they provide complementary information regarding stress. The ability and outcomes of EEG and HRV would allow us to diagnose and monitor the progress of treatment or therapy, performance, learning, and decision making. Another contribution of this study can be Stress Management (SM) by the HRV and EEG data as inputs for treatments applications including meditation studies, bio-feedback training, attention disorder, attention deficit hyperactivity disorder (ADHD), depression and anxiety disorders.

References:

1. Tania P. CMAPIB, 2017; 148:71-80.
2. Jesus M. SEN. 2018.

Acknowledgments Eyad Talal Attar was supported by the King Abdul-Aziz University (KAU) Doctoral Scholarship Academic for Ph.D. Program at Florida Tech.

Forming Libraries of Magnetic Multicore Nanoparticles with Tunable Dimensions and their Biomedical Applications

Zhen Xiao[1], Qingbo Zhang[1], Vicki L. Colvin[1,2]
1. Department of Chemistry, Brown University.
2. Center for Biomedical Engineering, School of Engineering, Brown University.

Introduction: The development of magnetic nanomaterials has been of great interest over the past decades. Their unique magnetic properties have fascinated many researchers, and the applications of their properties has engaged diverse areas of biomedicine and engineering, including magnetic resonance imaging, drug loading and targeted delivery, and hyperthermia agents for cancer treatment. To further make their applications more practical and higher performances, enhanced susceptibility of the nanomaterials is desired. Here we report a systematic approach for forming libraries of multicore iron oxide nanoparticles (MIONs) with superior susceptibility, to meet the demand of different optimization of magnetic properties.

Methods: The multicore iron oxide nanoparticles (MIONs) are synthesized through a hydrothermal method. In a typical synthesis, 540 mg $FeCl_3 I6H_2O$ is first dissolved in 20 mL ethylene glycol, followed by 250 mg PAA, 1200 mg urea and 2500 mg water. The mixture is transferred to an autoclave and then heated at 185 °C for 6 hours. The average cluster size of the nanocrystal is 25.0 ± 2.0 nm and primary particle size is 4.0 ± 0.5 nm as measured by transmission electron microscope (TEM). The amount of water and reaction temperature is controlled to form different MIONs with tunable dimensions. The magnetic performance is evaluated by vibrating-sample magnetometer (VSM).

The as-synthesized MIONs are purified and then subject to surface modification for fluorescent dye labeling and cell culturing. Polyethylene glycol ($PEG\text{-}NH_2$) and nitrodopamine are grafted on poly(maleic anhydride-alt-1-octadecene) PMAO to form PMAO-PEG-Dopa. The functionalized copolymer is then attached onto the MIONs through dative bonds between iron oxide and dopamine. DiI is then attached to the PMAO functionalized MIONs.

Cellular uptake of clusters: U2OS cells were seeded in a 4-well glass chamber. The lysosomes were labelled with Green Fluorescent Protein (GFP) by incubating the cells with CellLight Lysosomes-GFP overnight. Clusters labelled with DiI were added to the medium and incubated overnight. The nuclei were stained with Hoechst. The images were obtained using DeltaVision fluorescence microscope.

Results: The MIONs are aggregates of small nanocrystals (primary particles) as shown by TEM. We have synthesized a library of multicore iron oxide nanoparticles whose primary particle size and cluster dimension can be independently controlled by the reaction temperature and mass of added water, respectively. Both the cluster size and primary particle size can vary over a wide range. The MIONs are superparamagnetic as indicated by VSM. At low applied magnetic fields on the order of Gauss, the magnetization of the multicore sample is much greater than that of the single-core nanoparticle, suggesting a superior susceptibility.

The as-synthesized MIONs, with a cluster size of 25 nm and primary particle size of 4 nm, is very stable in pure water but are prone to aggregation in cell culturing media with high ionic strength and crosslinking agents such as Ca^{2+}. However, after surface modification with PMAO-PEG-Dopa, the MIONs colloidal stability is greatly enhanced. The introduction of PEG also facilitates the biocompatibility of the MIONs.

The PMAO functionalized MIONs are then labeled by DiI. The color of the DiI labeled MIONs solution is between that of the original MIONs solution and DiI solution, indicating the successful attachment of DiI onto the MIONs. Fluorescence spectra also confirm such observation, as the fluorescent pattern of the DiI labeled MIONs is pretty similar to that of DiI solution, while the original MIONs shows no significant fluorescence.

The DiI labeled MIONs are subsequently tested in U2OS cell. The centering blue area is the nucleus of the cancer cell. The green dots are the lysosome while the red dots, distributed within the cell, are the DiI labeled MIONs. The MIONs successfully entering the cell membranes directed by external magnet indicates their promising applications for biomedicines.

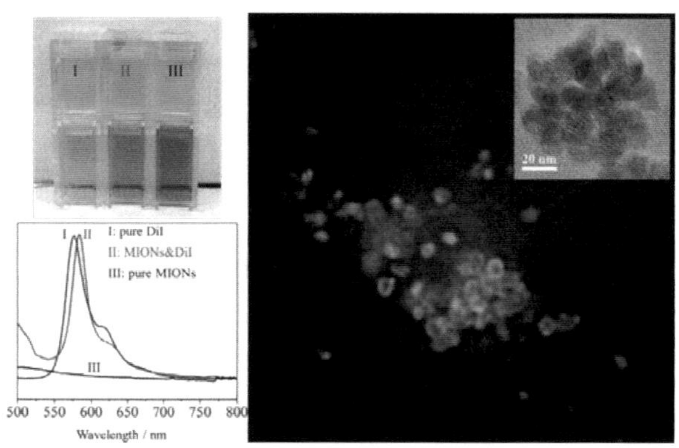

Figure 1. Result for cell culturing and the typical structure of a MION sample. The red dots represent the DiI labeled MIONs within the cell.

Conclusions: We have developed a synthetic strategy for forming libraries of biocompatible multicore iron oxide nanoparticles of tunable dimensions. These MIONs could successfully enter the cancer cells after surface modifications and labeling with florescent dye. Our preliminary experimental results shown here are very promising for biomedical applications, and a systematic investigation is underway.

References:

1. Sen, T. Adv. Drug Deliv. Rev. 2011;63:24-46.
2. Yin, Y. Chem. Soc. Rev. 2012;41:6874-6887.

Effect of Composition on Nanomechanics of Dental Adhesives

Z. Alabdali[1,2], J. Lynch, Ph.D.[1], A.B. Mann, Ph.D.[1,3]
[1] Materials Science and Engineering Department, Rutgers University
[2] Materials Engineering Department, University of Technology-Iraq
[3] Biomedical Engineering Department, Rutgers University

Introduction: Dental caries and the resultant cavities are amongst the most common diseases in the World. To fill the cavities with a composite restoration requires cavity preparation, application of an adhesive system, and then application of the composite [1]. Dental adhesives contain monomers, photoinitiator, and solvent, which are polymerized by light cure [2]. The adhesive forms a chemical and mechanical bond between the composite and the tooth. Ideally, the adhesive penetrates into the tooth surface, forming a hybrid, interfacial layer with the enamel and dentin [3]. Many factors influence the mechanical properties of the hybrid layer, including surface roughness of the teeth, wettability of the surface, and the composition of the adhesive [4]. This study examines the effect of adhesive system composition, Diurethane dimethacrylate (DUMA) in methyl methacrylate (MMA), on the nanomechanics of the adhesive and hybrid layers in bovine teeth.

Method: The adhesive was made from high molecular weight monomer DUMA (436909, 470.56 g/mol Mw.), diluent (low molecular weight) monomer MMA (M55909, 100.12 g/mol Mw.), initiator camphorquinone CQ (124893, 166.22 g/mol Mw.) and solvent ethanol (48075, 46.07 g/mol) all from Sigma Aldrich, USA. The restorative composite applied on the adhesive was Z100 shade A1 purchased from 3M ESPE.

The adhesive system was prepared by mixing 30 /40 /50 /60 /70 wt.% DUMA in MMA with 1 wt.% CQ and 9 wt. % ethanol. The mixtures were contained in amber glass and homogenized at room temperature using a magnetic stirrer (Corning, USA) at a speed of 3000 rpm for 1 hr within a dark room. Fifteen bovine molar teeth were cut using a low speed saw and stored in Hank's buffer solution. The teeth were prepared according to conventional clinical procedure. For the first step, Acid Etchant gel with 37% phosphoric acid was used to etch the teeth surface and increase surface roughness. For the second step, the adhesive system was applied on the tooth surface using a brush and light cure for 60, 50, 40, 30, and 20 sec for each composition of DUMA in MMA. For the third step, the restorative composite was applied on the adhesive layer and cured by light cure (wavelength 420-480nm) for 20 sec. Nanoindentation (Hysitron TriboIndenter™, USA) with a Berkovich diamond tip was used to measure reduced elastic modulus (Er) and hardness (H), in the adhesive layer and in the hybrid layer on both enamel and dentin. Nanoindents were conducted using displacement control up to 800 nm deep in a 5 x 5 pattern spaced 10 μm apart.

Results: Reduced elastic modulus (Er) and hardness (H) of the adhesive as a function of wt.% UDMA in MMA, is shown in Fig. 1. For the hybrid layers the highest value of hardness was obtained for the enamel and dentin at 50 wt.% of UDMA as shown in Fig. 2.

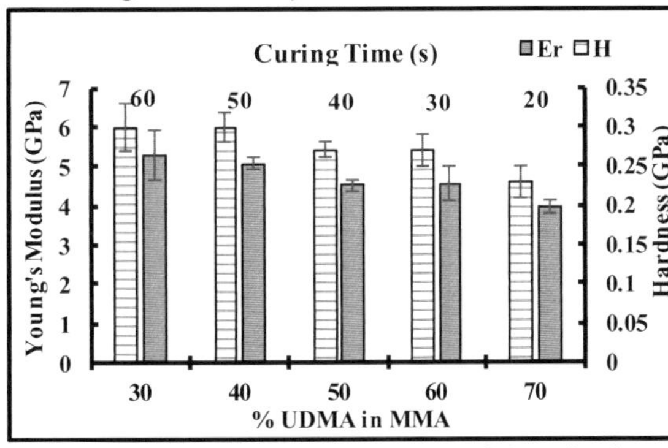

Figure 1. Mechanical properties of the adhesive

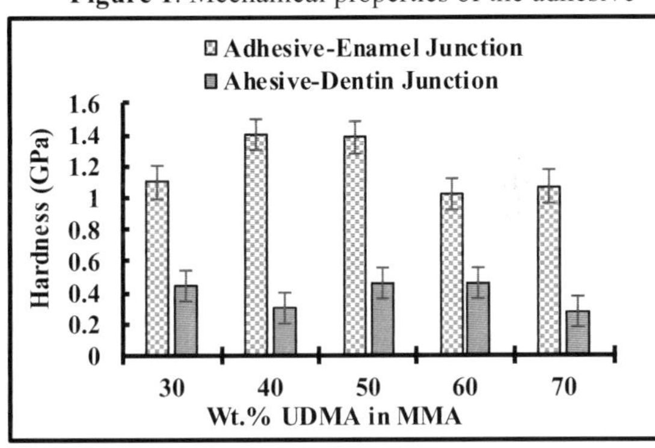

Figure 2. Hardness of the hybrid layers

Conclusions: Shorter curing times impact the mechanical properties as there is a decrease in the degree of conversion of carbon-carbon double bonds to single bonds. The shorter curing occurs when there is an increased weight percentage of the high molecular weight monomer in the adhesive. The poor conversion of double bonds gives shorter polymer chains. The result is similar to that found by S.A. Harahap et al. 2018 [5]. Hardness results on the hybrid layers show that the optimum concentration is 50wt.% UDMA in MMA. This is attributed to the viscosity at this concentration giving the best penetration into the tooth surface and the formation of a strong hybrid layer with both enamel and dentin. Future work will investigate rheology of each concentration to determine viscosity, and Raman spectroscopy will be used to investigate the hybrid layer structures.

References:

1. Keith H.S. chan et al. Materials.2010;3:1228-1243.
2. Eshrak Sofan et al. Annali di Stomatologia. 2017; VIII (1):1-17.
3. Bo Huang et al. Dental Materials. 2018; 34:1253-12162.
4. Ella A. Naumova et al. Dental Materials. 2016; 35(3) :345-352.
5. S.A. Harahap et al. Physics. 2018; Conf.Ser.1073 052020.

Acknowledgements: Iraqi Ministry of higher Education and Scientific Research (MoHESR).

3C-3

Biogenic metallic nanoparticles. A nanometric trojan horse approach

David Medina Cruz, Jungian Chen, Thomas J. Webster.
Chemical Engineering Department, Northeastern University, Boston (MA)

Introduction: Antimicrobial resistance to antibiotics (AMR) and cancer and two of the main concerns that the healthcare system should face nowadays. Current drugs and antibiotic treatments are becoming ineffective or have plenty of drawbacks and side-effects. Therefore, new alternatives are needed. Furthermore, traditional synthesis of nanomaterials is subjected to several disadvantages, such as the production of toxic-by-products and harsh conditions. Therefore, green nanotechnology is presented as a suitable answer, allowing the generation of nanostructures in a cost-effective and environmentally-friendly approach, with enhanced biocompatibility compared to their physicochemically-synthesized counterparts.

Methods: Pathogenic bacteria –both regular and antibiotic-resistant phenotype, such as *Staphylococcus aureus* and Methicillin-resistant *Staphylococcus aureus*- and human cells -both cancer and healthy phenotypes- were used for the synthesis of metallic nanoparticles: from monometallic formulations, such as gold (Au), platinum (Pt), palladium (Pd) and selenium (Se), to bimetallic structures, such as gold-palladium (AuPd) or gold-platinum (AuPt).

Bacteria and cells were cultured in the presence of metallic salts under standard conditions until the generation of nanoparticles, that was followed using microscopy and spectrophotometric techniques. After an in-depth characterization using FTIR, XRD, XPS, and TEM, the nanoparticles were used as antimicrobial and anticancer agents using colony counting unit assays and MTS assays, respectively, as well as characterized regarding the inhibition mechanism.

Results: Pathogenic bacteria were used for the synthesis of bacteriogenic metallic selenium nanoparticles. These agents were employed as suitable agents with antimicrobial activity against the same bacteria that synthesized them, showing low cytotoxicity for human dermal fibroblasts (HDF) cells, as well as an enhanced anticancer behavior towards glioblastoma and melanoma cells. A dose-dependent antibacterial and anticancer effect was found.

The synthesis of metallic nanomaterials using cancer and healthy cells –human dermal fibroblasts (HDF), Human fetal osteoblasts (HFOB), glioblastoma (GLIO), and astrocytes (AST) cells- was reported. Pure metal nanoparticles –palladium or platinum- and bimetallic structures –such as gold-platinum- are readily synthesized using the cells, and after purification, they are used as anticancer agents. A dose-dependent antibacterial and anticancer effect was found.

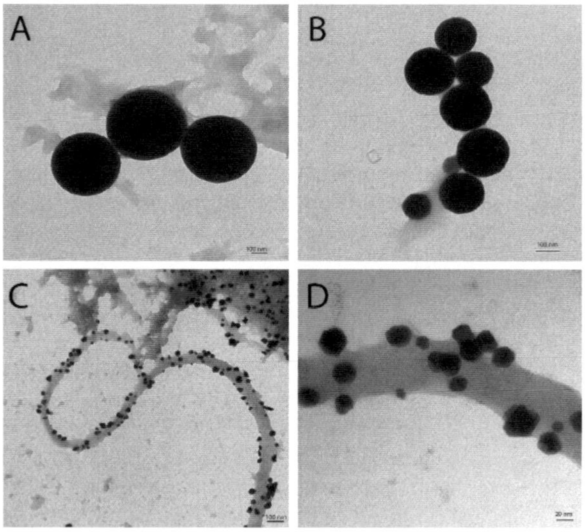

Figure 1. Selenium nanoparticles made by bacteria (A, B) and bimetallic Au-Pt nanoparticles made by melanoma cells (C,D).

Conclusions: Microbiological agents are successfully used as a synthetic biofactory for the generation of metallic nanoparticles of different compositions with biomedical properties. Therefore, they are presented as a suitable approach for the synthesis of nanomaterials in a green fashion, overcoming the limitations of traditional nanotechnology.

Acknowledgements: To the chemical engineering department at Northeastern University.

A Nanoscale Drug Delivery Platform with Controlled Drug Release Modulated by Aptamer Engineering

Robert J. Mosley[1], Ricky J. Whitener[1,2,3], Matthew V. Talarico[1], Jacek Wower[2,3], Mark E. Byrne[1,3]
[1] Biomimetic & Biohybrid Materials, Drug Delivery Laboratories, Department of Biomedical Engineering, Rowan University
[2] RNA Biochemistry Laboratories, Department of Animal and Dairy Sciences, College of Agriculture
[3] US Dept of Education GAANN Graduate Fellowship Program in Biological & Pharmaceutical Engineering

Engineering therapies with high selectivity and specificity for cancer cells can produce more effective treatments for patients while reducing off target toxicity and negative side effects. Aptamers are nucleic acid species that are formulated to fold into unique 3D structures that bind with high specificity to a target molecule. There has been much interest in using aptamers as targeting components on a nanoparticle platform to deliver drugs or assist diagnostics. Such a platform is advantageous as it increases the economical use of aptamers and allows for many of the benefits inherent in nanoparticle systems such as stability, bioavailability, and high payload capacity. Utilizing nucleic acids as a functional component also provides advantages. Base pairing mechanisms allow for controlled assembly of nucleic acid macromolecules. Double stranded nucleic acid structures can also be used to carry certain therapeutics, such as the anthracycline antibiotics that naturally intercalate into the DNA double helix (e.g., Daunomycin, Doxorubicin, Neomycin). Thus far, no nucleic acid aptamer has shown the capabilities of controlling drug loading or release. We report here a nucleic acid conjugated gold nanoparticle (AuNP) platform with the ability to carry a very high therapeutic payload and potential for controlled loading and release of the cancer therapeutic Daunomycin.

First, short oligonucleotides (ANC) with a 5' thiol group are conjugated to the surface of a gold nanoparticle at high density. Each of these ANC strands is engineered with polyT spacers on the 5' end to reduce interactions with both neighboring strands and the gold surface of the AuNP. The ANC strands therefore extend outward from the AuNP surface and act as linkage points for complementary nucleic acid molecules. This link between the AuNP-ANC platform and complementary nucleic acids provides a double stranded section capable of loading Daunomycin molecules. This drug binding section can be extended by applying the Klenow fragment of DNA polymerase I, increasing drug loading capacity. Additionally, the specific sequence used to load the drug can be tailored during oligonucleotide synthesis. One can then exploit the natural affinity of Daunomycin to various nucleotide combinations to modulate drug release.

Radioactive labeling confirmed binding of around 101 (+/-8) anchor strands per nanoparticle. We also show that variations in initial ANC concentration can be used to further control the amount of DNA oligonucleotides that we can conjugate to our platform, with observation of up to 106 individual DNA conjugates when using a large excess of ANC. Four unique DNA sequences were conjugated to separate AuNP-ANC platform to monitor drug loading and release. All four constructs revealed loading of over 1,000 Daunomycin molecules per AuNP,

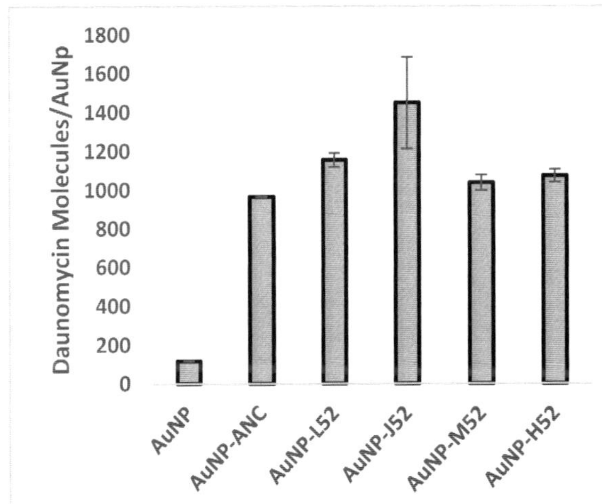

Figure 1 Daunomycin loading capabilities of four unique nanoparticle platforms.

with slight variances depending on the drug binding sequence. Dialysis membrane release studies revealed unique release profiles for each platform for up to 48 hours. Cell viability studies using MCF-7 breast cancer cells show that each of our AuNP platforms is more efficient at killing cells than equivalent amounts of free Daunomycin. We also employed a series of common endocytotic inhibitors to determine the methods of internalization our AuNPs exploit. The AuNPs are observed to enter the cell by a few different methods, with preference for lipid-raft endocytosis.

This platform employs gold nanoparticles, which exhibit little to no toxicity, easy modulation of size and polydispersity, and natural optical properties that have been used for tumor imaging and diagnosis. Thus, the capabilities include tumor targeting, imaging, and drug delivery. The size of the completed AuNPs stays between 16-100nm, allowing them to exploit the enhanced permeability and retention effect in tumors while staying large enough to avoid leakage into healthy vasculature. This platform also holds the potential to regulate drug loading and release using the aptamer itself, a feat that has yet to be seen in literature. Modifying the amounts of double stranded regions on our AuNPs, either through aptamer concentration or Klenow application, along with tailored drug binding sequences allows us to modulate the behavior of the loaded Daunomycin molecules. Current work in our lab involves optimizing our AuNP-ANC platforms for use with characterized aptamers, to prove that these novel, multifunctional particles can preserve aptamer structure – and therefore function – and remain stable for *in vivo* applications.

Application of disturbed fluid flow in a three dimensional cerebral bifurcation model

Nesrine Bouhrira, Brandon DeOre, Peter Galie.
Department of biomedical engineering, Rowan University, Glassboro, NJ

Introduction: Disturbed blood flow disrupts the shear stress profile exerted on the endothelium, but the effect of altered shear stress on the integrity of the blood-brain barrier (BBB) is not fully understood. The complexity of factors within the *in vivo* brain and the difficulty associated with modulating blood flow regimes prevent interrogation of the effect of altered shear stress on the BBB in animal models. Here, using a 3D in vitro model of a cerebral bifurcation consisting of a co-culture of endothelial cells, smooth muscle cells and astrocytes, we investigated the effect of disturbed blood flow on the integrity of the BBB. Combining microparticle image velocimetry with computational fluid dynamics provides the ability to apply distinct regions of both altered shear stress and fully developed flow to an endothelial lumen and to interrogate its effects on BBB integrity using immunocytochemistry and permeability testing.

Methods: Soft lithography was used to generate a microfluidic device with a flow inlet oriented at a 45-degree angle from the axial direction of a 3D vascular model (Fig. 1A). A collagen/hyaluronan composite hydrogel polymerized within inside the device was seeded with normal human astrocytes (NHA), an annulus layer containing human coronary artery smooth muscle cells (hCASMC), and a lining of GFP-labeled human cerebral endothelial cells (hCMEC/D3), all immortalized cell lines (Fig. 1B). A dual-stage, recirculating pulsatile pump controlled by an Arduino microcontroller was used to apply continuous perfusion to the vessel. A computational model validated by microparticle image velocimetry (µPIV) determined the flow rate necessary to produce consistent altered shear stress profiles. Following 24 hours of perfusion, the device was fixed with 4% paraformaldehyde and incubated with phalloidin to label actin, DAPI, and anti- zonula occludin-1 (ZO-1), a tight junction marker that localizes to cell-cell junctions in the BBB.

Results The CFD model predicted flow separation at the near wall of the cell-seeded vessel for an input flow rate of 4 mL/min, indicated by the recirculation zone measured with µPIV at the near wall (Fig. 1C). At this location, the velocity profile was no longer parabolic and shear stress varied along the wall. The model predicted that flow would reattach to the wall approximately 2.2-mm downstream of the vessel inlet. Three different geometries were tested using µPIV to validate the computational model. Flow characterization indicated that small changes in the channel geometry due to the hydrogel gelation and needle placement caused large deviations in the flow environment. Therefore, each of the geometries was input into the computational model first to determine the flow rate necessary to create repeatable flow profiles. As Figure 1D indicates, repeatable shear stress profiles were produced using this method. The stagnation point of the flow (where shear stress was null) occurred at approximately the same axial distance from the bifurcation for all geometries (0.543-mm +/- 0.029-mm). For all geometries, the shear stress in the

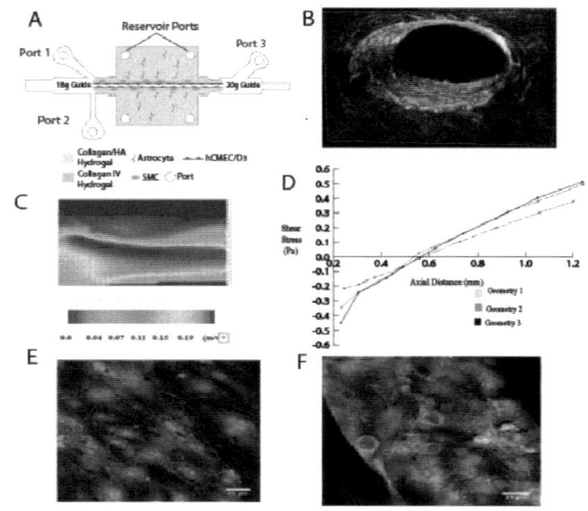

Figure 1: A) Schematic of the microfluidic device detailing the location of the ports and the needle guides used to create the layered cell-seeded vessel. B) A 3D reconstruction of the vessel.. C) Velocity vectors measured by µPIV. D) Shear stress profiles for three different geometries). E-F) ZO-1 staining in cells exposed to physiological (E) and non-physiological (F) shear stress (Scale = 25 µm).

fully developed region remained consistent as (0.835 Pa +/- 0.056 Pa). The effect of disturbed fluid flow on the integrity of the BBB was assessed by labeling ZO-1 within the endothelium lining the vessel in the region of altered shear stress. Figure 1E demonstrates an increased localization of ZO-1 to the cell-cell junctions of hCMEC/D3 exposed to fully-developed, laminar shear stress. This contrasts with the partial localization of ZO-1 to cell-cell junctions in endothelial cells exposed to altered shear stress (Fig. 1F). Permeability testing was conducted to interrogate the barrier function of the vessel at different points in the flow regime. Our results indicate the vascular permeability is higher in the locations of disturbed flow.

Conclusions: Overall, the cerebral bifurcation model described here provides a robust platform to study the response of brain vasculature to complex fluid mechanics. Unlike previous in vitro models designed to exert disturbed flow on endothelial cells, the current model preserves the three-dimensionality of the in vivo environment. Future work will focus on the effect of disturbed flow on endothelial gene/protein expression within this model, specifically related to the mechanism by which the endothelial cells sense and respond to altered shear stress in the context of aneurysm formation and BBB breakdown.

References:

High Throughput Identification of Synthetic Polymers with Globular and Protein-Like Features

Rahul Upadhya, N. Sanjeeva Murthy, PhD, Shashank Kosuri, Matthew Tamasi, Adam J. Gormley, PhD
Department of Biomedical Engineering, Rutgers University, Piscataway, NJ 08854

Statement of Purpose:
Synthetic mimicry of globular proteins is an ambitious goal in the fields of single-chain polymer nanoparticles and supramolecular chemistry.[1] All studies trying to achieve this goal have done so using time-consuming and traditional techniques that severely restrict experimental throughput and compositional exploration. Here, we use a new oxygen-tolerant photoinduced electron-transfer-reversible addition-fragmentation chain transfer (PET-RAFT) polymerization technique, developed in our lab with collaborators, to synthesize polymers with low polydispersity (PDI < 1.2) in open-air conditions and higher throughput than ever before possible.[2] We designed a library of over 300 polymers with varied hydrophobicity, charge, and functionality, then searched for a globular structure using high throughput small-angle X-ray scattering (SAXS). We hope to use this structure-function information for *de novo* design of synthetic proteins.

Methods:
Materials: RAFT agents with initiator Zinc tetraphenylporphyrin (ZnTPP) (Tokyo Chemical Industry Co., Ltd.) were used to simultaneously degas and polymerize monomer solutions in 96-well plates.[2] The backbone monomers utilized were either hydrophilic, hydrophobic, neutral, or charged. For backbones onto which PEG was tethered, acrylic acid N-hydroxysuccinimide ester (Sigma Aldrich) was incorporated into the backbone followed by dibenzylcyclooctyne-NH_2 (DBCO-NH_2) (Click Chemistry Tools) to enable click chemistry. Spin purification was done with Sephadex G-25 Superfine Resin (GE Healthcare Life Sciences), and PEG-N_3 (Sigma Aldrich) was clicked onto polymer backbones to assess shifts in size and conformation.

Polymer Synthesis: Light-initiated polymer synthesis was done in 96-well plates in open air conditions with a RAFT agent:initiator ratio of 100:1 and final monomer concentration of 1 M in DMSO. Molecular weight and polydispersity was quantified by size-exclusion chromatography and multi-angle light scattering (SEC-MALS). DBCO incorporation was assessed by UV spectroscopy at a wavelength of 310 nm.

SAXS: SAXS experiments were conducted at National Synchrotron Light Source II (NSLS-II, Brookhaven, NY) Beamline 16-ID for Life Science X-ray scattering on a Pilatus 1M detector. Samples were measured at 5 mg/mL in PBS/1% DMSO with 5-second exposures relative to buffer blanks. Averaging of the five exposures and buffer subtraction was completed.

Results:
We hypothesized that a variety of sizes and conformations can be characterized by SAXS due to modifications in monomer arrangement on the polymer backbone and the presence of PEG that is tethered to the backbone. Figure 1 shows the Kratky plot of several representative polymer backbones, including sample 195, a polymer backbone with a mixture of hydrophilic and hydrophobic monomers, and samples 189 and 396, which have the same backbone composition with PEG tethered. The peak in the Kratky plot indicates protein-like folding.

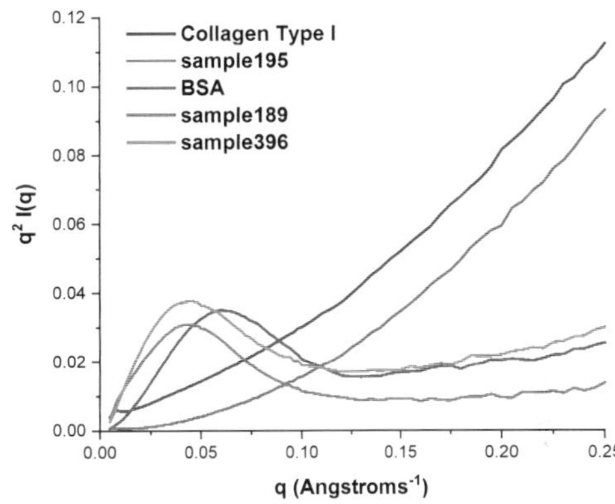

Figure 1: Kratky plots of extended (Collagen type I and sample 195) and compact (bovine serum albumin BSA, sample 189, and sample 396) macromolecules. Most polymer tested were intrinsically disordered, while three had protein-like folding behavior.

Conclusions:
Of the over 300 polymers tested, 3 or roughly 1% exhibited folding behavior. Such identification would not have been possible without our novel high throughput synthesis and characterization process. There is a fine balance in hydrophobicity and complexity necessary for folding conditions. Hydrophilicity allows the polymer to be soluble in water, while hydrophobicity drives globular assembly. Tethering PEG, which interacts with the solvent, allows for increased folding of the polymer by shielding hydrophobic monomers. This information will be used to design the next library of single-chain folded nanoparticles and ultimately synthetic proteins.

References: **[1]** Pomposo, JA, et al. "How Far Are Single-Chain Polymer Nanoparticles in Solution from the Globular State?" ACS Macro Lett 2014, 3(8).
[2] Gormley, AJ, et al. "An Oxygen-Tolerant PET-RAFT Polymerization for Screening Structure-Activity Relationships." Angew. Chem. Int. Ed. 2018, 57(6).

Acknowledgements:
Research was supported by the National Institute of General Medical Sciences of the National Institutes of Health under award number T32 GM008339.

4A-3

Solution Spun Protein-Based Polymer Fibers for Biomedical Applications

Xiao Hu
Department of Physics and Astronomy, Department of Biomedical Engineering,
Rowan University, Glassboro, NJ 08028, USA. (hu@rowan.edu)

Introduction: Protein-based fibers such as silk and corn fibers are a masterpiece of nature whose remarkable mechanical properties have been important materials in textiles, medical sutures, and armor since ancient times. Thanks to its incredible high strength and extensibility, as well as their combination of biocompatibility and biodegradability, protein fiber materials have found applications in various fields such as tissue regeneration and controlled drug delivery. Primarily produced by electro-spinning, it is possible to make protein fibers without the compatible solvents. However, the electro-spinning process still has low output efficiency and is limited by its dielectric constant. Here, we report a method for the production of protein micro and nanofibers (e.g. silk fibroin and corn zein fibers) from an adjustable concentric nozzle. This blow spinning technique consists of the extrusion of protein solution into an evaporative atmosphere producing fiber diameters similar to those generated by electro-spinning. This strategy will avoid the biologically toxic solvent systems needed for wet- and electro-spinning. Fourier transform infrared spectroscopy (FTIR), Temperature-modulated differential scanning calorimetry (TM-DSC), thermogravimetric analysis (TG) and scanning electron microscopy (SEM) were used to demonstrate the structural, morphological changes as well as the thermal properties of these fiber materials. Model drugs of varying solubility, molecular weight, and charge were then incorporated into the fibers, and the release profiles of these drugs were monitored over 96 hours and compared to their release from protein films to elucidate the effect of geometry on drug delivery kinetics. The yielded fibers can be used in the fabrication of protein-cell constructs for various pharmaceutical and biomedical applications in the future.

Methods: FTIR analysis of different protein fibers and films was performed using a FTIR spectrophotometer (TENSOR 27, BRUKR Co. Ltd, German), equipped with a deuterated triglycine sulfate detector and a multiple reflection, horizontal MIRacle ATR attachment (using a Ge crystal, from Pike Tech). The morphologies of the samples were observed by SEM (LEO 1530 VP, ZEISS Corporation, Japan). In addition, the protein samples (each about 5 mg) were encapsulated into Al pans and heated in a TA Instruments (Q100, TA Instruments Co. Ltd, USA) DSC, with purged dry nitrogen gas flow (50 ml·min^{-1}), and equipped with a refrigerated cooling system. TG Analysis (Hi-Res TGA 2950, TA Instruments Co., USA) was also used to measure weight changes of samples with increasing temperatures. Model drugs of varying solubility, molecular weight, and charge were then incorporated into the protein fibers and films prior to spinning or casting. The drug release study of protein samples was performed at 37 °C in PBS (phosphate buffered saline) solutions.

Results: FTIR, SEM, and DSC were performed on protein (zein and silk) nanofibers and films before and after release of drugs. Structural analysis revealed that the incorporation of model drugs into the fibers caused the secondary structure of the protein to transition from a network of random coils to a more alpha helical structure. Upon release, the protein fiber reverted to its original random coil network, resembling an elastic carrier with a controllable, reversible transition. This effect occurred in most nanofiber samples, but was not observed in film samples. In addition, thermal analysis indicated that fibers can protect the drug molecules in high temperature above 160 °C, while drugs within films will degrade below 130 °C. These findings can likely be attributed to the mechanical infiltration of the drug molecules into the ordered structure of the fibers during their solution fabrication. The slow release from fiber samples can be attributed to this biophysical interaction, illustrating that release is dictated by more than diffusion in protein-based carriers.

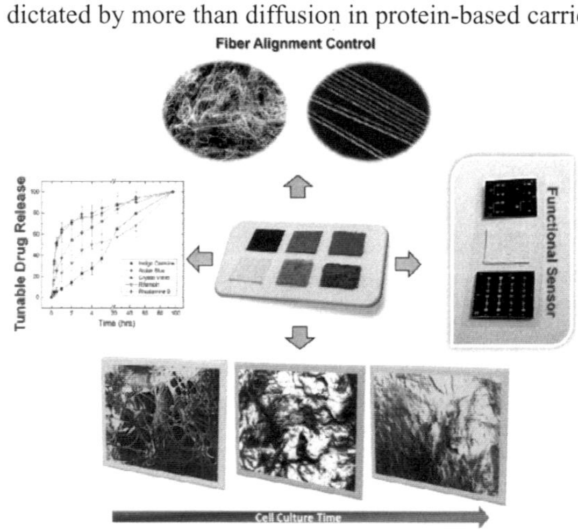

Figure 1. Air-Spun protein fiber materials loaded with different model drugs have tunable drug release rate, controllable fiber alignment, perfect tissue biocompatibility, and can be fabricated with sensor modulus to become a functional health care device.

Conclusions: In this study, air-spun protein (corn zein or silk) nanofibers and films were able to reversibly store a variety of model drugs. Incorporation of drug into fibers was shown to have significantly improved effect on the thermal stability of the drug molecules, while films can not prevent their thermal degradation/melting at high temperature. Understanding this interaction in more detail can lead to more efficient design of protein-based drug delivery systems.

References:

1. Hu X. et al. Mater. Today 2012;15:208-215.
2. Meinel L. et al. Adv. Drug Deliv Rev. 2012;64:1111-1122.

Adaptation of Liquid Handling Robotics for High Throughput Customizable RAFT-Polymerization

Matthew Tamasi, Shashank Kosuri, Rahul Upadhya, Adam J. Gormley
Department of Biomedical Engineering, Rutgers University, Piscataway, NJ 08854

Statement of Purpose:
Reversible addition-fragmentation chain transfer (RAFT) polymerization enables precise control over molecular weight and polydispersity and allows for investigation of complex polymer architectures with bioactive functionality (Yeow, Chapman et al. 2017). Traditionally, this oxygen-intolerant polymer chemistry is done in batch processes that are technically challenging and time consuming to complete. Recently, we have shown that oxygen-tolerant porphyrin-catalyzed photoinduced electron transfer RAFT (PET-RAFT) polymerization enables combinatorial preparation of bioactive polymers in 96 or 384 well plates (Gormley, Yeow et al. 2018).[1] To further expand on this unique capability and make polymer chemistry accessible to the non-expert, we have programmed a Hamilton Microlab STARlet liquid handling robot to perform high throughput polymer chemistry and bioconjugation based on a user-friendly graphical user interface (GUI). This automation allows for the reliable and simple creation of customizable polymers for a variety of biological applications.

Methods:
PET-RAFT polymerizations where carried out by a Hamilton Microlab STARlet liquid handling robot using dimethyl sulfoxide (DMSO) as the primary solvent and photo initiator zinc tetraphenylporphyrin (ZnTPP) with various RAFT agents. A user friendly GUI was programmed to enable the non-expert to make polymers with just 1 hour of training. Stocks of pre-aliquoted monomers, chain transfer agents (CTAs), and other reagents were prepared for easy reaction setup based on computer prompts after polymer library design. To initially validate the automation, all polymers were homopolymers consisting of a variety of acrylates, methacrylates, acrylamides, and methacrylamides. The Microlab STARlet was programmed to add required reagent volumes to 96 well plates at a final concentration of 1 M monomer and [CTA] : [ZnTPP] of 100:1. CTA and ZnTPP concentrations were varied to create varying degree of polymerizaiton (DP) from DP 100 to DP 500. Characterization of the polymers to obtain molecular weight and polydispersity information was carried out by gel permeation chromatography (GPC) and multi-angle light scattering (SEC-MALS).

Results:
More than 300 polymers have been synthesized and characterized using this liquid handling platform for RAFT polymerization. Data presented below demonstrates precise DP control of *N*-dimethylacrylamide(DMA) with 2-(dodecylthiocarnonothioylthio)propionic acid as the CTA (Figure 1). Shifts in molecular weight correspond with intended elongation of the polymer chain and number of monomer units.

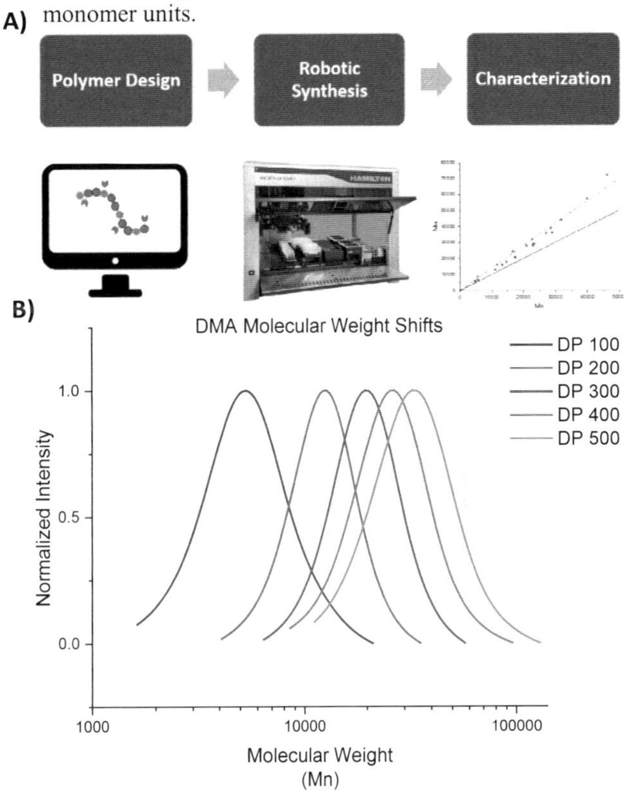

Figure 1. A) Polymer synthesis work flow. **B)** Polydispersity distributions of same DMA samples.

Conclusion:
We have tested over 300 variations of homopolymers with an ever-expanding library of monomers and CTA agents. Our results suggest that the adaptation of liquid handling robotics for high throughput synthesis of polymers is uniquely adaptable to PET-RAFT due to its reaction control in low volumes and oxygen tolerance. By leveraging a robotic system for polymer synthesis, we open the door to creating a robust and reproducible platform for the rationale design of biofunctional polymers for the non-expert.

References:
Gormley, A. J., J. Yeow, G. Ng, O. Conway, C. Boyer and R. Chapman (2018). "An Oxygen-Tolerant PET-RAFT Polymerization for Screening Structure-Activity Relationships." Angew Chem Int Ed Engl **57**(6): 1557-1562.
Yeow, J., R. Chapman, J. Xu and C. Boyer (2017). "Oxygen tolerant photopolymerization for ultralow volumes." Polymer Chemistry **8**(34): 5012-5022.

High Throughput Screening of Random Heteropolymers that Stabilize Enzymes

Matthew Tamasi[1], Shashank Kosuri, Adam J. Gormley
Department of Biomedical Engineering, Rutgers University, Piscataway, New Jersey

Statement of Purpose:

Random heteropolymers have recently been shown to be particularly strong enzyme stabilizers due to suitable matching of polymer-protein physicochemical characteristics.[1] These complementary interactions help safeguard enzymes from denaturing by forming a chaperone-like shell around them. However, discovering suitable polymer formulations on a case-by-case basis is highly time consuming due to the number of possible polymer sequences and the manual nature of synthesizing and testing each polymer design. Here, we describe a semi-automated and high throughput technique to synthesize and test large libraries of random heteropolymers for enzyme stabilization. This is achieved using a Hamilton Microlab STARlet liquid handling robot which has been designed to make these random heteropolymers using a novel oxygen tolerant PET-RAFT technique recently developed in our lab and collaborators.[2] Then, we tested the ability of these polymer libraries to stabilize the activity of a model enzyme, glucose oxidase (GOx).

Methods:

PET-RAFT polymerizations were conducted in a 96 well plate in dimethyl sulfoxide (DMSO).[2] Four different types of monomers were selected to impart different characteristics to the polymer backbone. The exact monomer names are omitted in this abstract, but will be available upon presentation. Monomer A was chosen for its hydrophobic nature, monomer B for its hydrophilicity, monomer C for its ability to form hydrogen bonds, and monomer D to impart cationic charge. Once polymerized, they were serially diluted in DMSO and added to sodium acetate buffer (50 mM, pH 5.1) in 96 well plates to reach final desired concentrations. Later, they were mixed with 1 μg/mL stock solution of GOx in equal volumes and allowed to sit for 30 minutes on a plate shaker. Then the 96 well plate was sealed with a plate sealing film and heated to 70°C for 1 hr to denature the enzyme. Negative controls were included in the same plate to compare the polymer's ability to protect the enzyme from thermal denaturation. GOx activity was assessed using an assay containing glucose, tetramethyl benzidine (TMB) and horseradish peroxidase (HRP). Baseline GOx activity was determined by preparing a stock protein solution in 50 mM sodium acetate buffer at pH 5.1 and applying the colorimetric assay.

Results

In just one month, approximately 270 polymer compositions were tested with varied molecular weight, hydrophobicity, polarity and charge. Data presented here is a set of four polymers tested with select domains eliminated to determine the protective effects of each monomer. When the charge domain was completely removed the polymer lost its ability to protect the enzyme. Polymers with 10% charge, on the other hand, had much better ability to protect the enzyme from heat. The best polymer had a composition of 70% A, 20% B, 0% C, and 10% D which retained native enzyme activity up to 40 percent even after 70°C heating for an hour. These results confirm that the design and composition of complex heteropolymers plays a key role in their ability to safeguard the enzyme from adverse conditions.

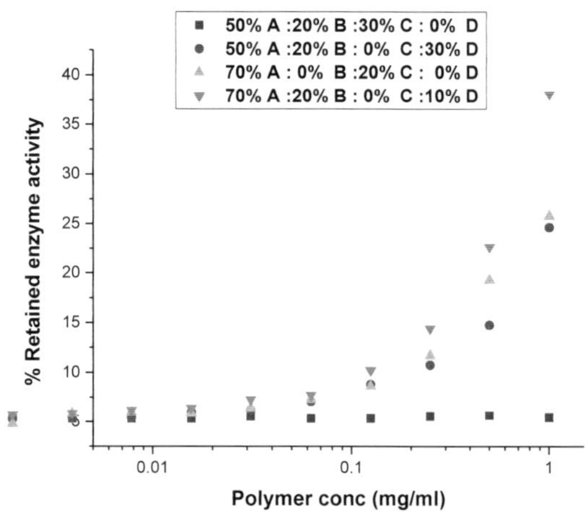

Figure 1. Effect of different polymer chain domain compositions on enzyme protection. Overall results from many more sets of data not shown suggests that small changes in polymer composition have dramatic effects on their ability to stabilize enzymes.

Conclusions:

We have tested 180 polymers with varied chain length, composition, charge and structure. These initial studies show promising signs that quickly varying the characteristics of polymers helps identify compositional patterns that lead to stable enzyme formulations. Our next set of experiments will be to characterize these polymers and understand the mechanism through which they stabilize GOx in more relevant conditions. Then, we will apply this approach to other valuable enzymes in need of advanced formulation.

References:

1. Panganiban, B., *Science*, *359*(6381), 1239-1243.
2. Gormley, A. J., Angewandte Chemie International Edition, 57(6), 1557-1562.

A Facile Method of Protein Crystallization using Gold Nanoparticles

Hyewon Kim[1], Xiaoting Guo[1], Becka Padgett[2], and Vicki L. Colvin[1,2]
1. Department of Chemistry, Brown University.
2. Center for Biomedical Engineering, School of Engineering, Brown University.

Introduction: Functionalized nanoparticles can serve as biosensors, drug carriers, and nanoprobes, making them widely applied in biological catalysis, immunization and drug delivery.[1] Protein crystals are important first step to understand molecular structure of proteins. Protein crystallization is also utilized widely for pharmaceutical industry in order to purify proteins from mixtures of proteins. Therefore, facile crystallization of protein is crucial developing a molecular understanding of disease and biology. Here, we present how gold nanoparticles (AuNPs) with a certain surface coatings can promote protein crystallization. The early stage of nucleation and the interaction between AuNPs and lysozyme is monitored using size exclusion chromatography (SEC) and UV-VIS spectroscopy.

Methods: Gold nanoparticles (AuNPs) were synthesized based on modified Frens method.[2] Briefly, 50ml of 1mM Gold(b) Chloride Trihydrate ($HAuCl_3 \cdot 3H_2O$; Aldrich, St. Louis, MO) reduced with 1mL of 0.25 M Sodium Citrate Tribasic Dihydrate (Aldrich, St. Louis, MO) at 100 °C. The solution kept stirring for 15 mins until the color becomes ruby red. The average size of AuNPs were 13.0 ± 1.0 nm based on TEM analysis.

Citrate-capped AuNPs were functionalized with polyethylene glycol (PEG) of different molecular weight by adding 3000 molar excess of PEG to AuNP solution and reacting overnight.

Size exclusion chromatography (SEC) was performed on Agilent HPLC 1100 series (Agilent, Santa Clara, CA). A pL-aquagel-OH 60 (25.0 x 300 mm, 15 μm, HPLC column) designed for hydrophilic polymer with Mw: 200K-10,000K was used for testing AuNPs-lysozyme conjugate in aqueous solution. A diode-array detector (DAD) was connected to monitor lysozyme at 280 nm and AuNPs at 526nm. Testing Condition was always kept constant for each sample: Flow: 1ml/min; Pressure: 40 bar; Injection volume: 25.0 μl; Column thermostat temperature: 30 ; RID unit temperature: 30 . The solution of AuNP-PEG and lysozyme was injected every thirty minutes and the eluent was detected using DAD.

The AuNP-PEG and lysozyme solution under crystallization condition was characterized using UV-VIS spectroscopy (Cary 5000). The solution was monitored for 24 – 48 hours and measured concentration change of AuNPs and lysozyme using absorbance.

Co-crystallization of PEGylated AuNPs and lysozyme is performed using batch crystallization method. 0.1 M sodium acetate with 1 M NaCl at pH 4.6 is used as crystallization buffer. 4-5 nM of AuNPs and 4-10 mg/mL lysozyme was mixed with the buffer and left at 10 for 24 h.

Results: The early stage of nucleation process was monitored using SEC. The retention time of single AuNP was around 8 to 9 min, and that of lysozyme was around 11 min. Fig. 1 shows the two different AuNPs that we tested: AuNPs functionalized with 10K molecular weight PEG (Au-10K), and AuNPs functionalized with 1K molecular weight PEG (Au-1K). As shown in Fig. 1 left, Au-10K shows the peak-shift of lysozyme towards smaller retention time within an hour, meaning that aggregation of lysozyme is formed in very early stage of crystallization. In addition, single Au-10K peak (red lines) disappears and broaden towards to smaller retention time as well. This indicates AuNPs also forms an aggregation meaning that they are incorporating in nucleation process. On the other hand, when Au-1K is mixed with lysozyme under crystallization condition, the peak of single AuNPs does not change over crystallization process (Fig.1 right).

0
0
0
k

As a result of this nucleation, we found that Au-10K incorporate into protein crystals, becoming a co-crystal of AuNPs and lysozyme, whereas, Au-1K remains in solution where lysozyme crystals are grown. We also found only Au-10K promotes crystallization process by changing nucleation time 3 times faster than that of native lysozyme.

Conclusions: We have developed a new facile process of protein crystallization utilize AuNPs. We have found that when AuNPs are coated with PEG larger than 5K molecular weight, they incorporate into protein crystals and facilitate crystallization.

References:

1. Gagner JE, Shrivastava S, Qian X, Dordick JS, Siegel RW, J Phys Chem Lett. 2012; 3(21): 3149-3158.
2. Frens, G. Nature Physical Science. 1973; 241:20.

4B-4

Alignment of Human Cardiomyocytes through Nano-Wrinkles on Shape Memory Polymers

Sarah Moore, Shiyang Sun, Chenyan Wang, Plansky Hoang, James Henderson Ph.D, Zhen Ma Ph.D
Department of Biomedical and Chemical Engineering, Syracuse University
Syracuse Biomaterials Institute

Introduction: Cardiomyocytes construct the cardiac muscles, which are the reason behind the blood pumping function of the heart. The alignment of cardiomyocytes is critical to maintaining the physiological functions of the cardiac muscle tissues, while disorganized alignment will lead to heart diseases. To better mimic the *in vivo* cardiac structure, researchers have been searching for new biomaterials and fabrication tools to align cardiomyocytes in a variety of culture systems. In previous research, it has been found that different surface topography could strongly influence the alignment of muscle cells [1]. More importantly, the topographic surface could promote the alignment of sarcomeres, the contractile apparatuses, within the cardiomyocytes [2]. In our project, we used shape memory polymers (**SMPs**) to form nano-wrinkles on the surfaces for cardiomyocyte alignment.

Materials and Methods: SMPs were fabricated with 95 wt% tert-butyl acylate (**TBA**) and 5 wt% butyl acylate (**BA**). These 95:5 TBA:BA SMP samples are heated and put under a 40% strain. The strained SMPs were then spin coated with polyelectrolyte multilayer (**PEM**) film. PEM film consisted of a serial layers of neutrally charged polyethyleneimine, negatively charged poly(allylamine) hydro-chloride, and positively charged poly(sodium-4-styrene sulfonate). Once PEM film was spin-coated onto the SMPs, the SMPs were heated to their glass transition state making the SMPs recover to their original shape (Figure 1a). Shape recovery of SMP would produce nano-wrinkles on the surface of PEM film. Analysis of the nano-wrinkles was completed by atomic force microscopy (**AFM**). The PEM-coated SMPs were sterilized under UV light and coated with gelatin for cell seeding. Human induced pluripotent stem cell derived cardiomyocytes (**hiPSC-CMs**) were then seeded onto the PEM-coated SMPs. hiPSC-CMs alignment with nano-wrinkles was analyzed with immunofluorescent staining.

Results and Discussion: We used the AFM to study the topographic feature of the surface of PEM-coated SMPs. Based on the AFM images, we found that parallel wrinkles were able to form on the surface (Figure 1b). We have successfully grown hiPSC-CMs on the PEM-coated SMPs, and found the alignment of cell body and nuclei parallel to the wrinkles and sarcomeres perpendicular to the wrinkles (Figure 1c). Currently, we are investigating the influence of the thickness of PEM layers on the wrinkle formation and cardiomyocyte alignment.

Conclusion: We demonstrated the formation of nano-wrinkles on the surface of PEM-coated SMPs after shape recovery. Additionally, we showed the alignment of hiPSC-CMs with nano-wrinkles on the surface, which mimics the physiological structure of the heart.

Acknowledgments: This work was supported by NSF (EBMS-1804875), CUSE grant, Nappi Research Scholar Fund, Lush Prize, and Syracuse Undergraduate Research Award. We would also like to acknowledge Dr. Ariel Ash-Shakoor from the Henderson Lab for training the polymer fabrication.

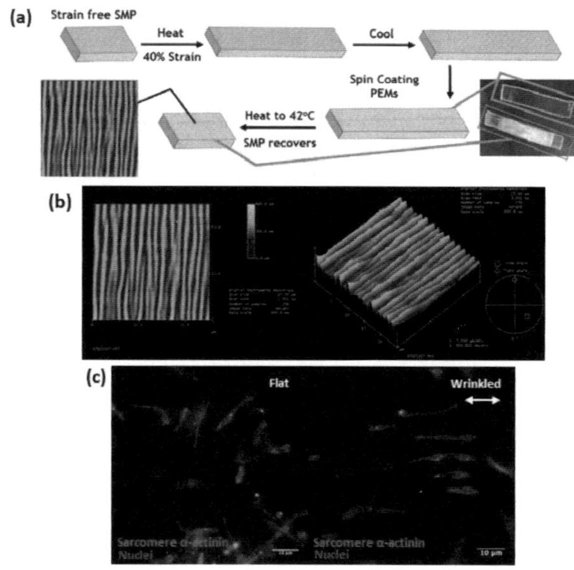

Figure 1. (a) Schematic of straining of SMPs and spin-coating of PEM film to form nano-wrinkles on the surface. (b) AFM image of nano-wrinkles. (c) Immunofluorescent staining of hiPSC-CMs grown on flat surface and wrinkled surface of PEM-coated SMP. The arrow indicates the wrinkle direction.

References: [1] Luna, Jesus Isaac, et al. "Multiscale biomimetic topography for the alignment of neonatal and embryonic stem cell-derived heart cells." Tissue Engineering Part C: Methods 17.5 (2011): 579-588. [2] Grosberg A, Alford PW, McCain ML et al (2011) Ensembles of engineered cardiac tissues for physiological and pharmacological study: heart on a chip. Lab Chip 11:4165–4173

Platelet-Derived Growth Factor Fused with Elastin-Like Polypeptides for Pressure Ulcer Healing

Suneel Kumar[1], Mehma Kaur Chawla[1], Henry Hsia[2], Kyle Quinn[3], Rick Cohen[1], Martin Yarmush[1], Francois Berthiaume[1]
[1]Department of Biomedical Engineering, Rutgers, The State University of New Jersey, Piscataway, NJ 08854
[2]Department of Plastic Surgery, Yale University, New Haven, CT 06510
[3]Department of Biomedical Engineering, University of Arkansas, Fayetteville, AR 72701

Introduction: Pressure sores or ulcers (PUs) are a common debilitating complication of spinal cord injury (SCI) and tend to occur in soft tissues (i.e. skin and underlying fat and muscle) located around bony prominences where body weight is concentrated while the patient is sitting or lying[1]. PUs causation is a multifactorial process associated with a number of extrinsic (pressure, shear, duration of loading, temperature, moisture, hygiene) and intrinsic factors (general physical conditions, the morphology of bone and tissues, muscle tissue mechanical properties)[2]. Our animal model of PU after SCI showed a novel system to investigate skin wound healing in SCI and shows that skin wound healing is delayed in SCI animals[3-4]. Furthermore, the data support the hypothesis that impaired blood flow may be responsible for this observation (Figure 1). In the present study, we wish to test this hypothesis by introducing angiogenic peptides at the wound site to show that promoting local angiogenesis can reverse the impaired wound healing. For this purpose, we developed fusion proteins whereby the angiogenic platelet-derived growth factor (PDGF) is attached to elastin-like polypeptides (ELPs) using our standard method.

Methods: The full-length encoding sequence for human PDGF was fused with 50 repeats of ELP encoding cassette via linker sequence as per our published protocol[5]. The expression vector containing PDGF-ELP was transformed into E. coli. The bacteria were grown in Terrific Broth media and protein expression was induced by adding 0.5 mM IPTG. Cells were lysed by sonication and the fusion proteins were purified by inverse transition cycling (ITC) three times alternatively at 4°C and 40°C. As a control, ELP proteins alone were also expressed and purified. Fusion proteins were characterized by SDS-PAGE gel and western blot. The ELP, PDGF-ELP, and PDGF biological activity is tested in a dose-dependent manner via cellular response assays (e.g. proliferation, and in vitro angiogenesis) *in vitro*.

Results: Mouse model data of PU suggested the slow healing[3-4] and decrease blood flow (Figure 1) in SCI animals as compared to control non-SCI animals. We hypothesis here to develop the stable form of PDGF fusion protein that can stimulate the local angiogenesis in the skin wound. Therefore, we successfully develop, purify and characterized the PDGF-ELP fusion protein. Physical characterization using SDS-PAGE and western blot result confirmed the presence of PDGF-ELP fusion proteins (~37.2 kDa) in the purified products. *In vitro* assay using endothelial cells showing the dose-dependent proliferation effect.

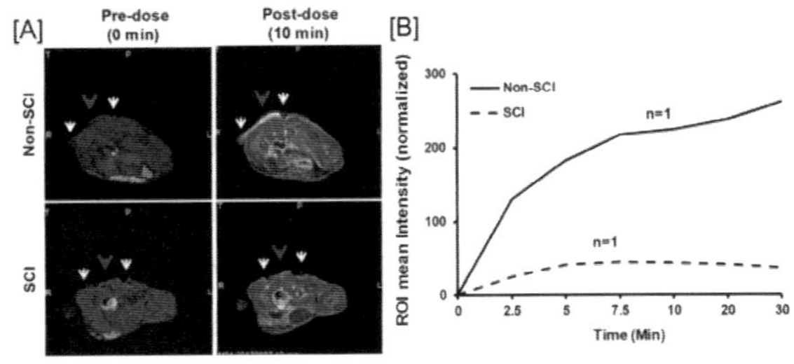

Figure 1. Dynamic contrast-enhanced MRI (DCE-MRI, T1 weighted scan) of blood flow in the wound area. Wounds were imaged at 3 days post-wounding. (A) The left column is the background image before the infusion of the contrast agent. The right column is image 10 min after contrast agent administration. White arrows point to edges of the wound, red arrow points to the center of the wound. (B) Signal intensity in the region of interest (ROI) corresponding to the wound area as a function of time

Conclusions: We have developed and characterized PDGF-ELP fusion proteins that could form nanoparticles at physiological temperature. These fusion proteins can be easily purified from the bacterial culture by ITC, thus enabling rapid and inexpensive purification. Our results suggest that PDGF-ELP could stimulate the local angiogenesis in skin wound thereby enhance the wound healing in SCI-related PU. Future studies will investigate the stability of this fusion protein in wound environment as well as in different *in vitro* system of angiogenesis before use in the animal model.

References:

1. Rappl LM. Int. Wound J. 2008;5:435-444.
2. Thomason SS et al. Adv. Skin Wound Care. 2016;29:269-276.
3. Kumar S et al. J Neurotrauma. 2018;35(6):815-824.
4. Kumar S et al. J. Vis. Exp. (Pending Publication), e58188, In-press (2019).
5. Yeboah A et. al. J Control Release. 2016;232:238-247.

Acknowledgments: This work was partially supported by the New Jersey Commission on Spinal Cord Research (CSCR15IRG010) and the U.S. Department of Defense (SC160029). We thank Sean O' Leary from the W.M. Keck Center for Collaborative Neuroscience, Rutgers for technical assistance.

Biodegradable PCL-PGA-beta TCP Scaffolds for Bone Tissue Engineering

Alok Kumar, Seyed Mohammad Mir, Carlos H Leon, Xiaojun Yu.
Biomedical Engineering Department, Stevens Institute of Technology, Hoboken, NJ, USA 07030.

Statement of Purpose: More than 2.2 million surgeries are performed each year worldwide to treat the bone defects [1]. Synthetic scaffolds engineered to promote the vascularization and support bone ingrowth have shown to be a possible alternative to the natural grafts due to its unlimited biological-risk free supply. Therefore, in the present work, biodegradable scaffolds with an ability to generate the pores for vascularization have been fabricated. An attempt has been made to obtain the mechanical properties comparable to host tissue by tailoring the composition and processing conditions. In this study, biodegradable polymers, polycaprolactone (PCL) and poly (glycolic acid) (PGA) along with Beta tri-calcium phosphate (βTCP) were used in various ratios to make the composites. The composites made of PCL and PGA with no pores are expected to provide the mechanical stability at the defect site. However, progressive degradation of PGA leads to the generation of pores within the composite. These pores with interconnected geometry are theorized to help in the vascularization and also will provide the space for bone ingrowth.

Methods: Three different composites, 25PCL-75PGA-20^2 TCP, 50PCL-50PGA-20^2 TCP, and 75PCL-25PGA-20^2 TCP were prepared by dissolving PCL, PGA, and beta TCP in 1,1,1,3,3,3-Hexafluoro-2-propanol (HFIP) at room temperature. The mixture was dried in air, followed by vacuum drying for complete removal of moisture and an organic solvent. The dried powder was treated with liquid nitrogen and then crushed to reduce the powder particle size. The powder was dried again in vacuum and then compressed in a PTFE mold, followed by heating for 10 min at 190 °C to make a cylinder of 12 mm length and 6 mm diameter. The as-prepared samples were cleaned in 70% ethanol and dried in vacuum. Dried samples were studied by scanning electron microscope (SEM) for the surface characterization. X-ray diffraction (XRD) and energy dispersive spectroscopy (EDS) were used for the phase analysis. The density of the samples was measured using the liquid displacement method. For the mechanical testing, cylindrical samples were tested at room temperature at a speed of 1.3 mm/min using the universal testing machine. Sterilized samples were immersed in 1×PBS and kept on a shaker at 37 °C for 2, 4, 6, and 8 weeks. The results of this study were statistically analyzed, and data were presented as the mean ± standard error of mean.

Results: The 25PCL-75PGA-20^2 TCP, 50PCL-50PGA-20^2 TCP, and 75PCL-25PGA-20^2 TCP samples were characterized by a dense structure of density ~0.6, ~0.8, and ~1.3 g/cm^3, respectively, which are comparable to the density of cancellous bone (0.1–1.0 g/cm^3) [2]. The SEM carried out in the secondary electron mode revealed microporous surface (Figure 1). The surface was characterized by a rough topography and roughness increases with the amount of PGA in the samples.

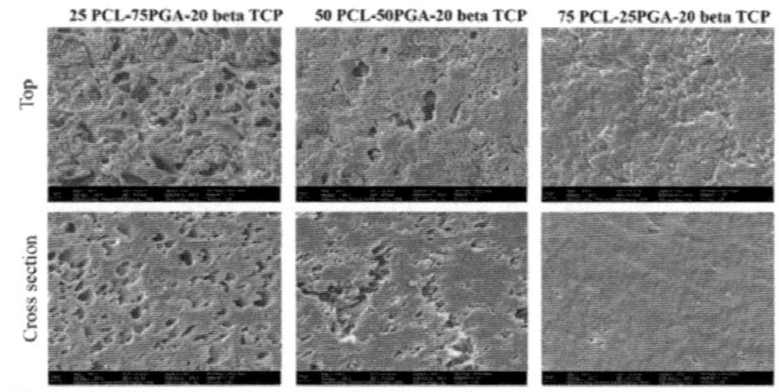

Figure 1: SEM micrograph showing the surface topography of the PCL-PGA-βTCP samples.

A comparison of XRD results with the existing literature and ICDD database confirmed the presence of PGA, PCL, and ^{2} TCP. Additionally, the EDS analysis confirmed the uniform distribution of βTCP (Ca, P, and O) in the PCL-PGA-βTCP samples. The mechanical testing of scaffolds showed a significant effect of PCL and PGA ratio on the yield strength with the highest value (~40 MPa) in case of 25PCL-75PGA-20^2 TCP samples.

The dissolution study was carried out to evaluate the stability and degradation behavior of the developed samples in 1×PBS under shaking condition. Results revealed a faster degradation of PGA than PCL. The degradation of PGA led to the formation of pores in the samples. As hypothesized, these pores are expected to help in the neovascularization and bone ingrowth, *in vivo*.

Conclusions: We have theorized that the dissolution of PGA from PCL-PGA-^{2} TCP samples leads to the formation of pores, required for the vascularization and bone tissue ingrowth. However, initial samples without no pores are expected to withstand the compressive load and help in the bone regeneration process.

Results showed the compressive strength of as prepared samples comparable to the cancellous bone. Furthermore, results confirmed the formation of pores in the samples due to faster degradation of PGA from the PCL-PGA-^{2} TCP samples. The initial study has supported the suitability of PCL-PLGA-βTCP scaffolds as a potential material for the reconstruction of bone defects.

References:

1. Leppik L, Scientific Reports. 2018; 8: 6307
2. Murugan R, Composites Science and Technology. 2005; 65:2385-406

Acknowledgements:

The work was partly supported by the National Institute of Biomedical Imaging and Bioengineering of the National Institutes of Health (award number R01EB020640), and the Assistant Secretary of Defense for Health Affairs, through the Peer Reviewed Medical Research Program under Award No. W81XWH-16-1-0132.

Title of Abstract: Release of Naloxone for Long-term Management of Opioid Addiction

Authors: **Ruhi Naik**[1], **Kayla Mitchell**[1], **Zain Siddiqui**[1], **Vivek Kumar**[1,2,3], **PhD**
[1]Department of Biomedical Engineering, [2]Department of Chemical and Materials Engineering, New Jersey Institute of Technology, 07103, [3]Department of Restorative Dentistry, Rutgers School of Dental Medicine, 07103

Introduction: Treatment for opioid addiction is both lacking and in high demand. According to the CDC, since 2000, the rate of drug overdoses has increased 137% with a 200% increase in overdose deaths involving opioids such as heroin and pain relievers [2]. Current treatment methods for opioid overdose are generally delivered intramuscularly through an injection, intranasally or sublingually. All of these treatments contain naloxone, a specific kind of drug that competitively binds to the same mu-opioid receptors in the brain as heroin does, reversing the effects of heroin overdose [1]. Overall, naloxone and suboxone are current effective treatments for overdose and, with consistent administration, can slowly wean addicts off of heroin, a common opioid. However, drug addicts often do not cease opioid use or do not stick to their dosing schedules; many patients who leave the clinic or rehabilitation center still take opioids. In order to truly combat opioid addiction, improve treatment consistency and help prevent overdose for patients receiving rehabilitation treatment for their addiction, an extended release form of naloxone should be developed. This project outlines a delivery platform that will slowly release naloxone over a period of three to four weeks in the form of an implantable and easily injectable hydrogel.

Methods: Multidomain peptide (MDP) hydrogels were formulated to release the drug naloxone over an extended period of time. 2 different MDP were investigated based on overall charge of peptide. We ran our typical barrage of chemical and mechanical characterizations, described here [3]. To create drug the hydrogels for release assays, 10 mg of E2 and K2 were dissolved in 1 mL of 10 mg/mL naloxone, 1 mg/mL naloxone, and 0.1 mg/mL naloxone to create a total of six different hydrogels. 1.5 mL microcentrifuge tubes, one set for the three conditions of both E2 and K2. For each condition, 200 μL of the hydrogel was added to the bottom of the microcentrifuge tube, followed by adding 1 mL of 1X PBS to the microcentrifuge tube. Before aliquoting and replacing the release media, 1X PBS, each tube was centrifuged at 200 g for 2 minutes. 1 mL of the release media was aliquoted without disturbing the hydrogel at hours 1, 2, 4, 8, 12, and 24 and stored at 4°C. The results were read using the NanoQuant on a Tecan M200 Infinite Pro microplate reader at 230 nm. The 1 mL aliquots were replaced with 1X PBS. 1 mL aliquots of release media were performed at 36 hours, 2, 3, 4, 5, 6, 7, 9, 11, 13, 15, 18, 21, 24, 27, and 30-day time points. The removed aliquots were stored in 1.5 mL microcentrifuge tubes and kept at 4°C. The concentration of naloxone released was determined by using the linear trend line equation from the naloxone standard curve below (Figure 1), where y is the absorbance and x is the concentration (mg/mL). The amount of drug released is calculated as follows: Amount of drug released (mg) = [(Concentration (mg/ml)) * (Dilution factor) * (Volume of dissolution medium = 0.1 mL)] / 1000. The percentage of drug released is calculated by dividing the amount of drug released (mg) by the dosage (mg), and multiplying by 100.

Results:

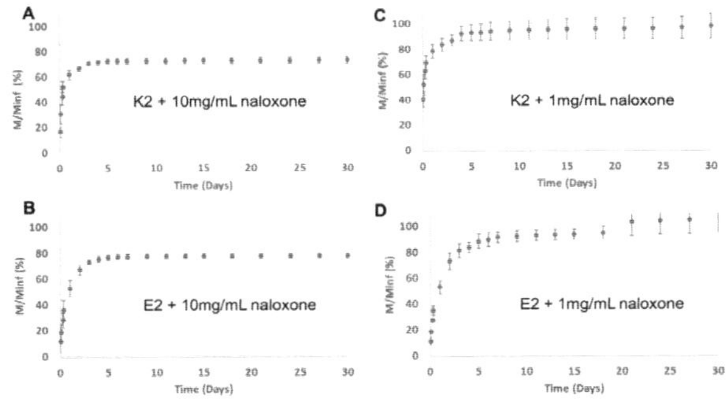

Figure 1: Release of naloxone at 2 different concentrations (A&B – 10mg/mL) and (C&D – 1mg/mL).

Positively charged K2 scaffolds showed faster release of positively charged naloxone, while negatively charged E2 scaffolds had attenuated release. MDP did not show significant differences in ultrastructure and mechanical properties. However, the majority of naloxone release occurred in the first 48 hours regardless of peptide type. However, naloxone continued to release through the MDP hydrogel for several more days. Most of the drug should be released within the first 24 hours and then continue to steadily release naloxone through the gel. We expect to see naloxone steadily release over the rest of the 21 days, through both hydrogels. The design of the MDP hydrogel will be able to address the issue of opioid overdoses in patients prone to using opioids more effectively than the currently available treatments. We expect to distinguish which hydrogel formulation (K2, E2, etc.) is optimal for release of naloxone for over a 30 day period. After gathering results from this trial, we will be able to establish the sufficient amount of naloxone to be loaded into the optimal hydrogel formulation to be effectively released over the three to four week time period. If the prolonged release of naloxone through our MDP hydrogels is observed, then this will provide evidence for the future development and analysis of long term naloxone release in an opioid rodent model.

Conclusion. The long release form of naloxone will be able to stay in the body for three to four weeks, buffering against potential opioid overdose. Additionally, this treatment may improve addiction management: longer drug activity reduces cravings and dose frequency.

References:

1. Gianluigi T., Francesco E. P., Gaetano D. C. Science. 1997; 276(5321), (2048-2050).

2. Rudd R. A. et al. 2016; Am J of Transpl. 16(4), 1323-1327.
3. Wickremasinghe, N. C., Kumar, V. A., & Hartgerink, J. D.2014;Biomacromolecules, 15(10), 3587-3595.

Significant Associations of Urinary Essential Elements and Autism Spectrum Disorder

Fatir Qureshi[1,3], James Adams[2], Devon Coleman[2] and Juergen Hahn[1,3,4]
[1]Department of Biomedical Engineering, Rensselaer Polytechnic Institute, Troy, NY, United States
[2]School of Engineering of Matter, Transport and Energy, Arizona State University, Tempe, AZ, United States
[3]Center for Biotechnology and Interdisciplinary Studies, Rensselaer Polytechnic Institute, Troy, NY, United States
[4]Department of Chemical and Biological Engineering, Rensselaer Polytechnic Institute, Troy, NY, United States

Introduction: Autism spectrum disorder (ASD) is a neurodevelopmental condition which results in impairments in social behavior and communication. While the underlying cause of ASD remains unknown, the current understanding points towards ASD being a result of complex interactions between environmental and genetic factors. Growing attention has been given to ASD as its national prevalence has been increasing steadily, and it is estimated to now affect approximately one out every 59 children [1].

While environmental factors can be too numerous to look at in their entirety, one approach that is indicative of some environmental factors uses composition analysis of urine collected from individuals diagnosed with ASD. Although there has been a substantial body of research that has been devoted to the analysis of heavy metals in urine content among individuals with ASD, only a few studies have been performed that examine the essential element content. Thus, the purpose of this investigation was to comprehensively assess the association between urinary element compositions and the presence of ASD in children and, separately, of mothers of children with ASD.

Methods: 47 children and 59 mothers were recruited for this study. Children were required to be between ages 2-5 and were characterized as either typically developing (TD) or having ASD. Mothers included in this study were required to have children that met these criteria. The group assignment for children was based on autism diagnosis verified by the ADI-R (Autism Diagnostic Interview-Revised). For the children assessed, 21 had been diagnosed with ASD (children ASD cohort) and 26 developing typically (children TD cohort). Usage of supplements containing folic acid and/or vitamin B12 was grounds for exclusion for mothers in this study. Additionally, mothers were expected to not be pregnant. Among the mothers included, 30 were assigned to the mothers ASD group and 29 to the mothers TD group based upon their child's ADI-R diagnosis status. Urine samples (mostly first-morning) were collected from 106 participants and the samples were analyzed for 18 essential and 20 toxic elements by Doctor's Data. Urine samples were normalized by creatinine to account for differences in dilution, and creatinine was measured using the Jaffe method. Statistical analysis was performed using univariate analysis for each element. While no correction was performed for multiple hypothesis testing of the univariate results, Fisher discriminant analysis (FDA), a multivariate technique, was utilized to characterize samples as belonging to the ASD or TD group based solely on their urinary element composition. "Leave-one-out" cross-validation was used for FDA to ensure statistical independence of results.

Results: Five essential elements were significantly lower in the ASD child group compared to the TD child group from the univariate analysis. Specifically, zinc (-23.4%, p-value=0.0438), selenium (-26.4%, p-value=0.0061), molybdenum (-33.5%, p-value=0.0053), phosphorous (-33.5%, p-value=0.0019), and sulfur (-35.8%, p-value=6.02E-05) levels were found to be lower in the child autism group. In contrast, chromium levels were found to be significantly greater in concentration within the autism child cohort (+20.8%, p-value=0.04).). Additionally, different concentrations of the toxic metals mercury (+553%, p-value=0.03), tungsten (-35.7%, p-value=0.0476) and tin (-48.05%,p-value=0.0030) were observed in the autism group of children Among the mothers, there were no significant differences between groups, except for a slightly higher copper concentration in the ASD mothers (+9%, p-value=0.0347). FDA analysis was carried out on both the mother and child cohorts, with notable results for the child cohort but not for the mother cohort.

	Actual ASD	Actual TD	
Predicted ASD	TP = 18	FP = 8	PPV = 0.857
Predicted TD	FN = 3	TN = 18	NPV = 0.692
	TPR = 0.308	TNR = 0.143	

Table 1: Confusion matrix of FDA analysis performed on the children cohorts using significant essential and toxic elements. True positive rate=86%, true negative rate=69%,

Conclusions: Although there have been other studies which have sought to analyze the concertation of certain essential elements in urine between children with ASD and those that are TD, this study represents one of the most comprehensive analyses performed in terms of the breadth of elements examined. It was shown that there are statistically significant differences between the ASD and TD children cohort's concentrations for nine specific elements. Notably, the concentrations of five essential elements in the urine of the children with ASD were lower than for their typically developing peers. This is particularly noteworthy as 46% of the children in the ASD cohort and only 24% of the TD cohorts were taking multivitamins/minerals. The univariate analysis performed on the mother cohort revealed little significant difference between those that had children with ASD and those that did not. Lastly, the use of multivariate techniques such as Fisher discriminate analysis further illustrated the differences between the element concentrations in the urine of the children in the ASD and TD groups.

References:

[1] J. Baio, et al, MMWR Morb Mortal Wkly Rep, 2018 Nov 16, 67(45): 1279. Available: 10.15585/mmwr.mm6745a7

Developing an Inexpensive Powered Myoelectric Prosthetic Arm for Persons with Amputation

Ryan Rattazzi[1], Emad Haque[1], Ricardo Whitaker[1], Nicole Baldassini[1], Sergei Adamovich, Ph.D.[1], Ghaith J. Androwis, Ph.D.[1,2*]
[1]New Jersey Institute of Technology – Biomedical Engineering Department, Newark, NJ, USA
[2]Kessler Foundation, The Center for Mobility and Rehabilitation Research, West Orange, NJ, USA

Introduction: 2.1 million people in the USA live with a lost limb [1]. With 185,000 amputations occurring annually [2], the total number of people with amputation in the US is projected to double by year 2050. Amputation can be the result of several clinical complications or diseases, with 54% due to vascular disease (e.g. diabetes and peripheral arterial disease), trauma (45%) and cancer (less than 2%) [1]. Upper limb amputation account for 35% and of those amputation, 70% are distal to the elbow[1]. There are several issues associated with the majority of commercially available prosthetics devices. Multiple challenges are faced by amputees starting in the recovery, rehabilitation, adaptation and management of daily life activities (ADL). Among these challenges is the costs, and perhaps is considered as the greatest cause of prosthesis and rehabilitation-related outcomes discrepancies for persons with amputation [3, 4]. Unmet rehabilitation services has been reported by 20% of nonmilitary amputees, and mostly due to their inability to cover associated costs [4]. The average price of a prosthetic arm is around $30,000, making it difficult for many amputees to afford a prosthetic. In addition, most prosthetics devices boast a mechanical look, without consideration to the natural appearance or sensory feedback of the amputated body limb (e.g. feel, color, and sensation feedback to the user). Therefore, the aim of this development project is to fabricate a prosthetic arm (i.e. below elbow amputees) that will remedy these issues, while considering to produce a similar look and feel to human skin, thereby, improves morale of the user, at a total estimated cost of less than 10% of the market average. This, indeed, will enable persons with amputation from any background or financial status to afford such needed prosthetic device.

Features of the developed prosthetic device: With cost being one of the greatest obstacle for many persons with amputees [4], we plan to develop a more affordable alternative to commercially available prosthetic arms. The developed device will be open-source, electromyography (EMG)-controlled forearm muscles in two degrees of freedom (DOF) at the wrist, includes a powered grasp motion, and has a cover that mimics the look and feel of human skin with impeded capacitive sensors.

Methods: Our prosthetic has been designed in Creo Parametric 5.0, and 3D printed entirely in Onyx (Nylon & chopped Carbon Fiber) filament. Design includes three joints in the fingers, two in the thumb, and spheroid joint at the wrist to enable two 2DOF. The skeleton of the prosthetic is comprised of 20 total pieces printed individually. We have utilized Arduino Nano to acquire the needed EMG data and control 6 Actuonix L16R linear servo motors. Servo motors will be housed in the forearm, and connected to the wrist and hand through the embedded holes of the prosthetic device via cables. EMG data is recorded by Myo Armband positioned on the forearm, the linear servo actuators and cables will adjust the angular position of the prosthetic's joint, responding to the intended motion direction generated by the user. **Real-time data analysis:** Six features will be extracted from the raw EMG signal to distinguish different movements and activate the appropriate actuators. IMU data (also recorded by Myo Armband) will further be utilized to create a more accurate profile of each movement. The EMG parameters are mean amplitude, slope changes, zero-crosses, variance, frequency, and summation of amplitude differences.

Fig 1 Myo Gesture Control

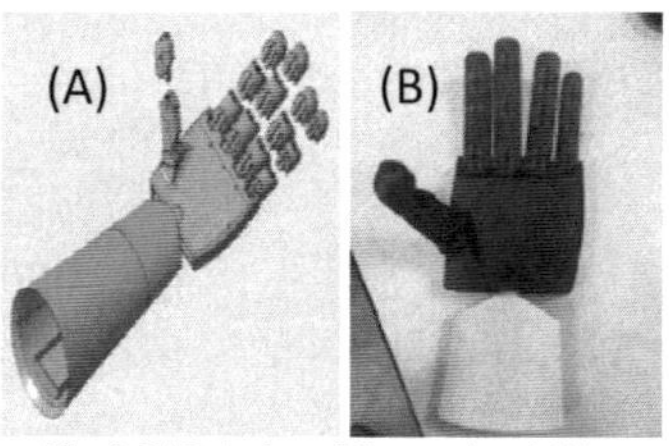

Fig. 2 CAD design of the arm-hand (A), and a 3D printed prototype of the design

Appearance and look of the developed prosthesis: The skin-like cover will be fabricated from PlatSil-0020 Gel Silicone. This is attained by dip casting the PlatSil-0020 Gel Silicone on a 3D printed hand model.

Results: The current state of our project is a working model of the hand printed in onyx, along with an initial prototype of the forearm attachment printed in ABS. Signal characterization for different poses of the 3 degrees of motion is complete, and electronics and coding in order to properly activate the actuators is underway. In addition, an initial casting of the outer skin layer has been completed.

Conclusions: Overall, our product has the potential to reach high levels of success, as there are no similar products with the same powered DOF at the price range we are targeting (less thank $2,000). Additionally, our product has the potential to not only give customers their missing motion and the ability to accomplish ADL, but also improve confidence and emotional levels.

References

1. Ziegler-Graham, K., et al., *Estimating the prevalence of limb loss in the United States: 2005 to 2050*. Archives of physical medicine and rehabilitation, 2008. **89**(3): p. 422-429.
2. Owings, M.F. and L.J. Kozak, *Ambulatory and inpatient procedures in the United States, 1996*. Vital and health statistics. Series 13, Data from the National Health Survey, 1998(139): p. 1-119.
3. Pasquina, C.P.F., A.J. Carvalho, and T.P. Sheehan, *Ethics in rehabilitation: access to prosthetics and quality care following amputation*. AMA journal of ethics, 2015. **17**(6): p. 535.
4. Resnik, L., et al., *Advanced upper limb prosthetic devices: implications for upper limb prosthetic rehabilitation*. Archives of physical medicine and rehabilitation, 2012. **93**(4): p. 710-717.

Instrumented Platform for the Quantitative Assessment of Human Balance Control

Karen Ayoub, Karina Dsouza, Swathi Pavuluri, and Carlotta Mummolo
New Jersey Institute of Technology (NJIT), Newark, NJ

Introduction: A lack of balance control puts individuals at risk of falling, especially those with osteoporosis, multiple sclerosis, and Parkinson's disease, as well as victims of stroke and traumatic brain injury. In 2014, Medicare costs associated with falling reached $31 billion.[1] In addition, balance disorders significantly reduce individuals' quality of life. To alleviate this, balance training and rehabilitation therapy are critical. Research on reliable clinical exercises is still far from reaching a consensus on a "golden standard" practice[2], due to the insufficiency of existing outcome measures for quantifying a person's state of balance. This work introduces an instrumented platform and a new computational element for quantifying balance: the Balance Stability Boundary (BSB)[3]. If a person's center of mass (COM), for a given stance, is inside of the BSB, then the state is defined as balanced; otherwise, the state is unbalanced, and the person must change his/her foot placement in order to restore a static posture. The BSB criterion is a novel direct measure of static and dynamic balance performance, allowing for selection of more meaningful, subject-specific target states in rehabilitation exercises. In addition, the use of visual display of BSB and trajectories during the balance tests makes the proposed instrumented platform an interactive human-computer interface. With these advantages, the proposed instrumented platform based on the BSB concept will contribute to filling the gap of a much needed "golden standard."

Methods: This project has three essential objectives. Objective 1 is to numerically construct the BSB for a sagittal plane model of a human subject in double stance[3]. Objective 2 is to prototype a portable instrumented platform to capture the subject's COM motion via Kinect camera and to detect changes in the subject's foot placement though socks instrumented with contact sensors. Objective 3 is to use the BSB as a reference "map" to assess the subject's state of balance at all times and create novel balance rehabilitation exercises. During the exercise, the subject will perform general balancing tasks, such as reaching a target state from a rest state, and vice versa. A computer interface collects, synchronizes, and processes sensors data in real-time. The BSB concept and the instrumented platform are used to design and conduct experimental balance tests with *rest* and *target* COM states. During the tests, the subject will visualize on a screen the start and target states, the real-time COM state captured by the motion system and the BSB map.

Results: The BSB results for one test subject (healthy human) are generated in figure 2. The area within the BSB is the region of "balanced states", while outside of BSB is the region of "unbalanced states." For each COM position in the anterior-posterior direction, the BSB quantifies the maximum allowable COM velocity perturbation (along the same direction), such that the subject can maintain the double stance posture while reaching a final static equilibrium. Based on this map, the start (triangle) and target (star) states are for a balance test are designed. The platform captures no contact changes when the subject's COM is reaching target 1, showing the capability to return to the rest state without compromising his/her balance (i.e., without stepping). Since target 2 is outside of the BSB, the platform captures an inevitable contact change during the second balance test, in which the subject must move his/her feet to ensure not to fall.

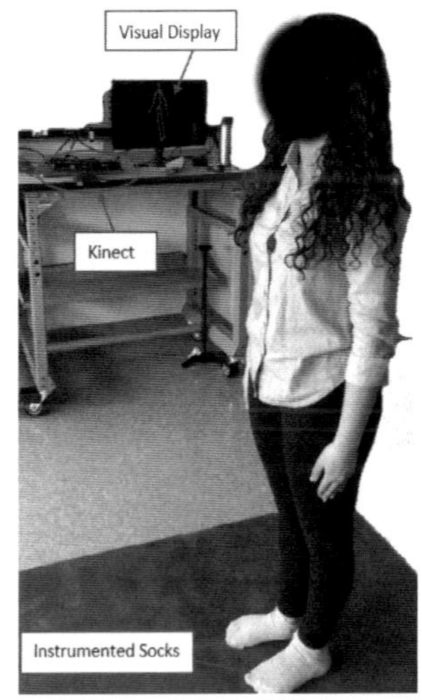

Figure 1: Experimental Platform Setup

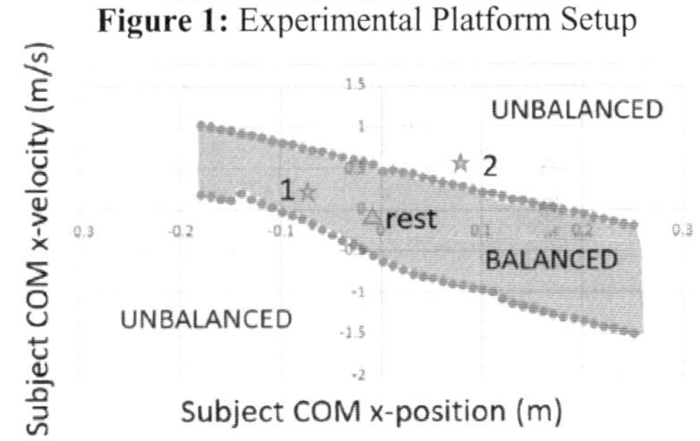

Figure 2: Anterior-posterior BSB in double stance shown as the shaded region. Inside this, the subject is balanced. Outside this, the subject is unbalanced.

Conclusions: In future work, a novel index will be formulated (based on the location of the subject's state relative to their BSB) in order to quantify the subject's margin of balance and severity of unbalance.

References:

1. Sozen T. et al. *Eur J Rheumatol.* 2017; 4(1): 46–56.
2. Sibley KM. *ArchPhysMedRehabil.*2015;96(1): 122-132.
3. Mummolo C. *Front Neurorobot.* 2018 Oct 15;12:62.

Acknowledgements: The authors acknowledge the Undergraduate Research and Innovation (URI) Program at NJIT for supporting this research through the 2018 URI Student Seed Grant - Phase I.

Biomechanical Threshold of in vivo Neonatal Brachial Plexus after Stretch Injury

Rachel Magee (1), Virginia Orozco (2), Sriam Balasubramanian (2), Maria Delivoria-Papadopoulos (3), Anita Singh (1).

(1) Department of Biomedical Engineering, Widener University, Chester, PA, USA, (2) School of Biomedical Engineering, Sciences and Health Systems, Drexel University, Philadelphia, PA, USA, (3) Department of Pediatrics, College of Medicine, Drexel University, Philadelphia, PA, USA.

Introduction: Neonatal brachial plexus palsy (NBBP) is a stretch injury to the brachial plexus (BP) during the birth process, resulting in varying degree of paralysis. No data exists on the biomechanical responses of the complex BP when subjected to stretch. The study reports mechanical responses of the *in vivo* neonatal BP when subjected to stretch.

Methods: Four neonatal piglets (3-4 days old) were anesthetized and the brachial plexus was exposed using an axillary approach. The institutional animal care and use committee preapproved all techniques. Once exposed, various segments of the BP including the nerve, chord, and divisions were identified and labeled. BP nerves were then cut at distal end and anchored to the testing setup shown in Fig.1.

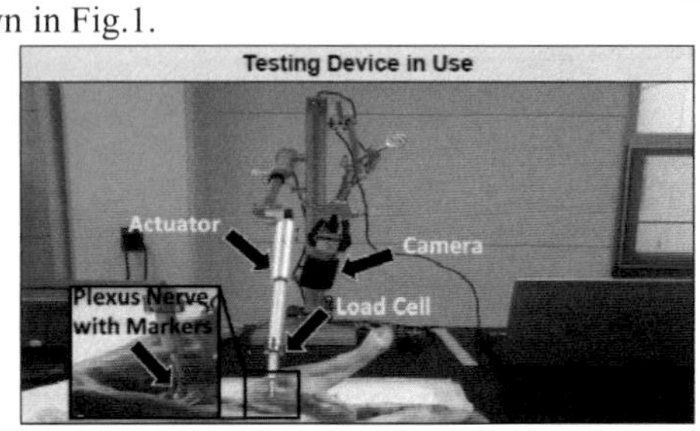

Figure 1: Mechanical Testing Device.

To measure the real-time strain on the various BP segments during stretch, black acrylic paint markers were placed on the entire length of the BP shown in Fig. 2.

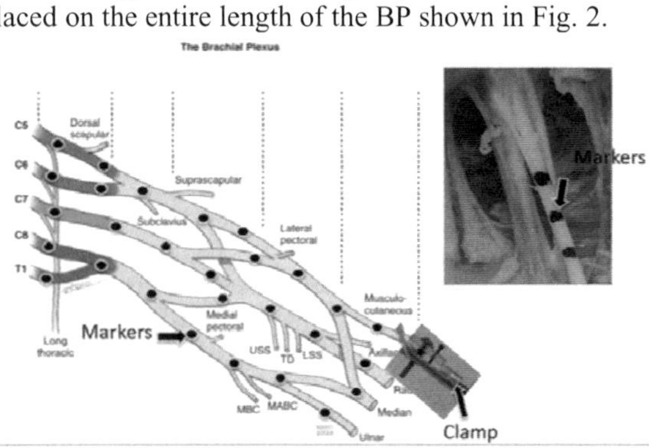

Figure 2: BP with Markers.

A high-speed video camera (Basler acA640-120uc camera, PA), placed directly over the tested segments was used to monitor the motion of the markers and determine the actual tissue strain at any time point. The BP segments were stretched at the assigned rate of 500 mm/min until complete failure occurred in any segment of the BP. The test was performed for 20 nerves. The load and displacement data was acquired at a sampling rate of 1000 Hz. Kinovea was used to record the displacement between marker points as recorded from the camera. Fig. 3 shows how the displacement was converted from an XY plane view to a 3D view. Maximum strain was calculated from this displacement.

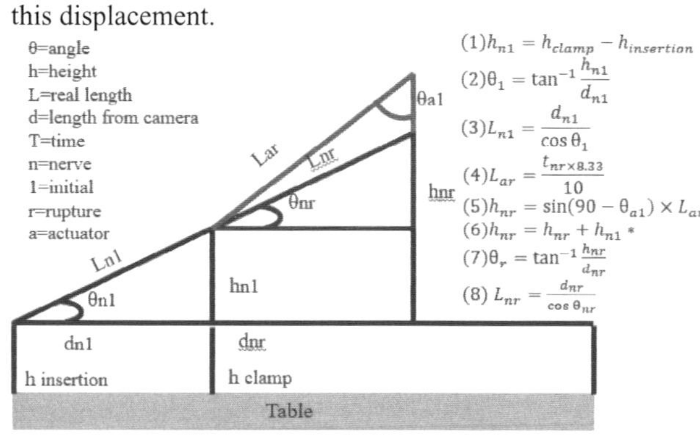

Figure 3: Data Conversion.

Results: Failure occurred all along the length of the tissue including near the clamp and near the foramen (Fig 4). Calculated Stress and strain values are shown in Table 1. The maximum strain value was found to be 29.0 ± 8.0% and the maximum load was found to be 14.64 ± 4.81N.

Table 1: Mechanical Data of BP Segments

	Brachial Plexus Complex
Rate	500 mm/min
N	14
Failure Load (N)	14.64 ± 4.81
Strain	0.29 ± 0.08

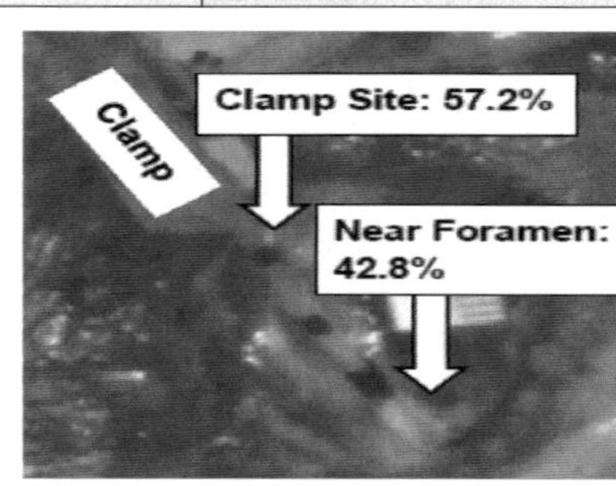

Figure 4: Area of Rupture

Conclusions: This study is the first to report *in vivo* biomechanical responses of neonatal brachial plexus and the knowledge of biomechanical properties obtained from this study will help characterize the mechanism of injury that could allow clinicians to develop strategies that minimize the occurrence of NBPP and guide treatment strategies.

Acknowledgements: This study was supported by the Eunice Kennedy Shriver National Institute of Child Health and Human Development of the National Institutes of Health (R15HD093024).

Protecting the Integrity of Surfactant-Stabilized, Oxygen Filled Microbubbles

Brian E. Oeffinger[†], Purva Vaidya[†], Iman Ayaz[†], John R. Eisenbrey[^], and Margaret A. Wheatley[†]
[†] *School of Biomedical Engineering Science and Health Systems, Drexel University, Philadelphia, PA, 19104.*
[^] *Department of Radiology, Thomas Jefferson University, Philadelphia, PA 19107*

Introduction: Ultrasound contrast agents consist of a gas filled, stabilized microbubbles, usually consisting of phospholipid, polymer or surfactant shells. We are developing a surfactant-stabilized microbubble platform with a shell composed of sorbitan monostearate (Span 60) and water soluble vitamin E (α -tocopheryl polyethylene glycol succinate, abbreviated as TPGS), which we have named SE61. This is a second-generation agent in which we have employed TPGS to replace the less biocompatible Tween 80 in the original agent, ST68. Both are formed via sonication in the presence of a perfluorocarbon (PFC) gas. For clinical use it is essential that the platform is stable under storage at room temperature, and we have developed the ability to freeze-dry these microbubbles using glucose as a lyoprotectant. This also allows for the bubble to refilled with a gas of choice, and our group has successfully investigated these microbubbles ($SE61_{O2}$) for oxygen delivery to hypoxic tumors.[1,2] Upon moving from the original ST68 platform to $SE61_{O2}$ we observed a loss of effectiveness during room temperature storage of freeze-dried samples. We therefore investigated the shelf-life of $SE61_{O2}$ and the effect of increasing the concentration of glucose lyoprotectant.

Methods: Microbubbles were fabricated based on a previously reported method, then freeze-dried.[1] Briefly, surfactant mixtures of TPGS and Span 60 with sodium chloride in PBS were continuously sonicated while purged with a steady stream of PFC gas. Microbubbles were washed and separated, then diluted 1:1 by volume with one of three lyoprotectant solutions resulting in the following concentrations (w/v): 1.8% glucose-PBS solution, 1.8% glucose-water solution, and a 5.0% glucose-water solution. Four mL aliquots of microbubbles in lyophilization vials were frozen by being placed into a -20°C bath, then lyophilized for 18-20 h. At the end of the cycle, the vials were sealed under vacuum. Prior to acoustical testing, lyophilized samples were filled with oxygen, introduced via a needle through the stopper, and then reconstituted with 4mL of solution

Acoustical testing was conducted using a custom-built acoustic setup, which closely mimics *in vivo* conditions,[3] consisting of a pulsed A-mode ultrasound system with a 5 MHz transducer focused through an acoustic window of sample vessel submerged in a water bath (37°C). Cumulative dose response curves, signal returned to the transduced as a function of microbubble dose, were constructed by pipetting increments of $SE61_{O2}$ into the sample chamber. To examine the stability of $SE61_{O2}$ while being exposed to an ultrasound beam, a single dose was insonated over a 10 min period with readings taken every minute with data normalized to allow for comparison.

Bubble counting was performed using a flow cytometer, LSRII (BD Biosciences, San Jose, CA), and the sizes of microbubbles were determined by dynamic light scattering measured using a Zetasizer Nano ZS (Malvern Inst., Worcestershire, UK). Differential scanning calorimetry (DSC) scans of dried product were conducted using a T.A. Instruments Q2000 (New Castle, DE) differential scanning calorimeter. Dried SE61 samples were cooled to -20°C then heated to 80°C at a rate of 10°C/min using approximately a 3 mg sample.

Results: $SE61_{O2}$ created with 1.8% glucose-PBS and 1.8% glucose-water had significant reductions in dose and time response curves after 7 days storage at room temperature. These reductions were accompanied by reductions in size and total microbubble concentrations. Increasing the concentration to 5.0% glucose-water resulted in stabilization of both the dose and the time response curves after 7 days. Changes in bubble size and total microbubble concentration were also not significant. These results are summarized in Table 1. This indicates that the 5.0% glucose-water lyoprotectant is better at stabilizing microbubble acoustical properties by preventing the destruction of the microbubble population.

Table 1: Effect glucose concentrations on stability

		Accoustical Testing		Size	Flow Cytometery
		Dose Response (dB) at 580 µl/mL	Time Response Half-life	Z Average (µm)	Total Microbubbles/mL
$SE61_{O2}$ - 1.8% glucose-PBS	Day 0	21.4 ± 0.5	2 min	2.55 ± 0.53	17.8 ± 1.2 E+07
	Day 7	14.1 ± 1.6	1 min	1.79 ± 0.23	9.6 ± 0.4 E+07
$SE61_{O2}$ - 1.8% glucose-water	Day 0	17.6 ± 1.3	2 min	2.17 ± 0.40	22.8 ± 1.9 E+07
	Day 7	6.8 ± 1.8	0.5 min	1.39 ± 0.26	15.2 ± 0.9 E+07
$SE61_{O2}$ - 5.0% glucose-water	Day 0	17.4 ± 1.3	2 min	1.47 ± 0.22	15.3 ± 2.4 E+07
	Day 7	16.1 ± 0.6	2 min	1.42 ± 0.16	11.8 ± 0.7 E+07

Increasing the glucose concentration from 1.8% to 5.0% was also shown by DSC to lower the melt temperature (Tm) associated with TPGS in freeze-dried microbubbles from between 24.4 and 31.3°C, above room temperature, to between 11.6 and 19.5°C, below room temperature. This reduction in Tm prevents the TPGS in the microbubble from transitioning from a liquid crystalline phase to a more solid gel phase, which we believe leads to the better shelf-life stability.

Conclusions: We have shown that the shelf stability of $SE61_{O2}$ microbubbles can be enhanced by increasing the glucose concentration to 5.0%. This is due to the ability of glucose to lower the Tm of TPGS, keeping it in a liquid crystal phase at room temperature. In the future, samples of $SE61_{O2}$ should be frozen in a 5.0% glucose-water solution to maximize bubble stability during drying and storage.

References:

1. Eisenbrey, JR. et al. Int J Pharm. 2015; 478: 361-367.
2. Eisenbrey, JR. et al. Int J Radiat Oncol Biol Phys. 2018; 101: 88-96.
3. Wheatley, MA. et al. Ultrasonics 2006; 44: 360-367.

Acknowledgements: Drexel University STAR program, NIH R01 EB026881 and R25 EB013038, Dr. Christopher Li and Mark Staub of the Department of Materials Science and Engineering, Drexel University (DSC), Drs. Amir Yarmahmoodi and Lei Yu from Thomas Jefferson University's Flow Cytometry Core Facility.

Multimodal Foot-Ground Contact Interaction in Human Postural Stability

Giulia Vicentini[1] and Carlotta Mummolo[2]
[1]New York University, Brooklyn, NY, USA; [2]New Jersey Institute of Technology, Newark, NJ, USA

Introduction: During various balancing and locomotion tasks, the foot structure undergoes a complex multimodal contact interaction with the ground, allowing the human body to move with remarkable agility and stability [1]. Implementing an accurate model of the human foot is essential to generate realistic whole-body motion in computational biomechanics applications. In postural stability analysis, the balance recovery action from a perturbed posture would benefit from intermittent heel and toe lifting, as compared to the case of fixed flat-foot contact. Balancing trajectories that bring the system's center of mass (COM) from a perturbed state (position and velocity) to a rest state (upright static posture) should be generated to investigate the maximum allowable COM velocity perturbation along a given direction [2]. In this study, a unified approach is proposed to generate the balancing trajectories of a human posture model that recovers from extreme COM perturbed states in the antero-posterior direction, while exploiting the multimodal nature of the foot-ground contact interaction.

Methods: A six-link planar mechanism is used to represent the human posture in sagittal plane (Fig. 1). The left and right lower-body joints are modeled in parallel, through one equivalent joint for hip, knee, ankle, and toe motion, connected by the respective links. The upper body is modeled by the HAT body segment (head-torso-arms), connected to the hip by a spine joint and a pelvis segment. A segmented foot model is proposed, in order for the system to establish various contact modes with the ground (Fig. 1).

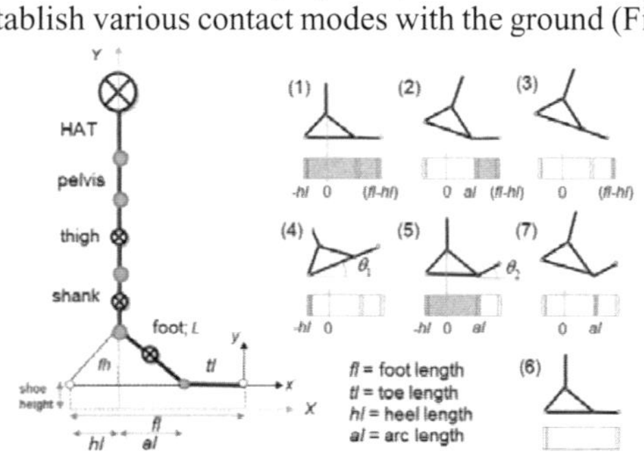

Figure 1. (left) Human posture model in sagittal plane (right) Contact modes and associated COP feasible region, relative to global frame $\{X,Y\}$

A contact mode detection algorithm maps at each time instant the values of foot and toe angles into to corresponding contact modes and associated lower and upper bounds for the center of pressure (COP) global position (Fig. 1). The discrete contact modes are approximated by two surface equations that provide continuous values for the COP lower and upper bounds, for any given pair of foot and toe angle values. The contact interaction is modeled through one equivalent resultant force applied to the system through the toe link, at varying COP positions relative to the toe local frame $\{x',y'\}$, to simulate the actual pressure distribution from heel to toe. The system dynamics is expressed both in joint space and in the COM space. The equations for the COM dynamics are used to calculate the ground reaction forces and their point of application, i.e., the COP. A constrained optimization problem is formulated to find optimal joint trajectories and actuator torques to balance the system from an initial COM position and maximum X velocity to the final rest position. The joint and COM motion resulting from the optimization are subject to several kinematics and kinetics constraints. *Kinematics Constraints:* (*i*) each joint is subject to anatomical ranges of motion; (*ii*) contact mode 6, i.e., flying phase, is excluded among the achievable modes by constraining the toe joint position on an arc centered on the heel and vice versa; (*iii*) ground penetration is prohibited; (*iv*) a specified COM initial position and the desired COM final state. *Kinetics Constraints:* (*i*) each joint actuation is limited by human joint strength data from the literature; (*ii*) the normal and tangential ground reaction forces are applied at the COP and must satisfy unilateral and friction constraints, respectively; (*iii*) at all times the COP global position must be within the lower and upper surface values resulting from the mode detection algorithm. The cost function is the maximum velocity along the antero-posterior direction that the COM can sustain, such that a final rest posture is achieved.

Results: The benefit of a multimodal contact interaction for postural stability is demonstrated by comparing the maximum allowable COM velocity perturbation in case of segmented foot (0.788 m/s in the anterior and -0.411 m/s in the posterior direction) with that achieved in case of fixed flat-foot (i.e., contact mode 1, 0.483 m/s in the anterior and -0.284 m/s in the posterior direction). Balancing trajectories from anterior COM velocity perturbations are reported in the two cases (Fig. 2).

Conclusions: The multimodal foot-ground contact schedule naturally arises from optimization results, leading to greater balancing capabilities in the human postural model.

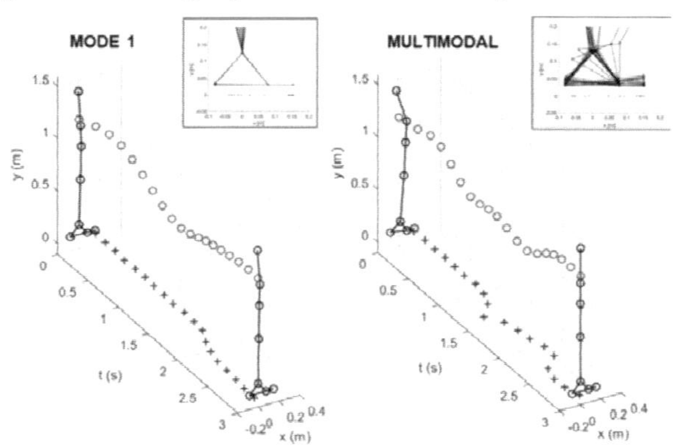

Figure 2. Balancing trajectories from initial COM position and maximum velocity perturbation in the anterior direction to final upright static equilibrium. System in contact mode 1 (left) and in multimodal contact interaction (right).

References:
[1] Ku P.X., et al., PLoS One. 2012; 7(7), p.e41539.
[2] Mummolo C., et al., Journal of Mechanisms and Robotics. 2018; 10(2):021009.

Effects of Walking Forces on Shoe Wear Rate

Sarah L. Hemler, Jessica R. Sider, Kurt E. Beschorner
Department of Bioengineering, University of Pittsburgh, Pittsburgh, USA

Introduction: Slips that result in falling are a major cause of injury in the workplace. In 2016, falls accounted for 26% of all injuries in the workplace in the U.S. and cost over $16 billion [1]. Previous research has shown that the ability for shoes to prevent slips and falls degrades over time [2] and that the amount of tread on the shoe outsole may be an important factor in determining slip risk [3]. Preliminary research suggests that shoes may wear down at different rates per individual [4]. However, the factors that contribute to these variations in shoe wear are unknown. Furthermore, one metric utilized to assess an individual's risk of slipping at the shoe-floor interface is the required coefficient of friction (RCOF) measured during walking [5]. As RCOF is specific to an individual's gait, this metric may be a good indicator for the rate of shoe wear.

Methods: In this study, twelve participants (9 males, 3 females; age = age = 44 ± 12.2 years, weight = 91 ± 13 kg) wore two types of slip-resistant shoes with different outsole tread while walking over force plates (AMTI, Watertown, MA, 02472). The peak RCOF was evaluated for each subject and shoe. Participants then wore the pairs of each type of shoe, equipped with a step tracker, for one month at a time. At baseline and after each month of wear, rubber silicone molds of the shoe heels were generated with the shoes at a sagittal plane angle of 17° (Figure 1). The heel tread molds were filled with water and the change in mass per each month of wear was calculated. The volumetric difference between the measurements at baseline and after wear was normalized to the distance the shoes were worn and termed as the wear rate [mm^3/km]. Only the first month of wear in which the subjects walked 100 km or more were included. As such, data for both shoes were included for three participants and data for one pair of shoes were included for eight participants.

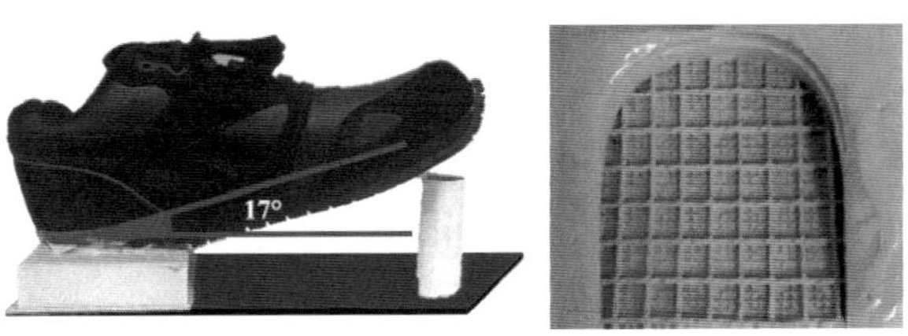

Figure 1. Silicon rubber mold set-up with shoe set at sagittal plane angle of 17° (left) and resultant heel tread mold (right).

Repeated measures ANOVA with incomplete data using maximum likelihood method was utilized to assess the effects of RCOF on wear rate. This method is suitable for repeated measures with incomplete data [6]. Using this method, the dependent variable was the wear rate, which was log-transformed to achieve normal residuals, and the independent variables were shoe type, peak RCOF, and bodyweight.

Results: Peak RCOF ranged from 0.092 to 0.225. The wear rate was significantly affected by the peak RCOF ($p=0.0211$), but not by the shoe type ($p=0.7302$) or the bodyweight ($p=0.8521$). Two subjects largely drove the trend with both low RCOF and wear rate.

Conclusions: The results from this research provide preliminary evidence that an individual's peak RCOF may be a good indicator for determining the rate at which shoes will wear down. Furthermore, this research did not support a strong influence of bodyweight on shoe wear rate. As such, the distance that was walked and walking behaviors may be more important than anthropometric measures and shoe type when considering shoe replacement schedules. Given the two types of shoe tread and small population, the results of this study should be monitored as more data becomes available.

References:

1. Liberty Mutual Research Institute for Safety (2017). Liberty Mutual Workplace Safety Index.
2. Verma SK, et al. Ergonomics, 57, 1919-1926, 2014.
3. Singh & Beschorner.
4. Hemler SL, et al. (2019). "Changes in under-shoe traction and fluid drainage for progressively worn shoe tread". App Ergo, In review.
5. Beschorner KE, et al. Gait & Pos, 48, 256-260, 2016
6. Gornbein JA, et al. Stat Methods in Med Res, 1, 275-295, 1992.

Acknowledgements: This work was funded by a grant from the National Institute for Occupational Safety and Health (NIOSH R01 OH 010940) and the National Science Foundation (NSF 16-104).

Effects of diabetic therapeutic footwear and traditional athletic footwear on gait: a cross-sectional investigation

Kelly Poretti, Dr. James Peters, Dr. Anita Singh.
Widener University.

Introduction: The number of individuals living with diabetes has increased in recent years and is expected to reach 366 million by the year 2030 [Juster-Switlyk *et al.*, 2016]. Secondary complications such as neuropathy, characterized by loss of protective sensation, put these individuals at an increased risk of developing diabetic foot ulcers (DFU), which are the leading cause of lower extremity amputations in this population [Crews *et al.*, 2016]. Therapeutic footwear is one of the many preventative measures used to reduce the risk of DFU, through the unique structure that allows for offloading of at-risk locations on the foot; however, patient compliance is low, with 58% of patients wearing their custom footwear during less than 60% of their total daily walking hours [Crews *et al.*, 2016, Bus *et al.*, 2016]. There are various reasons for lack of compliance, including appearance and discomfort [Crews *et al.*, 2016]. Although it is not known how therapeutic footwear affects joint mobility and gait, it is known that limited joint mobility increases the risk of DFU by increasing plantar pressure and shear stress [Crews *et al.*, 2016].

The purpose of this pilot study was to investigate the effects of therapeutic footwear and traditional athletic footwear on known risk factors for diabetic foot ulceration.

Methods: IRB approved all procedures. A preliminary sample of one diabetic, neuropathic individual (Male, 79) and one non-diabetic, non-neuropathic individual (Female, 77) were recruited for this study. Data collection was performed at Widener University's motion capture lab, utilizing the 16 camera Qualisys Miqus set up to measure the dynamic range of motion. A total of 18 markers were placed on the participants' lower limbs in accordance with the 2002 ISB recommendation for defining joint coordinates of the ankle [Wu et al., 2002]. Each subject completed six walking trials each while wearing therapeutic footwear and athletic footwear. The 3D marker data were processed using MATLAB (Mathworks, MA, USA) to determine the maximum plantarflexion, dorsiflexion, inversion, and eversion for each limb for each trial. Statistical comparisons were made with a two-way repeated measures ANOVA (± = 0.05) for each ankle range of motion measurements using shoe type and subject condition (diabetic and non-diabetic) as factors.

Results: Ankle range of motion values generally differed between subject conditions but not between footwear types (Table 1). Significant differences ($p < 0.05$) between subject conditions were observed for inversion and plantarflexion. No significant differences were found between subject conditions for eversion or dorsiflexion. Significant differences ($p < 0.05$) between footwear types were only observed for dorsiflexion (Table 2).

Table 1. Average dynamic ROM values.

Group	Footwear	Inversion	Eversion	Plantarflexion	Dorsiflexion
Diabetic	Athletic	16.8 (1.6)	-6.4 (1.9)	28.9 (5.4)	-1.2 (1)
	Therapeutic	17.8 (2.4)	-4.4 (4.2)	26.3 (5.3)	-15.8 (4.4)
Control	Athletic	13.9 (1.5)	-5.5 (3.4)	20.7 (3.9)	-8.7 (5.2)
	Therapeutic	13.3 (3.8)	-5.1 (6)	18.8 (2.3)	-13 (3.4)

Table 2. P-values.

	Inversion	Eversion	Plantarflexion	Dorsiflexion
Footwear	0.8445	0.4910	0.2227	0.000007
Group	0.0019	0.9428	0.0003	0.1459

Conclusions: The significant differences observed between subject conditions were expected: diabetes has been shown to cause limited joint mobility. The significant difference observed for dorsiflexion between footwear types could be attributed to differences in the structures and material properties of the footwear.
In conclusion, it appears that there are no significant differences in the risk factors for diabetic foot ulceration between the two footwear types used in this study. This indicates that further research is required to understand the relationship between gait and diabetic foot ulceration which could inform the design of improved therapeutic footwear.

Acknowledgements:
2018 New Balance Footwear Research Award for partial funding of this study.

Chester Community Physical Therapy Clinic for assistance in data collection.

References:

1. Caputo G.M. et al. New Eng J Med. 1994;13:854-860.
2. Crews, R.T. et al. Diabet/met res rev. 2016;8:791-804.
3. Juster-Switlyk K. et al. F1000 Res. 2016;5:738.
4. Wu G. et al. J Biomech. 2002;4:543-548.

Trophoblast Enrichment Using Surface Adhesion Properties

Sumaiya Sayeed[1], Christina M. Bailey-Hytholt[1], Anita Shukla, PhD[1], Anubhav Tripathi, PhD[1].
[1]Brown University, Providence, RI, USA.

Introduction: Prenatal diagnosis allows for the identification of genetic abnormalities and can be performed with the acquirement of trophoblast genetic content. While current diagnostic methods, such as chorionic villous sampling and amniocentesis, extract whole cells, they are invasive, carrying a 2% risk for fetal loss [1]. Cell-free fetal DNA circulating in maternal plasma provides a non-invasive method for obtaining fragments of fetal DNA and can screen for chromosomal abnormalities, but it requires follow-up confirmation by invasive diagnostic procedures [2]. Whole trophoblast cells can be collected from the cervix [3], and these cells, the extravillous trophoblasts (EVT), appear at a frequency of about 1 in 2000 cervical cells [4] for a total of approximately 1000 EVT in a 20-mL cervical swab sample [5]. We have designed a device and enrichment process for these fetal cells from the cervical sample of maternal cells, mucus, and extracellular proteins using their intrinsic surface adhesion properties and varied incline at which the cells are flowed.

Methods: Cervical swab samples in PreservCyt® (Hologic, Marlborough, MA) from women in their 5th to 20th week of pregnancy were obtained, and JEG-3 cells were cultured and suspended in PreservCyt® as an extravillous trophoblast model. A channel device (Fig 1a) was designed using Adobe Illustrator and made using laser-cut poly(methyl methacrylate) directing cell flow. To test their differential surface adhesions properties, the cells were flowed on a slide that consists of a channel and collection chamber. To optimize the flow process, we tested the effect of a five-minute incubation before inclined flow, the angle of incline (25°, 50°, 75°), and the material of the slide (glass, polystyrene). These parameters vary velocity of fluid flow, the static settling vs. dynamic convection mobility of the cells, and material interactions. The cells adhered on the slides were counted using a hemocytometer, 4',6-diamidino-2-phenylindole (DAPI) stain, and human leukocyte antigen G (HLA-G) stain. Statistical analysis was performed using two-way Anova with Tukey post-hoc analysis, $\alpha = 0.05$.

Results: Using polystyrene slides and a 5 min incubation prior to inclined flow, we observed a statistically significant difference in JEG-3 vs. cervical adhesion for all angles of incline tested (Fig 1b). After the incubation and inclining to 25°, we observed the greatest % adhesion of JEG-3 cells at 76 $\pm$ 3% with cervical cells adhering at 5 $\pm$ 1% (Fig 1b). After the incubation and inclining to 50°, the frequency of JEG-3 cells in the cervical sample population increased 31x. The significance of the 5 min incubation was observed when performing the inclined flow without prior incubation, which resulted in less JEG-3 adhesion to the surface. Next, we performed the inclined flow on glass slides. Overall, there was a less significant frequency increase of JEG-3 cells compared to the cervical, with the highest at 4x in the 50° flow without prior 5 min incubation. Thus, the polystyrene material is important for ideal enrichment. Incubating both the JEG-3 and cervical cells together on polystyrene with prior incubation and inclined flow at 50°, we observed a ~400% enrichment of the JEG-3 cells on the surface using HLA-G staining.

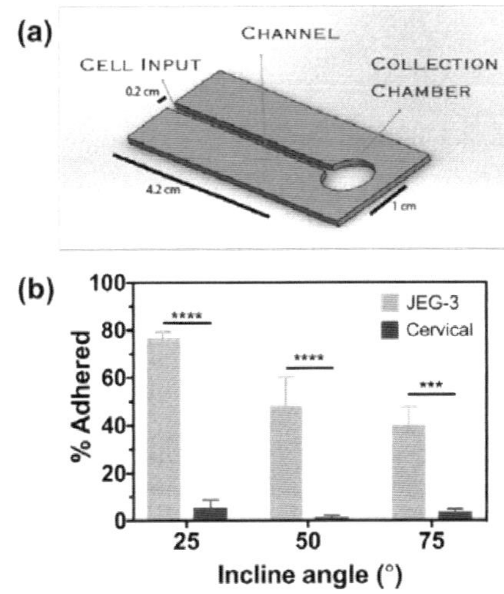

Figure 1. (a) Channel flow device, (b) Percent of JEG-3 and cervical cells adhered to polystyrene surface with 5 min prior incubation. Statistics were performed with n=3, two-way Anova with Tukey post-hoc analysis, $\alpha = 0.05$, *$p<0.05$, **$p<0.01$, ***$p<0.001$, ****$p<0.0001$.

Conclusions: Cells experiencing inclined flow for the purpose of JEG-3 cell enrichment and isolation provides a new method of collecting rare cells in a heterogeneous cell and mucus sample. Here we provide an optimized device and process for performing such cell flow studies and analyze the effects of surface material, prior incubation, and angle of incline on cell adhesion and enrichment. We find that the optimal conditions are using polystyrene surface with a prior incubation at 25° and 50 ° inclines. This approach to enriching JEG-3 cells paves the way for label-free cell enrichment techniques that can allow for cells to be separated based on their interactions within a dynamic fluid medium above varied surface materials. This is especially useful when cells, such as the extravillous trophoblast, are found in small quantity amidst other types of cells in outnumbering quantity. Using this method allows for samples to be tested without changing sample chemical characteristics (ie. by adding nanoparticles or reagents) and using only a lab-on-chip device.

References:
1. Imudia AN. Hum Reprod. 2009;9:2086-92
2. Alfirevic Z. Cochrane Database Syst Rev. 2014;3:1-105
3. Moser G. Hum Reprod Update. 2018;24:484-96
4. Imudia AN. Fertil Steril. 2010;93:1725-30
5. Bolnick JM. Fertil Steril. 2014;102:135-42

Acknowledgements: Sumaiya Sayeed acknowledges the Karen T. Romer Undergraduate Teaching and Research Award from Brown University and PerkinElmer, Inc. CBH is supported by a National Science Foundation Graduate Research Fellowship.

Wnt and Nodal Signaling Pathways Influence Chiral Bias in Human Embryonic Stem Cells

Kathryn E. Worley, PhD and Leo Q. Wan, PhD.
Rensselaer Polytechnic Institute, Troy, NY.

Introduction: Various signaling pathways have been shown to be influential in the establishment of left-right (LR) asymmetry.[1-3] Animal embryos are typically used to study effects of the asymmetric expression of pathway components on developmental outcomes, but few have directly observed the influence these pathways have on cellular organization, especially on intrinsic cellular biases (or cell chirality).[2,4] We aim to shed some light on how specific signaling pathways could influence cellular organization during development through observation of changes in the intrinsic cellular chirality of human embryonic stem cells (hESCs) using our 3D chirality assay. Based on our recent work examining the role of hESC differentiation on cell chirality[5], together with previous studies, we settled on a list of pathways including Nodal, BMP, Wnt, and FGF that were likely to influence intrinsic cellular chirality.

Methods: The hESC, RUES2 cell line, was procured through the Rensselaer Center for Stem Cell Research (RCSCR) and cultured in MEF-CM (RCSCR). Cells are clump passaged and allowed to attach overnight, fresh MEF-CM containing the drug is then applied for either 12, 24 or 72 hrs, with medium changes every 24 hrs containing fresh drug. The drugs include: activin A (PeproTech, 100 ng/mL), SB431542 (Tocris, 10 μM), LDN193189 (Sigma, 200 nm), IWP2 (Tocris, 5 μM), Wnt3a (R&D Systems, 500 ng/mL) and FGF4 (PeproTech, 500 ng/mL). Cells are then single cell passaged, seeded into the 3D Matrigel bilayer system and time lapsed for 2-4 hrs at 1-5 min intervals using a Keyence BX-X700 microscope.[5] Analysis of single cell rotation is done in ImageJ, cells are classified into five groups, clockwise (CW), counterclockwise (CCW), no rotation, complex rotation and migrating. The binocdf function in MATLAB was used to evaluate if the ratio of CW or CCW rotating cells significantly differed from 50%. Statistical significance was determined at a confidence level of 0.05. Error bars in all figures represent standard error.

Results: Analysis of hESCs in our 3D bilayer system pretreated with activators or inhibitors of key signaling pathways indicates a role for Wnt and nodal signaling in modulating cellular chiral bias. Untreated hESCs were found to exhibit no bias in rotation direction which is illustrated in Figure 1A. Pre-treatment of hESCs for 12 hrs with the nodal signaling activator, activin A, results in an increase in CCW biased cells, but this increase is lost at the 24 and 72 hr time points. Pre-treatment with the nodal inhibitor, SB431542, has the opposite effect, causing more CW biased cells for 12 and 24 hrs of pre-treatment, (Figure 1B). Activation of the canonical Wnt signaling pathway using Wnt3a showed a statistically significant increase in CCW biased cells for 12, 24 and 72 hr treatment groups (Figure 2A). Inhibition of Wnt signaling using IWP2, a porcupine inhibitor, caused an increase in CW biased cells for the 12, 24 and 72 hr treatment times (Figure 2B). BMP pathway inhibition using LDN193189 and FGF pathway activation using FGF4 showed no significant differences in chiral bias similar to untreated hESCs.

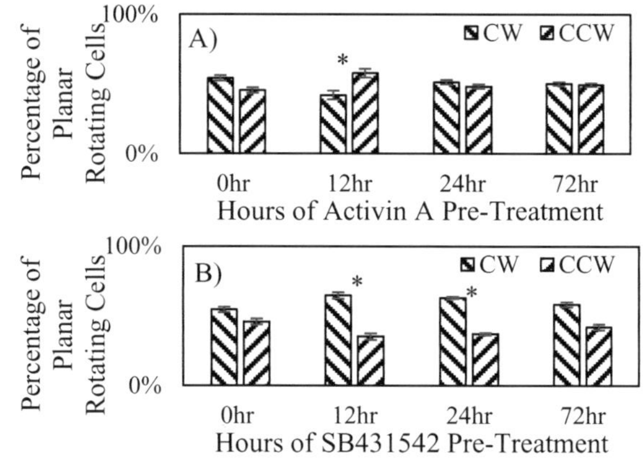

Figure 1. Activation and inhibition of the nodal signaling pathway alters biased rotation of hESCs. A) activin A, a nodal signaling activator B) SB431542, a nodal signaling inhibitor. * $p < 0.05$

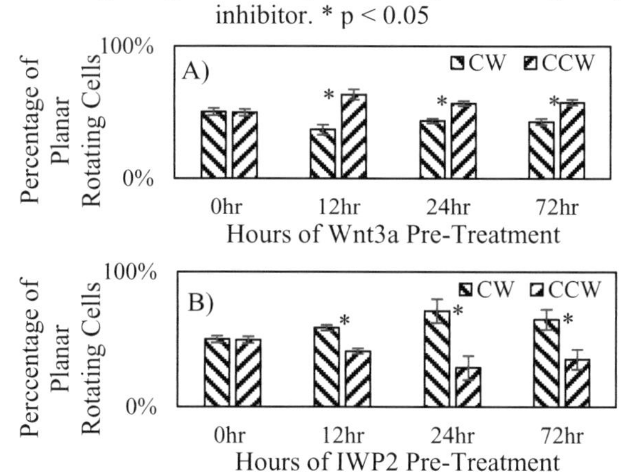

Figure 2. Activation and inhibition of canonical Wnt signaling alters biased rotation of hESCs. A) Wnt3a, a Wnt activator, B) IWP2, a Wnt inhibitor. * $p < 0.05$

Conclusions: Using our 3D bilayer chirality assay, we show that nodal and Wnt signaling play a role in altering the inherent chiral bias of hESCs. It connects cellular chirality with fundamental developmental pathways already known to regulate developmental LR asymmetry. Future works will examine the mechanisms behind this bias in cellular rotation.

References:

1. Kitajima K. Dev. Biol. 2013;380:222-232.
2. Veerkamp J. Dev. Cell. 2013;24:660-667.
3. Neugebauer J.M. Dev. Biol. 2014;386:123-134.
4. Mentink R.A. Dev. Cell. 2014;21:188-201.
5. Worley K.E. Stem Cells Int. 2018;1848605.

Acknowledgements: Funding support from National Institutes of Health (OD/NICHD DP2HD083961), National Science Foundation (CAREER CMMI-1254656), and March of Dimes (MOD 5-FY14-111). Leo Q. Wan is a Pew Scholar in Biomedical Sciences (PEW 00026185), supported by the Pew Charitable Trusts.

Constitutively Active Ras Alters Cellular Chirality and Actin Structure of Mammary Epithelial Cells

Amanda S. Chin, PhD, Kathryn E Worley, PhD, and Leo Q. Wan, PhD.
Rensselaer Polytechnic Institute, Troy, NY.

Introduction: The spatial occurrence of some diseases is left-right (LR) asymmetric in paired organs and tissues. Studies of international populations have shown that different types of cancer have confirmed LR asymmetric laterality, including tumor development in the kidneys, lungs, ovaries, testes and breast.[1] Elevated levels of Ras-proto-oncogene protein product occur in approximately 50% of breast carcinomas.[2] The Ras superfamily contains several members that play a role in increasing migration potential.[3] There is growing evidence of chirality playing a role in tissue morphogenesis[4,5], however, any effect on cancer laterality has yet to be seen. In order to expand our understanding of primitive mammary gland development and breast carcinogenesis, we investigate the impact of Ras activation on the chirality and actin structure of human mammary epithelial cells in a 3D culture platform.

Methods: Michigan Cancer Foundation 10a (MCF10a) are an immortalized, nontransformed human mammary epithelial. MCF10AT1 (AT1) is a cell line derived from MCF10a that overexpress Ha-Ras. MCF10a and AT1 cells were gifts from Dr. Jason Herschkowitz's lab at University at Albany-SUNY. The cells are seeded into the 3D Matrigel bilayer system[3], allowed to grow overnight and time lapsed for 4 hrs at 5 min. intervals using a Keyence BX-X700 microscope. Salirasib (Sigma-Aldrich) treatment was applied overnight at varying concentrations. ImageJ was used to evaluate cell rotation.[5] Cells are classified into: clockwise (CW), counterclockwise (CCW), no rotation, and complex rotation. The binocdf function in MATLAB was used to determine if the ratio of CW or CCW rotating cells significantly differed from 50%. Error bars in all figures represent standard error. Phalloidin 568 (Invitrogen) and DAPI (ThermoFisher) were used to label actin and the nucleus, respectively. Images were taken using a confocal microscope, Leica TCS SP8 STED.

Results: Human mammary epithelial bicells (MCF10a) cultured in our 3D chirality assay demonstrated a CW chiral bias, as seen in Figure 1.

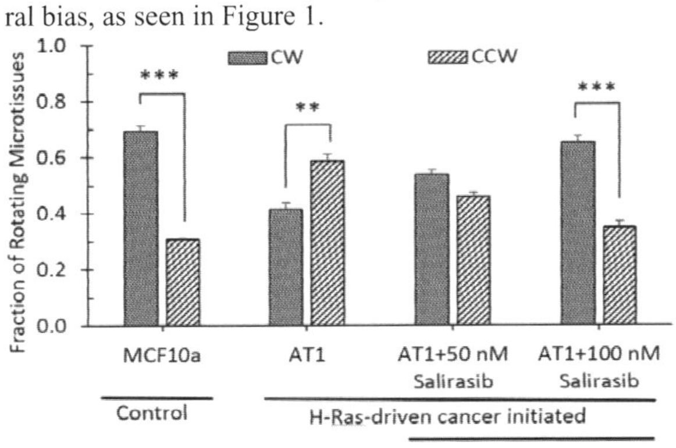

Figure 1. Constitutively active H-Ras reverses the chiral bias of MCF10a cells. H-Ras inhibition of the pre-malignant AT1 cells, by Salirasib treatments, restored the directional bias of rotation to that of MCF10a cells. $^{**}p < 0.01$; $^{***}p<0.001$.

The H-Ras overexpressing line, AT1, exhibited a bias in rotation direction (CCW) opposite of the MCF10a line (Figure 1). This difference in bias was restored back to CW through treatment of the AT1 line with the Ras inhibitor salirasib (Figure 1). The actin structure is visually different between the MCF10a and AT1 cell lines. Both lines have filopodial protrusions along the cell membrane (Figure 2). However, bicells of the AT1 line, have dense actin at the cell-cell interface and appear to have denser and longer actin filopidal structures when compared to the "normal" human mammary epithelial line.

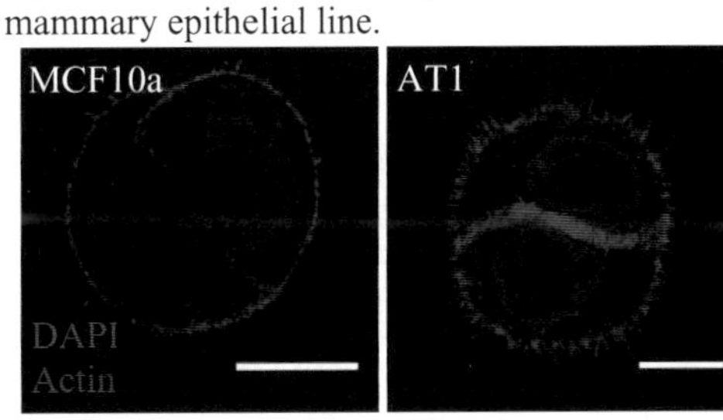

Figure 2. The actin cytoskeletal and actin-dense filopodial structures of MCF10a and AT1 bi-cells. Scale bars are 10 μm.

Conclusions: Using our 3D chirality assay, we found that the human mammary epithelial cells line (MCF10a) and its Ras-oncogene-transfected counterpart (AT1) displayed opposite rotational biases. Additionally, substantial differences in actin structure between the two cell types were observed. The "normal" chirality was rescued through treatment with the Ras inhibitor, salirasib, providing a functional link between cellular chirality and Ras, thus demonstrating that cell chirality could be a potential biomarker for breast cancer. Future works will explore specific downstream mechanisms likely related to actin structure as well as other known oncogenic pathways.

References:

1. Roychoudhur R. CCC. 2006;17:655-662.
2. Basolo F. Mol. Carcinog. 1991;4:25-35.
3. Izdebska M. Biomed. Res. Int. 2018;4578373.
4. Wan L.Q. PNAS. 2011;108:12295-12300.
5. Chin A.S. PNAS. 2018;115:12188-12193.

Acknowledgements: Funding support from National Institutes of Health (OD/NICHD DP2HD083961), National Science Foundation (CAREER CMMI-1254656), and March of Dimes (MOD 5-FY14-111). Leo Q. Wan is a Pew Scholar in Biomedical Sciences (PEW 00026185), supported by the Pew Charitable Trusts.

The Effect of Zinc on the Osteogenic Differentiation of Mesenchymal Stem Cells

Jennifer Moy, Ateka Khader, and Treena Livingston Arinzeh
Biomedical Engineering, New Jersey Institute of Technology, Newark, NJ 07102

Introduction: Zinc is an essential trace element known to impact bone formation [1] and has been shown to stimulate collagen production in osteoblast cells [2]. However, the effect of zinc on the osteogenic differentiation of mesenchymal stem cells (MSCs) has been limited. MSCs play a significant role in fracture healing and are thought of as a promising cell source for bone tissue engineering strategies. This study investigated the osteogenic differentiation of human MSCs in cultures supplemented with zinc. Changes in cell growth and osteogenic markers were examined over time.

Methods: Human MSCs were seeded at 2x10^4 cells/cm^2 in 96-well tissue culture plates and cultured for up to 14 days in either standard growth media (DMEM, 10% fetal bovine serum, 1% antibiotic-antimycotic) or osteoinductive media (standard growth media supplemented with 100 nM dexamethasone, 0.05 mM ascorbic acid and 10 mM ²-glycerophosphate). Zinc was added to each media condition at the following concentrations: 0, 10^{-6}, 10^{-5}, 5x10^{-5}, 10^{-4}, or 10^{-3} M. Samples were collected and analyzed for cell growth using the PicoGreen ® dsDNA assay (Invitrogen) and alkaline phosphatase activity based on the enzymatic conversion of para-nitrophenyl phosphate (p-NNP) to para-nitrophenol (p-NP). Calcium was quantified with the QuantiChrom Calcium Assay kit (BioAssay Systems). Collagen was quantified through hydroxyproline content with a hydroxyproline assay kit (Sigma-Aldrich). Cell morphology (F-actin phalloidin stain, Invitrogen) and collagen type I (rabbit polyclonal anti collagen I, Abcam) were evaluated by fluorescent microscopy. Additional samples were stained with the following: Alizarin Red S to evaluate matrix mineralization. Data analysis was performed using factorial ANOVA and post hoc Tukey HSD at $p<0.05$ significance level (SPSS-22).

Results: For general media, cell number was significantly higher for media containing 10^{-4} M zinc by day 14 (figure 1a). Alkaline phosphatase activity was significantly higher for 10^{-4} M zinc by day 14. Collagen production increased with zinc concentration up to 10^{-4}M, but was not statistically significant. Calcium content increased significantly with increasing zinc concentration by day 14. Increased collagen type I and calcium deposition were present in the zinc supplemented general media groups when compared to 0 M control (Figure 1c and d respectively). In osteoinductive media, cell number was significantly higher for media containing 5 x 10^{-5} M and 10^{-4} M zinc in comparison to the control group (0 M) by day 14 (Figure 2a). Alkaline phosphatase activity increased with increasing zinc concentrations (Figure 2b). Collagen production was significantly higher for 5 x 10^{-5}M and 10^{-4}M when compared to the control. The cells in zinc supplemented media appeared to produce more collagen type I than the 0 M control group (Figure 2c). Calcium was statistically significant for 10^{-4}M when compared to the 0M control group on day 14. Increased calcium deposition in the zinc supplemented groups also appeared to be present as compared to 0 M control (Figure 2d).

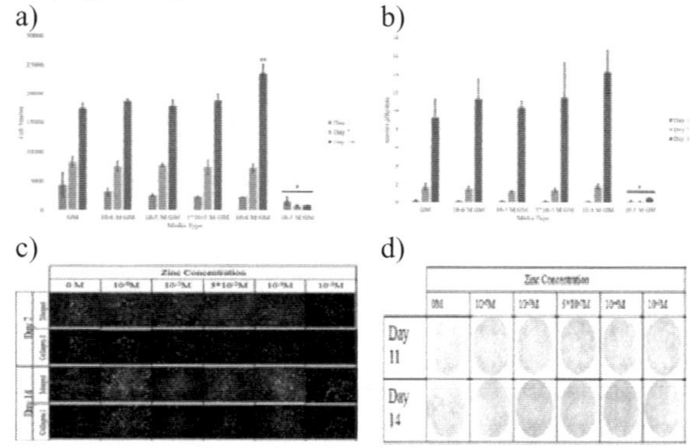

Figure 1: a) cell number, b) alkaline phosphatase activity, c) fluorescent microscopy images of nucleus (blue), actin (red), and collagen type I (green) in general media, and d) optical images of alizarin red stain. Magnification = 20x. * $p< 0.05$ as compared to other media conditions. ** $p<0.05$ as compared with all other media conditions for time point.

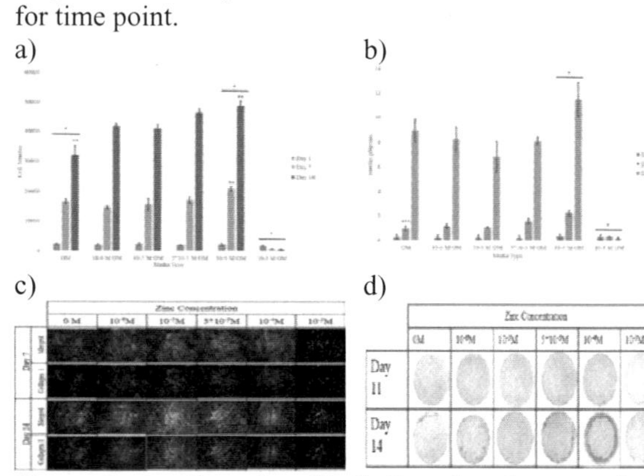

Figure 2: a) cell number, b) alkaline phosphatase activity, c) fluorescent microscopy images of nucleus (blue), actin (red), and collagen type I (green) in osteogenic l media, and d) optical images of alizarin red stain Magnification = 20x. * $p< 0.05$ as compared to other media conditions. ** $p<0.05$ as compared with all other media conditions for time point. *** $p<0.05$ as compared with all media types except 10^{-6}M

Conclusions: Zinc supplemented media enhanced cell growth and the osteogenic differentiation of human MSCs. This study demonstrates that zinc may be a potent supplement for enhancing the osteogenic differentiation of MSCs. Future studies will focus on the potential use of zinc for bone tissue engineering strategies.

References:

[1] Wey, A, J. Orthop. Res., 2014; 32, 834-41
[2] An, S, Biol Trace Elem Res, 2017; 175: 112-21

Acknowledgements: The authors would like to thank funding from the National Science Foundation (#1610125).

Kinetics of MSC-Based Enzyme Therapy for Immunomodulation

Alexandra Burr, Dr. Biju Parekkadan
Department of Biomedical Engineering, Rutgers University, Piscataway, NJ

Introduction: Mesenchymal stromal cells (MSC) demonstrate innate and regulatory immunologic functions and have been widely explored for cell therapy applications. Mechanisms by which MSCs achieve therapeutic effects are in many ways unknown, with appropriate dosing and circulation time as continuing issues[1]. One immunosuppressive function of MSCs is through ectoenzyme expression of CD73 and CD39 which cooperatively hydrolyze inflammatory, extracellular adenosine triphosphate (ATP) to anti-inflammatory adenosine. Extracellular ATP has a key role in autoimmune and inflammatory diseases, which therapeutic MSCs have the potential to harness[2]. It has been shown that in chronic inflammatory diseases, ATP is released by cellular apoptosis and is recognized as a danger signal. For T-lymphocytes, excess ATP has been shown to either induce apoptosis or cause proliferation, depending on the level of ATP exposure[3]. For monocytes, co-stimulation of ATP results in pro-inflammatory IL-1^{2} secretion[4].

Methods: *In vitro* experiments were performed to determine the hydrolysis rates of ATP by MSCs. Through kinetic modeling from experimental results, the rate at which a single cell can metabolize ATP was determined. Based on these rates, the ability of MSCs to downregulate inflammatory signaling pathways was validated with respect to two different mechanisms: extracellular ATP stimulates lymphocytes to suppress proliferation and induce apoptosis and with co-stimulation, it stimulates monocytes to release pro-inflammatory IL-1^{2} . MSCs were co-cultured with immune cells using transwell inserts and compared to immune cell only groups.

Results: Hydrolysis of ATP was efficiently modeled by first-order enzyme kinetics. By exposing different cell populations to various doses of ATP, we plot the hydrolysis rate for these doses and non-linear regression shows a strong correlation of fit to Michaelis Menton kinetics (Fig 1A). As shown in Fig 1B, the maximum hydrolysis rate (Vmax) is plotted as a function of normalized population size and has a strong linear correlation. From this we derive that for in vitro culture, the rate at which a single cell can hydrolysis ATP is 8.9 nmol/min. To confirm the ability of MSCs to modulate a physiologic response, cocultures with immune cells were exposed to various doses of ATP. For lymphocyte-MSC cocultures, the lymphocytes were stimulated to proliferate and dosed with ATP which would in turn prevent proliferation and cause cell death. Using the model, it was determined that the delivered dose MSCs could hydrolyze approximately 572 nmol/hour (~1 mM) of ATP. As shown in Fig 1C, MSCs are able to mitigate the effects of ATP at 1 mM which restores proliferation in lymphocytes. As the ATP dose exceeds what MSCs can effectively hydrolyze, the proliferation decreases. Similarly, when MSCs are cocultured with monocytes, they are able to reduce the effects of ATP and therefore the secretion of IL-1^{2} which is typically induced by ATP stimulation.

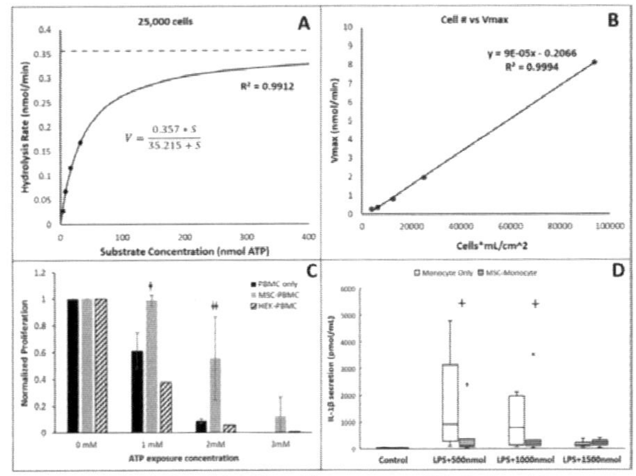

Figure 1: **Modeling MSCs as Enzymes to Modulate Responses to ATP in Immune Cells. (A)** MSC hydrolysis of ATP over time follows first order enzyme kinetics with $R^2>0.99$ **(B)** Plotting the Vmax of hydrolysis for various population sizes of MSCs is shown to be linear, hypothesizing the maximum rate of hydrolysis on a per cell basis is 8.9 nmol/min. **(C)** Verification of MSC ability to restore proliferation in T-lymphocytes that have been stimulated and exposed various doses of ATP. **(D)** MSCs prevent pro-inflammatory IL-1^{2} secretion by monocytes in the presence of ATP.

Conclusions: In addition to demonstrating that MSC hydrolysis of ATP can be modeled by enzyme kinetics, we show that the in a dose-dependent manner, MSCs can modulate the immune response with relation to three different mechanisms. Such use of these models may allow for better dosing predictions for MSCs to be used therapeutically for chronic inflammatory diseases such as rheumatoid arthritis, diabetes, pancreatitis, and other systemic inflammatory response syndromes. For the first time, the effect of MSCs on ATP hydrolysis in immune cell response is quantitatively analyzed on a cell-molecule basis by modeling the hydrolysis as an enzyme-substrate reaction. The results also give insight into MSCs' dynamic response mechanisms to ameliorate effects of extracellular ATP whether it be through positive or negative regulation.

References:

[1] Le Blanc K, Ringden O: Immunomodulation by mesechymal stem cells and clinical experience. J Intern Med. 2007, 262: 509-525. 10.1111/j.1365-2796.2007.01844.x.

[2] Trautmann, A. (2009). Extracellular ATP in the immune system: more than just a "danger signal". *Sci. Signal.*, *2*(56), pe6-pe6

[3] Savio, L. E., de Andrade Mello, P., da Silva, C. G., & Coutinho-Silva, R. (2018). The P2X7 Receptor in Inflammatory Diseases: Angel or Demon?. *Frontiers in pharmacology*, *9*, 52.

[4] Di Virgilio, F. (2007). Liaisons dangereuses: P2X7 and the inflammasome. *Trends in pharmacological sciences*, *28*(9), 465-472.

Acknowledgements: Research report in this manuscript was supported by the National Institute of General Medical Sciences of the National Institutes of Health under award number T32 GM008339.

5C-1

A Modular Cell-Cell Adhesion Toolbox for Engineering Novel Multicellular Systems

Nathaniel Borders, Timothy Wannier, PhD George Church, PhD
Department of Genetics, Harvard Medical School

Introduction: Multicellularity confers a wide array of valuable properties to organisms, including a defined 3D-structure, compartmentalization, and patterning. Recently, the engineering of synthetic multicellular systems has led to a greater understanding of developmental principles and generation of spatially organized living materials.[1] Natural multicellularity is directed by protein mediated cell-cell adhesion, but only one synthetic toolbox with three hetero-specific binding pairs has been reported.[1] Desired complexity in future synthetic multicellular systems will undoubtedly require a significantly larger set of modular cell-cell adhesion proteins.

We have assembled and characterized a toolbox of protein pairs that when expressed on the surface of a bacterial cell specifically and orthogonally drive cell-cell adhesion. This toolbox will form a basis for future bioengineering efforts toward artificial multicellularity, enabling the next generation of synthetic biomaterials.

Methods: A library of 83 potential binding proteins was compiled and expressed onto the surface of *Escherichia Coli.* Cell-cell adhesion was assayed through quantification of macroscopic aggregation with pairwise mixing of stationary phase library members and comparison of the optical density (OD600) of the top 25% of culture tubes after 0, 4, and 18 hours.

In addition, to gain the ability to assay $>10^5$ cell-cell adhesion pairs, we have optimized an intracellular tripartite split-GFP (sGFP) assay that relies on protein-protein interaction to reconstitute GFP.[2] Potentially interacting proteins are fused to the two beta strand subunits of GFP, S10 and S11, have been severed from the main barrel of GFP subunits S1-9. If the proteins bind, the three disparate GFP parts associate, producing fluorescence, which is read out via plate reader.

Results: Through use of an aggregation-based assay we have identified three sets of heterospecific cell-cell adhesion proteins (Fig. 1, lower OD600 indicates binding). When mixed together, strains expressing protein pairs that bound one another were measured to fall out of solution more quickly than the strain alone or in combination with off-target binders. The first and second sets are composed of an easily engineerable binding motif called the α-helical coiled coil (CC). Currently, we have identified three anti-parallel (Fig. 1a) and two parallel (Fig. 1b) orthogonally binding pairs of coiled coils. Additionally, we found that extracellular display of the reported covalent binders SpyCatcher-SpyTag but not SnoopCatcher-SnoopTag also mediates cell-cell adhesion (Fig. 1c).[3,4]

To more rapidly expand the set of protein adhesion tools, we optimized the *E. coli* based tripartite split-GFP (sGFP) assay. Through several iterations of linker design, we have increased the stringency of our results and validated our system by comparison with the aggregation-based assay (Fig. 1d, higher arbitrary fluorescence units indicate binding). Specifically, the strongest four binding pairs (highlighted in green) are also found in the aggregation assay.

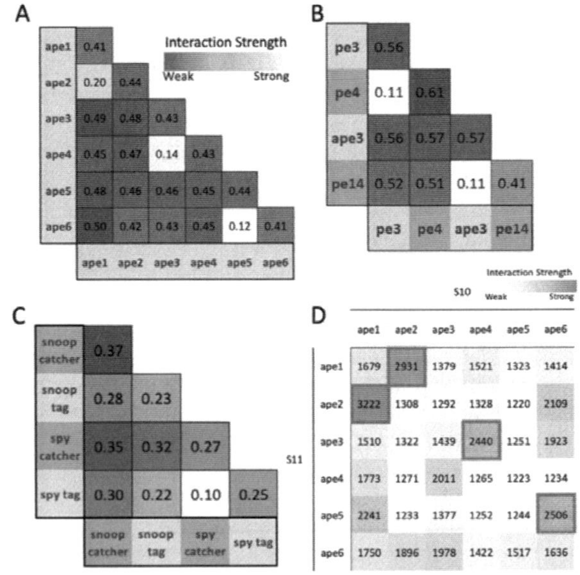

Conclusions: The toolbox presented within this abstract expands the known set of synthetic cell-cell adhesion proteins by three-fold, which already greatly increases the potential for synthetic multicellular engineering. As multicellular systems continue to increase in complexity more orthogonal synthetic parts will be needed to compensate. With this in mind, we have designed and begun to build four 1024-member libraries of coils that we will assay against each other in the sGFP assay for upwards of 16 million potential CC interacting pairs. Library designs were informed by CCs that performed well in both the aggregation and sGFP assays. Once construction of the large sGFP libraries has been completed, each library will be flow-sorted, those CC pairs that bind most strongly will reconstitute GFP fluorescence and be sorted for, and CC pair sequences will be extracted from captured cells through next generation sequencing. Binding pairs that prove most orthogonal will be synthesized and further assayed for binding and orthogonality at the cell surface via the settling assay.

References:

1. Glass D. S. & Riedel-Kruse I. H. Cell. 2018;174(649-658)
2. Cabantous S. et al. Sci. Rep. 2013;3(2854)
3. Zakeri B. et al. Proc Natl Acad Sci. 2012;109(690-697)
4. Veggiani G. et al. Proc Natl Acad Sci. 2016;113(1202-1207)

Acknowledgements: We would like to thank the DARPA ELM program for funding this work. We would also like to thank the Wyss Institute of Harvard and The Department of Genetics at Harvard Medical School for hosting this research. Additionally, we would like to thank all members of the Church lab, including Emma Taddeo and Andrew Irish for administrative tasks.

Model Based Approach To Measuring Human Endothelial Arterial Function

Gary Drzewiecki, PhD. & Kristin Distefano, M.S., Rutgers University, N.J.
P. Lenehan, M.D., Everist Health, Ann Arbor, MI.

Introduction: Prior studies have shown the brachial artery flow mediated dilation (BAFMD) is a good indication of arterial endothelial function. Poor endothelial function subsequently indicates the presence of cardiovascular disease. The value of detecting poor endothelial function is that it precedes advanced cardiovascular disease by months and in some cases by several years [1]. A quantitative measurement of arterial endothelial function begins by measuring the resting arterial diameter of typically the brachial Artery. The artery is then caused to vasodilate by inducing flow mediated dilation. Ultrasonic imaging is typically employed to measure the vessel's diameter. Percent dilation is then calculated relative to the resting diameter. PD (percent dilation) greater than 7% is generally considered healthy. While this test is noninvasive and repeatable there are a few drawbacks that prevent its acceptance. First, the use of ultrasound requires some skill to measure the artery diameter accurately. Secondly, although the measurement is repeatable with some care for an individual there is often measurement variability due to unknown factors in a population of subjects. For example, arterial diameter is affected by other factors besides endothelial function. For example, vessel diameter is a known function of blood pressure. In this article, we report on the direct effect of blood pressure on PD. We further report on the use of an arterial model-based approach to measuring percent dilation, PD.

Methods: The method of cuff cFMD was applied to obtain the brachial artery diameters in 6 normal undergraduate students in comparison to Ultrasound (uFMD) cFMD has the advantage of being able to alter the subjects Arterial Transmural Pressure over a wide range while uFMD measures only at the subject's mean arterial pressure (MAP). General reagents (e.g. buffers, deionized water) need not be included if they are not unique to the work. cFMD begins with measuring the subject's baseline Pressure – Lumen area curve. A 5-minute period of blood flow occulusion is the applied to cause FMD. Then a second Pressure - lumen Area curve is recorded to represent the dilated condition. Lumen area recordings were obtained and calibrated using the Cuffsoft instrument provided by Applied Processor and Measurements, Williamsville, NY. This device permits the measurement of arterial diameter over the range of blood pressure from systolic to diastolic. This permits a direct analysis of the influence of blood pressure on PD. A modeling approach was then applied to obtain PD as a function of blood pressure. The arterial model of Drzewiecki [2] was applied. A simplified version of the model relates trans mural pressure to lumen area as follows.

$$\boldsymbol{Pt = a\left(e^{\frac{b(A-A_b)}{A_b}} - 1\right) - E\left(\left(\frac{A_b}{A}\right)^n - 1\right) + P_b} \qquad \textbf{(1)}$$

a, b,n E and A_b, are parameters that are calibrated to the individual subject's data.

[mmHg]

Results: PD was calculated from the change in vessel diameter from the r baseline rest condition to the FMD condition for all values of transmural pressure for all subjects. A single subject is shown in fig.1. The model data results are shown for the rest and the vasodilated FMD condition. Using the data from all subjects, the chi- squared test revealed an error of less than 5%. indicating that the model can successfully represent all data recorded. PD as determined from the model id provided for subject 4 in fig 2.

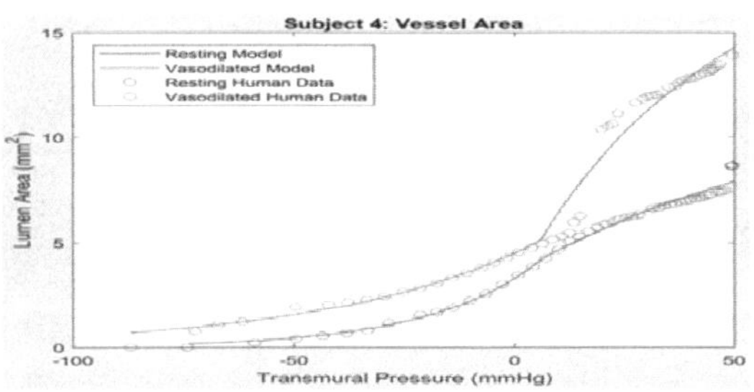

Figure1. Pressure-Area data and model result for a single subject. The solid curve is the model result after parameters were calibrated.

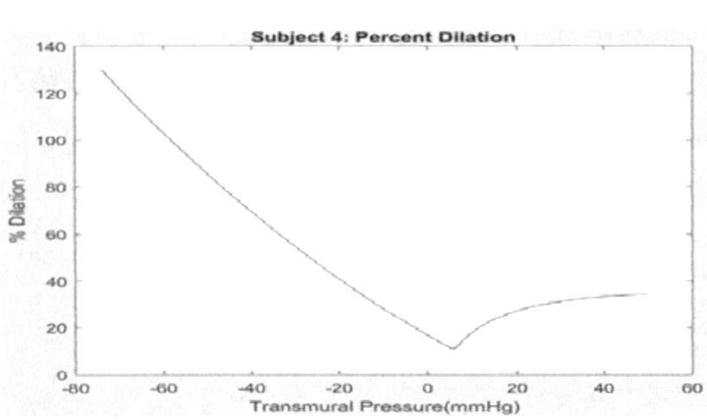

Figure 2. PD data for the subject in figure 1. for all transmural pressures

Conclusions: It was first revealed that our vessel model is capable of modeling all of our subjects and over a full range of blood pressure from systolic to diastolic, as shown in figure 1 with less than 5% error. This supports the conclusion that both the baseline condition and FMD conditions can be modeled accurately by eq. (1). Once calibrated, our model was used to find the functional relationship between blood pressure and PD. Referring to figure 2, it is clear that PD is affected by blood pressure as we originally proposed.

Our results confirm that the FMD test should be normalized to the subject's blood pressure to be generally valid as a true indicator of endothelial function.

References:

1. Hamburg et al. Circulation,104;22;2673-2678,2001

2. Drzewiecki, G. et al. Am J Phys. 1997;4:H2030-H2043.

Acknowledgements

The authors wish to acknowledge their collaborations with the everist Health Angiodefender™ team towards the development of an automatic instrument for the evaluation of true endothelial function. We also acknowledge the technical aid of Applied Processor and measurements in the support of our measurements

Maddox Components of Vergence

Sebastian N. Fine[1], Elio M. Santos, Ph.D.[1], Mitchell M. Scheiman, O.D., Ph.D.[2], Tara L. Alvarez, Ph.D.[1]
Department of Biomedical Engineering[1], New Jersey Institute of Technology, Newark, NJ
Department of Optometry[2], Salus University, Elkins Park, PA

Abstract – **The Maddox components of vergence eye movements (tonic, accommodative, disparity, and proximal) have been studied piecemeal. Hence, the combinational contribution of each component remains unknown. This study quantifies baseline data for those with normal binocular vision so that it will serve as a reference to compare how patients with binocular vision dysfunctions may differ. Data were acquired from 25 binocularly normal subjects using novel instrumentation. For healthy controls, the disparity cue most significantly contributes to the overall vergence response. Understanding the contributions of each Maddox component has the potential to uncover the underlying neural mechanism by which therapeutic interventions may lead to the reduction of visual symptoms.**

I. Introduction

Oculomotor dysfunctions, such as convergence insufficiency (CI), inhibit normal binocular vergence function. CI affects nearly 16 million Americans and is associated with symptoms of eyestrain, blurred vision (Fig. 1), and fatigue when engaged in near work. Office-based vergence and accommodative therapy with home reinforcement (OBVAT) leads in successful remediation of vision symptoms for 73% of patients and remains effective at least one-year post therapy.[1] Nevertheless, this is accomplished by generalized treatment sessions, which treat all ocular vergence components as dysfunctional. Consequently, establishing normative metrics for vergence eye movements is crucial for baseline comparisons to determine therapy efficacy. Hence, this study objectively determines normative data metrics for the Maddox components of vergence [2] in binocularly normal subjects.

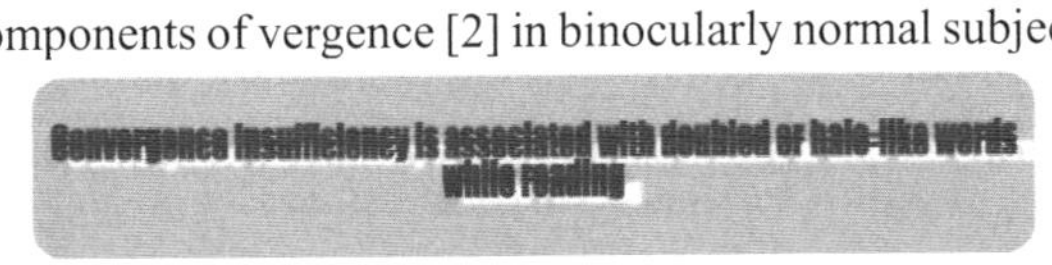

Figure 1. The inability to correctly converge causes inconsistencies in focus, leading to words doubling and "haloing" while reading.

II. Methodlogy

A total of 25 naïve binocularly normal subjects (19 males, 6 females) ages 18 to 35 were enrolled. Subjects were centered into a novel haploscope (Figure 2A) and presented with three different visual stimuli (Figure 2B) which stimulated 2.5° (monocular) and 5° (binocular) vergence responses. The following visual stimuli were used: a Maltese cross for accommodation, a Gabor patch to diminish accommodation and stimulate disparity, looming a stimulus to induce proximal responses, and a Gabor ring to stimulate peripheral disparity contributions. Nine unique vergence cue movements, each containing multiple trials, were recorded and analyzed using a custom developed MATLAB software.

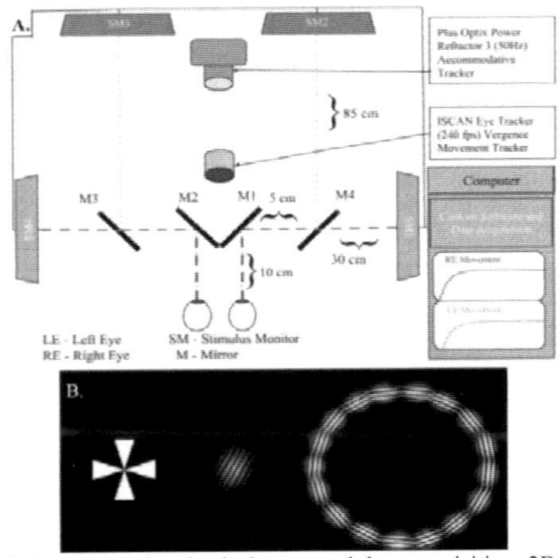

Figure 2. A presents the physical setup and data acquisition. 2B displays the 3 stimulus types (left to right, Maltese cross, Gabor patch, Gabor ring) shown to subjects during trials within the haploscope.

III. Results

Table 1 presents two key metric's statistics for the nine single and multiple Maddox cue combinations studied.

Cue	Peak Velocity (Deg/s)	Final Amplitude (Deg)	N
All Cues	29.7±6.9	5.0±0.3	24
Accommodation Only	12.5±3.6	2.8±0.5	18
Proximal Only	7.9±3.6	1.3±0.7	19
Disparity Only	28.0±4.9	4.8±0.4	25
Proximal & Disparity	28.9±6.8	4.9±0.5	25
Proximal & Accommodation	12.4±3.6	2.7±1.3	20
Disparity & Accommodation	30.7±6.9	5.0±0.3	24
Proximal & Disparity	28.1±7.3	4.8±0.5	25
All Cues (Post)	27.5±7.2	4.9±1.0	25

Table 1. Mean convergent peak velocities and final amplitudes for each Maddox cue. N is the number of subjects that data could be analyzed from.

IV. Discussion

Across all combinations it was observed that responses that contained the disparity cue induced higher velocities, more accurate final amplitudes, and contained less omitted movements compared to responses which did not contain the disparity cue. If found to be compromised in CI patients, disparity reinforced therapeutic treatments may have heightened effectiveness and further improve CI suffers' visual comfort.

Acknowlegements

This research was supported by, NSF MRI CBET 1428425 and NIH 1R01EY023261 to TLA.

References

[1] M. Scheiman, M. Rouse, M. T. Kulp, S. Cotter, R. Hertle, and G. L. Mitchell, "Treatment of convergence insufficiency in childhood: a current perspective.," *Optom. Vis. Sci.*, vol. 86, no. 5, pp. 420–8, May 2009.

[2] M. Morgan, "The Maddox Classification of Vergence Eye Movements./: Optometry & Vision Science, *Optom. Vis. Sci.*, vol. 57, no. 9, pp. 57–60, 1980.

The Impact of Vascular Ehlers-Danlos Syndrome Mutations on Collagen III Structure, Dynamics, and Function

Cody L. Hoop[1], Sonal Ghalawat[2], Allysa Kemraj[1], Baifan Wang[1], Jie Zhu[1], Madison Godesky[2], H. Rose Warren[2], David I. Shreiber[2], Jean Baum[1]

[1]Departments of [1]Chemistry and Chemical Biology and [2]Biomedical Engineering
Rutgers, The State University of New Jersey, Piscataway, NJ

Introduction: Vascular Ehlers-Danlos Syndrome (vEDS) is the most severe of the EDS connective tissue disorders. A life-threatening disease, it leads to arterial, uterine, and intestinal rupture and an average life expectancy of less than 50 years. Most vEDS cases are due to a single glycine substitution (G→X) in collagen III in vascular walls. In this work, we investigate how these mutations may impact collagen III structure and dynamics and its critical interactions with binding partners.
The interaction of collagen with the cell receptor, integrin ±2² 1, plays a pivotal role in cell adhesion and platelet activation. Its ±2I-domain (±2I) specifically recognizes the GROGER sequence in collagen III through a metal-mediated interaction with the collagen Glu. This binding domain lies in a region where vEDS mutations are highly prevalent. To understand how vEDS mutations impact collagen III structure and dynamics and its ability to bind ±2I, we use a multi-tiered approach combining sequence analysis, biological adhesion assays, biophysical methods, and molecular dynamics (MD) simulations.
Methods: *Sequence Analysis:* The Ehlers-Danlos Syndrome Variant database (https://eds.gene.le.ac.uk) was used to analyze the frequency and distribution of different amino acid substitutions for glycine in 601 reported vEDS cases. *Peptide design:* Collagen mimetic peptides (CMPs) incorporating a native ±2I binding motif ($G^{237}ROG^{240}ER$) and three naturally occurring vEDS mutations (G240A, G240R, and G240V) were designed to elucidate the impact of different glycine substitutions on the conformation, dynamics, and function of collagen III. *Peptide characterization:* Circular Dichroism (CD) temperature scans were employed to measure the melting temperatures of each CMP. Peptides were selectively ^{15}N-labeled at G16, X19 (mutation site), and G28 to characterize residue-specific effects to the triple helix backbone by nuclear magnetic resonance (NMR) spectroscopy. Disruptions to the hydrogen bonding network were assessed by temperature-dependent chemical shift perturbations. Dynamics of the peptide backbone were measured by $\{^1H\}$-^{15}N NOE experiments. *CMP-±2I complex characterization:* MD simulations (500 ns) were used to evaluate atomic-level interactions of ±2I with the WT CMP and mutants. Stabilities of specific interactions were assessed by monitoring distances between residues. *Adhesion Assays*: The relative Mg^{2+}-dependent adhesion of recombinant ±2I to the CMPs was determined with ELISA. Human mesenchymal stem cell (hMSC) adhesion to substrates coated with CMPs or control materials was assessed 1 h after cell seeding, and morphology was evaluated after 24 h.
Results: Strikingly, more than 60% of all reported G→X vEDS mutations are Gly-to-charged residue substitutions, which are especially destabilizing to collagen's canonical triple helix conformation. CD and NMR of CMPs that include the GROGER ±2I binding site or one of three natural vEDS mutations show that, while the overall triple helix secondary structure is preserved, substitutions destabilize hydrogen bonding and enhance ps-ns dynamics local to the mutation. MD simulations reveal that G→X mutations cause asymmetric diameter expansions of the triple helix near the mutation site. These substitutions decrease cell spreading and hinder integrin ±2I binding, trending with triple helix stability. Moreover, incorporation of a Val mutation, which shows the largest decrease in ±2I adhesion, destabilizes the fundamental Mg^{2+}–±2I coordination required to mediate its interaction with collagen.

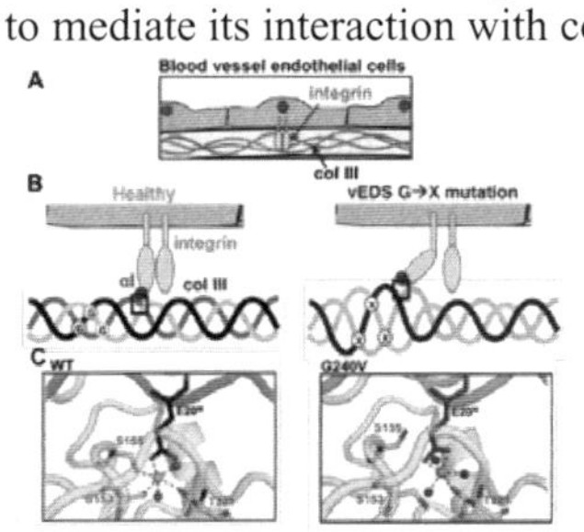

Figure 1. Perturbation of collagen III–integrin interactions with vEDS mutations. A) Collagen III in blood vessel walls interacts with integrin B) Natively (left), collagen III folds into a triple helix of three polypeptide chains. Glycines are involved in the critical hydrogen-bonding network that stabilizes the triple helix and ±I domain integrin binding. G→X mutations (right) disrupt the triple helix, with an expansion of the diameter even beyond the mutation site. C) MD simulations reveal that the hindrance of the integrin ±2I–collagen III triple helix interaction in the case of a G→V mutation may be due to a disruption of essential ±2I–Mg^{2+}.

Conclusions: These results unveil atomic-level insights into how single pathological G→X mutations impact the collagen III–integrin ±2I interaction. The inhibition of the interaction trends with the destabilization of the triple helix. Furthermore loss of Mg^{2+} coordination with ±2I in the case of vEDS mutations may contribute to deficient collagen III–±2I adhesion in the blood vessels. Novel insight into this sequence-structure/dynamics-function relationship sets the foundation for new drug therapy techniques to combat collagen disorders.
Acknowledgements: This work is funded by American Heart Association Post-doctoral Fellowship 17POST33410326 to CLH, NIH grant GM 45302 to JB, and the NSF REU in Cellular Bioengineering NSF EEC 1262924 to DIS, which supported HRW.

Effects of Orthoptic Treatment on Convergence Insufficiency

Ayushi Sangoi[1], Elio M. Santos, Ph.D.[1], Mitchell M. Scheiman, O.D., Ph.D.[2], Tara L. Alvarez, Ph.D.[1]
Department of Biomedical Engineering[1], New Jersey Institute of Technology, Newark, NJ
Department of Optometry[2], Salus University, Elkins Park, PA

Abstract – **Convergence insufficiency (CI) is a common binocular vision problem and while randomized clinical trials report an effectiveness of 73%, the underlying mechanism by which Office-Based Vision Therapy (OBVT) is not completely understood. The rate of phoria adaptation to a 6 base-out and 6 base-in prism at near (40cm) was compared before and after 12 hours of OBVT. After therapy, CI patients had an increased rate of phoria adaptation compared to baseline measures for both base-in and base-out prism adaptation experiments. The base-in prism adaptation exhibited a greater change compared to the base-out adaptation.**

I. Introduction

Convergence insufficiency (CI), the reduced ability to coordinate the eyes while performing near work such as reading, affects 4-9 % of the population [1,2]. Typical symptoms include blurred or double vision, eyestrain, headaches, and the inability to concentrate when conducting near work. Vision symptoms in patients with CI can be remediated with Office-Based Vision Therapy (OBVT). Randomized clinical trials show that OBVT has an effectiveness of 73% in CI patients. However, the underlying neural mechanism by which OBVT is remediating symptoms is not completely understood. This study is designed to test the hypothesis that vision therapy will lead to CI patients having an improved ability to perform phoria adaptation.

II. Methodology

This analysis investigated 25 CI patients who were diagnosed by one of the co-authors (MS) who is an optometrist. CI was diagnosed as a near point of convergence greater than 6 cm, reduced positive fusional vergence of less than 15 ” , and exodeviation greater than 4 prism diopters at near than at distance or failing Sheard's criteria. [1] The phoria levels were measured subjectively using the flashed Maddox rod test.

A 6 ” base-in (BI) prism or a 6 ” base-out (BO) prism was inserted in front of an eye to conduct a BI or BO phoria adaptation experiment at near (40 cm away from the subject). The details of this method are described in our prior publication.[2] The phoria adaptation experiment was 7 min in duration. Processing and analysis of the data were performed using a custom MATLAB script. The rate of adaptation is calculated as the change in magnitude divided by the time constant of the exponential curve of best fit.

III. Results

Figure 1 shows the average rate of adaptation for the base-in and base-out phoria adaptation experiment for the CI subjects pre and post OBVT. The rate had a more dramatically visible change in the base-in prism tests post OBVT compared to baseline. The base-in had almost an 80% increase in rate with the treatment, while the base-out test had a 10% increase.

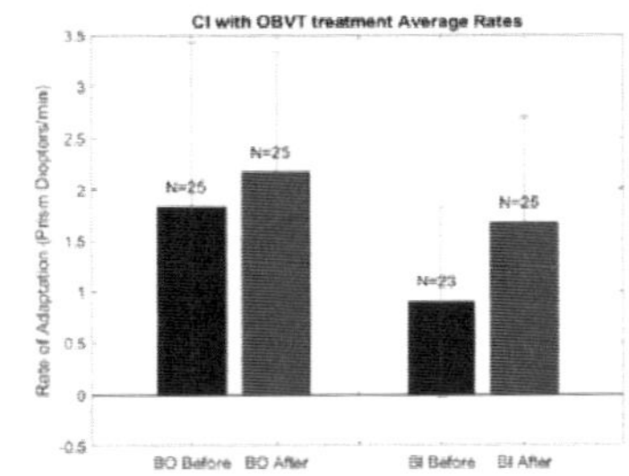

Fig. 1. Average rate of adaptation for BI and BO for CI subjects at baseline (blue) and after vision therapy (red) before values in blue and after values in red showing the base-out averages on the left and the base-in averages on the right.

Figure 2 shows the average phoria adaptation levels for the 25 CI patients to the BI phoria adaptation experiment before (blue) and after (red) OBVT. The magnitude of change and the rate of phoria adaptation was greater after OBVT. The slope is steeper in the CI subjects when comparing before and after measurements.

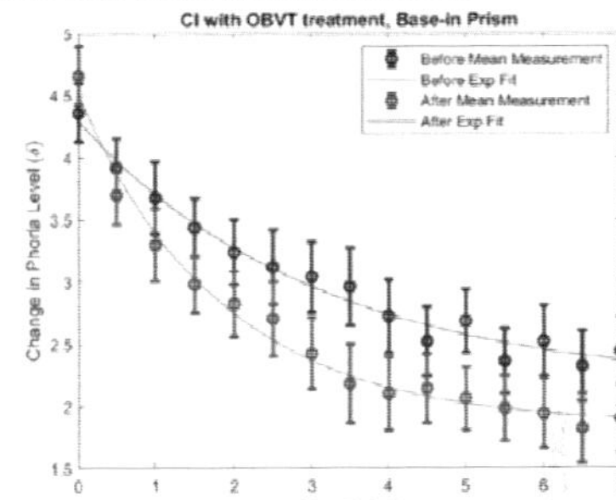

Fig. 2. Average change in phoria levels for base-in prism for CI patients. The averages and exponential fit before treatment are shown in blue, while the averages and exponential fit after treatment are shown in red. The SEM is also depicted as bars for each point.

IV. Discussion and Conclusion

Prism-induced phoria adaptation substantially changed post OBVT compared to baseline measurements for patients with CI. These results support that one mechanism by which vision therapy may lead to a reduction in vision symptoms is an improvement in the slow fusional vergence system to enhance a CI patient's ability to adapt to near or far space.

Acknowledgements

This research was supported by NIH 1R01EY023261 to TLA.

References

[1] M. M. Scheiman, H. Talasan, G. Lynn Mitchell, and T. L. Alvarez, "Objective Assessment of Vergence after Treatment of Concussion-Related CI: A Pilot Study," *Optom. Vis. Sci.*, vol. 94, no. 1, pp. 74–88, Jul. 2017.

[2] E. M. Santos, C. Yaramothu, and T. L. Alvarez, "Comparison of symmetrical prism adaptation to asymmetrical prism adaptation in those with normal binocular vision," *Vision Res.*, vol. 149, pp. 59–65, Aug. 2018.

Controlled Synthesis and Solubility Characterization of Polymer-Peptide Conjugates for Biomedical Applications

Reem Eldabagh, Shashank Kosuri, Matthew Tamasi, Rahul Upadhya, Adam J. Gormley, Ph.D.
Department of Biomedical Engineering, School of Engineering, Rutgers University, Piscataway Township, NJ 08854

Introduction: Regenerative medicine entails the regrowth of lost or damaged cells, tissues, or organs. This often occurs in response to proteins known as growth factors. Naturally-derived growth factors are costly, difficult to produce, immunogenic, and may be unstable. Therefore, synthetic peptides are favored over their biological counterparts and those that mimic growth factors are produced for regenerative medicine. These peptides, however, may easily be broken down by proteases or lysosomes, among other intracellular species. Conjugating synthetic peptides to suitable polymers protects these peptides from degradation. The conjugation also allows for multivalency, via the attachment of multiple peptides to a single polymer, enhancing bioactivity and regenerative effectiveness. As such, polymer-peptide conjugates are preferred for the delivery of proteins to their biological targets. Previously in the lab, polymer-peptide conjugates were synthesized and administered to cells, yet no cellular response was observed. Insufficient characterization of the conjugates disabled conclusive structure-activity relationship studies. Thus, controlled methods of synthesis using click chemistry and solid phase support are being developed. The strain-promoted azide alkyne cycloaddition (SPAAC) reaction and bifunctionalized polymer enable this controlled synthesis.

Methods: Homopolymers of methyl methacrylate (MMA), methyl acrylate (MA), butyl methacrylate (BMA) butyl acrylate (BA), hydroxyethyl methacrylate (HEMA), and hydroxypropyl methacrylate (HPMA) were synthesized via the oxygen-tolerant PET-RAFT synthetic method. A prospective study of aqueous solubility was completed by qualitatively assessing the critical solubility points of MMA, MA, BA, BMA, HEMA, and HPMA homopolymers using the CKX41 Inverted Olympus Microscope. The solubility assessments were tested again using Zetasizer Nano from Malvern, to measure dynamic light scattering (DLS). These tests were initially conducted in DMEM cell media and later in PBS. The polymeric molecular weights were determined using the Agilent Technologies GPC-SEC. The study of aqueous solubility was carried out to choose an appropriate polymer for the backbone of polymer-peptide conjugates for biomedical applications, which must be soluble in aqueous media. The bifunctionalized polymer was synthesized by initial reduction to form a thiol, followed by the click thiol-ene reaction, and azide addition in a nucleophilic substitution reaction (**Scheme 1**). The aim of bifunctionalization is to create polymers of known molecular weights with reactive functional groups on either end, able to be clicked to other such functionalized polymers to synthesize polymers of well-defined molecular weights. The realization of this would result in controlled synthesis of materials with well-defined molecular weights, enhancing the quality of any future structure-activity relationships.

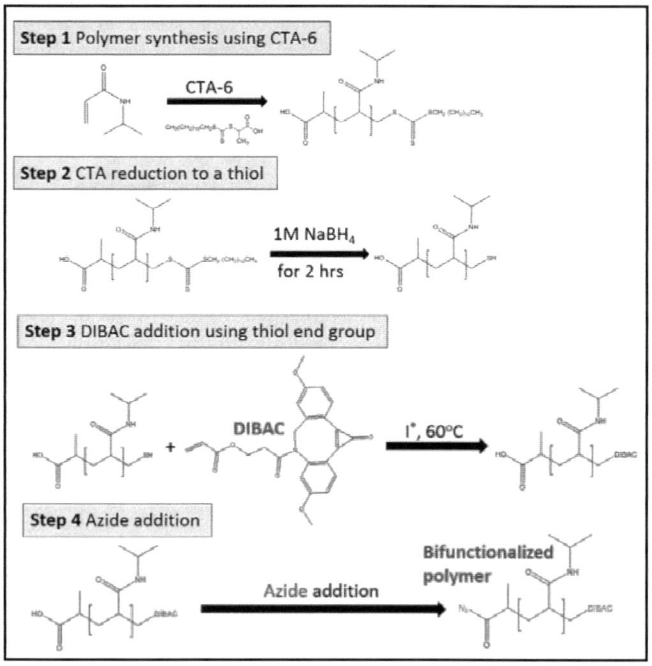

Scheme 1: Bifunctionalization of N-isopropylacrylamide

Results: Homopolymers of acrylate monomers with more polar groups (like HPMA and HEMA) were expected to be more soluble than others. The solubility tests in DMEM cell media contradicted these expectations, with DLS data supporting the findings. These tests were repeated in PBS, and the results more closely followed the expectations. To the polymer, the alkyne was successfully attached, but the azide was not. Instead, an apparent dimerization was observed by GPC-SEC.

Conclusions: The aqueous solubility of the polymers in decreasing order were as follows: HPMA, HEMA, BA, MA, MMA, BMA. The increase in molecular weight was not from dimerization. Further studies are necessary to test the most soluble polymers as backbones in polymer-peptide conjugates, and to successfully synthesize and verify the synthesis of the bifunctionalized polymer.

References:

1. Gormley AJ. Angew. Chem. Int. Ed. 2018;57:1557-1562.

Acknowledgements: Thanks is due to all members of the Gormley lab, Dr. David Shreiber, Dr. Evelyn Erenrich, and the grant that made this work possible:
The National Science Foundation, Grant #NSF EEC 1559968
REU Site: Cellular Bioengineering - From Biomaterials to Stem Cells

Bioinspired Vascularized Polymers for the Delivery of Bioactive Compounds at Surfaces

Kayla Marquis[1], Benjamin Chasse[1], Caitlin Howell[1,2]
[1]Department of Chemical and Biomedical Engineering, University of Maine, Orono, Maine
[2]Graduate School of Biomedical Science and Engineering, University of Maine, Orono, Maine

Introduction: Similar to how the veins in a leaf deliver nutrients throughout a plant, channels can be added to a material to temporally and spatially control the delivery of bioactive compounds to a surface.[1] The channels allow for the targeted delivery of a chosen compound. Applications may include the studying of the healing or toxic effects of various chemicals on organisms, the antifouling applications used for a ship's hull to prevent the attachment of sea creatures, the isolation of desired genes from a culture, or the separation of bacteria in bacterial multi-cultures. By varying the system geometry and drug concentration, the growth patterns of biofilm on the surface can be controlled. The purpose of the research is to derive predictive modeling variables via experimental designs to produce highly controllable and optimizable bioactive compound delivery systems for a broad range of applications.

Materials and Methods: Nutrient agar with empty embedded channels was fabricated using previously reported methods.[2] Then, *Escherichia coli K12* was streaked over the surface of the system using a cotton swab, and the channels were injected with the antibiotic gentamicin sulfate at varying concentrations. The antibiotic was then allowed to diffuse to the surface of the agar and affect the growth of the bacteria at the surface. The system was incubated overnight at 37°C, the inhibition zone was imaged and measured using ImageJ. These data were then correlated with the concentration of the antibiotic and used to define the dose response curve, which was integrated with computational models made with COMSOL with inputs of initial concentration and the experimentally determined diffusion coefficient.

Results: To define the necessary variables for developing a predictive model, a simple system was first constructed[2]. **Figure 1A** shows images of the experimental system with bacteria compared to the theoretical model (left), and the dose response curve by graphing the inhibition zone (mm) versus the antibiotic concentration (μg/mL) (right). **Figure 1B** shows the dose response curve, which was used to predict the formation of a more complex pattern, in this case an "M", with a target inhibition zone of 2.53 mm. The experimental inhibition zone had an average of 2.70 mm with a standard deviation of 0.32 mm.

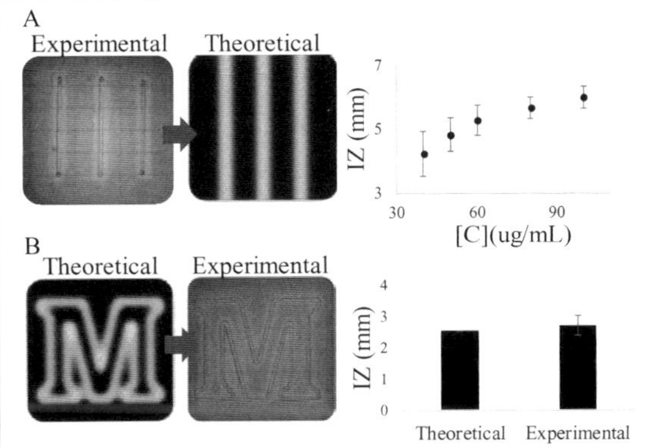

Figure 1: Image A shows a correlation between the inhibition zones of *E. coli K12* and the antibiotic concentrations. Image B shows the minimum inhibitory concentration with the dose response curve to spatially control the growth of *E. coli K12*. Graph B compares the theoretical and experimental inhibition zones. "IZ" is defined as inhibition zone.

Conclusions: Predictive models have been developed in COMSOL using known diffusion coefficients, experimentally determined minimum inhibitory concentrations and dose curves. We are currently in the process of increasing the accuracy of the predictive modeling system by using liquid media infused with antibiotics to determine the minimum inhibitory concentration instead of using a plating method. We are also increasing the adaptability and control of the system by orienting the channels in three dimensions within the matrix.

Citations:

1) Howell C, et al. *ACS Applied Mat. Int.* 2014;6:13299-13307.
2) We W, et al. *Adv. Mat.* 2011;23:H178-H183

Acknowledgements:

This work was supported by the University of Maine System Research and Reinvestment Funds, the University of Maine Vice President for Research, and the Maine Space Grant Consortium Fellowship Program.

Synthesis and Characterization of Glycosaminoglycan Mimics for Cartilage Repair Applications

Vincent, R., Hammond, W., Collins, G., Arinzeh, T.
Department of Biomedical Engineering
New Jersey Institute of Technology

Introduction: Damaged cartilage, or osteoarthritis (OA) can lead to severe pain, loss of mobility, swelling and decreased quality of life [1,2]. Currently, the regeneration of articular cartilage tissue remains a challenge. Often surgical techniques result in the formation of mechanically inferior fibrocartilage which only temporarily alleviates the symptoms of OA[3]. During cartilage development, chondroitin-6-sulfate and heparin sulfate are the predominant glycosaminoglycans (GAGs) present [4] and one of their functions is to bind growth factors, which is determined by their structure [5]. Our lab has demonstrated the potential of partially sulfated cellulose (pSC) as a GAG mimetic, with degree and position of sulfation similar to chondroitin-6-sulfate, for cartilage repair [7]. We also seek to investigate other structural features in GAG mimetics. Cellulose has glucose monomer units linked by beta glycosidic linkages. However, to more closely mimic the structure of heparin sulfate, a GAG mimetic derived from starch, which has glucose monomers linked by alpha-glycosidic linkages, will be prepared. Similarly, a GAG mimetic derived from chitin, which is composed of N acetyl glucosamine, also will be prepared since the N acetyl glucosamine group is present in naturally occurring GAGs. These structural changes may impact stability and growth factor binding. We synthesized and characterized the products of the chemical sulfation of cellulose, chitin and starch.

Materials and Methods: ***Synthesis of Sodium Cellulose Sulfate:*** Partially sulfated cellulose (pSC) was prepared by modifying a previously published technique [6]. Briefly, a sulfating reagent was prepared by the dropwise addition of chlorosulfonic acid to dimethylformanide (DMF) under argon purging. Vacuum dried microcrystalline cellulose was added to anhydrous DMF swollen prior to sulfation. The sulfating agent was added dropwise to the swollen cellulose while stirring to form crude cellulose sulfate. The crude compound was then isolated and purified. ***Synthesis of Starch Sulfate:*** Partially sulfated starch (pSA) was prepared by modifying a previously published technique [6]. Briefly, a sulfating reagent was prepared by the dropwise addition of chlorosulfonic acid to pyridine under argon purging. After the reaction is completed, the agent emulsion is heated to 70^{0}C. Vacuum dried soluble starch (Sigma) was added to the agent and allowed to react forming the pyridinium salt of starch sulfate. The reaction was then quenched using sodium
hydroxide to a pH of 5-6, forming the sodium salt. The crude pSA was then isolated and purified. ***Synthesis of Chitin Sulfate***: Partially sulfated chitin (pSCH) was prepared by modifying a previously published technique [6]. Briefly, a sulfating reagent was prepared by the dropwise addition of chlorosulfonic acid to pyridine under argon purging. After the reaction is completed, the sulfating agent emulsion is heated to 85^{0}C. Chitin powder (alpha Aesar) was added to the agent and allowed to react forming the crude pyridinium salt of chitin sulfate. The reaction was then quenched and titrated to a pH of 5-6 using sodium hydroxide forming the sodium salt. The crude pSCH was then isolated and purified. ***Characterization:*** The presence of sulfate substitution was confirmed by Fourier transform infrared spectroscopy (FTIR). The sulfur content (%S) of the purified products were determined by acid digestion and barium sulfate precipitation. Proton NMR was utilized to identify impurities on the compounds (Brucker 500 Hz Rutgers, Newark).

Results: FTIR of all sulfated compounds showed a broad peak located at 1222 cm^{-1} which corresponds to S=O stretching, and primary sulfate peaks at 822 cm^{-1} and shoulder at 1000 cm^{-1}[7]. This indicates that all the compounds are sulfated. For glucose, the sulfation reaction kinetics favors the sulfation of primary hydroxyls, which is at the 6-carbon. Trace amounts of impurity were also identified by proton NMR (HNMR) in all samples showing small amounts of sodium acetate, (DMF) and pyridine. The mass percentage of sulfur was measured by barium sulfate precipitation and all samples were confirmed to have a 0.3 to 0.5 sulfate groups per glucose residue. Future characterization will include molecular viscosity, number and weight average molecular weights of the reaction products.

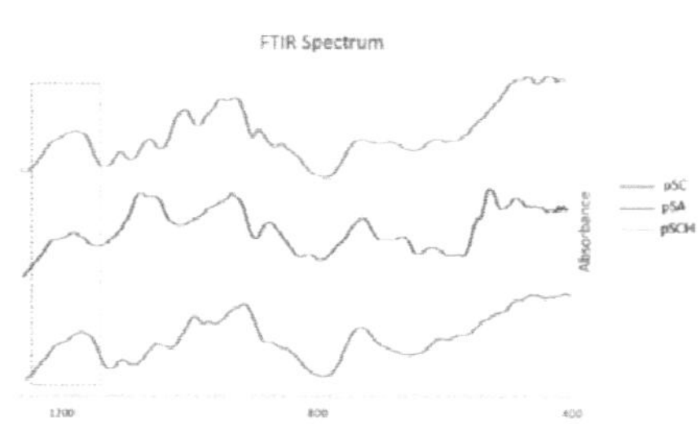

References:

[1] The Basic Science of Articular Cartilage: Structure, Composition, and Function Alice J. Sophia Fox, MSc,* Asheesh Bedi, MD,† and Scott A. Rodeo, MD

[2] Magnussen RA, et al. Clinical Orthopaedics and Related Research 2008;466:952-962.

[3] Kuo CK, et al. Current Opinion in Rheumatology 2006;18:64-73.

[4] Lash JW et al. Journal of Experimental Zoology 1974;189:127-131.

[5]Forsten-Williams K, et al. Annals of Biomedical Engineering 2008;36:2134-2148.

[6] Barnes, C., et al. Tissue Engineering 13, 1593, 2007.

[7] Huang G. et al. Tissue Engineering, 2017. 23(15-16): p. 1011-1021.

Acknowledgements: This work was supported by the NSF Center for Engineering MechanoBiology (CEMB) CMMI: 15-48571

Zein Protein Fibrous Matrices for Promoting Cell Adhesion and Osteogenic Differentiation

Apurva Limaye, Jessica Cardenas Turner, Treena Livingston Arinzeh.
Department of Biomedical Engineering, New Jersey Institute of Technology, Newark, New Jersey.

Introduction: Tissue engineering approaches using scaffolds are utilized to mimic native extracellular matrix (ECM) to support tissue formation. Plant derived proteins are renewable, abundant and have low immunogenic effects. Zein, a protein found in corn, has been studied as a potential scaffold due to its cytocompatibility, antibacterial properties and biodegradability[1,2]. In the following studies, we evaluated electrospun zein scaffolds for cell adhesion and growth using human mesenchymal stem cells (hMSCs). These results were compared to gelatin scaffolds, which is denatured collagen and well established for its favorable cell adhesion properties[3]. Electrospun fibrous gelatin and zein scaffolds were evaluated for cytocompatibility, stiffness and hydrolytic stability. The osteogenic differentiation of MC3T3-E1 pre-osteoblast cell line was evaluated on zein scaffolds. MC3T3-E1 cells were cultured in both standard growth media and induction media and evaluated for presence of osteogenic specific markers.
Methods: Fabrication of Scaffolds: 24% (w/w) gelatin (bovine type B) in 60/40 acetic acid/water solutions and 30% (w/w) zein in 80/20 ethanol/water solutions were prepared for electrospinning. Solutions were electrospun using standard conditions and crosslinked. **Cell Culture:** Human mesenchymal stem cells (hMSCs), passage 5, were seeded at 32,000cells/cm^2 onto gelatin and zein scaffolds, cut into discs and cultured in complete growth medium (DMEM, 10% FBS, 1% antibiotic/antimycotic) for up to 11 days. MC3T3-E1, passage 5, were cultured up to 21 days on zein scaffolds at 30,000 cells/cm^2 in either complete growth medium (GM; DMEM, 10% fetal bovine serum, 1% antibiotic/antimycotic) or osteogenic induction medium (OM; GM supplemented with 0.05mM of ascorbic acid, 10mM of β-glycerophosphate, and 100nM of dexamethasone). **Cell Growth:** Quanti-iT PicoGreen® dsDNA reagent was utilized to quantify cell number at days 1, 7 and 11. Confocal imaging of hMSCs seeded on scaffolds was conducted at day 1 or day 7 and stained with Phalloidin (actin). **Osteogenic Differentiation Studies**: Cell morphology and secretion of matrix proteins associated with osteogenic differentiation (osteocalcin) were evaluated with confocal imaging by immunostaining (anti-osteocalcin), rhodamine phalloidin (cytoskeletal actin filaments) and DAPI blue (cell nuclei stain). Matrix mineralization was quantified by the calcium assay kit following the manufacturer's protocol (BioAssay Systems) and evaluated by SEM with energy dispersive X-ray spectroscopy (EDX).
Results: Cell number increased on zein scaffolds at later time points as compared to gelatin. Confocal images demonstrate viable cell attachment and standard hMSC morphology on both gelatin and zein scaffolds. Mechanical properties of the two scaffolds were evaluated by conducting tensile testing. The elastic modulus was 195 kPa±47 kPa (zein) and 194 kPa ± 53 kPa (gelatin). Hydrolytic stability studies of these scaffolds also exhibited similar percent weight loss of 7.4%±3.6% (zein) and 6.7%±1.0% (gelatin) over 42 days suggesting stability in hydrolytic conditions.

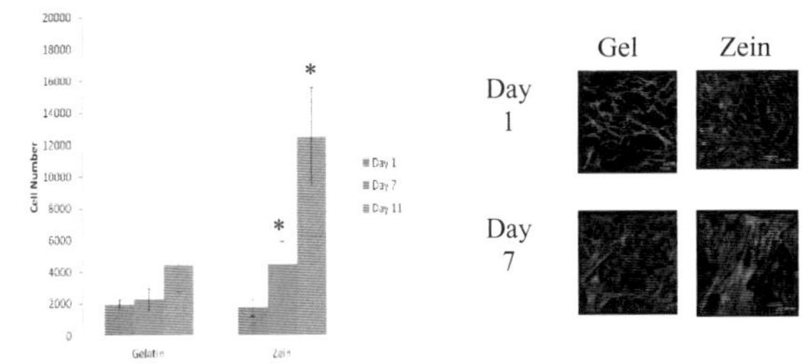

Figure 1. hMSC growth on zein scaffolds compared to gelatin scaffolds, *p<0.05. Confocal imaging showing hMSCs attachment on gelatin and zein scaffolds with actin staining (green).

Osteogenic potential of MC3T3-E1 cells was observed when seeded on zein scaffolds. Confocal imaging of MC3T3-E1 cells on zein scaffolds in OM conditions expressed osteocalcin by day 21. Mineral deposits by the MC3T3-E1 cells were observed using SEM at day 21 on zein scaffolds. Calcium deposition was significantly higher for cells seeded on zein scaffolds than on TCP control in GM conditions suggesting osteogenic potential.

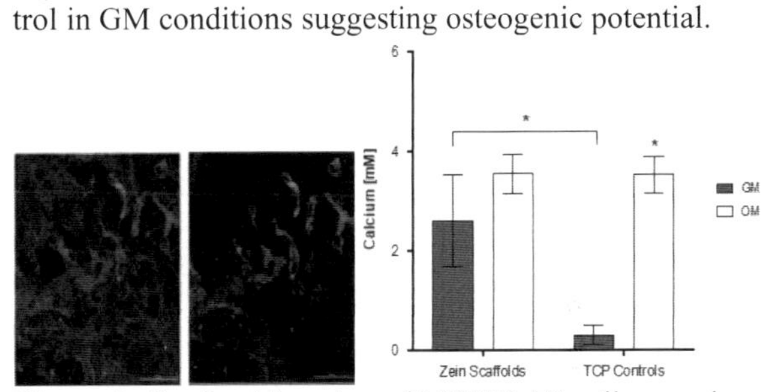

Figure 2. Confocal imaging of MC3T3-E1 cells on zein scaffolds (day 21 in OM conditions) showing osteocalcin (green) and actin filaments (red). Calcium quantification on scaffolds and TCP controls. * indicates statistical significance (p<0.05).

Conclusions: These results suggest that zein, a plant-derived protein, may be a suitable alternative to the animal-derived protein gelatin and demonstrates its potential as a promising scaffold for bone tissue engineering.

References:

1. Jiang Q. Acta biomaterialia. 2010;6(10):4042-51
2. Shukla R. Industrial crops and products. 2001;13(3):171-92
3. Hoque ME. Polymers Research Journal. 2015;9(1):15

Acknowledgements: This work supported by the Center for Engineering MechanoBiology (CEMB), an NSF Science and Technology Center, under grant agreement CMMI: 15-48571

Stop Go Flow Encapsulation

Scott McKirgan, BS, Ayesha Aijaz, PhD, Biju Parekkadan, PhD.
Department of Biomedical Engineering, Rutgers University

Introduction:Microencapsulation is a technique that can be used for cell transplantation, organ replacement and the controlled delivery of drugs[1]. Allogeneic and xenogenic pancreatic islet transplantation also utilizes cell microencapsulation however, microencapsulation scale-up remains a challenge. Compared with extrusion-drip methods, high-throughput capsule production by emulsification techniques is easier because they are not limited by scale. Herein, we present a T-junction device that produces droplets that are polymerized under white light. Continuous flow encapsulation will help meet the product demand to treat more patients a year and decrease the price of manufacturing considerably.

Methods: A photopolymerization emulsion encapsulation method is used in which mixing an oil and pre-polymer solution creates the capsule shape. Photopolymerization was achieved with a pre-polymer, co-initiator, photoinitiator, comonomer, oil, and a light source. A metal halide lamp with liquid light guide (Fiber Light MH -100) is used to initiate free-radical based light polymerization. Poly (ethylene glycol) diacrylate (PEGDA) is used as the macromer backbone, Triethanolamine (TEOA) co-initiator, Eosin Y the photoinitiator, Pluronic Acid, 1-vinyl-2 pyrrolidinone (NVP) to accelerate gelation kinetics in a HBS Solution[2]. The oil phase is composed of mineral oil with acetophenone/NVP. These solutions were implemented into peristaltic pumps.
Using peristaltic pumps the polymer solution (1.3 mL/min) was pumped into a T junction along with the oil solution (144mL/min) to form droplets. The pumps were then stopped and the droplet formation was exposed to white light with the Fiber Light MH-100 for 15 seconds to polymerize and crosslink the PEGDA and form a capsule.
Optimization of solutions using current method of encapsulation have also been implemented to determine the chemicals are needed to form different types of couples.

Results:
Capsules were created using a peristaltic pump stop go flow with a light exposure time of 15 seconds. (Figure 1A, B).
Using all chemicals in the solutions mentioned previously the capsules created have crosslinked core and shell. Removing Triethanolamine (TEOA) created a capsule with crosslinked shell and liquid core was created by removing Triethanolamine (TEOA) (Figure 1D).

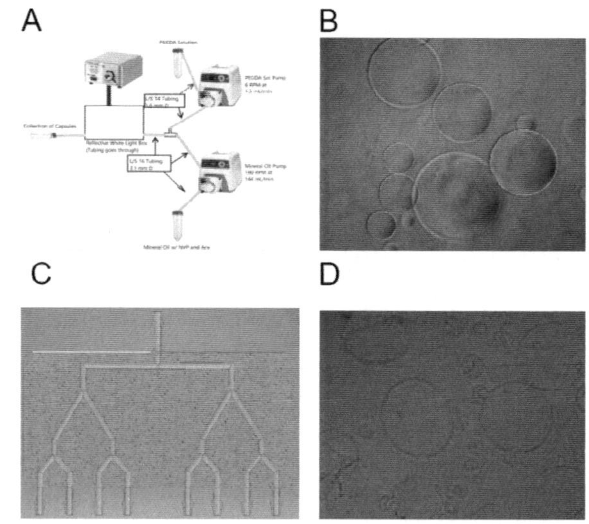

Figure 1 A: Stop Go Flow Encapsulation Setup, B: Capsules from Stop Flow Encapsulation, C: 3D printed device is used to divert the flow, D: Capsules created without Triethanolamine (TEOA).

Conclusions: Capsules were created using the stop go flow encapsulation model with a light exposure time of 15 seconds. In the future this method of encapsulation can be implemented to create a continuous flow system at a large scale the flow rates must be high enough to scale up the encapsulation but slow enough for the mixed solutions to get enough light exposure for polymerization. High flow rates during mixing phase at the T junction need to be slowed down enough after to get long light exposure. A 3D printed device could be used to divert the flow into multiple different outlets to slow down the mixed solution for long enough light exposure. (Figure 1C).
Capsule formation using different chemicals can be used for different capsules for specific cells. This is a customizable option for the continuous flow in the future.

References:
1. G J. Lim. Prin of Reg Med 2008; 954-966
2. 1.Hao, Y. 2014; 34(3): 269-273

Acknowledgements: TechAdvance grant as funding source.

The biomechanics of distal colon and rectal wall and its implication in visceral sensation and hypersensitivity

Background: Visceral pain is the cardinal symptom of irritable bowel syndrome (IBS), afflicting 13% of the US population and costing healthcare ~$30bn annually. Current drug treatments for visceral pain have significant side effects that often outweigh the analgesic benefit. Unlike other types of pain, visceral pain has a unique biomechanical component: it is mechanical distension/spasm of hollow visceral organs – not heat, cutting or pinching – that reliably evokes pain from these organs. Encoding of mechanical stimuli (i.e. mechanosensation), relies first on a biomechanical process, namely transmission of bulk mechanical deformation in visceral organs to local stress/strain distribution in microns-thick sensory nerve endings (i.e., afferent endings) in the organ. However, this biomechanical process of mechanosensation has never been mechanistically understood. In this study, we systematically studied the biomechanical properties of mouse distal colon and rectum (colorectum) using combined biaxial tissue testing and nonlinear imaging method on different layers of the colorectum to reveal the bulk and micro-mechanical environment of sensory nerve endings.

Methods: The ~30mm distal colon and rectum were harvested from male C57BL/6 mice, placed in phosphate buffered saline containing L-type Ca^{2+} blocker (nifedipine), protease inhibitors and antibiotics (penicillin and streptomycin), and stored at 4°C overnight to completely relax smooth muscle activities. The cylindrical colorectum is then cut open, pinned flat, and separated into mucosal and muscular layers by fine blunt dissection at the submucosal space. Tissue squares of 6x6mm, either mucosal, muscular or combined layers were harvested from along the longitudinal locations of the colorectum, which were subjected to bi-axial tissue testing via a servo-controlled two-channel force actuator (Model 300D, Aurora Scientific) that records both bulk tissue displacement and force. In addition, the tissue surface was speckled by carbon black powder and video recorded during the deformation to allow imaging processing and calculation of the locale mechanical strain. Immediately after biaxial testing, some tissue samples were subjected to non-linear imaging (second harmonic generation, [SHG]) to reveal collagen fiber orientations throughout the thickness of the tissue via two-photon fluorescent microscopy (Zeiss LSM 780).

Results: From bulk biaxial mechanical tissue testing, the Cauchy stress versus stretch ratio plots indicate that the mechanical strength is comparable between the circumferential and axial directions for both colonic and rectal tissues. Compared with the muscular layers, mucosal layers are less stiff but do provide significant load bearing strength in contrast to what is predicted from their folded villi-crypt structures. The SHG imaging data revealed a significant network of collagen fibers at the basal layer of the mucosa and submucosa, which likely subserves the unexpected load-bearing feature of the mucosal layer.

Conclusions and significance: Our biomechanical studies of distal colon and rectum have discovered the unexpected mechanical strength of the mucosa, especially the submucosal region, which abundant extrinsic and intrinsic nerve endings reside in. Further studies on the biomechanics of distal colorectum will nicely complement the neurophysiological approach to synergistically advance our mechanistic understanding of colorectal afferent neural encoding and sensitization in disease conditions of IBS. In addition, studying the biomechanics of colorectum will likely guide the next-generation drug development that targets biomechanical factors to treat visceral pain, which could mechanistically avoid off-target side effects to the CNS because mechanosensation arises in the peripheral organs.

P2-7

Synthetic Tear Duct Drainage System

Kyle Fehn, Alexis Rader, Robert Gettens, Christopher Thiagarajah.
Western New England University, Department of Biomedical Engineering, Springfield, MA 01119.

Significance: Disgruntled patients with blocked tear ducts need a reliable and comfortable way of draining tears to the nasal cavity because the current Jones Tube is brittle, uncomfortable, and prone to migration. The Jones Tube was developed by Lester T. Jones and Gunther Weiss in 1961. Lester Jones was the lead designer and Gunther Weiss was a professional glass blower [1]. Today, Gunther Weiss Scientific Blowing Company manufactures the majority of all Jones Tubes [1]. A Jones Tube is implanted after canalicular obstruction. Most commonly this obstruction is due to trauma from either a dog bite or other injury. The Jones Tube exists in standard sizes. The diameter of a Jones Tube ranges from 3.0mm to 5.0mm, increasing in .5 mm intervals. The length of the Jones Tube ranges from 9mm to 40mm [1]. Many designs already exist. Some Jones Tube have a 130° angle present while some have a suture hole allowing for the surgeon to suture the Jones Tube to the corner of the eye. Frosted glass has been used as a way to combat migration however, it has proven ineffective [2]. Flanges have also been added to combat migration. This has been seen to prevent extrusion but not intrusion [3].

Design Description: Through the ideation process, a final design concept was chosen for prototyping. The final concept (Fig. 1) is a combination of three concepts that were previously generated.

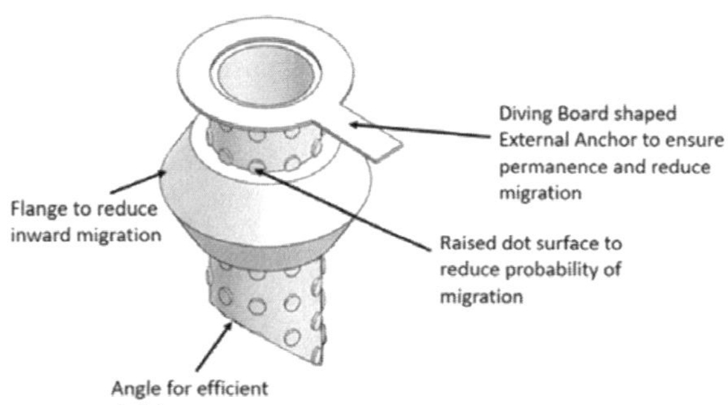

Figure. 1. Final Concept

The model (Fig. 1) shows the shape, interface, and surface of the device. Other aspects that affect the design include the material that will be used to create the device. The material for this device was chosen by testing the hydrophobicity of several candidates. This test helped aid in the choosing of silicone as the material this device will be made from (Table 1). Further research into the manufacturing process that will be used to make the device is needed. Initial product sterilization methods will most likely be gamma radiation. Instructions will be included with the device. The instructions will include patient care, handling, and storage. The instructions and device will be packaged in Wipak®. Wipak® is the most common packaging for sterilized medical devices and tools.

Personalized Fit: One option is to allow for personalized fit using imaging and additive manufacturing techniques. A CT scan can be used to enhance the outcome of the surgery. The scan from the patient can be used to create a 3D model using a software package of 3DSlicer® and MeshMixer® (Fig. 2). This 3D model can be used to determine proper placement of the implant and custom sizing for that patient's specific anatomy. This would also aid in surgery preparation to ensure the surgeon knows exactly where the implant is going and which size tube to use. This path can be used or standard sizes of jones tubes can be created and the current method of surgery will not have to be modified. Further testing and design are needed before this can be finalized.

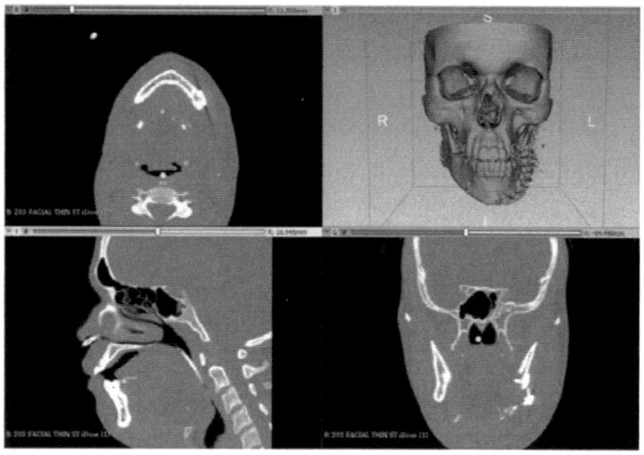

Figure. 2. 3DSlicer® Screenshot depicting the CT scan and 3D Model created through choosing specific parts of scan.

To prototype the device, three materials were chosen based on the production cost and speed. For the first processing step, polylactic acid is used in 3D printing. In the next step, silicone (PDMS) is used as a mold. Silicone is again used to create the prototype.

Design Criteria: The final concept was developed using the FDA's waterfall design process. User needs were taken into consideration. One user need is that patients found Jones tubes uncomfortable and corrective surgeries were needed to combat migration. To address these user needs, the device must be made from a softer plastic and designed to prevent movement. The materials studied had to be hydrophobic in order to effectively drain tears. The materials tested were silicone, polyethylene (HDPE), polyethylene (LDPE), and acrylic. The contact angle method was used to test the materials for hydrophobicity. The design specifications were based on the customer needs (Table 1). The specs were created using industry standards. Each specification has a metric, unit and value associated with it. The metric values are based on the current glass Jones tube. These specifications were involved in creating concepts and then deciding on the final one.

Table 1. Specifications of Drainage System

Category	#	Metric	Units	Value
Biocompatibility	1	International Standard ISO 10993-1, "Biological evaluation of medical devices - Part 1: Evaluation and testing within a risk management process" for Implant in Tissue	Binary	ISO 10993-1:2009 recommended endpoints
Material Properties	2	Diameter	mm	3.0-6.0
	a	Length	mm	8.0-40.0
	b	Stress (Ultimate)	N/m^2	>210
	c	Strength	MPa	>62
	d	Lightweight	g	~50
	e	Durometer (Softness)	Unitless	<60
	f	Contact Angle	Degrees	>90
	g	Roughness	µm	>.05
Sterilization	3	Sterilized packaging	Binary	Pass/Fail
Cost	4	Manufacturing Cost	USD	<100
	a	Medical Cost (Covered by Insurance)	USD	<5-7k
Recovery Time	5	Time	Weeks	<2-3

Results: As described in the design criteria, the material for this device was chosen by testing the hydrophobicity. This test helped aid in the choosing of using silicone as the material this device will be manufactured out of (Table 2).

Table 2. Contact Angle using Two Different Liquids

Material	Distilled Water (°)	Artificial Tears (°)
HDPE	70.77	43.1
LDPE	58.11	40.55
Silicone	92.13	73.14
Acrylic	70.90	48.29

Future Implications: This design will drain the tears of the patient effectively without causing physical or emotional harm. This ensures the customer needs and specifications are met. The device is currently in the prototyping phase. Drain rate will be measured through testing to see if the concept effectively drains tears. Further research into the manufacturing process to make the device is needed. Sterilization methods will be further studied based on silicone. Further work will be done to verify and validate the device.

References:

[1] E. A. Design, "Jones tubes: what are they, how do they work and what's the right tube for you?," Lester Jones Tube. [Online].
[2] L. Bagdonaite and A. R. Pearson, "Twelve-Year Experience of Lester Jones Tubes—Results and Comparison of 3 Different Tube Types," Ophthalmic Plastic and Reconstructive Surgery, vol. 31, no. 5, pp. 352–356, 2015.
[3] G. B. Bartley and R. O. Gustafson, "Complications of Malpositioned Jones Tubes," American Journal of Ophthalmology, vol. 109, no. 1, pp. 66–69, 1990.

Acknowledgments: The design team would like to thank Greg Gagnon of Baystate Medical Center for collaboration with image processing and fabrication techniques.

P2-8

Effect of Bifurcation Angle on Red Blood Cell Lingering and Partitioning

Andrew Pskowski. Jeffrey D. Zahn
Department of Biomedical Engineering, Rutgers University, Piscataway, NJ

Introduction: Blood flow in the microvasculature provides the surrounding tissues and organs with the majority of nutrients and oxygen needed to survive. Because of this large role, an understanding of the dynamics of blood flow in these vessels, which range from 4-100 µm in diameter is important. We previously reported RBCs "lingering" at bifurcations in microchannels where lingering is characterized by the transit time needed for the red blood cell (RBC) to perfuse through the bifurcation. RBCs with transit times greater than 1.25 the average transit time are considered to "linger", often deforming and remaining pinned against the bifurcation for milliseconds (figure 1A-D). Lingering RBCs have been shown to cause perturbations in hematocrit both upstream and downstream of the bifurcation. These perturbations can have implications for local oxygen delivery to surrounding tissues. Here we build upon that research by investigating the role of both bifurcation angle and RBC position upon lingering and partitioning.

Materials and Methods: *Sample preparation*: Blood was collected through finger prick from healthy consenting volunteers into a heparinized tube (Ram Scientific, Nashville, TN) and washed 3 times in phosphate-buffered saline (PBS). The sample was diluted in PBS to reduce the hematocrit to either 20% or 1% for testing. Hematocrit was confirmed by hemocytometer. *Experimental Setup:* Y-shaped bifurcating microchannels were fabricated using traditional soft lithography techniques. Devices were designed with bifurcations of 60°, 90°, 160°, 180°, and 300°. Prior to perfusion devices were incubated in 1% bovine serum albumin for 1 hour to prevent non-specific cell binding. RBCs were pipetted into the reservoir at the entrance of the device. A negative pressure gradient generated by water columns on the device outlets enable independent flowrate control in the daughter channels. Images were captured at 40× magnification using a high-speed camera and analyzed via a custom MATLAB script.

Results and Discussion:
Cell partitioning (ratio of cells in daughter channel to feeder channel, N*) is not significantly affected by bifurcation angle. The frequency of cells lingering *(n)* is significantly greater in the 160° bifurcation than the 60° and 90° as seen in figure 1B (two sample t-test, $p<0.05$). We believe this is due to greater wall surface area found in the bifurcation of the 160° degree bifurcation compared to the smaller bifurcations and the flatter wall reducing the reliance on RBC position to cause cells to linger. As Q* (flowrate ratio of daughter channel to feeder channel) increases *n* decreases due to more RBCs moving away from the center as they enter the bifurcation.

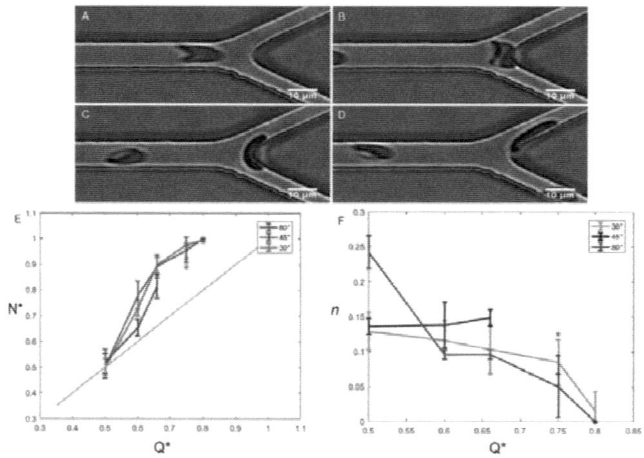

Figure 1 (A) RBC entering bifurcation (t=0). (B) RBC beginning to linger (t = 3 ms). (C) RBC lingering (t = 6.7 ms). (D) RBC moving off into daughter channel (t= 20.3 ms). (E) Angular dependence of N* vs Q* and (F) lingering frequency vs Q*.

Conclusions: The study demonstrates that there is a correlation between bifurcation angle and frequency of lingering cells. Further studies will involve angles of 180° and greater as well as the continued use of shear focusing to investigate the role of cell position in this phenomenon. These studies can provide insight into what tissues this phenomena may take places, as well as the effects on tissue oxygenation.

References: (1) Balogh, P. *Biophys. J.* 2017, 113:2851-2826. 12

Acknowledgements: This work supported under National Science Foundation NSF CBET 1353918

Injectable Solution of Epigallocatechin Gallate as a Potential Preventative Measure against Cartilage and Connective Tissue Damage

Mary Pat Reiter[1], Adrian B. Mann, PhD[1,2], Joseph W. Freeman, PhD[1].
Rutgers University, Piscataway – Department of Biomedical Engineering[1], Department of Materials Science and Engineering[2]

Introduction: Cartilage and connective tissue damage can be the result of any number of conditions including genetic disorders, athletic injury, and degenerative diseases. The focus of this research was for treatment of osteoarthritis (OA) of the articular cartilage in the knee and the sprains and strains of connective tissue in patients with genetic disorders such as Ehlers-Danlos syndrome. OA is a debilitating disease defined by degradation of joint tissue resulting in significant pain and hindered joint mobility diminishing patients' quality of life. OA's impact on articular cartilage is both marked by areas of thinned cartilage and exposed bone and a decrease in the mechanical properties of the remaining cartilage. Sprains and strains are a serious concern for injury-prone patients especially those with connective tissue disorders since such conditions cause joint instability and reduced tendon and ligament strength making healing difficult. Few treatments exist for people who have OA or connective tissue disorders making development of prevention and treatment methods a critical goal of healthcare, tissue engineering, and biomechanics research. Green tea's effects on the human body have been studied for decades and sworn by for millennia, but recent research has shown that its most abundant catechin, Epigallocatechin Gallate (EGCG), is responsible for many of its benefits. EGCG is a polyphenolic compound and has been shown to alter the levels of inflammatory cytokines as well as cross-link with collagen, a major component of articular cartilage and connective tissues. These studies aimed to test the theory that cross-linking EGCG with collagen can increase the mechanical properties of articular cartilage and connective tissue.

Methods: All subjects of the nanoindentation study were female Lewis rats weighing approximately 350 grams each. In the experimental group, three rats of approximately were injected daily in the lateral side of the hindleg knee cavity for five days with a 300 μg/25μl solution of EGCG (Sigma-Aldrich) in saline in each right knee and 25 μl of saline in each left knee. For the control group, three rats underwent no injections. All subjects were sacrificed on Day 7, two days following the final injection. Right and left hindlegs were frozen in saline. Articular cartilage on the tibial plateaus were evaluated with a TriboIndenter by Bruker/Hysitron (Eden Prairie, MN) with a fluid-cell tip while submerged in phosphate-buffered saline (PBS) using a load-controlled function and quasi-static indentation. Data analysis was completed with the Oliver-Pharr method to determine elastic modulus, hardness, and stiffness. For the tensile study, rat tendons were extracted from frozen tails and preserved in PBS. Experimental groups were exposed to 0.06% (wt/v) EGCG in PBS for 2.5 hours and rinsed while control groups were kept in PBS. Tendons were evaluated with an Instron 5800 (Norwood, MA). Data was analyzed with MATLAB to determine elastic modulus and ultimate tensile strength.

Results: In the cartilage study, the treated right knees showed a statistically significant increase in the properties of elastic modulus, hardness, and stiffness compared to the respective untreated left knees of each subject (±<0.05). The means of the ratio of the right legs to left legs for Group 1 were statistically significant for all mechanical properties (±<0.05), (*Figure 1*).

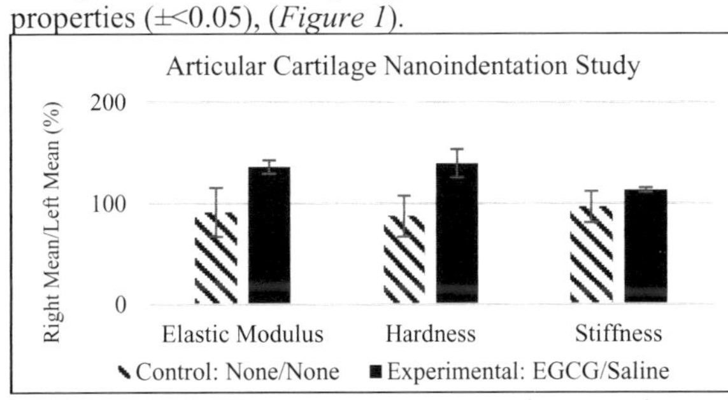

Figure 1. Nanoindentation study comparing experimental and control groups

In the tensile testing study, all experimental samples had statistically significant increases in means compared to the control groups except for test 3 for Elastic Modulus (±<0.05), (*Figure 2,3*).

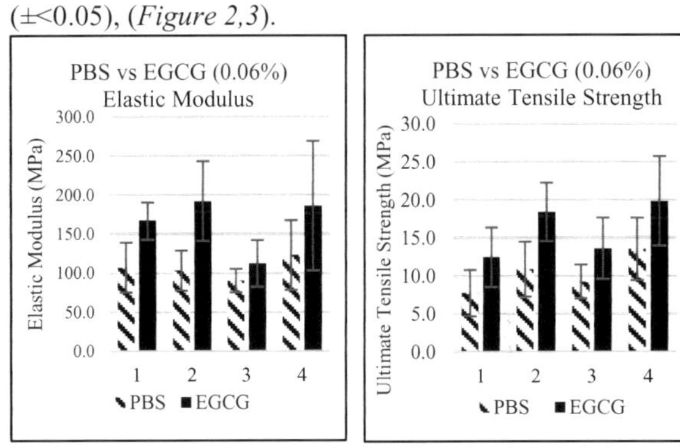

Figures 2, 3. Tensile testing data for Elastic Modulus and Ultimate Tensile Strength for PBS vs EGCG groups

Conclusions: In this study, we examined the effectiveness of a polyphenolic treatment in increasing crucial mechanical properties of healthy cartilage and connective tissue. For cartilage, we compared the findings for each leg within each subject as well as the relative increase of treatment for all the experimental subjects compared to subjects who had no injections. For tendons, we compared treated to untreated across individual vertebra and expanded the data by repetition. Our findings in conjunction with its established anti-inflammatory effects suggest that EGCG injections could be a potential preventative measure against cartilage and connective tissue damage. Future recommended studies include nanoindentation with a larger sample size, staining histological samples, and evaluating damaged tendons.

Acknowledgements: The authors would like to thank Shawn H. Ward, PhD and Michael Pellegrini for their assistance.

Effect of Counterface Surface Roughness on Tribological Rehydration of Articular Cartilage

Meghan E. Kupratis, B.S[1]., Margot S. Farnham, M.S.[1], David L. Burris, Ph.D.[2], Christopher Price, Ph.D.[1,2]
[1]Biomedical Engineering, University of Delaware [2]Mechanical Engineering, University of Delaware

Introduction: Articular cartilage comprises the load-bearing tissue of synovial joints and, in the absence of injury or degradation, can support smooth articulation over decades of use. A major interest in cartilage biomechanics is to understand how cartilage can maintain low friction values ($\mu<0.02$) under repetitive loading conditions.[1-3] Recently, our team rediscovered a unique *ex vivo* tribological testing configuration, the convergent stationary contact area (cSCA) that allows assessment of cartilage biomechanics under fully-sustainable and physiologically-consistent cartilage strain, strain recovery, hydration, and lubricity over hours of testing; through a mechanism we have termed 'tribological rehydration.'[4] Our Previous studies have used standard glass microscope slides as the sliding counterface. In order to further characterize cartilage biomechanics in the cSCA configuration, *the goal of this study was to investigate the effect of glass counterfaces of different surface roughnesses on cartilage tribomechanical outcomes.*

Methods: Tissue Specimens and Tribological Testing: 19mm osteochondral cores were removed from the femoral condyle centerline of mature bovine stifle joints. Explants were stored in PBS + protease inhibitors[5] and tested using a custom-built tribometer[6] in which explants were compressed and slid against a reciprocating glass slide. Glass Surfaces: Glass slides of three different roughness and asperity heights were tested in this study: super-polished quartz microscope slides (SPI Supplies, West Chester, PA), plain microscope slides (Fisher Scientific), and frosted microscope slides (Fisher). Testing Protocols: A tribological rehydration characterization scheme, with a repeated-measures design, was used for each testing group. It consisted of 30 min of static compression at 7N (~0.25 MPa), followed by 30 min of reciprocal sliding at 80 mm/s (~walking speed) under 7N compression. Characterization was first performed against a plain microscope slide, then repeated against polished, plain, and frosted slides, sequentially. Between sliding tests, explants free-swelled in PBS for two hours. After testing, damage was assessed using stereomicroscopy, and explant were bisected to measure cartilage thickness (h). Data Analysis: Deformation (δ), normal force (F_N), and friction coefficients (μ) recorded by the tribometer were analyzed using MATLAB. Measures of tissue strain ($\varepsilon = \delta/h$) were calculated, strain and friction magnitudes were analyzed at the beginning and end of active sliding, and strain recovery (i.e. tribological rehydration) during sliding was measured. Characteristic deformation rates were obtained from the time-dependent deformation data. Friedman's Test was used to identify statistically significant changes between the repeated tests.

Results: Deformation and friction responses were similar with sliding against regular and polished slides, and strain recovery (i.e., tribological rehydration) was observed under both of these conditions. The most pronounced changes occurred when explants were tested against 'rough' glass, where tribological rehydration was not observed. When sliding against rough glass, strain and deformation recovery were significantly impaired ($p<0.0001$). Equilibrium coefficients at the end of sliding were similar for all three surfaces ($\mu = 0.04 - 0.07$). However, there was a significant increase in start of sliding friction coefficient for rough glass ($\mu = 0.5288$, $p<0.0001$) compared to regular and polished surfaces ($\mu= 0.288$ and 0.289, respectively; $p=0.9996$). Surface roughness had no effect on the static deformation time constant ($p>0.1$); however, increased surface roughness had a significant effect on the sliding deformation time constant relative to regular and polished glass surfaces ($p<0.0001$), consistent with the lack of strain recovery observed (positive; **Fig. 1**).

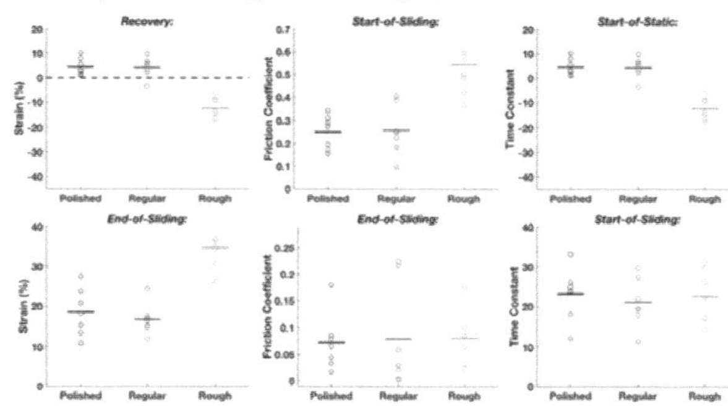

Figure 1: Sliding-induced strain recovery occurs with both polished and regular, but not rough glass; the static compression time constants were similar for polished and regular glass while explants reached characteristic deformation faster when compressed against rough glass; end-of-sliding deformation time constants show that strain recovery occurred for polished and regular glass while sliding against rough glass resulted in continued deformation

Conclusions: The results of this study illustrate that decreasing surface roughness of the glass counterface has no appreciable effect on the tribomechanical response of articular cartilage under the cSCA testing configuration. These findings suggest that regular microscope slides are adequate for assessing the tribological behavior of articular cartilage in the cSCA and that ultrapolished glass is not necessary for these experiments. Conversely, the use of frosted glass, which is of a greater asperity height, and thus, surface roughness, prevents cartilage from sustaining tribological rehydration, and induces clear tissue wear.

References: [1] Mansour, J.M., *Kines*, 2009. [2] Wright, V. et al., *J Anat*, 1976. [3] Ateshian, G.A. et al., *J Biomech*, 2009. [4] Moore, A.C. et al., *OA&C*, 2015. [5] Stewart, R.C. et al., *Cart*, 2016. [6] Moore, A.C. et al, *OA&C*, 2016. [7] Bonnevie, E.D., *JOR*, 2018.

Acknowledgements: This work was supported by the NSF BMMB [1635536]. All opinions, findings, and conclusions are those of the authors and do not reflect views of the NSF.

Biophysical Model of Ebola Virus Interactions with TIM Proteins

Xinyu Cui[1], Luke Wang[1], Chuqian Xiong[1], Nicole Lapinski,PhD[3], Xiaohui Zhang,PhD[1,2] and Anand Jagota,PhD[1,3]

[1]Department of Bioengineering, [2]Department of Mechanical Engineering & Mechanics, [3]Department of Chemical and Biomolecular Engineering, Lehigh University, 19 Memorial Drive West, Bethlehem, PA 18015, USA.

Introduction: Viruses often hijack normal physiological processes in order to be uptaken by the cell they infect. Therefore, the principal difficulty in designing therapies against viruses lies in the fact that blocking binding to a cell surface is likely also to deleteriously affect normal physiological processes. Our focus is on the Ebola virus (EBOV), which has posed a major threat to global health. It utilizes T-cell immunoglobulin and mucin domain (TIM) protein as its main receptors on the cell surface. The TIM protein family is composed of IgV and mucin like domain (MLD), where IgV can interact with phosphatidylserine (PS) on the outer surface of EBOV. Recent studies shows that for TIM-dependent EBOV entry, MLD of at least 120 amino acids is required.[1] We have two competing hypotheses for this requirement, that (a) sufficient length is needed for the flexibility of the TIM stalk as it seeks a binding partner, or (b) without sufficient length the electrostatic and hydration repulsion between the cell surface proteins and virus coat overcome the TIM/PS attraction. The main goal of the work is to (1) test these hypotheses with Brownian Dynamics simulations of a coarse-grained model of TIM/MLD/PS, and (2) develop a continuum model for EBOV-TIM cell interaction that is driven by its adhesion to cell surface receptors and resisted by membrane bending or tension. Available models, to our knowledge, have not been put to such use yet.

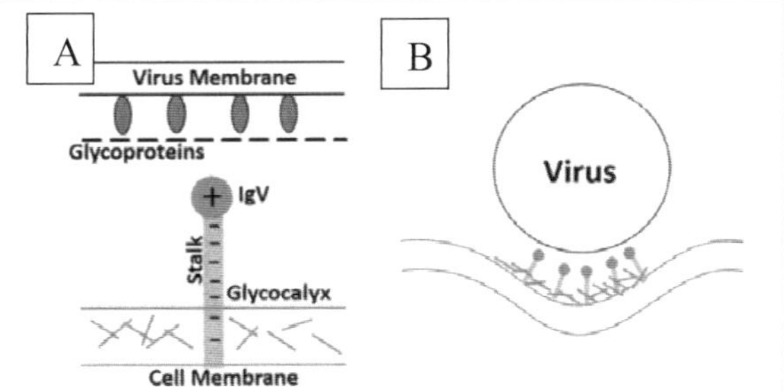

Fig. 8. Biomechanical Model for TIM-mediated viral attachment to a host-cell. The problem is being studied at two length scales: (A) At the intermolecular scale the adhesion between a single TIM-1 and the virus membrane is studied using a coarse-grained model with Brownian Dynamics. (B) The mechanics of whole virus attachment and internalization incorporates the interactions calibrated in (A) into a model at the scale of the viral particle (tens to hundreds of nm). At the whole virus length scale, the effects of membrane mechanics, specifically its bending rigidity and tension, are incorporated.

Methods: The Brownian Dynamics model is composed of PS plane, IgV and MLD beads connected serially. One additional dummy bead whose radius is much less than that of the IgV bead is placed between PS plane and MLD beads in order to set up the right force between PS plane and IgV bead. A Lennard- Jones potential is applied to the dummy bead and a time-dependent force is applied to the last MLD bead at the bottom. The simulation is built and run through LAMMPS (Large-scale Atomic/Molecular Massively Parallel Simulator) and result is analyzed by MATLAB and VMD (Visual Molecular Dynamics). The continuum model is built by deriving an expression for the total energy of the system as a sum of contributions from the tension and bending of the membrane, adhesion between the membrane and viral particle, and the potential of external force:

$$U_{Tot} = U_{Ten} + U_{bend} + U_{ad} - P\delta \quad (1)$$

The adhesion energy is proportional to the contact length while the deformation energies are obtained by solving the beam/membrane equation:

$$\kappa\frac{d^4w}{dx^4} - T\frac{d^2w}{dx^2} = F \quad (2)$$

where w is the deflection of the membrane, κ is the bending rigidity, T is the tension, and F is the distributed applied force. For given parameters, we compute the force-deflection relationship by minimizing the total system energy with respect to contact width.

Results: Results of Brownian Dynamics simulations show that the simulated pulling force increases logarithmically with loading rate, consistent with experimental measurements.[2] This supports the capability of our model to study single-molecule pull-off experiments. The continuum model reveals that adhesion contact width depends on two dimensionless parameters. When bending of the membrane dominates, the solution depends on a single parameter, $\alpha = \kappa/(2\rho WR^2)$. In particular, its value needs to be smaller than unity for any adhesion to occur.[2]

Conclusions: Our preliminary results establish the coarse-grained model as a promising tool for the study of binding between individual receptors and the viral surface. The continuum model shows how full-virus adhesion can be modeled, now including effects of membrane bending and tension.

References:

1. Moller-Tank S. J Virol. 2014;88:6702-13.
2. Matthew A.Dragovich et al. Sci. Rep. 2019;9(1):267

Acknowledgements: This work is supported in part by NIH grant R15AI133634 and NSF grant 1804117 to A.J. and X.Z.

Developing Methods to Validate a Subject-Specific Magnetic Resonance Based Finite Element Model to Predict Strain in the Femur

Bryan Choate, MS[1], Karen L. Troy, PhD[1]
[1]Department of Biomedical Engineering, Worcester Polytechnic Institute, Worcester, MA

Introduction: Osteoporosis affects over 10 million people in the United States[1], resulting in over 1.7 million fractures and $70 billion in direct healthcare costs[2]. Failure to obtain and maintain optimal peak bone mass is a primary and modifiable cause of osteoporosis[3]. Signals related to mechanical strain magnitude and rate are thought to elicit improvements in bone mass and structure, especially in growing adolescents. However, it is not possible to directly measure these signals in living humans. Currently, computed tomography (CT) based computational models have been validated to estimate bone strain. Unfortunately, CT is not suitable for all populations (e.g. adolescents) due to high exposure of ionizing radiation. Thus, the objective of this project is to establish methods to validate a magnetic resonance (MR) based subject-specific finite element (FE) model as a safer alternative to estimate bone mechanical behavior.

Methods: This project was broken up into two distinct sequences: (1) computational development (image acquisition, image segmentation, FE model creation, and FE model simulation) and (2) experimental development (fixture design, fixture fabrication, fixture verification, speckle pattern formation for digital image correlation (DIC)).

Computational Development: Sample MR image sequences were created in collaboration with the University of Massachusetts Medical School using a 3T Philips Achieva Whole-body scanner. Bovine femora were imaged *in vitro* either wrapped in water saturated paper towels or submerged in water filled polyethylene bags. A volunteer (BC) was imaged to assess *in vivo* image quality. Consistency of image segmentation using Mimics 18 (Materialise; Leuven, Belgium) was evaluated by calculating the coefficient of variation (CV) from 3 repeated volume measures of segmented regions (integral, cortical, and trabecular) from a single scan at 3 subregions. A segmented three-dimensional (3D) object was created and meshed with 10 node tetrahedral elements for finite element (FE) modeling. Material properties (elastic modulus and Poisson' ratio) were assigned as either a function of bone density or set as typical values. Lastly, axial compression was simulated using replicated experimental boundary conditions using Abaqus CAE 2017 (Simulia; Providence, RI).

Experimental Development: To ensure model generalizability, five mechanical tests (four-point bend, axial compression simulated stance, simulated fall, and axial torsion) were designed based upon previously reported fixtures[4-6]. Fixture components were fabricated using standard subtractive manufacturing techniques (lathe and mill) and made out of 6061 aluminum for its versatility and corrosive resistant properties. Additional verification of the test setups was completed using an injection-molded femur (SAWBONES; Vashon, WA) by applying low forces (~100 N). To measure surface strain during testing, 3D DIC will be used. White enamel spray paint, black blow markers, and black enamel spray paint where used to create test speckle patterns on porcine ribs to determine their robustness for cadaveric bone specimens.

Results: *Computational Development:* The best MR image sequence used a voxel size of 0.5 mm^3, repetition time (TR) of 5.696 ms, echo time (TE) of 2.272 ms, fat-shift of 0.869 pixels, and pixel spacing of 0.446 pixel/mm. A combination method was found to be highly repeatable (Table 1) and visually most effective at segmenting bone from in vivo MR scans. Simulated axial compression of 1000 N produced nominal maximum principal element strains between 0 and +1000 ¼μ, which is within the physiologic range.

Table 1. Variance of Image Segmentation

	Coefficient of Variation		
	Integral	Cortical	Trabecular
Proximal	2.40%	5.02%	1.47%
Mid	0.33%	0.36%	0.33%
Distal	0.01%	0.16%	0.08%
Total	2.74%	5.54%	1.88%

Experimental Development: Speckles generated with blow markers were water soluble and not durable when compared to black enamel spray paint. Simulated fall and axial torsion fixture verification tests revealed fixture design deficiencies that resulted in an excessive bending moment and assembly difficulty respectively.

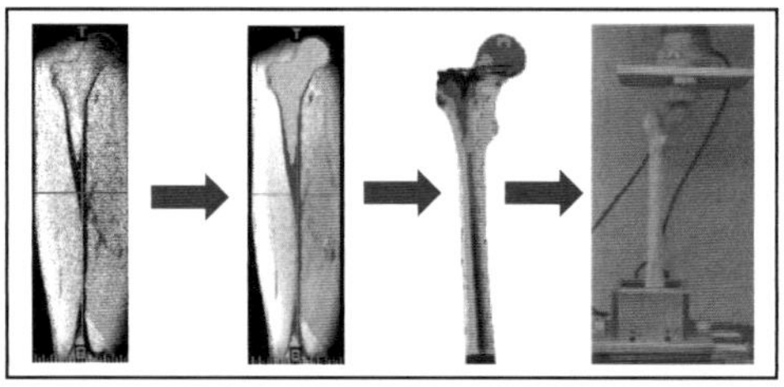

Figure 1. (a) Image, (b), Segmented, (c) FE, (d) Test

Conclusions: Currently, prototype methods for MR image acquisition, image segmentation, FE model creation/simulation, experimental testing, and speckle pattern application have been developed. Future work involving cadaver functional testing using 3D DIC, cadaver model creation and simulation, and the statistical analysis of resulting surface principal strain from both tests will need to be completed for full model validation.

References:

1. Wright NC et al. J Bon Min. 2014;29(11):2520-2526
2. Watkins-Castillo et al. 2014
3. Hernandez CJ et al. Osteo Int. 2003;14(10):843-847
4. Gardner MP et al. Ann of BME. 2010;38(3):613-620
5. Cristofolini L et al. J Biomec. 1996;29(4):525-535
6. Nishiyama KK et al. J Biomec. 2013;46(7):1231-1236

Acknowledgements: The authors would like to thank Shaokuan Zheng, PhD at the University of Massachusetts Medical School for his help in developing a suitable MR image sequence for our application.

Biomechanical characterization of TIM protein–mediated Ebola virus–host cell adhesion

Chuqian Xiong, Matthew A. Dragovich, Nicole Fortoul, Anand Jagota, Wei Zhang, and X. Frank Zhang

Lehigh University, Bethlehem, PA, 18015

Introduction: Since the most recent outburst, the Ebola virus (EBOV) epidemic remains one of the world's public health and safety concerns. EBOV is a negative-sense RNA virus that can infect humans and non-human primates, causing hemorrhagic fever. It has been proposed that the TIM (T cell immunoglobulin mucin like) protein family act as the major cell surface receptors for the PS (phosphatidylserine) which is expressed on the EBOV capsid. Here we characterize the mechanical strength of the TIM-PS bond that allows the virus-cell pair to resist external mechanical perturbations. This study utilizes single-molecule atomic force spectroscopy to quantify the specific interaction forces between two important candidates TIM-1 or TIM-4 and the following binding partners: PS, EBOV virus like particle (VLP), and EBOV glycoprotein/vesicular stomatitis virus pseudovirion (VSV). We have developed a simple model for virus–host cell interaction that is driven by its adhesion to cell surface receptors and resisted by membrane bending (or tension), showing how single molecule adhesion measurements relate quantitatively to the mechanics of virus engulfment by the cell.

Methods: Cantilever preparation/ coverslip preparation: To functionalize AFM cantilevers (MLTC, Bruker Nano) with PS, the cantilever was first silanized with 3-(trimethoxysilyl)propyl methacrylate to obtain surface thiol-groups. Phosphatidylserine(PS)-PEG-Maleimide (NANOCS) was coupled to the cantilever (3). VLPs (1, 2) of EBOV, or VSV were immobilized onto a (3-aminopropyl)-triethoxysilane sinalized AFM cantilever (MLTC, Bruker Nano) using a hetero-bifunctional polyethylene glycol (PEG) cross-linker, Acetal-PEG-NHS (Institute of Biophysics, Johannes Kepler University), according to the protocol developed by Dr. Hermann J. Gruber, Johannes Kepler University (http://www.jku.at/biophysics/content). Soluble recombinant TIM-1 or TIM-4 (R&D Systems) was attached to the sinalized glass coverslips using the same crosslinking approach. Functionalized cantilevers and glass surfaces were stored in PBS (3 × 5 min) and used for AFM experiment within 8 hours. **Single-molecule force measurements:** All single-molecule force measurements were conducted using a custom-designed AFM apparatus. AFM measurements were collected at cantilever retraction speeds ranging from 0.19 to 0.63 μm/s to achieve the desired loading rate (200-4,000 pN/s) or 1.53 to 0.63 μm/s. **Statistical Analysis**: For each pulling speed, over 500 force curves were recorded, which yielded 40 to 200 unbinding forces. Curve fitting was performed using IGOR Pro or Origin software by minimizing the chi-square statistic for the optimal fit. a p-value less than 0.05 considered to be statistically significant.

Results:

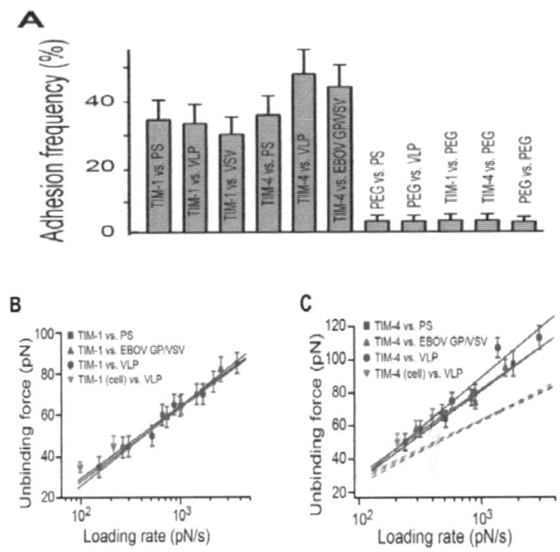

Interaction specificity was shown by the adhesion frequency measurement under the same measurement conditions. Fig. A shows a significant decrease in adhesion when either the TIM or its binding partner (PS, VLP or VSV) was absent, confirming that the vast majority of the recorded unbinding force stemmed from specific TIM-ligand interactions. The biophysical properties of TIM-ligand interactions were studied by the means of a dynamic force spectrum (DFS) (Fig B). The DFS is the plot of most probable unbinding force as a function of the loading rate. The loading rate is obtained by multiplying the system's spring constant and the pulling speed of each force curve. Fig. B and 2C show that the unbinding force of the TIM-1 and 4–PS, and TIM-1 and 4–VLP bonds increased linearly with the logarithm of the loading rate, ranging similarly from 40 to 80 pN and 40 to 100 pN over a loading rate of 200 to 4,000 pN/s, respectively. In order to confirm our findings on a cell surface, HEK293T cells were transfected with both TIMs. Under similar loading rates, the interaction of cell-bound TIM-1 and TIM-4 with VLP was mechanically comparable to that when purified soluble TIMs were covalently coupled to glass cover slips (Fig. B, C). A more detailed analysis of the biophysical properties of TIM-1 and TIM 4–ligand interactions were conducted by fitting the acquired DFS data to the Bell-Evans model and Dudko-Hummer-Szabo model.

Conclusions: The study shows the biomechanical parameters crucial for Ebola attachment to host cells. The study has demonstrated experimentally that TIM-EBOV interactions are mechanically comparable to adhesion molecule (e.g., selectin)–ligand interactions. Through a simple mechanical model, we further demonstrate how molecular binding parameters determine whether adhesion is sufficient for viral encapsulation. The study may provide new information to aid in the development of new antiviral therapeutics for the prevention and treatment of EBOV disease.

References:

1. Rhein BA, Brouillette RB, Schaack GA, Chiorini JA, & Maury W (2016) Characterization of Human and Murine T-Cell Immunoglobulin Mucin Domain 4 (TIM-4) IgV Domain Residues Critical for Ebola Virus Entry. *Journal of virology* 90(13):6097-6111.
2. Rankl C, et al. (2008) Multiple receptors involved in human rhinovirus attachment to live cells. Proc Natl Acad Sci U S A 105(46):17778-17783.
3. Moller-Tank S, Kondratowicz AS, Davey RA, Rennert PD, & Maury W (2013) Role of the phosphatidylserine receptor TIM-1 in enveloped-virus entry. *Journal of virology* 87(15):8327-8341.

Acknowledgements: This work was supported in part by a Lehigh Faculty Innovative Research grant, an NIH grant (1R15AI133634), and an NSF grant (CBET-1804117) to A.J. and X.F.Z.

Biomechanics of Arteries under Cyclical Pulsations in Hypertension and Drug Delivery

Yueya Ge[1], Mehmet Kaya[2], Vignesh Balasubramanian[2], Peter L.M. Kerkhof[3] and John K-J. Li[1].
[1]Rutgers University, [2]Florida Institute of Technology and [3]VU Amsterdam, the Netherlands.

Introduction: High blood pressure has been recognized as the most common risk factor of cardiovascular diseases, which are the leading cause of death. There is a strong correlation of blood pressure (P) and mechanical properties of arteries, as characterized by vascular compliance and stiffness, peripheral resistance and characteristic impedance. For this reason, an accurate estimation of these parameters is necessary for better interpretation of arterial system function in hypertension. The classic Windkessel model is most popularly used in the clinical setting to describe arterial system function. This model assumes a constant arterial compliance (C) throughout the entire cardiac cycle, although it has been known that compliance is a function of pressure. A recently modified Windkessel, also known as the Li model, which incorporates a pressure-dependent nonlinear compliance (C(P)) component has shown that arterial compliance continuously varies with time, i.e. not showing a constant value under cyclic pulsations[1]. We explore the impact of this new model in quantifying mechanical properties of arteries in hypertension and subsequent vasodilator drug therapy.

Methods: Previously collected experimental data on anesthetized, ventilated mongrel dogs were utilized. Experiments were performed with Rutgers IUCUC approval. Hypertension was induced with *i.v.* methoxamine (2-5 mg/ml) and vasodilator treatment was through *i.v.* nitroprusside (50 μg/ml) infusion. Computational modeling based on the Li model provided the pressure-dependent nonlinear compliance (C(P)) data, along with aortic characteristic impedance and peripheral resistance and the Windkessel based constant compliance C. Comparison was made between measured data and the prediction, where a root mean square error (RMSE) was used as the metric of accuracy.

Results: Both Windkessel and Li models predictions are fairly accurate in early systolic stage, whereas the nonlinear model generates closer estimations from end-systolic to diastolic period. The averages of RMSE of nonlinear pressure predictions in normal, hypertension and vasodilator conditions are lower than those with Windkessel model. This outcome indicates that incorporating a continuously varying compliance can significantly improve the reliability of three-element Windkessel model in predicting hemodynamic activities. Further analysis of the C(P) vs pressure loop showed that the area of each loop varies in accordance with the prevailing condition: larger with vasodilator and smaller area under hypertension.

Conclusions: Modeling facilitates the exploration of physiological information from limited sources. Only simultaneously measured pressure and flow are required for the Windkessel model in providing several crucial characteristics of the arterial system, giving a simplified approach to obtaining gross feature of arterial system function. The dynamic association between pressure and vascular compliance employed in this study expressed an

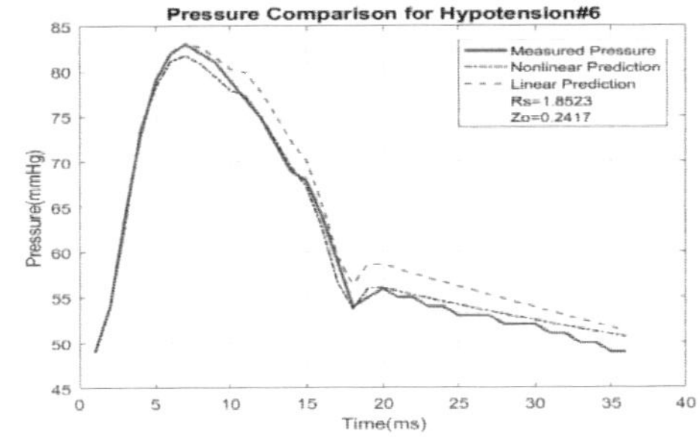

Figure 1. Comparison of Windkessel model and Li-model predicted blood pressure pulsations with the measured pressure.

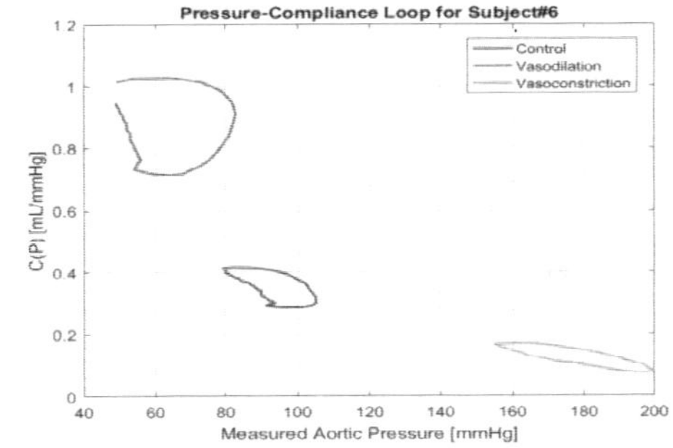

Figure 2. Li-model based C(p)-P loops during cyclic pulsations, while comparing normal (blue), hypertension (yellow) and vasodilator treatment (red).

excellent discernibility during each stage of systole in prediction. Meanwhile, the concept of pressure-compliance loop provides a novel interpretation of the structural mechanical properties of arterial wall interacting with underlying pulsatile pressure-flow relationship[1,2]. One of the marked observations is that during systole and diastole, pressure could reach the same value while compliance differed. This result highlights a speculation, which was often omitted, that mechanical properties of vessels alter from systole to diastole. In other words, the vascular elastic stiffness varies throughout the cardiac cycle even under the same pressure level. Increased vascular stiffness or decreased large vessel compliance impedes cardiac ejection. Thus, in hypertension, the significantly reduced compliance-pressure loop area reflects constricted blood vessel lumen and the beneficial effect of vasodilator is to allow a greater increase in blood flow through increase in arterial compliance, as seen in the resultant larger C(P) vs P loop area. Future evaluation with human data will allow the beneficial effect of drug treatment to be continuously monitored with the present approach.

References:

1. Li, J.K-J. *Dynamics of the Vascular System: Interaction with the Heart.* World Scientific, 2018.

2. Kaya. et al. *Computers in Biology and Medicine.* 2018;99: 98-106.

Pericellular Collagen VI Reorganization After Painful Temporomandibular Loading in the Rat

Melissa Franklin[1], Megan Sperry[2], Evan Phillips[1], Eric Granquist[3], Beth Winkelstein[2], Michele Marcolongo[1]
[1]*Materials Science and Engineering, Drexel University,* [2]*Bioengineering,*[3]*Oral & Maxillofacial Surgery, University of Pennsylvania, Philadelphia, PA*

Introduction: Temporomandibular Joint (TMJ) disorders are the second most common source of orofacial pain[1]. A subset of patients with TMJ disorder develop osteoarthritis (OA), characterized by intraarticular inflammation and cartilage degeneration[1]. Structural changes in the cartilage extracellular matrix (ECM) have been previously identified with OA, with earliest changes occurring in the pericellular matrix (PCM) of chondrocytes[2]. The PCM is rich in the protein collagen VI and a host of proteoglycans, which serve as support structures and biomechanical transducers for chondrocytes to the ECM. PCM collagen VI acts as a mechanical tether and cytoprotective layer surrounding chondrocytes, and it is believed that it may be disrupted in joints with OA[3]. We hypothesized that the pericellular structure of TMJ fibrocartilage may be altered in a model of repeated overloading of the TMJ that is painful. In this study, we characterized collagen VI expression in a rat model of TMJ pain, using one group with repeated 2N loading (non-painful) and another with repeated 3.5N loading (painful), which induce different levels of inflammation along with variable orofacial pain[4].

Methods: All procedures were approved by the IACUC at the University of Pennsylvania. Adult female Holtzman rats were exposed to mechanical loading of the jaw under isoflurane anesthesia at 2N (induces sensitivity that resolves) and 3.5N (induces persistent sensitivity) for 1 hour/day for 7 days[1]. On day 8 or 15, rats were anesthetized, perfused with phosphate buffer solution (PBS) and fixed by 4% paraformaldehyde perfusion. TMJs (n=5/group) exposed to 3.5N were harvested at day 8 as well as day 15 and 2N-loaded TMJs were harvested only at day 15. Experimental groups were compared to adult female rats that did not receive loading (n=5). TMJs were stored in 30% sucrose in PBS and later decalcified with 0.25M EDTA for 3 weeks at 4 . Tissue was embedded in OCT compound, sagittally sectioned at 18μm, and thaw-mounted onto slides. Immunohistochemistry (IHC) was performed, incubating sections with primary antibody against collagen VI (1:1250; Fitzgerald) overnight at 4 . After washing, sections were incubated with biotinylated horse anti-rabbit secondary (1:1,000; Vector) for 30 minutes, developed using 3,3-diamimobenzidine (DAB), and cover-slipped. Two sections from each TMJ sample were imaged at 40x using widefield microscopy. Collagen VI expression was quantified by densitometry using a custom MATLAB script. The width of the PCM was measured manually using ImageJ, using collagen VI as the guidance marker. One-way ANOVA and post-hoc Tukey tests were performed to determine differences in percent collagen VI expression between experimental groups. Results were verified with immunofluorescence confocal microscopy (63X magnification) with fresh 5μm sagittal sections acquired from normal, 2N, and 3.5N TMJs from day 15.

Results: The percent positive pixels for collagen VI increased in 3.5N-loaded TMJs on day 15 (12.99±1.32) over both normal (8.51±1.09; ^p=0.013) and day 8 levels (9.34±0.50; *p=0.048) (Figure 1A). No changes were detected in 3.5N-loaded TMJs at day 8 compared to normal levels (p=0.837). The percent of collagen VI increased in 2N-loaded TMJs on day 15 (16.40±1.38) over normal (#p=0.004) (Figure 1B). Yet, no changes in collagen VI were identified between the 2N and 3.5N day 15 TMJs (p=0.148). The PCM widths for 3.5N-loaded TMJs (2.99μm±0.08) and 2N-loaded TMJs (3.28μm±0.11) both indicated expansion of pericellular region over normal widths (2.00±0.06, p<0.0001).

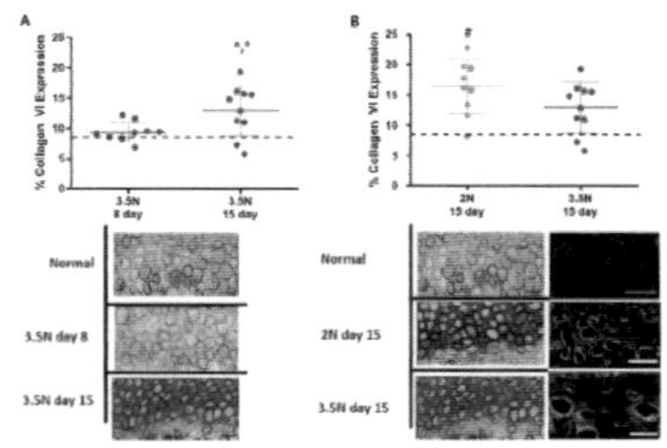

Figure 1. ***(A-top)*** Collagen VI increases at day 15 and is unchanged at day 8 relative to normal (blue dashed line) ***(A-bottom)*** respective DAB-IHC images ***(B-top)*** Collagen VI increases in both 2N and 3.5N-loaded TMJs compared to normal at day 15. ***(B-bottom)*** respective DAB-IHC and IF-IHC images (scale =25μm).

Conclusions: This study identified increased collagen VI expression and PCM expansion in both TMJ pain models at day 15 that are not observed in normal TMJs. Structural change in the PCM surrounding TMJ chondrocytes occurs regardless of loading protocol. Although changes in cartilage structure were previously identified only in TMJs exposed to 3.5N loading[5], this study shows there are pericellular structural changes also with 2N loading which may precede global changes in the joint. Increases in the expression of collagen VI and pericellular expansion may arise because chondrocytes are attempting to maintain normal tethering to the cartilage ECM when exposed to overloading conditions. However, the pericellular expansion does not exceed 4μm, which is characteristic of degenerative OA[2]. PCM thickness of less than 4μm suggests that 2N and 3.5N loading does not induce advanced OA pathology at day 15. As changes were only found at day 15, they indicate that there may be a potential time window prior to permanent cartilage structural change. Although pain, inflammation, and catabolic factors are upregulated by 3.5N TMJ loading on or prior to day 8[1], there are no structural changes to the PCM at this time point. This study bridges previously determined inflammatory and behavioral outcomes of this model to detailed pericellular structural modulation. Additionally, our investigation establishes a motivation to molecularly engineer the PCM of TMJ fibrocartilage with therapies that preserve the region's structural integrity and biomechanical properties.

References: **[1]** Kartha, S. *J Oral Maxillofac Surg*. 2016. **[2]** Guilak, F. *Matrix Biology*. 2018. **[3]** Godwin, A. *Acta Biomaterialia. 2017.* **[4]** Sperry, M. *Nature Sci Reports*. 2018.**[5]** Sperry, M. *Biomed. Eng. Society*. 2017. **Acknowledgements:** This study was funded by the Sharpe Foundation, Oral and Maxillofacial Surgery Foundation, and Penn Dental's Schoenleber Fund. The authors also acknowledge the Penn Microscopy Core and Penn Center for Musculoskeletal Diseases for equipment use.

Glycosaminoglycan Mimetic Aligned Fibrous Scaffolds Promote Neurite Extension and Myelination

Sharareh Hashemi[1], Patrice Maurel[2], and Treena Livingston Arinzeh[1]
[1]Biomedical Engineering, New Jersey Institute of Technology, Newark NJ
[2] Department of Biological Sciences, Rutgers-The State University of New Jersey, Newark NJ

Introduction: Spinal cord injury (SCI) is a devastating heterogeneous neurological condition with no effective treatment at the present time. Tissue engineered scaffolds mimicking the native extracellular matrix (ECM) may be a promising strategy to promote axonal growth. Glycosaminoglycans (GAGs), specifically chondroitin sulfate (CS), can either inhibit or promote axonal growth depending upon the degree and pattern of sulfation. We have developed a GAG mimetic, sodium cellulose sulfate (NaCS), which can be tailored to have varying degree of sulfation similar to native GAG. Schwann cells (SCs) are of interest to be used in combination with this scaffold since they secrete neurotrophic factors stimulating neuron survival and extension of axons. This study fabricated, characterized and evaluated NaCS scaffolds consisting of aligned fibers for supporting SC growth and neurite extension.

Methods: Aligned fibers were prepared by electrospinning a solution of 24% (w/v) of gelatin (obtained from Bovine skin, Sigma) with 0.25% (w/w) GAG mimetics. Partially sulfated sodium cellulose (pNaCS), which mimics chondroitin sulfate-C (CS-C) where the sulfate groups are on the 6th carbon position of the glucose monomer unit and has ~0.5 sulfates per glucose monomer unit, was prepared according to previously reported protocols [1]. A fully sulfated sodium cellulose (fNaCS), with sulfate groups on the 2, 3, and 6th carbon position of the glucose monomer unit and ~3 sulfates per glucose unit, was commercially available. For electrospinning aligned fibers, a grounded rotating mandrel was used as a collector. The scaffolds were then crosslinked using EDC/NHS. Fiber morphology and alignment of the scaffolds were evaluated using SEM. GFP labeled SCs were seeded onto scaffolds at 6.3x 10^4 cells/cm^2 and cultured for up to 14 days to evaluate cell growth and survival. Elongation of SC processes (n=4) within 24 hrs was observed by confocal microscopy. Purified dorsal root ganglion (DRGs), from the spinal cord of rat embryos (E-16), were seeded onto scaffolds with or without SCs. After 3 days in culture, neurite extension was evaluated by confocal microscopy and the length was measured using ImageJ. SC myelination was investigated by purifying dissociated DRGs on scaffolds for 2 weeks followed by seeding SCs and maintaining the culture for 3 more weeks. The appropriate growth media for each study was prepared as previously reported [2].

Results: The scaffolds had a uniform morphology with an average fiber diameter of 700 nm with greater than 90% alignment maintained after crosslinking (Fig-1). Significantly longer processes were observed for SCs seeded on fNaCS scaffolds in comparison to the other scaffold groups after 24 hrs. Significantly longer neurite outgrowth were determined for explant DRGs in co-culture with SCs on fNaCS scaffolds in comparison to the other groups and for DRGs cultures without SC (data not shown). More myelination was observed on both Gel/pNaCS and Gel/fNaCS in comparison to Gel (Fig-2), where Gel/fNaCS appears to have the greatest myelination.

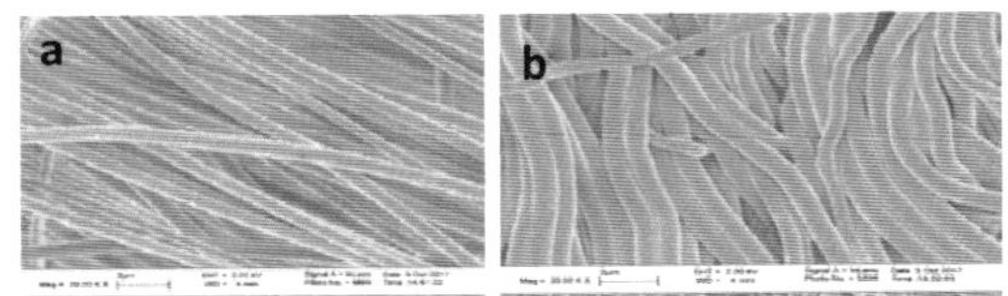

Figure 1: SEM image of Gel with 0.25% fNaCS (Mag. 20,000X), before (a) and after (b) crosslinking.

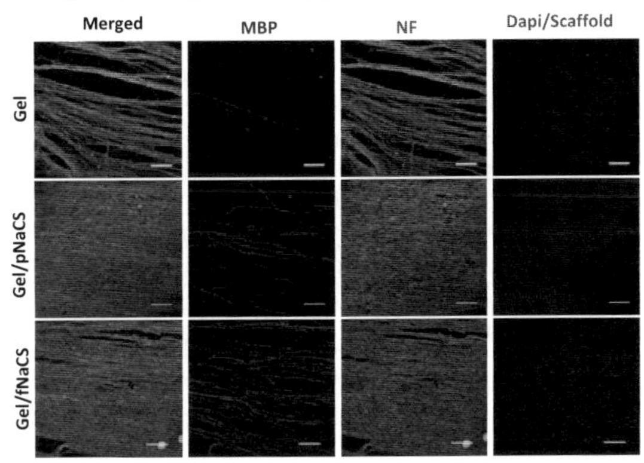

Figure 2: Myelination study: dissociated DRGs are co-cultured with exogenous SCs for up to 36 days. Stained for myelin basic protein (red), neurofilament (green) and nucleus (Dapi-blue). Fibers autofluoresce in blue. Mag: 20 X, Scale bar: 250 um.

Conclusions: A combination of SCs and GAG-mimetic scaffolds can be used as a novel approach to promote and direct axonal growth. The significant neurite growth for SCs seeded on fNaCS scaffolds suggests that neurite outgrowth can be promoted, and it may depend on the degree of sulfation. Similarly, myelination was promoted on the GAG-mimetics demonstrating their potential for neural repair. A combination of SCs and scaffolds consisting of GAG-mimetics, derived from cellulose, may be a novel approach to effectively promote and direct axonal growth.

References:

[1] M.R. Portocarrero Huang G, et al, Tissue Eng Part A 23(17-18) (2017) 1011-1021.
[2] S. Wu, M.S. et al. J Neural Eng 15(5) (2018) 056010.

Acknowledgements: This work was supported by the NJ Commission on Spinal Cord Research.

P2-19

Uric Acid Decreases Cell Death and Reactive Oxygen Species following Glutamate-induced Excitotoxicity in Organotypic Slice Culture

Salman Khaliq, BS and Bonnie L. Firestein, PhD
Departments of Cell Biology and Neuroscience and Biomedical Engineering,
Rutgers University

Introduction: Glutamate-induced excitotoxicity (GIE), resulting from secondary spinal cord injury, leads to increased production of reactive oxygen species (ROS) (Becker, 1993; Ryan et al., 1997). Damage elicited to neurons by glutamate overstimulates N-methyl-D-aspartate (NMDA) receptors, which causes an influx of calcium ions into the cell, catalyzing production of ROS. Uric acid (UA) has been demonstrated to protect neurons from excitotoxicity by reducing glutamate induced oxidative stress (Scott et al., 2005). Previous studies conducted on dissociated spinal cord cultures lack intrinsic connectivity and cell types, such as oligodendrocytes and microglia. In this study, we optimized organotypic spinal cord cultures and exposed them to GIE. Dose dependent effects of UA were elucidated by quantifying cell death using propidium iodide and ROS using CellROX™ Green Reagent.

Methods: Spinal cord slices were dissected from rats at embryoic day 16 (E16) in Gey's balanced salt solution supplemented with kynurenic acid and glucose. A dorsal incision was made in the embryos to isolate spinal cords, which were then chopped into 350μm slices using a McIlwain tissue chopper. The slices were placed into 35mm glass bottom petri dishes previously coated with 0.1% poly-D-lysine for 24 hours and 0.1mg/ml laminin for 2 hours at 37°C and 5% CO_2. The slices were cultured for 8 days *in vitro* (DIV) in Neurobasal A medium supplemented with B-27, GlutaMAX, and gentamycin.
Locke's buffer (NaCl, 154 mM; KCl, 5.6 mM; CaCl2, 2.3 mM; MgCl2, 1.0 mM; NaHCO3, 3.6 mM; glucose, 5 mM; Hepes, 5 mM; pH 7.2) was used as vehicle for glutamate and uric acid (UA) treatments. On DIV8, the slices were treated with 3mM glutamate for 1 hour and different concentrations of UA for 24 hours at 37°C and 5% CO_2.
For assessment of oxidative stress, phase contrast and fluorescence micrographs were taken under a 4x objective using the EVOS FL Microscope (ThermoFisherScientific) following the incubation of slices in 3 μM propidium iodide (PI) for 30 minutes and 5 μM CellROX™ Green Reagent for 1 hour.

Results: Addition of UA decreased glutamate-induced ROS production in a dose-dependent manner (**Figure 1**).
In addition, glutamate-induced cell death was attenuated by UA treatment, similar to that seen for UA-promoted decreases in glutamate-induced ROS production (data not shown).

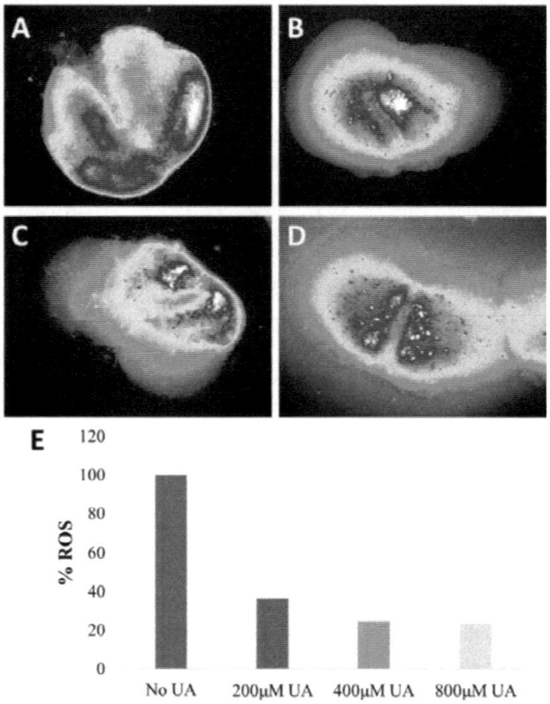

Figure 1. Representative heat maps demonstrating ROS in organotypic spinal cord slices following 3 mM glutamate treatment and post-treatment with (A) vehicle (control), (B) 200 μM UA, (C) 400 μM UA and (D) 800 μM UA. (E) Quantitation of %ROS from heat maps.

Conclusions: In this study, UA concentration was optimized to achieve the greatest attenuation of ROS and cell death following GIE. In the future, UA delivery via a nanofiber scaffold will be developed to provide a sustained supply of UA to *in vitro* organotypic spinal cord cultures, and ultimately *in vivo*.

References:
1. Becker BF, Free Radic Biol Med. 1993;14(6):615-31
2. Ryan M, Ann Clin Biochem. 1997;34 (Pt 6):688-9
3. Scott GS, Proc Natl Acad Sci USA. 2005; 102:3483–3488.

Acknowledgements: This project is funded by grant #CSCR14IRG005 from the New Jersey Commission on Spinal Cord Research (to BLF).

The Effect of Phase-Specific Optogenetic Stimulation on Memory Recall in Mice

Benjamin Lahner[1], Melanie H. Quick[1], Bahar Rahsepar[1], Jad Noueihed[1], John A. White, Ph.D.[1]
[1]Department of Biomedical Engineering, Boston University, Boston, MA

Introduction: Memory and its driving mechanisms are relevant to a wide variety of clinical and physiological applications. Better understanding of memory could lead to treatments of debilitating neural disorders or clinical techniques to improve memory. Theta waves (slow sinusoidal oscillations) are hallmarks of memory encoding and retrieval in the hippocampus. Disruptions of theta wave propagation limits the creation of new memories and retrieval of old. A prominent theoretical model proposes the phase of theta separates the state of encoding new information from recalling stored memories [1]. Efforts have been made to understand the specifics of theta's role in memory encoding and recall, however, understanding the functional role of theta wave phase suffers from a lack of a real-time approach. The post-processing of theta waves eliminates computational difficulties but stymies the possibility of memory manipulation and therefore prevents causal investigation of their role in memory processing. Noisy real-time local field potentials (LFP) and the quasi-periodic nature of theta waves deter accurate detection of peaks and troughs without excessive latency. The algorithm developed addresses the specific features of theta waves, allowing for accurate peak and trough detection within the constraints of real-time computation. The provided platform will increase the specificity by which theta waves and other memory processes can be explored.

Methods: All algorithm development was completed in MATLAB R2018a (Mathworks Inc., Natick, MA). The algorithm was tested causally using sample LFP data collected in previous experiments. Algorithm performance was measured by comparing peaks and troughs identified by the algorithm with manually labeled peaks and troughs used as truth. Five randomly selected, independent, 10,000 millisecond sections of the LFP were manually labeled. After offline bandpass filtering from 4 to 10 Hertz, the algorithm detected peaks and troughs on the same randomly selected ten second sections. The real-time algorithm uses first and second derivative testing to determine all minima and maxima. The algorithm then utilizes the frequency of theta as a threshold latency and a minimum power threshold to classify the minima or maxima as a true peak or trough. Progression with this algorithm into real-time experimentation requires implementation in Real-Time eXperiment Interface (RTXI), a platform developed for real-time experimental manipulations [2]. During *in vivo* experimentation, memory neurons will be stimulated optogenetically at peaks or troughs of theta in mice and the animal's memory recall will be evaluated [3]. The performance of the RTXI module created from the algorithm during *in vivo* experimentation can be assessed via offline analysis.

Results: From the 50 seconds of data analyzed, 396 maxima and 395 minima were measured. For peak detection, the accuracy was 92.2%, and the misclassification rate was 7.8%. The True Positive Rate (TPR) was 93.7%, the False Negative Rate (FNR) was 6.3%, the False Positive Rate (FPR) was 19.6%, and the True Negative Rate (TNR) was 80.4%. For trough detection, the accuracy was 93.4%, and the misclassification rate was 6.6%. The TPR was 94.3%, the FNR was 5.7%, the FPR was 13.0%, and the TNR was 87.0%.

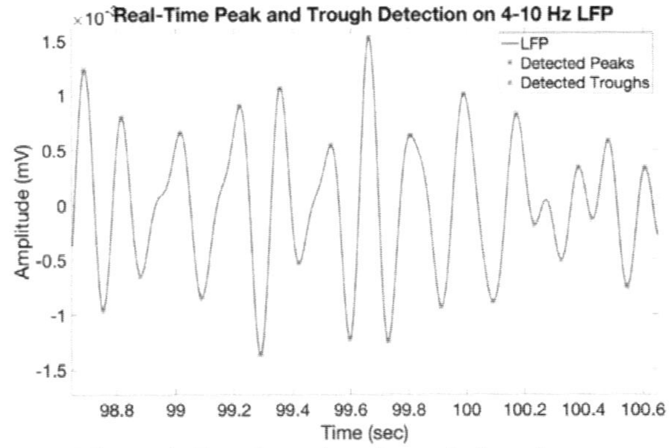

Figure 1. Sample peak and trough detection.

Conclusions: Implementation in a causal environment with sample LFP signals yielded accuracies of 92.2% and 93.7% and misclassification rates of 7.8% and 6.6% for peaks and troughs, respectively. These performance results demonstrate the algorithm overcomes barriers to accurate peak and trough detection in theta waves, namely noise, quasi-periodicity, and real-time computation limitations. The algorithm is suitable for *in vivo* experimentation in combination with RTXI and optogenetic stimulation. In integrating real-time manipulations of memory neurons with behavioral observations, the algorithm will help inform understanding of theta phase with the state of memory encoding versus retrieval. From these results, the neurophysiology of phase-specific stimulation on memory recall can be studied with increased specificity. Furthermore, the results of the experiment will help yield a better understanding of the neural circuitry behind larger memory processes. The experimental paradigm created through this project will enable more detailed dissection of the physiology behind memory processing and offer useful methods to study and modify theta waves in real-time. Building a foundation for real-time memory stimulation opens the door to memory creation and deletion, brain wave-based therapies, and better probes into the mechanisms of memory-based disorders.

References:

1. Hyman JM. et al. J Neurosci. 2003;23:11725-11731.
2. Patel YA. et al. PLoS Comput Biol. 2017;13:e1005656.
3. Josselyn SA. et al. Nature Rev Neurosci. 2015;16:521-534;

P2-21

Antioxidant Nanoparticle Films for Improving Deep Brain Recording

Victoria Vafaee, Yue Hu, Elaina Atherton.
Brown University.

Introduction: Brain recording devices are small (e.g. ~ few mm) electrical sensors that allow researchers and doctors to monitor and measure neural electrical activity deep in the brain. While used frequently in research applications in primates and humans, they are also used in clinical practice. Brown's BrainGate project pioneered this research when it used these sensors to translate neural signals associated with people thinking about movement into actual movement of their prosthetic limbs (Nuyujukian et al., 2018). While brain recording devices can take a variety of forms, in this work we study small (1 mm) microelectrode.

Although promising, this technology is limited because the brain recognizes these chronically implanted devices as foreign bodies. Once this happens, dense tissue forms around the sensitive electrode and blocks the weak signals from surrounding neurons from being detected. As a result, recording is no longer possible and patients must have additional surgeries to implant new electrodes. This process happens over months. Biological studies have shown that the electrode stimulates an immune response which culminates in the deposition of dense fibrotic scar tissue around the electrode. This is the reason the devices stop working in most patients.

In this research we investigate whether a new kind of nanoparticle made from ceria (cerium and oxygen) can be used to extend the lifetime of the recording devices by diminishing scar formation in the brain. A central part of our strategy is to reduce highly reactive oxygen species (ROS) in and around the recording electrode. These ROS are the signaling molecules that initiate the inflammatory process, leading to scar formation. Biological studies using genetically modified cells have shown that if the ROS generation process is shut down, then no scar will form (Shroff et al., 2014). Unfortunately, there is no practical way to apply genetically modified cells. Instead this work looks at whether a nanoparticle known as an "oxidant sponge" can be applied to solve this problem.

Methods: Our work will address whether or not thin films of nanoparticles can be formed on the microelectrode casing and released over days into physiological buffers? Our approach will be to generate polymer-nanoparticle composites using layer-by-layer manufacturing. We are currently using an in house made device that is controlled and automated by an Arduino, to dip the glass slides into the two film solutions. When immersed in physiological fluids these films come apart and release the nanoparticles. Given the chronic nature of the disease, we need to engineer a very slow release process that will be too slow to study in the lab. Modeling will help fill this gap using basic drug release and degradation kinetics to fit our release curves, and ultimately inform methods that would lead to the slowest release possible. One teammate used the solutions that were generated from the drug delivery experiments to examine whether the reactivity of the ceria is preserved. Their study of the anti-oxidant properties fed back into the modeling of the thin films degradation.

Results: Optimization of the drug dispensing film is found to be controlled by the concentration of cerium nanoparticles in each layer. It can be achieved by changing dipping parameters, and controlling the number of cerium nanoparticles released with each released layer. These parameters also effect the rate at which the film itself disassembles. The release of cerium nanoparticles from 10-layer films show that 25% of the loaded cerium nanoparticles were released in the first week. There is a estimated total release at 1 month. The rate of degradation and drug release is also found to be related to film thickness, where the film thickness increases as the number layers increases. We expect that a 40-layer film will have a lot larger loading and slower release than a 10-layer film.

Conclusions: Parameters of the layer-by–layer manufacturing can be altered to create an optimized drug release process. It is still difficult to determine the exact antioxidant requirements of the tissue. The target concentration will be determined by the therapeutic window of the cerium nanoparticles still to be measured. In vivo testing is going to be the final process for optimizing the drug delivery parameters.

References:

1. Nuyujukian P, Albites Sanabria J, Saab J, Pandarinath C, Jarosiewicz B, Blabe CH, Franco B, Mernoff ST, Eskandar EN, Simeral JD, Hochberg LR, Shenoy KV, Henderson JM. Cortical control of a tablet computer by people with paralysis. PLoS ONE. 2018; 13(11). doi.org/10.1371/journal.pone.0204566
2. Shroff A, Mamalis A, Jagdeo J. Oxidative Stress and Skin Fibrosis. Curr Pathobiol Rep. 2014; 2(4):257-267. doi.org/10.1007/s40139-014-0062-y

Acknowledgements:
Dr. David Borton
Dr. Vicki Colvin
Dr. Hyewon Kim

Microphysiological Modeling of Adipose Tissue for High-Throughput Applications

Michael Struss, B.S. Evangelia Bellas, Ph.D.
Department of Bioengineering, Temple University, Philadelphia, PA, United States

Introduction: Adipose tissue is a highly vascularized endocrine organ, dependent on its vasculature to maintain whole body metabolic homeostasis[1]. Inadequate vascularization leads to tissue hypoxia ultimately resulting in metabolic dysfunction[2]. Traditionally, adipose tissue is studied in vitro, using 2D and 3D cultures, but falls short from a lack of spatiotemporal organization of adipocytes and endothelial cells that carry out this endocrine signaling. Microphysiological models are tools for modeling the spatiotemporal complexity of biological systems in vitro, but are often limited in utility as they are often incompatible with traditional culture ware. Here, we introduce a novel, miniaturized adipose tissue (µAT) model compatible with standard culture ware and thus amenable to high-throughput platforms.

Methods: The µAT devices are fabricated using polydimethylsiloxane (PDMS) casted from a 3D printed template and bonded to a 15 mm circular glass coverslip using air plasma. To accommodate high-throughput culture plates (e.g. a 96 well plate), the miniaturized µAT device is 6 mm in diameter and 750 microns in height. The µAT device is designed with a central hydrogel reservoir, to contain adipocytes encapsulated within a collagen type I hydrogel, a microchannel guide which serves as a precursor for the vessel fabrication, and two media reservoirs which allow media to flow through the vessel, across the device (*Figure 1A*). A stainless steel rod (120-micron diameter) is inserted through the microchannel guide and the hydrogel reservoir is then filled with a 2mg/mL collagen type I (Corning) hydrogel with adipocytes (8 million cell/mL). The hydrogel is crosslinked and the stainless steel rod is removed leaving a microchannel void. The microchannel is then seeded with human umbilical cord vascular endothelial cells (HUVEC). The µAT devices are cultured on a rocker (BioRocker, Denville Scientific) for 7 days in order to expose the endothelial cells to physiological shear flow.

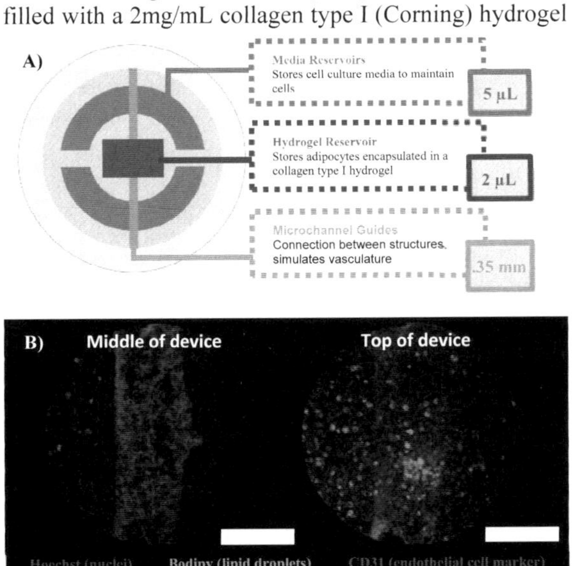

hydrogel. The microchannel guides (green) aid in the fabrication process of the endothelial cell vessel. B) Middle and top of device showing homogenous distribution of adipocytes and endothelial cell vessel.

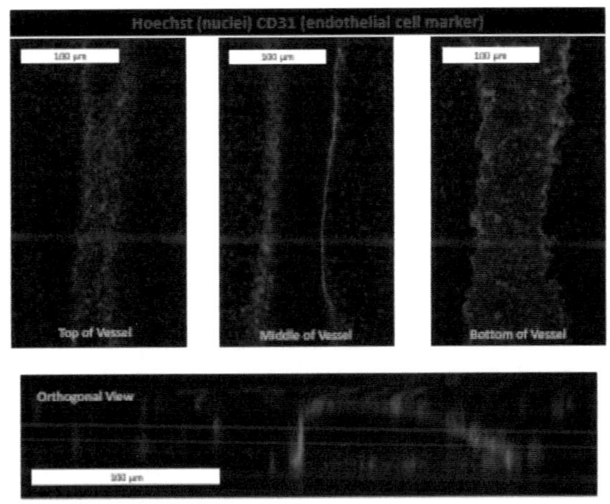

Figure 2: The confocal images represent the top, middle, bottom, and orthogonal view of the endothelial cell vessel. The features are visualized with Hoechst (blue, nuclei) and CD31 (red, endothelial cells).

Results and Discussion: The organization of cells within the µAT model was validated by confocal imaging (Olympus Fluoview). The homogeneous distribution of adipocytes inside of the hydrogel was using verified using Bodipy (lipid droplet stain, *Figure 1B*). The endothelial vessel formation was visualized with CD31 (endothelial cell junction) immunofluorescence. The CD31 staining demonstrated clear cell-cell junctions between endothelial cells and an open lumen cross-section through the vessel (*Figure 2*).

Conclusion: The proposed miniaturized adipose tissue (µAT) model provides a robust microphysiological platform for tissue engineering and disease modeling applications, which are otherwise not readily achievable or require too many resources. Our future work will include quantifying the permeability of the endothelial vessel with fluorescently tagged dextran and studying the cross talk between the adipocytes and the vascular network in models of disease, such as obesity and type 2 diabetes.

References:

1. Coelho M. et al. Arch Med Sci. 2013; 9(2): 191-200
2. Goossens GH. et al. Front Endocrinal. 2017; 6: 55

Acknowledgements: We acknowledge funding support from Dr. Bellas' startup funds (Temple University) and the NIH NIDDK Diabetic Complications Consortium DK07616 and DK115255 grants (to EB).

Antimicrobial Susceptibility Testing using Piezo Electric Plate Sensors

Pawan Rao [a], Dolores Conover [a], Wei-Heng Shih [b], Wan Y. Shih [a]
[a] School of Biomedical Engineering, Science and health Systems, Drexel University, Philadelphia, PA-19104, USA
[b] Department of Materials Science and Engineering, Drexel University, Philadelphia, PA-19104, USA

Introduction: In the U.S., each year 2 million people acquire serious infections with bacteria resistant to one or more of the antibiotics designed to treat their infections. The total economic cost of antibiotic resistance to the U.S. economy was estimated to be as high as $20 billion in excess of direct healthcare costs, with additional costs to society for lost productivity as high as $35 billion a year. In particular, nosocomial bloodstream infections (BSIs) including catheter-related BSIs, are important causes of patient morbidity and mortality. The annual incidence of BSIs is >750,000 cases with a 30% mortality rate, and the proportion of antibiotic-resistant bacteria causing BSIs is increasing. Increasing rates of multidrug-resistant bacteria require use of broad-spectrum antibiotics to avoid inadequate antibiotic coverage, but these drugs are often expensive compared to narrow-spectrum antibiotics, and their use promotes bacterial resistance and the infection with *Clostridium difficile*, a serious healthcare-associated infection that costs billions a year. Rapid initiation of appropriate empiric antibiotic therapy is strongly associated with decreased mortality rates in BSIs since the risk of death from septic shock increases by >7% with every passing hour from the onset of shock until the start of targeted therapy. In contrast, inappropriate empiric antibiotic therapy and every hour delay in antibiotic administration after septicemia diagnosis have been associated with higher patient mortality. The goal of physicians and of all antimicrobial stewardship programs (ASPs) today is to place patients with BSIs and other infection types on targeted narrow-spectrum antibiotic therapy based on AST results as quickly as possible. Rapid delivery of definitive antimicrobial susceptibility testing (AST) results is key to allow minimal use of broad-spectrum therapy and effective targeted narrow-spectrum antibiotic regimen.
Current AST methods reply on the growth (or the lack of it) of isolated bacteria in different antibiotic concentrations to determine the minimum inhibition concentration (MIC) of the most desirable antibiotic to treat the infection. As such they typically take 5 hours to days to get the result depending on the methods and the bacteria strains.

Methods:
The goal of this study is to develop an instant AST (iAST) using a piezoelectric plate sensor (PEPS) developed in Shih and Shih's laboratory. What is unique is that PEPS is particularly suited to measure minute stresses generated by the metabolism of live bacteria, both mechanical and electrical due to its piezoelectric nature and its enhanced sensitivity. Minute random stresses both mechanical and electrical can shift the resonance frequency of a PEPS randomly, causing the top of the resonance peak to become blunt (or widened). Monitoring how the top of a PEPS resonance peak becomes less blunt (wide) with antibiotic offers a novel and instant AST.
Antimicrobial susceptibility testing was performed on Escherichia coli (E. coli) (ATCC25922) which is a Clinical and Laboratory Standard Institute (CLSI) standard quality control strain using Ampicillin as an antibiotic. Resonant frequency of width extension mode (WEM) peak was monitored in water (negative control), after coating with bacteria and various concentration of antibiotics. Before testing, PEPS was monitored in DI water for 5 minutes to get the negative control signal (without bacteria). It was then soaked with 100 μl of live bacteria suspension at 0.5 McFarland concentration for 15 minutes. PEPS was then rinsed in 100 ml DI to remove unbound bacteria and monitored for 5 minutes in DI water before immersing in a series of 2-fold increasing antibiotic solutions (e.g., 1, 2, 4, 8 and 16μg/ml) each for 5 min.

Results: Figure 1(a) shows the change in the resonant peak shape: widening of the peak by the presence of live bacteria and then narrowing by the antibiotics. Figure 1(b) shows the peak width measured 0.1 degree below the highest point of the peak versus time. As can be seen, at the MIC (4 mg/ml) or higher, the width of the peak stopped further narrowing. The obtained MIC is consistent with the literature value.

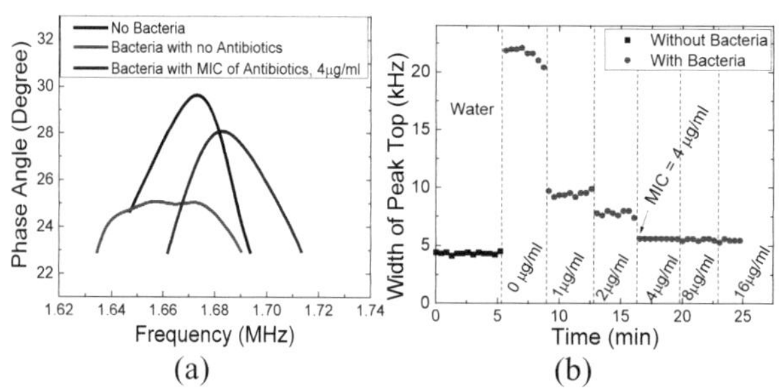

Figure 1. a) Phase angle and Resonance Frequency spectra of PEPS in water (no bacteria-black), after coating with bacteria-red and in MIC concentration (4μg/ml) of antibiotic solution. b) Width of Peak, 0.1 degree from top of the peak in water (no bacteria), after coating with bacteria and increasing concentration of antibiotics.

Conclusions: In this study we showed that a PEPS iAST can indeed accurately determine the MIC in 30 minutes.

Acknowledgment: This work was supported in part by the National Institute of Health Grants No. 1R41AI112224 and 1R41AI120445 and Pennsylvania Department of Health.

Sensitive detection of nucleic acids (DNA and RNA) using a novel biosensor electrochemical platform - ESSENCE

Yu-Hsuan Cheng, Zhenglong Li, Mahima Hariharan, Pedro Moura, and Sagnik Basuray
New Jersey Institute of Technology

Introduction: Electrochemical sensors are sensitive to the charge of a particular target molecule that results in a measurable detection signal. This can be measured (quantitatively or qualitatively) by several electrochemical, spectroscopic techniques such as electrical impedance spectroscopy (EIS), pulse voltammetry (PV), cyclic voltammetry (CV), among others. However, all these spectroscopic techniques have a serious disadvantage. The electrical double layer results in a parasitic capacitor that blocks/overshadows the signal of the attachment of the target molecule to the electrode, resulting in reduced sensitivity. The reduction in sensitivity of the sensor leads to false positives and false negatives. Here we show a new electrochemical sensing platform that uses shear-enhanced, flow-through, nanoporous, electrode (called ESSENCE) that significantly reduces the noise from the electrical double layer. ESSENCE is used here to detect DNA's and RNA's using Electrical Impedance Spectroscopy (EIS).

Method: The device consists of nonplanar interdigitated microelectrodes of sizes 10μm X 100μm with a pitch of 10μm separated by a microfluidic channel of 500μm X 100μm X 1 cm. Carboxylated Carbon Nanotubes (CNT) is surface activated using standard EDC-NHS chemistry. A short probe DNA/RNA oligo is attached to the CNT. The CNT with the attached DNA/RNA oligo is packed into the microfluidic channel between two non-planar interdigitated electrodes to form a nanoporous, flow-through electrode. FTIR spectroscopy is used to confirm all the attachments to the CNT.

The DNA/RNA oligo-functionalized CNT factionalized chip is connected to the Impedance Analyzer. An NE-300 syringe pump from New Era Pump Systems, Inc. (USA) is connected to the inlet of the chip. The chip is first washed with 1X (pH 7.4) PBS solution at a flow rate of 1 μl/min for 2 hours or until the EIS signal from the chip stabilizes. This stabilized EIS signal taken in 1X PBS is treated as the baseline measurement. Post-stabilization, the 1X PBS solutions is switched to the 1 M blocking agency, Ethanolamine 99%+ which is acquired from Alfa Aesar, for 90 mins. Post-blocking, the chip is washed with 1X PBS to remove the blacking agency. After the system is stabled, the settled Target-DNA\RNA flow into the system with 1 μl/min. Post-Target-DNA/RNA oligo passage, the Target-DNA/RNA solution is again switched back to the washing 1X (pH 7.4) PBS solution. The chip is again washed with 1X (pH 7.4) PBS solution at a flow rate of 1 μl/min until the EIS signal stabilizes. The difference between the EIS signal from the pre and post-PBS wash is interpreted as the Target-DNA/RNA capture signal. The capture signal is fitted to an equivalent circuit model.

Results: The Result show that in our ESSENCE sensor, the EIS signal shifts post target DNA/RNA binding to the CNT with the capture oligos. The limit of detection is the femtomolar range for DNA/RNA. The design of our sensor allows for multiple different types of configurations for taking the EIS signal. It also allows a novel cascade electrochemical technique that allows for detection of small molecules. Further, some of the configurations allow us to make a calibration curve for the DNA/RNA concentration over a large range of concentrations. ESSENCE biosensor has been used to detect (both qualitatively and quantitatively) p53 protein from a mixture of p53 and HER2.

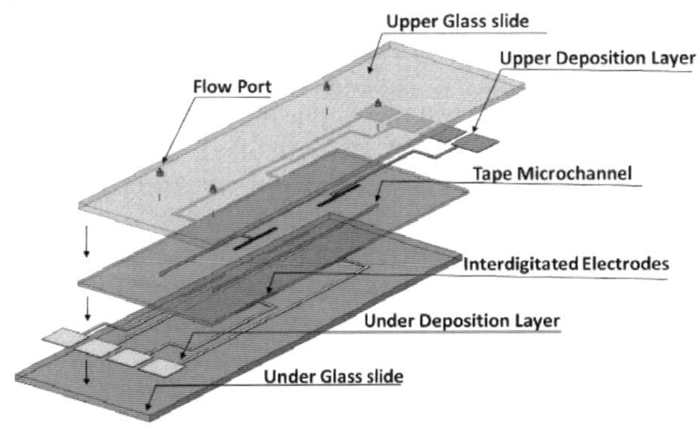

Figure 1. The structure of the ESSENCE microfluidic sensor showing the different elements

Conclusions: The packing of the CNT leads to a significant increase in convective transport that shifts the parasitic double layer capacitor to high frequency. Thus, it allows the observation of the charge transfer resistance, a measure of the binding of a DNA/RNA target to its corresponding DNA/RNA oligo, at a higher frequency. As the detection signal is seen at a higher EIS frequency, it allows for rapid detection of the EIS signal. Hence ESSENCE offers multiple advantages in contrast to current EIS devices like an increased signal to noise ratio, high sensitivity, rapid detection. This allows us to have a Limit of Detection (LOD) of ESSENCE in the femtomolar range. The chip design and the flow rate are being optimized to differentiate between 1 base mismatch, 2 base mismatch, 3 base mismatch, and complete mismatched DNA/RNA. This will allow us to detect single nucleotide polymorphisms (SNAPS) in both DNA and RNA.

References:

1. Basuray, S., Senapati, S., Aijian, A., Mahon, A. R., & Chang, H. (2009). Shear and AC Field Enhanced Carbon Nanotube Impedance Assay for Rapid, Sensitive, and Mismatch-Discriminating DNA Hybridization. ACS Nano, 3(7), 1823-1830. doi:10.1021/nn9004632
2. Mahon, A. R., Barnes, M. A., Senapati, S., Feder, J. L., Darling, J. A., Chang, H., & Lodge, D. M. (2011). Molecular Detection of Invasive Species in Heterogeneous Mixtures Using a Microfluidic Carbon Nanotube Platform. PLoS ONE, 6(2). doi: 10.1371/journal.pone.0017280

Acknowledgement: This research is supported by a National Science Foundation (NSF) Career grant (#1751759), "ASSURED electrochemical platform for multiplexed detection of Cancer Biomarker Panel using Shear-Enhanced Nanoporous-Capacitive Electrodes" to the Basuray Lab.

P2-25

Detection of anti-Tn antibody in serum with 1000-fold better sensitivity than enzyme-linked immunosorbent assay (ELISA) using Piezoelectric Plate Sensor

Song Han[1], Manoj Kumar Baskaran[1], Pawan Rao[1], Hassan Alotaibi[1], Alexandra Jednorski[1], Dolores Conover[1], Wei-Heng Shih[2], Wan Y. Shih[1]

[1]School of Biomedical Engineering, Science, and Health Systems, Drexel University, Philadelphia, Pennsylvania 19104 USA
[2]Department of Materials Science and Engineering, Drexel University Philadelphia, Pennsylvania 19104, USA

Introduction: Breast cancer (BC) is the leading cause of cancer for women. Current breast cancer screening relies on mammography to identify women with suspected lesions for further biopsy testing. However, mammography is not specific. Only 1 out of 5 mammograms with suspected lesions yield a positive cancer diagnosis, resulting in many unnecessary biopsies, anxieties, and worries for women. It would be desirable if there is a cancer-specific blood test that can be used to help determine if the suspected lesions in the mammograms are indeed cancer. However, none of the existing protein cancer markers are specific to cancer. As a result, none of the serum CEA, CA15.3, CA27.29 or circulating cancer cells (CTC) tests can sensitively and specifically detect BC. On the other hand, Tn antigen, GalNAc-O-Ser/Thr, a core monosaccharide antigen on mucins in 90% of human carcinomas that is exposed due to the abnormal truncation of glycans in cancer cells. Furthermore, Tn antigen elicits an immune response. Both Tn antigen and anti-Tn antibody could be found in the blood. However, enzyme-linked immunosorbent assay (ELISA) could only detect Tn antigen with <50% sensitivity due to the low concentration of Tn antigen in the circulation. Piezoelectric plate sensor (PEPS) is a new type of sensor developed in Shih and Shih's lab at Drexel University that exhibits unprecedented sensitivity due to its self-sensitivity-enhancing mechanism through binding-stress induced crystalline orientation switching. As a result, PEPS could detect the same target at a much lower concentration than ELISA. In this study, we intend to detect anti-Tn antibody spiked in bovine serum and compare its sensitivity with a comparing ELISA using exactly the same receptor and the same reagents. Furthermore, we will visually verify the detection of anti-Tn antibody on the PEPS surface with Tn-antigen-coated fluorescent reporter microspheres (FRMs) 6 μm in diameter.

Methods: PEPS were made of 8 μm thick lead magnesium niobite-lead titanate solid solution (PMN-PT) thin films with 150 nm thick gold electrode deposited on both sides of the PMN-PT film with a 100 nm chromium bonding layer by the thermal evaporator (ThermionicsVE90). They were cut in 2 mm x 0.5 mm strips using a wire saw (Princeton scientific). PEPS were then electrically insulated by repeatedly soaking in 0.1% 3-mercaptopropyltrimethoxysilane (MPS) (Sigma) solution in ethanol. Synthetic Tn antigen (Sigma) was immobilized on the PEPS surface as the receptor through a bifunctional linker SM(PEG_{12}) (Thermo Fisher) to detect anti-Tn antibody (Tn-218 IgM) (Genetex) spiked 1.25% bovine serum albumin (BSA) solution to mimic 4-fold diluted serum. The sensor surface was blocked with 1.25% BSA to minimize nonspecific binding. Comparison of ELISA was carried out using a clear amine-binding plate to directly immobilize Tn antigen on the plate.

Results: PEPS was able to detect anti-Tn antibody at a concentration as low as 1 pg/ml, which was further verified with the FRMs captured by the anti-Tn antibody bound on the sensor surface in a dose-responsive manner as shown in Figure 1a. In comparison, ELISA could not detect anti-Tn antibody at 1 ng/ml or lower, indicating that PEPS could detect anti-Tn antibody at a concentration 1000-fold lower than ELISA.

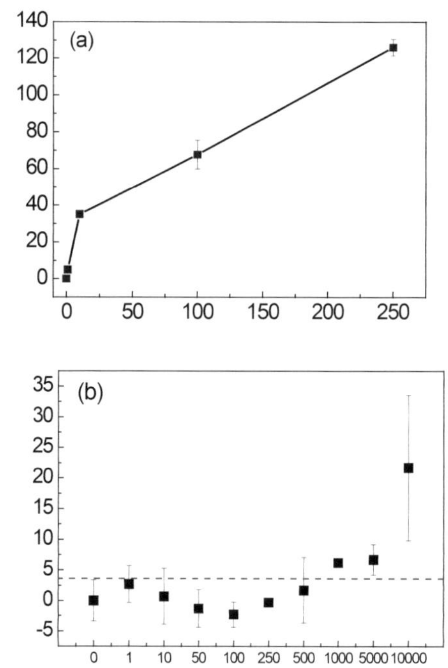

Figure 1: (a) Number of FRMs captured by PEPS for the anti-Tn antibody detection with 1-250 pg/ml and (b) results of comparing ELISA.

Conclusions: In this study, we have shown that PEPS could detect anti-Tn antibody at 1 pg/ml, a concentration 1000-fold lower than ELISA. This strongly indicates potential of using a PEPS anti-Tn antibody test as a sensitive companion blood cancer test to minimize unnecessary biopsies in mammography breast cancer screening.

Acknowledgment: This work was supported in part by the National Institute of Health Grants No. 1R41AI112224 and 1R41AI120445 and Pennsylvania Department of Health.

Optimizing Design of Electrochemical Bio-Sensor for Enhanced Sensitivity

Zhenglong Li, Yu-Hsuan Cheng, Sagnik Basuray
Department of Chemical and Material Engineering, New Jersey Institute of Technology

Introduction: Highly sensitive detection of biological targets (protein and deoxyribonucleic acid (DNA) among others) have attracted tremendous attention to many fields such as genomics, medical diagnostics, point-of-care, and so forth[1]. Although optical detection methods have high sensitivity, the need for professional knowledge, cumbersome techniques as well as the difficulty in miniaturizing has greatly limited their application. However, electrochemical impedance spectroscopy (EIS) has emerged as a promising candidate for rapid, economical, highly selective and label-free detection of a diverse range of analytes from chemical and environmental analytes to biological targets[2]. A more detailed analysis of the system can be obtained by observing the change in the resistance associated with charge transfer at the electrode surface. Carbon nanotubes (CNTs) are used in a multitude of applications from nanoelectronic devices, chemical sensors, gas sensors to solar cells and bio/analytical sensors. Our recent advance in a microfluidic-based sensor that combines EIS with CNTs has shown great promise in the sensitive detection of some analytes, unlike current optical sensing techniques[3].

Methods: Our sensor combines a shear-enhanced, flow-through nanoporous capacitive electrode made of non-planar interdigitated electrodes and CNTs, with a lab-on-a-chip electrochemical platform. Each chip has nonplanar interdigitated microelectrodes of sizes 10μm X 100μm with a pitch of 10μm separated by a microfluidic channel of 500μm X 100μm X 1 cm[4]. Oxidized CNTs are trapped in between the microelectrodes to form tightly packed CNTs. This non-planar interdigitated microelectrode with the tightly packed CNT becomes one of the electrodes of the EIS.

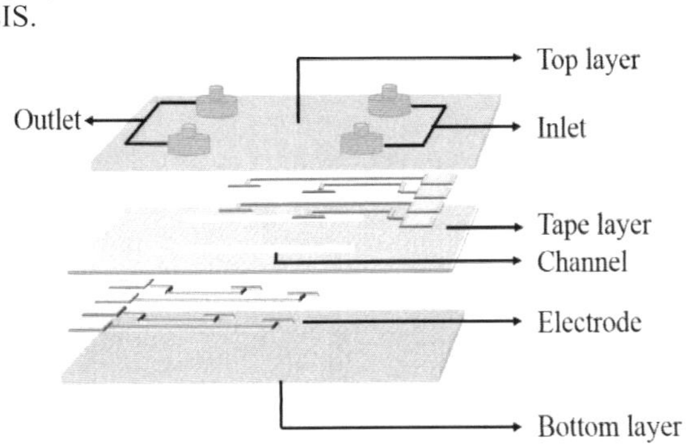

Figure 1. Schematic diagram of device structure

Results: By changing the distance between the two sets of electrodes in the horizontal direction, we observed that the sensitivity ($S = \frac{R_{CT-PBS} - R_{CT-DNA}}{R_{CT-PBS}}$) of the device increases with the increase of the distance between the electrodes, and eventually remains constant range, S= ~5%.

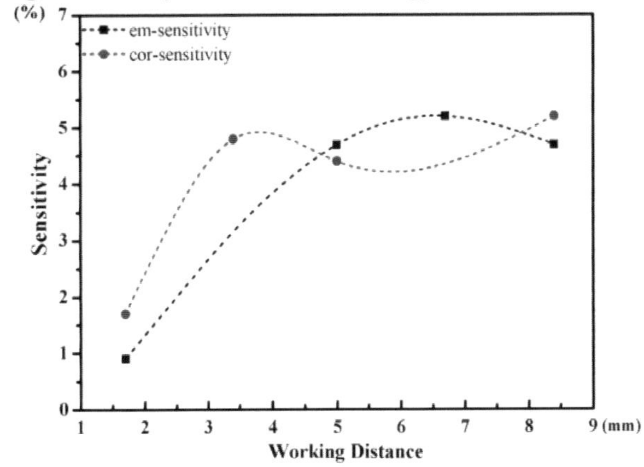

Figure 2. Experimental results analysis of the Sensitivity of the device

Conclusions: We wanted to determine if the sensitivity is a function of the channel architecture or it depends on the biomolecule. Further, the working distances between electrodes is changed also. By changing the distance between the two sets of electrodes in the horizontal direction, we observed that the sensitivity S increases with the increase of the distance between the electrodes, and eventually remains constant range. These allow designing an optimized channel dimension for high sensitivity of our sensor. Further, we are looking to see if there is any effect on the sensitivity due to the frequency of the electric field. Finally, we are also looking at the packing of the CNT in conjugation with channel architecture within the channel to explore selectivity.

References:

1. Pale ek E. Surface-attached molecular beacons light the way for DNA sequencing[J]. Trends in biotechnology, 2004, 22(2): 55-58..
2. Lisdat F, Schäfer D. The use of electrochemical impedance spectroscopy for biosensing[J]. Analytical and bioanalytical chemistry, 2008, 391(5): 1555.
3. Sengupta, S.; Battigelli, D. A.; Chang, H.-C. A Micro-Scale Multi-Frequency Reactance Measurement Technique to Detect Bacterial Growth at Low Bio-Particle Concentrations. Lap Chip 2006, 6, 682–692.
4. Basuray, S., Senapati, S., Aijian, A., Mahon, A. R., & Chang, H. (2009). Shear and AC Field Enhanced Carbon Nanotube Impedance Assay for Rapid, Sensitive, and Mismatch-Discriminating DNA Hybridization. ACS Nano, 3(7), 1823-1830. doi:10.1021/nn9004632

Acknowledgements: This research is supported by a National Science Foundation (NSF) Career grant (#1751759), "ASSURED electrochemical platform for multiplexed detection of Cancer Biomarker Panel using Shear-Enhanced Nanoporous-Capacitive Electrodes" to the Basuray Lab

Discovery of Novel Photoreceptor Gene by Single-Cell RNA-seq Data Analysis

Alexandria Pinto, Xin Ai and Li Cai
Department of Biomedical Engineering, Rutgers University, New Brunswick, NJ

Introduction: Topoisomerase II [2] (TOP2B) is an enzyme that participates in the unwinding of DNA, an important step which regulates transcription. Studies have established that the *Top2b* gene has an essential function in retinal cell differentiation. Here, we analyze the function of TOP2B in the retinal development using single-cell RNA sequencing (scRNA-seq) data. This data allows analysis of mRNA expression at an individual cell level in complex tissues with multiple cell types, such as the retina. The retina contains seven major types of cells – bipolar, Müller glia, ganglion, horizontal, amacrine, rod and cone photoreceptor cells. Retinal differentiation, e.g., cell fate determination, neurite growth and synapse formation, requires specific gene expression. In this study, we focused on the photoreceptor cells (rods and cones), which are responsible for converting incoming light into a signal that will be passed to the brain resulting in vision.

Methods: We utilize publicly available datasets GSE63473 [1], containing P14 mice retinal cells, and 10x single cell RNA-seq data from the developing mouse retina [2], containing E16, E18, P0, P2, P5 and P8 mice retinal cells. We performed scRNA-seq data analysis using Bioconductor packages, scater and MAST, as well as Seurat. A training set of cells for each time point was first isolated. From the training cells, we identify genes that are differentially expressed with Top2b expression and utilized the top 1000 highly variable genes for clustering individual retinal cells (**A**).

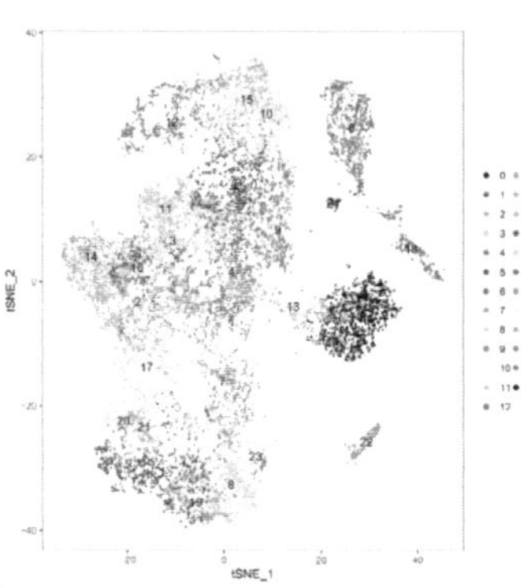

A

Clusters contained cells with similar gene expression with 25 marker genes per cluster. Genes used for clustering in the training set were then used to cluster all the cells in each developmental time point.

Results: The training set of cells prove that the method of using differentially expressed highly variable genes for clustering of retinal cells is feasible. The method is capable of separating out cones, rods bipolar, ganglion, and other cell types. Applying the same set of genes to all remaining cells resulted in the same type of separation. We found that *Fam19a3* is among upregulated genes, while corresponding to expression of *Opn1sw*, a cone marker **(B)**. *Fam19a3* expression was first detected at E16 and continues to be expressed within cone cells. After P8, *Fam19a3* is expressed with both cone cells and bipolar cells.

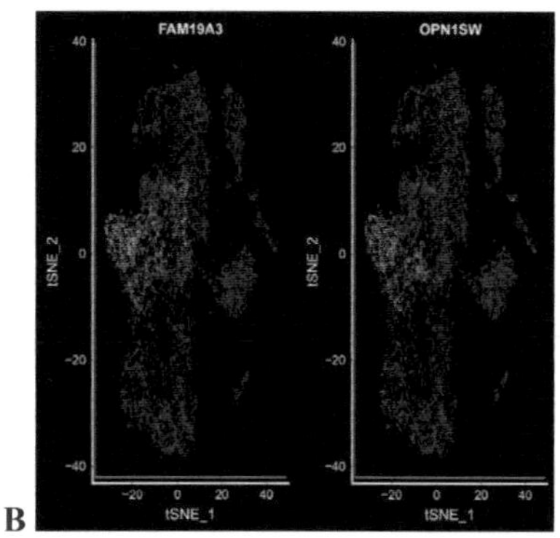

B

Conclusions: *Fam19a3* is a novel candidate gene for cone photoreceptor development. Since *Fam19a3* encodes for chemokine like molecule, it is likely that it plays an important role in synapse formation between cones and bipolar cells. Future studies of *Fam19a3* will determine its biological functions in retinal development using gain- and loss-off-function analyses.

References:

1. Macosko, E. *Cell* 2015, 161(5): 1202-1214.
2. Clark, B. *BioRxiv Preprint* 2018.

Friday, March 22, 2019

Schedule

9:00 am–4:00 pm	Registration
9:00 am–10:00 am	Breakfast
10:00 am–10:40am	Keynote Speaker from Cancer Institute of NJ—*LIVINGSTON HALL*
10:40 am–12:00 pm	**Design Competition, Business Pitch, and Undergraduate Research Presentations** (15 minute talks with 5 minutes for questions)

	Session 6A Design Competition *—LIVINGSTON HALL*	**Session 6B Business Pitch** *—201*	**Session 6C Undergraduate Research Talks** *—202*
10:40–11:00 am	6A-1:Design of a Digital Motion Analysis Device to Improve the Markup Procedure of Cerebral Palsy Subjects during Gait Analysis	6B-1:Kristen Labazzo Rutgers University	6C-1:Ronke Olabisi Rutgers University **Meaningful Research with Undergraduate Students.**
11:00–11:20 am	6A-2:Pulse Amplitude Ratio Based on a Preload-Afterload Normalization	6B-2:Flexible Sensors for Human Motion Analysis	6C-2:Validating Hydrogel Microencapsulated Insulin Secreting Cells for Wound Healing
11:20–11:40 am	6A-3:Design of Detection System for Inaccessible Venous Access Ports in the Pediatric Emergency Department	6B-3:Fast Tumor Spheroid Growth and Drug Testing in Microfluidic Device	6C-3:Fluid flow rate dictates the efficacy of low intensity anti-vascular ultrasound therapy in a microfluidic model
11:40 am–12:00 pm	6A-4:Bioreactor for Mechanical Stimulation of 3D Scaffolds for Tendon Tissue Engineering	6B-4:Auditory Assistive Laboratory Application for Blind and Visually Impaired Students	6C-4:Spatial Organization of Biochemical and Physical Properties in 3D-Printed Scaffolds

12:00 pm–2:00 pm	Lunch/Vendors and Sponsors/Poster Session

2:00 pm–3:20 pm **Parallel Platform Presentations: Session VI**
(15 minute talks with 5 minutes for questions)

	Session 7A Rehabilitation *—LIVINGSTON HALL*	**Session 7B Cancer Biology** *—201*	**Session 7C Biomedical Imaging II** *—202*
2:00–2:20 pm	7A-1:Raviraj Nataraj Stevens Institute of Technology	7B-1:Marcin Iwanicki Stevens Institute of Technology	7C-1:Spectrally Encoded Compressive Imaging For Improved Resolution in Fiber-Bundle-Based Endoscopy
2:20–2:40 pm	7A-2:Alex Zhou NJIT	7B-2:Detection of DNA Mutations in Extracellular Vesicles Derived from Patients with Malignant Ground Glass Opacities	7C-2:Environmentally-Controlled Approaches for Near Infrared Spectroscopic Assessment of Cortical Bone Water
2:40–3:00 pm	7A-3:Effects of Robotic Exoskeleton Gait Training on an Adolescent with Chronic Brain Injury	7B-3:Superparamagnetic Iron Oxide Nanocrystal Clusters for Cancer Therapy	7C-3:Short-Wave Infrared Spatial Frequency Domain Imaging For Non-Invasive Quantification Of Tissue Hydration
3:00–3:20 pm	7A-4:Modeling and Evaluation of an Admittance Controlled Hand Exoskeleton for Neuromuscular Rehabilitation	7B-4:Tunable Hydrogels to Investigate Breast Cancer Dormancy and Metastatic Relapse	7C-4:z

3:20–4:00 pm Conference Research Awards Ceremony Introduction of the 46th NEBEC hosts & Conference Closing—*LIVINGSTON HALL*

Abstracts

6A-1 **Design of a Digital Motion Analysis Device to Improve the Markup Procedure of Cerebral Palsy Subjects during Gait Analysis** 157
Haley Marion, Isabelle Rivas, Alexis Van Buren, Robert Sitbon, Andrea Kwaczala
Western New England University

6A-2 **Pulse Amplitude Ratio Based on a Preload-Afterload Normalization** 160
Rebecca Edgerton, Mackenzie Cash, Kareem Boura, Ying Sun
University of Rhode Island

6A-3 **Design of Detection System for Inaccessible Venous Access Ports in the Pediatric Emergency Department** 162
Hannah Blakely [1], Sumaiya Sayeed [1], Kelly Williams [1], Colette Bare [1], Sakina Sojar [2]
[1] Brown University, [2] Hasbro Children's Hospital

6A-4 **Bioreactor for Mechanical Stimulation of 3D Scaffolds for Tendon Tissue Engineering** 163
Brennen Zolnoski, Emily Fernschild, Devina Jaiswal
Western New England University

6B-2 **Flexible Sensors for Human Motion Analysis** 165
Sagar Doshi, Kaleb Burch
University of Delaware

6B-3 **FAST TUMOR SPHEROID GROWTH AND DRUG TESTING IN MICROFLUIDIC DEVICE** 166
Yaling Liu
Lehigh University

6B-4 **Auditory Assistive Laboratory Application for Blind and Visually Impaired Students** 169
Haoyang Chen, Aida Kupa, Junyu Lu, Pushpita Rahman
Drexel University

6C-2 **Validating Hydrogel Microencapsulated Insulin Secreting Cells for Wound Healing** 171
Shreya Soni, Zeiny Aubdoollah, Eddy Iturbide
Rutgers University, New Brunswick

6C-3 **Fluid flow rate dictates the efficacy of low intensity anti- vascular ultrasound therapy in a microfluidic model** 172
Brandon DeOre [1], Peter Galie [1], Chandra Sehgal [2]
[1] Rowan University, [2] University of Pennsylvania

6C-4 **Spatial Organization of Biochemical and Physical Properties in 3D-Printed Scaffolds** 173
Anne Behre, Paula Camacho, Hafiz Busari, Kelly Seims, Peter Schwarzenberg, Colin Bader, Hannah Dailey, Lesley Chow
Lehigh University

7A-3 **Effects of Robotic Exoskeleton Gait Training on an Adolescent with Chronic Brain Injury** . . . 174
Kiran Karunakaran [1,2,3], Naphtaly Ehrenberg [1,2], JenFu Cheng [2], Karen Nolan [1,2]
[1] Kessler Foundation, [2] Childrens Specialized Hospital, [3] New Jersey Institute of Technology

7A-4 **Modeling and Evaluation of an Admittance Controlled Hand Exoskeleton for Neuromuscular Rehabilitation** 175
Xianlian Zhou, Ashley Mont, Sergei Adamovich
New Jersey Institute of Technology

7B-2 **Detection of DNA Mutations in Extracellular Vesicles Derived from Patients with Malignant Ground Glass Opacities** 176
Yi Wen, Komal Abhange, Yuan Wan
The Pq Laboratory of Micro/Nano BiomeDx, Binghamton University-SUNY

7B-3 **Superparamagnetic Iron Oxide Nanocrystal Clusters for Cancer Therapy** 177
Qingbo Zhang [1], Zhen Xiao [1], Linlin Zhang [2], Sheng Tong [2], Gang Bao [2], Vicki Colvin [1]
[1] Brown University, [2] Rice University

7B-4 **Tunable Hydrogels to Investigate Breast Cancer Dormancy and Metastatic Relapse** 178
Shantanu Pradhan, John Slater
University of Delaware

7C-1 **Spectrally Encoded Compressive Imaging For Improved Resolution in Fiber-Bundle-Based Endoscopy** 179
John Dumas, Muhammad Lodhi, Batoul Taki, Waheed Bajwa, Mark Pierce
Rutgers, The State University of New Jersey

7C-2 **Environmentally-Controlled Approaches for Near Infrared Spectroscopic Assessment of Cortical Bone Water** 180
Jack Oswald, Ramyasri Ailavajhala, Nancy Pleshko
Temple University

7C-3 **Short-Wave Infrared Spatial Frequency Domain Imaging For Non-Invasive Quantification Of Tissue Hydration** 181
Christine Sahyoun [1], John Dumas [1], Yanyu Zhao [2], Matthew Applegate [2], Darren Roblyer [2], Mark Pierce [1]
[1] Rutgers, The State University of New Jersey, [2] Boston University

7C-4 **Diagnosis of Myocardial infarction in cardiac cine MR images using corr-entropy based endocardium detection and ventricle wall motion analysis** 182
Kavitha G [1], Muthulakshmi M [2]
[1] Associate Professor, MIT Campus, Anna University, [2] Research scholar, MIT Campus, Anna University

Posters

P3-1 **Targeted Tumor Drug Delivery by Magnetic nanoparticles ; An Invitro Study** 183
Negin Farzad [1], Peter klesczews, Christina Zito [1], Hashini Mohottala, Saion Sinha [1]
[1] University of New Haven

P3-2 **Design of a Standardized in vitro Model of Skeletal Muscle Regeneration for Implantable Microthread Scaffolds** 184
Erin Heinle, Emily Morra, Emily Mossman, Alyssa Paul, Meagan Carnes, Catherine Whittington, George Pins
Worcester Polytechnic Institute

P3-3 **Inflammatory Secretome of Macrophages Treated with Hemoglobin-Haptoglobin Complexes** .186
Paulina Krzyszczyk [1], Kishan Patel [1], Maurice O'Reggio [1], Kristopher Richardson [2], Martin Yarmush [1], Andre Palmer [1,2], Francois Berthiaume
[1] Rutgers University, [2] The Ohio State University

P3-4 **Nanoparticle Containing V domain of sRAGE for Diabetic Chronic Wounds** 187
Hwan June Kang
Rutgers University

P3-5 **Engineered Extracellular Vesicles Derived from Human Umbilical Cord Mesenchymal Stem Cells for Skin Rejuvenation** . . . 188
Komal Abhange, Yi Wen, Yuan Wan
The Pq Laboratory of Micro/Nano BiomeDx, Binghamton University

P3-6 **Alginate Encapsulation for Bupivacaine Delivery and MSC Co-therapy** . . . 189
Mollie Davis [1], Xiomara Perez [1], Ileana Marerro Berrios [1], Charles Rabolli [1], Rene Schloss [1], Joel Yarmush [2], Martin Yarmush [1]
[1] Rutgers University, [2] New York Presbyterian- Brooklyn Methodist Hospital

P3-7 **A Thermoreversible and Photoactive Collagen-Based Scaffold for Tissue Engineering Applications** . . . 190
Yoliem Miranda Alarcon, Dorota Jazwinska, David Shreiber
Rutgers University

P3-8 **Nanofibrous nerve conduits pre-seeded with bone marrow stromal cells and pre-cultured in bioreactors for peripheral nerve regeneration** . . . 191
Gan Zhou, Xiaojun Yu
Stevens Institute ofTechnology

P3-9 **Investigating the use of low-cost bioprinting for the fabrication of complex tissue scaffolds** . . . 192
Robert Warren [1], Carolina Leynes [2], Joseph Freeman [1]
[1] Rutgers University, [2] University of Texas Rio Grande Valley

P3-10 **Intelligent Rock Climbing Shoes That Provide Haptic and Interactive Feedback to Users with Lower Leg Prosthetics** . . . 193
Gianna Morrongiello [1,2], Lauren Bartlett [1,2], Meaghan O'Connell [1,2]
[1] Ying Sun, [2] Jiang Wu

P3-11 **Catechins Inhibit Toxin Activity by Changing Secondary Structure of Toxin** . . . 195
Joanne Huang, En Hyung Chang, Angela Brown
Lehigh University

P3-12 **Collagen Type-I Antibacterial Hydrogel for Wound Healing Applications** . . . 196
Dorota Jazwinska, Yoliem Miranda-Alarcón, David Shreiber
Rutgers University

P3-13 **Effect of Muscle Cell Differentiation on AAV Transduction Efficiency** . . . 197
Kedar Patel, Alexandra Burr, Biju Parekkadan
Rutgers University - New Brunswick

P3-14 **Significant Glycosaminoglycan Deposition in the Aorta of a Murine Model of Hutchinson-Gilford Progeria Syndrome Associates with Microstructural Alterations and Compromised Biomechanical Properties** . . . 198
Yuki Kawamura [1], Kristin Zimmerman [2], Demetrios Braddock [2], Jay Humphrey [1,3], Sae-Il Murtada [1]
[1] Department of Biomedical Engineering, Yale University, [2] Department of Pathology, Yale School of Medicine, [3] Vascular Biology and Therapeutics Program, Yale School of Medicine

P3-15 **Kinetic Characterization of Nitrite reduction to NO by the Molybdopterin Enzyme mARC2** . . . 199
Eric Cecco [1,2], Jesus Tejero [1], Mark Gladwin [1], Courtney Sparacino-Watkins [1]
[1] Vascular Medicine Institute, [2] University of Pittsburgh

P3-16 **Methods for Encapsulating Cells into Hydrogel Sheets for Wound Healing** . . . 200
Eddy Iturbide, Zeiny Aubdoollah, Shreya Soni, Ronke Olabisi
Rutgers University, New Brunswick

P3-17 **Hearing Characterization Using Spoken Words** . . . 201
Mohammed Janoudi, Justin Jewell, Liam Stamp, Ying Sun, Jiang Wu
University of Rhode Island

Friday

P3-18 **Monitoring coral health via allosteric transcription factor biosensor technology** 204
Melanie Martinsen, Rachel Petherbridge, Kevin Lorch, Pulkit Mittal
Boston University

P3-19 **Establishing a Three-Dimensional Collagen Based Cell Culture of Human Monocytes** 205
Blair Landon [1], Yun Jiao [1], Laura Simpson [1], Jennie Leach [1], Gregory Szeto [1,2]
[1] University of Maryland, Baltimore County, [2] Marlene and Stewart Greenebaum Comprehensive Cancer Center, University of Maryland School of Medicine

P3-20 **The Automation of the OculoMotor Assessment Tool** .206
Nabih Armout [1], Christopher Morris [1], Taiga Akiyama [1], Anthony Chirayath [1], John Vito d'Antonio-Bertagnolli, Chang Yaramothu, Tara Alvarez
[1] Vision Neural Engineering Laboratory of New Jersey Institute of Technology

P3-21 **Enhanced properties of liquid infused paper for bacteria handling and point-of-care diagnostics**. 208
Emily LeClair [1], Daniel Regan [1,2], Caitlin Howell [1,2]
[1] University of Maine, [2] Graduate School of Biomedical Science and Engineering, University of Maine

P3-22 **Rapid Assessment of Cellular Activity of Growing Tissue Engineered Constructs** 209
Sabrina Zouaghi, Mugdha Padalkar, William Querido, Rutvin Kyada, Shital Kandel, Nancy Pleshko
Temple University

P3-23 **Modulation of Covalently Crosslinked 3D Collagen Hydrogels Regulate Metabolic and Fibrotic Gene Expression** 210
Nikolas Di Caprio, Evangelia Bellas
Temple University

P3-24 **The Design of an Assisted Intravenous Transfer System for Patient Ambulation** 211
Scott Selig, Lillian Margolis, Laura Parra
University of Rhode Island

P3-25 **Assistance Calling Device For Advanced Stage Amyotrophic Lateral Sclerosis Patients** 213
John Furey, Andres Gutierrez, Chadwick Dotson-Jones
Drexel University

P3-26 **Glucose Sensitivity in Engineered Adipose Tissue Model of Type 2 Diabetes** 214
Alexandra Abruzzo, Michael Struss, Evangelia Bellas
Temple University

P3-27 **The Effect of Macrophage Infiltration on Adipose Tissue Inflammation** 215
Eleanor Caston, Evangelia Bellas
Temple University

P3-28 **The Effect of Vascularization on Adipose Tissue Glucose Metabolism** 216
Jennifer Hammel, Golnaz Anvari, Michael Struss, Evangelia Bellas
Temple University

P3-29 **Exploring New Parameters of The K562 Cytotoxicity Assay**. 217
Riya Patel, Lauren Timmens, Biju Parekkadan [1]
[1] Rutgers University

P3-30 **Angiogenic response to IL-4 eluting coatings in mesh tissue explants**. 218
Hannah Geisler [1], Alexis Nolfi [1], Aimon Iftikhar [1], Bryan Brown [1,2]
[1] University of Pittsburgh, [2] McGowan Institue for Regenerative Medicine

P3-31 **Gait Analysis Through Pressure Sensing**. 219
Stephanie Speaker, Patrick Riolo, Bailey Felix, Emerson Iannone, Miguel Gomez
Syracuse University

P3-32 **Deep Tissue Injury (DTI) Detector** 220
Eric John, Amanda Joyce, Meenakshi Patel, Michael Neidrauer, Richard Huneke, Michael Weingarten, Wei-Heng Shih, Wan Shih
Drexel University

P3-33 **An Acoustic Bioreactor for High Throughput Chemotherapeutic Screening on Perfused Three Dimensional Tumors** 221
Kayla Van Buren[1], Matthew Carey[1], Ben Ross-Johnsrud[2], Bart Lipkens[2], Anthony English[1]
[1] Western New England University, [2] FloDesign Sonics Inc.

P3-34 **Neonatal Real-Time Photoplethysmogram Heart Rate Monitoring System During Resuscitation** . 222
Alex Gray, Luke McConnaghy, Brandon Williams
University of Rhode Island

P3-35 **A 2-Photon Approach to Imaging the Spatial Relationship of Collagen and Nerve Fibers in MMP-13 Inhibited and Uninhibited Subcutaneous White Adipose Tissue** 224
Mitchell Harling, Karissa Tilbury, William Breeding
University of Maine

P3-36 **Design of a Bone Morphogenetic Protein Based Carapace Repair Device** 225
Sylvia Baeyens[1], Elizabeth Cummins[1], Daniel Kanter[1], Matthew Figucia[1], Aislinn Keane[2], Christopher Wilson[2]
[1] Boston University, [2] Bioventus LLC

P3-37 **Monitoring Self-Contained Breathing Apparatus Performance through Temperature and Physiological Recordings** 226
Thomas Totillo, David Cleary, Jake Duerwald, Eugene Chabot
University of Rhode Island

P3-38 **Neuron Emulator with Realistic Action Potentials Voltage-Clampable via an Electrophysiological Instrument** 228
Pascaline Uwase, Fatima Issa, Huafen Guo, Samuel Ng
University of Rhode Island

P3-39 **Monitoring the Patency of Pediatric Tracheostomy Patients' Airways using the TrachAlert Device** 230
Kenneth Poser[1], Yanique Spigner[1], Franzisca Komar[1], Matthew Iobst[1], Julianna Ricci[1], Sally Shady[1], Frank Castello
[1] Stevens Institute of Technology

P3-40 **Monitoring Balance Board Pressure during Ankle Rehabilitation** 231
Meghan Keenan, Aiden Keene, Tessa Arsenault
University of Rhode Island

P3-41 **In-theater Concussion Assessment through Reaction Time Evaluation Utilizing an Android Application** 233
Christian Chaneski, Duncan Gandley, Gabrielle McCormack, Caroline Shusta, Magdalena Slonski
Stevens Institute of Technology

P3-42 **Micro-catheters for Targeted Fluid Delivery in Neural Tissues** 234
Michael Fabiunke
Rutgers University

P3-43 **Body Powered Distal Interphalangeal Finger Prosthetic** 235
Ricardo Garcia, Christian Pignataro, Giovanna Nolan, Madison Taylor, Ashley Mont, Sergei Adamovich
New Jersey Institute of Technology

Friday

P3-44 **Otto-Mobile: The Modular Wheelchair for Children** 236
Samanatha Abate, Maria Catalane, Alina Zdebska
Syracuse Univeristy

P3-45 **Inter-user Variability Analysis for Manual Bone Surface Segmentation from Ultrasound Data** 237
Hridayi Patel, Ilker Hacihaliloglu
Rutgers University

P3-46 **Induced Heart Rate Variability Index Derived from the Valsalva Maneuver and Sudden Standing Up**. 238
Abdulrahman Alsasa, Nathan Guillemette, Dylan Young
University of Rhode Island

P3-47 **Soothing Patients with Autism in the Emergency Room - BusyBin**. 240
Jeff Ung, Craig Hughes, Faith Martin, Daniel Natale
Stevens Institute of Technology

P3-48 **Rapid Prototyping to Roll to Roll manufacturing of Microfluidic devices** 241
Chris Toothaker [1], Bailey Corless [1], Amber Boutiette [1], Caitlin Howell [1,2]
[1] University of Maine, [2] Graduate School of Biomedical Science and Engineering

P3-49 **Biomarker Predictors of the Development of Alzheimer's Disease** 242
Katherine Ward [1,2], Nathan Ackermann [1,2], Natalie Brunette [1,2], Jared Klug [1,2]
[1] Boston University, [2] Massachusetts General Hospital

P3-50 **Effects of Substrate Stiffness and Cell-Cell Contact Area on Stem Cell Signaling** 243
Sarah Furman [1], Sebastian Naranjo [1], Kirstene Gultian [1], Abigail Loneker [2], Rebecca Wells [2], Sebastian Vega [1]
[1] Rowan University, [2] Perelman School of Medicine, University of Pennsylvania

P3-51 **Creating an Assistive Beach Wheelchair Device**. 244
Gabriella Borodyansky, Jacqueline Chi, Justin Stowe
Stevens Institute of Technology

P3-52 **Development of an Electroactive Tissue Engineering Scaffold** 245
Ruchi Patel, Jeeba Thomas, Madara Dias
Rutgers University

P3-53 **Targeted Heating Gadget: A Heating Device for Treatment of Hypothermic Conditions**. 246
Daniel Cleary [1], Salma Alshafie [1], Melanie Caba [1], Emily Ferguson [1], Christopher Perry [2]
[1] Stevens Institute of Technology, [2] Long Island Jewish Hospital

P3-54 **Evaluation of endogenous neural stem cell activity in a mouse model of traumatic brain injury** 247
Rebecca Risman [1], Jeremy Anderson , Dr. Li Cai
[1] Rutgers University

P3-55 **Designing a high-throughput cell immobilizer for microscopy imaging in a monolayer** 248
Tina Ta, Chance Jackson
Boston University

P3-56 **Mechanical Tests on a Polyelectrolyte Complex (PEC)** 249
Hima Tallam, Brandon Mirabile
Rutgers University

P3-57 **Mid-Fidelity Simulator Model for Training Cesarean Sections in Kampala, Uganda** 250
Victoria Utria, Rachel Junod, Michelle Krach, Lyndsey Sbarro
Drexel University

P3-58 **Design of optimization procedure for detection of a breast cancer-specific surface molecule** .252
Virginia Tanner, Priya Gupta, Raadiya Qadeer
Drexel University

P3-59 **Mapping Functional Networks in the Brain Through Increased Adaptability and Comfort Levels in fNIRS Technology** .253
Elizabeth Tarangelo
Syracuse University

P3-60 **Classification For Visual Objects Categorization Using EEG** .255
Hongbo Du, Ismail Jouny
Lafayette College

P3-61 **Continuous Multi-Limb Monitoring Device for Correlation Between Blood Clot Formation and Bed Rest for Stroke Patients** .256
Sydney Robinson, Julia Donovan, David Edgar
University of Rhode Island

P3-62 **Novel Uterine Manipulator to Reduce Risk of Uterine Perforation and Infection Post-Surgery** 258
Courtney Evans, Kristine Bombardiere, Abigail Emerson, Gabriella Ferrara
Stevens Institute of Technology

P3-63 **Optical Density based Measurement of Cell Count within Polymeric Particles** .259
Dylan Trawinski, Ayesha Aijaz, Biju Parekkadan
Rutgers University

P3-64 **Non-Invasive Arterial Pressure Sensor under Low and High Deflections from Force Signals** .260
Angela Meseha
Rutgers University

P3-65 **Platform for Developing Beat-to-Beat Heart Rate Detection Algorithms Based on Photoplethysmogram** .261
Jack Cammarata, Jake Matte
University of Rhode Island

P3-66 **Classification of Tumor in Mammograms with Deep Convolutional Neural Networks** .263
Tianyu Zhu, Ismail Jouny
Lafayette College

P3-67 **3D Orientation and Adjustment System for Automated Venipuncture Device** .264
Zachary Lopez
Rutgers University

P3-68 **Novel Drug Delivery System using Anti-Angiogenic Peptides for Glioblastoma Multiforme** .265
Anna Mathew
New Jersey Institute of Technology

P3-69 **Android Based Prosthetic Training Using Augmented Reality** .266
Merci-Pauline Ujeneza, Anya Brian Duroha, Corvah Akoiwala, Eugene Chabot, Ying Sun
University of Rhode Island

P3-70 **Point of Care Ultrasound for Spine Imaging in Space Explorations** .268
Nina Ninua, Hedy Makris, Pragya Hooda, Daniel Lefever, Ilker Hacihaliloglu
Rutgers University

P3-71 **A Novel Approach to Enhancing Proprioceptive Feedback through the Control of Vibration Motor Frequency** 269
Shreya Anjaria [1], Brandon Herb [1], Eric Radbourne [1], Richard Schermerhorn [1], Becky Tucci , Vikki Hazelwood [1]
[1] Stevens Institute of Technology

P3-72 **Sound Asleep: A Novel Solution for Active and Passive Noise Control in the Neonatal Intensive Care Unit (NICU)** 270
Rachel Hjerpe [1], Christian Jensen [1], Reinout Van Landegem [1], Kristian Quevada [1], Paul Moyer [1], Sally Shady [1], Marwa Khalil [2]
[1] Stevens Institute of Technology, [2] Hackensack University Medical Center

Design of a Digital Motion Analysis Device to Improve the Markup Procedure of Cerebral Palsy Subjects during Gait Analysis

Haley Marion, Isabelle Rivas, Alexis Van Buren, Robert Sitbon, Kayla Van Buren, Andrea Kwaczala, Ph.D.
Western New England University

Significance: Motion analysis systems used during gait analysis are vital for treating Cerebral Palsy (CP) patients. CP is a group of neurological disorders that affect movement and balance due to abnormal muscle tone. Gait analysis is used to plan corrective surgeries or to identify efficacy of orthotic devices. The most common type of CP is Spastic Cerebral Palsy, which effects around 80% of those with CP [1]. Around 70% of these people can walk independently or with a walking aid, and are commonly followed by routine gait analysis [2].
The use of assistive walking aides can impede visualization of reflective markers by cameras mounted around the perimeter of the room. Additionally, there is some discrepancy in the level of accuracy at detecting hip rotation during gait, even in healthy controls [3]. The main aim of this project was to determine if CP physiology contributed to variance in markup and analysis for motion analysis and to design a digital tool that could wirelessly track hip joint motion without the use of infrared cameras. We hypothesize that controlling marker placement and digital markup of a subject will improve the reliability and repeatability of motion analysis when used on a CP patient with abnormal gait.

Design Description: The design of the *Joystick-Talkee Marker System* allows for simultaneous motion analysis data via classical Infrared (IR) camera technology as well as digital wireless data via chips embedded in the standard reflective markers. The Joystick is designed to allow accurate placement on the body using articulating arms to improve visualization and proper positioning on any subject regardless of anatomy. The final design will be required to have 1.5mm of accuracy and communicate at a frame rate of up to 120 fps to match the data processing capabilities of the VICON motion analysis system.
Combining both the Joystick and the Talkee marker designs would result in the *Joystick-Talkee Marker System* where both devices would be used in tandem. We aim to utilize biomechanical models of gait to characterize body motion such as the Plug-in Gait model configuration and will test the device for standard gait studies as well as functional movement screens to assess feasibility of the design (**Fig. 1**).

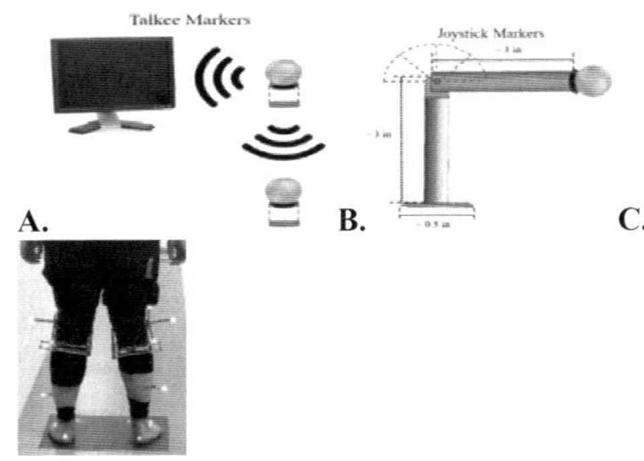

Figure 1. Simplified design of the Joystick and Talkee Markers. A.) The Talkee Markers communicate with each other and to local computer with a data acquisition software to compute position and distance between markers with respect to time. **B.)** The physical Joystick Marker can be placed on the subject in areas where markers become obscured . **C.)** The articulating, position-locking arm allows the reflective marker to be positioned at a known distance when attached to the subject's body to improve visibility during testing.

Design Criteria: This project arose from a clinical assessment problem identified in the Motion Analysis Center at Shriners Hospital for Children in Springfield, MA. The project sponsors were looking for a new tool to corroborate the data collected during gait analysis. The new device was designed to solve three main issues; 1. Improve the detection of markers even when used with walking assist devices, 2. Track hip joint center more accurately compared to the current data collection system, and 3. Increase the confidence between the clinical video data and gait data with a secondary data collection system that runs simultaneously.
Electrical System – Talkee Marker Design: The Talkee Marker is a wireless communication device which must provide a digital position of each marker reported to a computer station. The sensors embedded within the IR reflective markers must last 3 hours on battery life, communicate within 10 meters from the receiver, and accurately track marker position with a detection sensitivity of up to 0.5cm $\pm$ 0.15cm. The system must report values that match the current data collected at the Shriners Motion Analysis Lab with an acceptance criterion of α less than 0.025.
Mechanical System – Joystick Design: To conform to standard clinical markup procedures and work within the Motion Analysis Center, the physical device needed to be constructed within several constraints. The articulating arm of the device needed to be adjustable to provide an angle of the marker $180° \pm 1°$ relative to the axis of rotation. The

extension length could be no more than 12±0.5 cm with an articulating arm length no more than ~8 cm. The device should weigh no more than 450±5 grams to minimize wobble during dynamic movements and to remain adhered to the body throughout a dynamic motion testing trial and must stay attached for as long as 3 hours. The key design criteria of the Joystick Marker is that during functional movements, it needed to stay visible by the IR cameras for the duration of the examination and was not be obscured by abundant adipose tissue.

Clinical Testing - Verification and Validation Methods: A healthy control and a Spastic CP subject completed a gait analysis. Height and weight were collected and used within the data analysis process. Three retroreflective markers were taped onto the right leg of the subject at the; knee flexion axis, lateral malleolus, and greater trochanter. The marker placement was done by the same researcher each time. The markers were removed in between each data collection trial, and reapplied to determine intra-user variability due to difference in the placement of the marker. The trial consisted of 5 steps video recorded and analyzed with Kinovea video analysis software for three separate trials. Repeatability of marker placement was assessed using a crossed Gage R&R analysis. Differences between subjects was determined using a Student's t-test. Future testing will include visual gait data collected in the Motion Analysis Center at Shriners Hospital for Children compared to wireless data collected using new digital markers.

Results: A physical prototype has been developed and preliminary data collection of both a healthy control and CP subject have been collected using standard markers.

Talkee Marker Device: An electrical system has been built that powers the wireless communication devices. Software is being written to triangulate the markers in space with respect to time.

Joystick Marker Device: A prototype of the physical mounting system for the reflective markers has been built that allows standard reflective markers to be attached to a joystick design that holds them away from the subject's body. The final prototype will allow the articulating arm to lock into place at specific angles, which will then allow proper calculations and input into the standard VICON software. This information will allow the end user to easily triangulate the new reflective marker position to the original body landmark location. Mathematical code is still needed to determine the external marker location in space with respect to the human body attachment site.

Clinical Verification and Validation Study: The preliminary gait analysis of a healthy and CP subject revealed anatomical differences in the knee joint center. When joint angles were measured using Kinovea motion analysis software, the CP subject demonstrated a 47% smaller knee flexion angle at swing when compared to the healthy subject (Student's Two-sample t-test, p<0.05, **Fig.2**). The variance in knee flexion angle was similar in the two subjects (F-test, p>0.05, **Fig.2**). The CP patient had significantly different joint angles at each stage of gait compared to the expected outcome from the literature data (One-sample t-test, p<0.05). Whereas the healthy subject's joint angles were similar to the expected outcome.

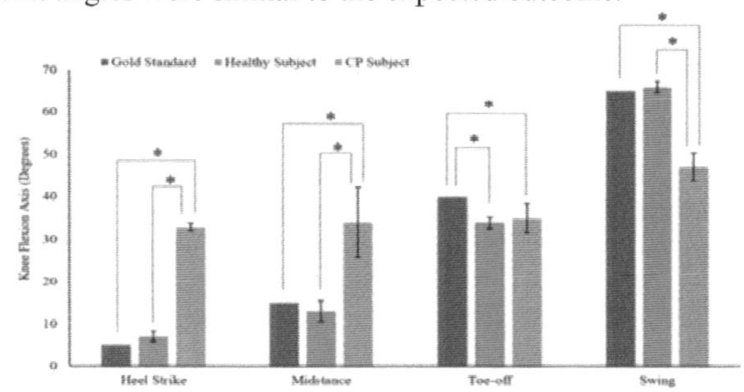

Figure 2. Knee Flexion Angles during the Gait Cycle. Knee angles were measured in a Healthy Control (Orange), a Spastic Cerebral Palsy (CP, Gray), and compared to a literature value (Blue) [4]. Angles were measured at heel strike, midstance, toe-off and swing in 2D videos analyzed via Kinovea motion analysis software. Data are represented as means +/- standard deviation for 5 steps measured across three trials, n=15/group. (Student's t-test, *p<0.05).

Repeatability of the markup procedure by a single operator revealed the measurement system was effective when used on the healthy subject but not on the CP subject. In the healthy control, the majority of the variance was due to the subject's inconsistency in walking, accounting for nearly 100% of the total variance. This suggests the measurement system itself had little variation and the operator applies the reflective markers consistently across trials where bony prominences were more obvious (**Fig. 3A**). However, in CP subjects the anatomical landmarks are more difficult to find and place on the body due to the irregular musculature and anteversion of the hips. In CP, the measurement system accounted for more than 30% of the variation (**Fig. 3B**). The inexperience of the operator contributed to the variation in the markup as well as in the operator in the more complicated case (CP subject). In order to reduce as many sources of variability as possible, the final gait analysis mark p procedure to validate the prototype will be conducted by a trained clinician.

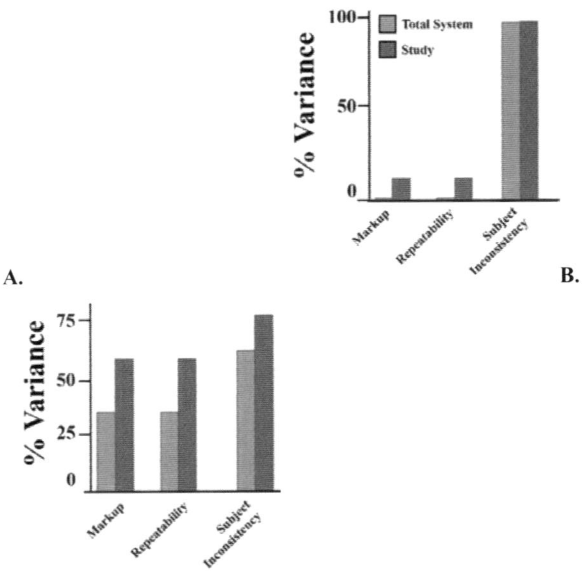

Figure 3. Analysis of Variance of the data collection system, markup procedure and during a gait analysis study for a healthy and CP subject. A.) The healthy subject yielded nearly 0 percent variation in the system, all variance was explained by the inconsistency in walking by the subject. **B.)** CP subject's pathology contributed 65% of the variation. More than 30% of the variation was contributed by the measurement system and demonstrated a lack of repeatability in the testing system.

Future Implications: The preliminary gait analysis study helped to demonstrate higher variability in data collected on a CP patient compared to a healthy control. This heled to characterize the need for a marker device that eliminates unnecessary variance caused by the data collection process. The goal of this design project is to develop a secondary data collection tool that can run simultaneously with standard IR reflective marker technology. Its purpose is to track marker location during gait studies conducted in motion analysis labs. This has the potential to help corroborate current clinical data to improve confidence in the gait study outcomes. However, for our device to be successful it must prove to be reliable and to be used with a high level of repeatability. This will require that the final prototype be evaluated using the Gage R&R analysis.
The proposed device design must meet all of the design criterion set forth by Shriners Hospital, and the outcomes of the collected data must match or exceed the quality of output data provided during a standard motion analysis gait study. The prototype will be tested simultaneously within the VICON motion analysis system. We will carefully study the intra-user variability to determine how operator, system and subject variability is developed within our new marker detection system.
Our focus will be to develop a marker system that reduces variation in order to better determine surgical planning and assess quality and fit of orthotic devices for CP patients.

The adaptable *Joystick Talkee Marker* is a device in development meant to address these issues. A device that can run wirelessly and without the use of IR camera technology could improve the data quality and expand its use to external motion analysis projects where instrumented labs are not available.

References:

1. Stern K. et al. Cerebral Palsy. 2018
2. CDC. Data and Statistics for Cerebral Palsy. 2018
3. Gorton G. et al. Kinematic. 2009; 29(3):398-402
4. Baliunas et al. Osteoarth. And Cart. 2002; 10, 573-579

Acknowledgements:

We would like to thank Shriners Hospital for Children in Springfield and the Motion Analysis Center for their guidance.

Pulse Amplitude Ratio Based on a Preload-Afterload Normalization

Rebecca Edgerton, Kareem Boura, Mackenzie Cash, Ying Sun
Biomedical Engineering, University of Rhode Island, Kingston, RI, USA

Significance: Several studies have utilized the Pulse Amplitude Ratio (PAR) as a standard of diagnosing potential hemodynamic irregularities. PAR can be defined as the measure of the amplitude change of the photoplethysmograph (PPG) waveform due to an external maneuver, divided by the baseline average pulse amplitude before the onset of the maneuver. Our study looks to combine the PAR of both the Valsalva maneuver, and a sudden standing up maneuver. The Valsalva maneuver can be defined as a forceful exhalation against a closed glottis. Doing this maneuver will greatly increase intrathoracic pressure, causing a myriad of hemodynamic changes. The Valsalva PAR has been used in several studies as a noninvasive correlation to deducing the pulmonary capillary wedge pressure, which is used to diagnose potential congestive heart failure [1,2].

A healthy Valsalva waveform should look like the one shown in *Figure 1*. The baseline is a steady and consistent waveform before the onset of the maneuver. Then, when the Valsalva begins a slight overshoot can be observed because there is an increase in arterial pressure. This is where the first red dot is seen in the image. After this point, the amplitude of each wave begins to slowly decrease as the systolic blood pressure decreases. This happens as the heart tries to manage the pressure within the thoracic cavity. The second red dot in the image shows where Valsalva is released resulting in a decrease in arterial pressure, at which point, there is a large overshot of PPG amplitude. Finally, the arterial pressure increases in response to the lack of impedance, and the wave then returns to the baseline amplitude.

In patients with high filling pressure, the Valsalva waveform will not see as great of a decrease in magnitude as pressure in the chest is already built up. This will result in a higher PAR. In extreme cases of heart disease, pressure will already be present in the chest caused by congestive fluid buildup in the cardiac chambers.

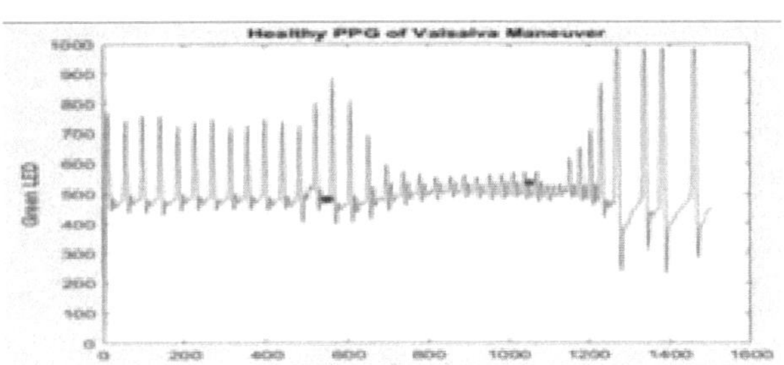

Figure 1 : A healthy PPG signal from a Valsalva Maneuver.

The standing up maneuver will be used to determine the effectiveness of the heart's autonomic response, and will be used as a scaling parameter for the Valsalva PAR. Shown in *Figure 2,* is a healthy sudden standing waveform. Upon the onset of standing, the gravitational shift creates a flow of nearly 500 ml of blood from the chest to the distensible venous capacitance system (ie venous pooling). The baroreceptors of the heart detect the change in pressure due to the blood volume shift, and an autonomic response occurs. In a healthy patient this response should amplify the afterload effects of the heart and then rapidly return to baseline. As seen within *Figure 2,* at around 600 samples the signal afterload increases then returns to the baseline. This event is indicated by the red dot. In a patient with a potential cardiac condition, the response will be drastic and the overall amplitude of the waveform will increase while the patient remains standing.

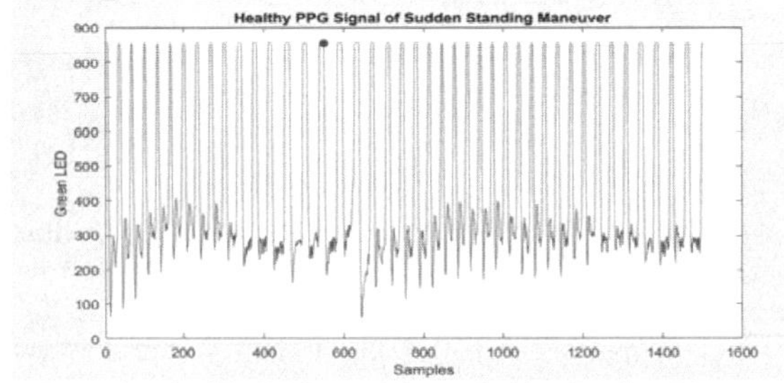

Figure 2: A healthy PPG signal from a Sudden Standing Maneuver.

Studies have shown a healthy PAR for the Valsalva maneuver to be below 0.55, while a healthy PAR for standing up should remain near 1.0, as a healthy autonomic response should have little change from the baseline average of sitting down. Our group believes incorporating the resulting PAR's of these two maneuvers will derive a new standard for comprehensive, noninvasive early heart failure detection.

Design Description: This study determines the PAR by utilizing a pulse sensor to read PPG, an Arduino Uno, and the MATLAB software suite. The team chose to use the PPG given its cost effective nature. The Arduino Uno was chosen because of how simple it is to implement with our PPG, as well as its ability to transmit information through the serial port. MATLAB was chosen because it also has the ability to communicate through the serial port with the Arduino hardware. MATLAB was also chosen because it is an extremely powerful software that contains all the necessary functions for PAR calculation within their signal processing toolbox, specifically the ability to find local minimums and maximums in PPG waveforms. One thing to consider when using serial communication with MATLAB is that the real time processing has a time delay.

The pulse sensor was embedded into a finger splint to make it easy to place on someone's finger and reduce risk of the sensor moving. This is hardwired to the Arduino Uno, which for convenience, has been implemented into a wrist strap. The Arduino is then plugged into a laptop via USB so that the sensor may communicate with MATLAB, and provide real time data on screen.

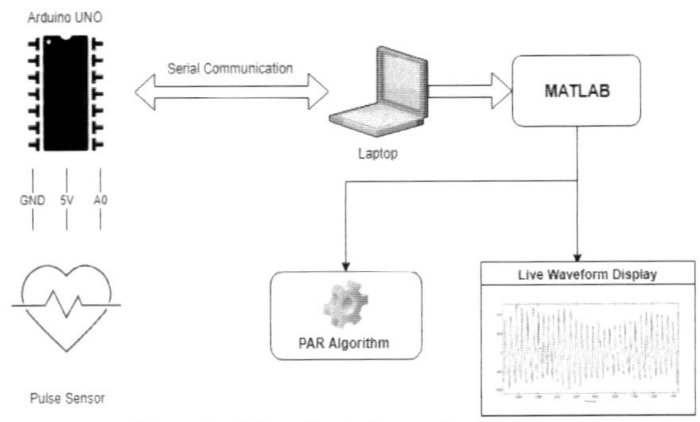

Figure 3: A flowchart of our schematic.

Design Criteria: The first step in finding this new combined ratio is to measure the PAR of the Valsalva maneuver. The PAR measurement from the Valsalva maneuver is taken three times to confirm accuracy of the reading, then the average of these three measurements is used in our subsequent calculations. Following this step, the PAR of the sudden standing maneuver is measured, again using the average from three maneuvers. The standing up PAR is normalized to see how far the recorded value differs from 1, meaning the variation in the waveform caused by the maneuver compared to the baseline waveform. This is calculated using *Equation 1*, where we subtract the standing PAR from 1, then subtract the absolute value of the result from 1 again. By normalizing this value, one can observe that the lower the resulting normalized standing PAR is, the greater the scaling factor is in increasing the original Valsalva PAR. The Valsalva PAR is then divided by the normalized standing PAR, shown by *Equation 2,* giving us a new scaled value that our group is calling the adjusted amplitude ratio (AAR).

Equation 1. $\text{Norm. Standing} = 1 - |1 - \text{Standing}|$
Equation 2. $\text{AAR} = \text{Valsalva} / \text{Norm. Standing}$

Since the PAR from the Valsalva maneuver is affected by both the preload and afterload of cardiac output, and the PAR from the standing maneuver is primarily affected by the afterload, dividing the PARs should give a value that may indicate the heart's health based solely on the preload.

Results: The normalized standing PAR that we calculate using *Equation 1* will always be less than or equal to 1.0. When we use this in *Equation 2*, the resulting AAR will always be greater than the Valsalva PAR. However, in a healthy heart, dividing the low Valsalva PAR value by an expected normalized standing PAR that is less than, but very close to 1.0, will result in a low AAR value, indicating low risk for heart disease. As an example, measurements of PARs were taken from ourselves, this data along with our AAR calculations are shown below in *Figure 4.*

A lower normalized standing PAR, or a greater Valsalva PAR are both indicators of potential heart failure. From our equation, a lower normalized standing PAR or a greater Valsalva PAR will both generate a higher AAR. The synergistic effect of observing the heart's response to both preload and afterload stresses is why our group believes that calculating the AAR can be a new method of assessing a patient's heart health.

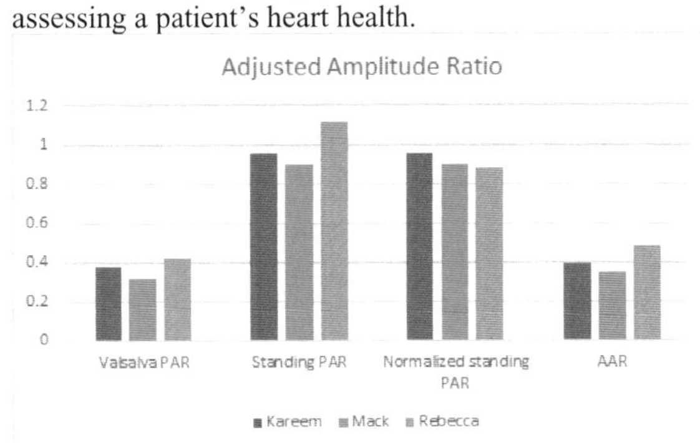

Figure 4: Measurements of the groups' PARs and calculated AAR values.

Future Implications: The technology necessary for this study was successfully developed and produced accurate results measuring the PAR of a healthy person under ideal conditions. While there are other diagnostic modalities out there, the group chose the PPG given that the PPG was the strongest choice given the noninvasive, simple, and cost effective nature of obtaining that signal. However, a possible complication that could arise while using a PPG signal is that reading may be inaccurate if the circulation is poor in the extremity being measured. Circulation can be negatively affected by the temperature, or by other potential health issues such as Raynaud's disease, or conditions affecting oxygen levels in the bloodstream.. In the end, the team decided that the positives of using a PPG signal were of a higher magnitude than the negatives.

The AAR alone is not meant to be used to determine specific heart issues. The innovation of this study is that the AAR could potentially be used as a metric to classify general heart health, or could be used as a preliminary modality to determine if a patient requires further evaluation. This procedure could be implemented in existing wearable devices (ie fitbit). In either case, further research would be required, specifically on unhealthy participants, to develop parameters for diagnostic purposes.

References:

[1] H.A. Silber, et al. *Am Jour Phys,* 302, 10, H2043-H2047, 2012.
[2] C. Gillard M.D., et al. *Clin Cardio,* 29, 3, 107-111, 2006.
[3] KM. McIntyre, et al. *NE Jour Med,* 327, 24, 1715-1720, 1992.
[4] J. Akatsu, et al. *Industrial Health,* 37, 1, 68-75, 1999.
[5] Galiatsatos, et al. *Am Jour Cardio*, 119, 7, 2017, 1053–1060.
[6] Gilotra, Nisha A., et al. *Am Jour Cardio*, 120, 6, 2017, 966–972

Design of Detection System for Inaccessible Venous Access Ports in the Pediatric Emergency Department

Hannah Blakely,[1] Sumaiya Sayeed,[1] Kelly Williams,[1] Colette Bare,[1] Sakina Sojar, MD.[1,2]
[1] Brown University, [2] Hasbro Children's Hospital

Significance: Totally implanted venous access ports (TIVAPs) improve the ability to quickly, safely, and routinely access the bloodstream for pediatric oncology patients. TIVAPs are composed of a reservoir covered by a silicone lumen with an attached catheter directly inserted into a major vein. Placement varies by patient, but TIVAPs are most frequently implanted in the upper chest (sub-clavicular or axillary, with access to subclavian or internal jugular vein) or in the arm (with access to the cephalic vein).[1] While TIVAPs reduce vein damage, patient pain, and anxiety related to frequent IV access, these improvements to quality of life do not come without risk. Infection rates associated with port access are estimated to be as high as 12%,[2] likely correlated with increased number of needle sticks to achieve access.[3] Standard treatment for port-related infection is rapid administration of antibiotics via the TIVAP.[4] When hospital oncology clinics are closed (i.e. evenings and weekends), pediatric cancer patients presenting with signs of infection are diverted to the emergency department (ED) for care and administration of antibiotics. However, relative to oncology nurses, nurses in the ED have less experience locating and palpating TIVAPs, and often have difficulty achieving access. This results in the need to perform multiple needle sticks to successfully achieve blood return, a metric of access. Not only can this process can be distressing and frustrating for patients and their families, it can sometimes result in the need for an oncology nurse to come perform the procedure, which significantly delays the patient's care in the ED. Therefore, there is a need to facilitate the access of implantable venous access ports of pediatric cancer patients in the ED, in order to reduce needle sticks, line infections, and time in the hospital. By improving the efficiency of port access, we can enhance the quality of care received by pediatric oncology patients in the ED, increase these patients' willingness to let unfamiliar medical practitioners and parents access their ports, and simplify stressful procedures for nurses.

Design Description: Our solution is an optics-based detection system utilizing a near infrared (NIR) light source to detect absorbance difference between the tissue and TIVAP. This signal will be read via our novel software system to identify the edges of the device, and to calculate the location of the lumen's center. Variation in port shadow area, as read by the device, will inform the relative angle between the device and port. This information will be conveyed to the healthcare provider via a simple user interface, with prompts to adjust the device positioning to confirm proper alignment. Finally, the device will facilitate successful port access by projecting an outline and center position of the port in red LED onto the patient's skin.

Design Criteria: Through discussions with medical providers we categorized our user needs and constraints into three categories: TIVAP access, ED workflow, and device compatibility. TIVAP access is constrained by the angle and location of the needle stick into the port's lumen. The device must facilitate access at the center of the lumen with minimal radial error, as well as at an angle perpendicular to the lumen. The device must integrate easily with established procedures and workflow in the emergency department to support nurses in access while minimizing patient anxiety. Finally, the variety of TIVAP geometries, implantation locations, and material properties must be considered. TIVAPs can have single or double lumens, can be composed of titanium or plastics, and can be implanted at varying depth and location by patient. Working with all of these variables constrains the design to ensure usability for all cases.

Results: To verify our technological developments, we will test the light source and camera device set-up using a model of the biological system, namely a TIVAP behind 1 cm of artificial tissue or non-human biological tissue. Successful access will be assessed via comparison of the indicated position relative to the exact center and perpendicular angle of the TIVAP (i.e. 'ideal' access position). The radial error and angular error will be quantified and compared between trials, varied TIVAP models, and individual practitioners operating the device. Separately, our team will manually verify software outputs to ensure accuracy, precision, and reliability of identified edges, port shape, and center positioning. Ongoing communication with clinicians and patients will validate device design, technological feasibility, and workflow integration. Additionally, we are developing a survey to quantify the perceived need for our proof-of-concept with the patients and parents of the pediatric emergency department who undergo port access.

Future Implications: Design components have not yet been integrated into a finalized device. The resolution and accuracy of the data obtained from a NIR light source is being quantified in a biological system and the software is being developed to produce precise results. With ongoing research and design optimization, our solution is being modified and better adapted to meet the identified unmet need.

References:

1. Lebeaux et al., *Lancet Infect Dis* 2014; 14: 146–59.
2. Sharp et al., *J Ped Surg* 2014; 49; 763-765.
3. Kurul et al., *Lancet Oncol* 2002; 3: 684–93.
4. Tabatabaie et al., *Am J Clin Oncol* 2017; 40; 94-105.

Acknowledgements: We acknowledge the Center of Biomedical Engineering at Brown University. We also acknowledge Professors Celinda Kofron, Vicki Colvin, and Anubhav Tripathi for their support.

Bioreactor for Mechanical Stimulation of 3D Scaffolds for Tendon Tissue Engineering

Brennen Zolnoski, Emily Fernschild, and Devina Jaiswal, PhD
Department of Biomedical Engineering, Western New England University, 1215 Wilbraham Road, Springfield, MA 01119

Significance:
Every year there are 16.4 million tendon and ligament injuries in the United States [1]. Since tendon exhibits poor self-repair capability, cases of tendinopathy are surgically treated which might result in a weaker fibrous tissue, susceptible to early failure. Tendon engineering is a powerful tool that aims at the replacement of damaged tissue with a mechanically stable tendon [2]. During normal physiological activities, tendon undergoes cyclic mechanical loading. This dynamic environment can be mimicked by bioreactors for *in vitro* tissue regeneration. Most of the commercially available bioreactors are expensive and focus on a single chamber unit that prevents testing multiple conditions in a single run. Flexcell™ works on monolayers cultured on cell culture membrane [3] whereas single chamber bioreactors designed for 3D scaffolds fail when multiple samples have to be run at similar condition for longer time

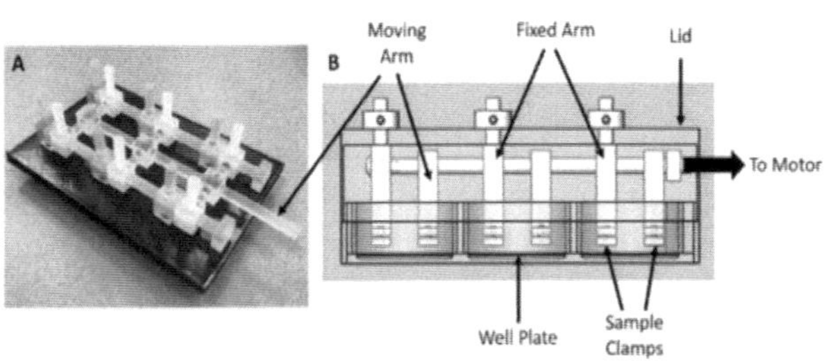

Figure 1. (A) 3D printed six-well plate mechanical stimulator for tendon. (B) The SolidWorks design of the overall stimulator with a custom lid. The stimulator consists of a movable and fixed arm to produce at least 4% strain.

points [4]. Other bioreactors are too bulky to be placed in a conventional incubator which amounts to increase in their fabrication cost [5]. There is a need for a high throughput bioreactor that works with 3D samples, is cost effective, user-friendly, and aids in testing multiple experimental conditions simultaneously. This study aims at design, fabrication, and validation of a six-well plate bioreactor for mechanical stimulation of scaffolds for tendon tissue engineering.

Design Criteria:
The device was created under the following main criteria: (a) it needs to mechanically stimulate tissue samples and biomaterials with a strain percentage ranging from 0-8%, (b) it should be compatible with an inverted microscope, (c) fit in a standard laboratory incubator, and (d) it needs to be compatible with a six-well plate. Other device needs include, the capability to test samples with lengths ranging from 5 to 20 mm, hold the sample tightly without causing slippage and being user-friendly. Lastly, considering certain device parts will come in contact with biological material during *in vitro* studies, the device must be biocompatible and follow Food and Drug Administration (FDA) and Biosafety Level 2 (BSL 2) regulations.
In order to fulfill the stimulation criteria, the device was conceptualized to have a moving and fixed arm in each well of the six well plate (Figure 2). The design was constrained to a 35 mm size limit for each arm mechanism due to the area of an individual well. The arms have a height of 29.5 mm which will fit under a custom-made lid compatible with the well plate. The custom lid is made from clear acrylic for compatibility with an inverted microscope. Each arm ends in a clamp that will hold the sample in place and these clamps are incorporated with 0.2 mm deep grooves in order to grip the samples. The dimensions of the clamps are 7 mm by 21 mm which were designed to maximize the size of the samples within the wells.
To make this a programmable device, a stepper motor was selected which could provide linear movement to the slider. A custom built MATLAB program would control Arduino giving users the flexibility to change the strain percentage applied to the samples.

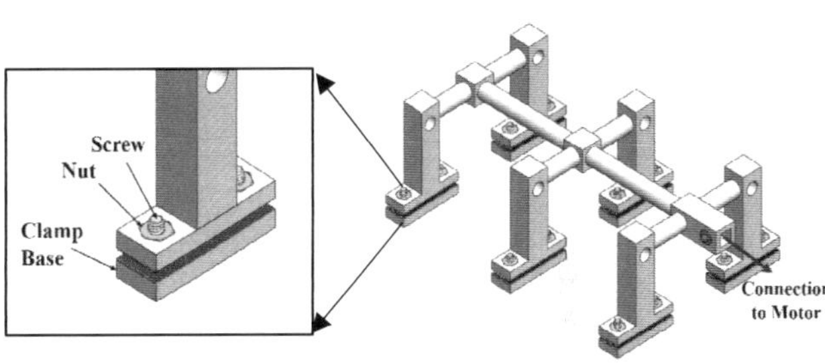

Figure 2. SolidWorks design for moving arm assembly showing six arms that would fit in a six well plate. The inset shows the zoomed in view of the clamp that would hold the sample.

Design Description:
The device was designed using SolidWorks as seen in Figure 1B. The design consists of a fixed and a moving arm to mechanically stimulate six samples at a time using a stepper motor (NEMA 17-size hybrid bipolar stepping motor) with 40 μm step size. The prototype fits a standard six-well plate and is compatible with a common laboratory incubator. The arms end in a clamp that holds the sample and prevents slippage during mechanical stimulation. COMSOL Multiphysics® Modeling Software (Solid Mechanics module) was used to simulate the force exerted on the poly(caprolactone) (PCL) scaffold during extension. A strain of 4%, representing the amount of strain exerted on the Achilles tendon in a human, was applied to the model [1]. PCL (Length: 10 mm, Width: 4mm, Thickness: 0.1 mm) was

created in the model by using following material properties: molecular weight of 80,000 g/mol, a density of 1.145g/mL, an elastic modulus of 429.1MPa, and a Poisson's ratio of 0.3 [6].

A study was conducted on stainless steel 304 to assess the degree of corrosion of screws and nuts that will be submerged in media in the bioreactor. Four samples of stainless steel were tested for a fourteen-day period. Through qualitative observation, it was determined that stainless steel did not rust and was acceptable for use in the bioreactor. For ease of use, the main component coming in contact with the biologics were 3D printed with nylon-12 plastic. This plastic was chosen because it is biocompatible and food safe for long-term use [7]. The rail, that the moving arm will slide along, will be fabricated with stainless steel to reduce friction. This part will be lubricated using a biocompatible grease to further reduce friction and deterioration.

Results:

The first 3D printed prototype is seen in figure 1A. This design was tested with a 6-well plate. The clamps mimic a standard Instron, which can accommodate samples with a size ranging from 5 mm to 20 mm. The movable shaft will be connected to the motor with a quick release mechanism that would make the setup user friendly for a cyclic stimulation schedule. Riboh et al. optimized the mechanical stimulation regime to intermittent cyclic 4% strain with 0.1Hz frequency and 1hr on/ 5hr off operation cycle. This regime improved cell proliferation and increased collagen production by 25% for adipose stem cells [8]. The actual operation conditions were simulated using COMSOL and the stress map for 4% strain is shown in figure 3. The maximum force exerted on the sample is 6.8 N which is within the force range given in literature for tendon stimulation studies [9].

Figure 3. Stress map of a PCL membrane after application of 4% strain.

Future Implications:

In conclusion, a user friendly bioreactor was designed and validation steps were followed to assess its feasibility for mechanical stimulation of tendon. In the future, the bioreactor will be used to conduct *in vitro* studies to understand cellular interaction with the scaffold. PCL will be used to fabricate electrospun aligned fibers to mimic tendon structure. The scaffold will be seeded with tenocytes and stimulated with the bioreactor designed in this work. Due to the inherent design features of the bioreactor, a full study consisting of control and treated samples could be run simultaneously to produce statistically significant data with minimum variability associated with each run. The stimulated samples will be further processed to quantify cell proliferation, collagen production and study tendon specific mechanosensors. Since the mechanical stimulation directly impacts mechanosensors of the cell, this programmable device can be used to test different stimulation regimes which can be correlated to cellular behavior.

One design criteria for this work was to be compatible with an inverted microscope. Even though the six-well plate has a clear bottom and the sample will be placed as close as possible to the base of the dish, the design is restricted to objective lens with long working distance. For the future, this aspect of the bioreactor can be improved for live cell imaging during mechanical stimulation by decreasing the overall height of the assembly. Another improvement for this design would include incorporation of a detachable, quick release clamp assembly. This design feature would make the device more user friendly and allow testing of a range of biomaterials and tissue samples. Lastly, the device can be incorporated with a force sensor to measure the forces that are exerted on the samples to make it a more precise device.

Overall, researchers can utilize this device in any laboratory setting to mechanically stimulate several samples *in vitro* as well as provide a controlled culture environment for better understanding of tendon mechanobiology.

References:

1. Wang, J. H., Guo, Q., & Li, B. (2012). *Journal of hand therapy*, *25*(2), 133-141.
2. Yin, Z., Chen, X., Chen, J. L., & Ouyang, H. W. (2010). *Expert opinion on biological therapy*, *10*(5), 689-700.
3. Matheson, L. A., Fairbank, N. J., Maksym, G. N., Santerre, J. P., & Labow, R. S. (2006). *Biomaterials*, *27*(2), 226-233.
4. Youngstrom, D. W., Rajpar, I., Kaplan, D. L., & Barrett, J. G. (2015). *Journal of Orthopaedic Research*, *33*(6), 911-918.
5. Lee, K. I., Lee, J. S., Kim, J. G., Kang, K. T., Jang, J. W., Shim, Y. B., & Moon, S. H. (2013). *Journal of Biomedical Materials Research Part A*, *101*(11), 3152-3158.
6. Eshraghi, S., & Das, S. (2010). *Acta biomaterialia*, *6*(7), 2467.
7. Abe, A., Kobayashi, S., Fuyamada, H., Yamaguchi, D., et al. (2013). *Journal of oral tissue engineering*, *11*(1), 42-50.
8. Riboh, J., Chong, A. K., Pham, H., Longaker, M., Jacobs, C., & Chang, J. (2008). *The Journal of hand surgery*, *33*(8), 1388-1396.
9. Saber, S., Zhang, A. Y., Ki, S. H., Lindsey, D. P., Smith, R. L., Riboh, J., & Chang, J. (2010). *Tissue Engineering Part A*, *16*(6), 2085-2090.

Sense-C | *Walk healthy again*

Sagar M. Doshi, Kaleb Burch, Dr. Jill S. Higginson, Dr. Erik T. Thostenson

Value Proposition:

We are developing nano-technology based low-cost, flexible sensors integrated into footwear to help physical therapists to utilize movement and gait data acquired in a realistic context to improve rehabilitation of patients after joint replacement surgery/injury

Problem:

- **Increasing patient population due to aging and obesity**
- **Lack of devices for measuring gait data to monitor rehabilitation**

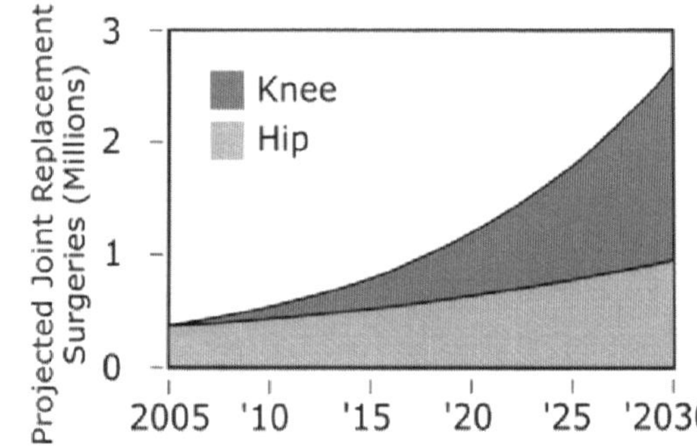

End User:

- **Physical Therapists: Monitor rehabilitation**
- **Athletes: Analyze and improve performance**
- **Researchers: Understand human movement**

Solution & Business Model

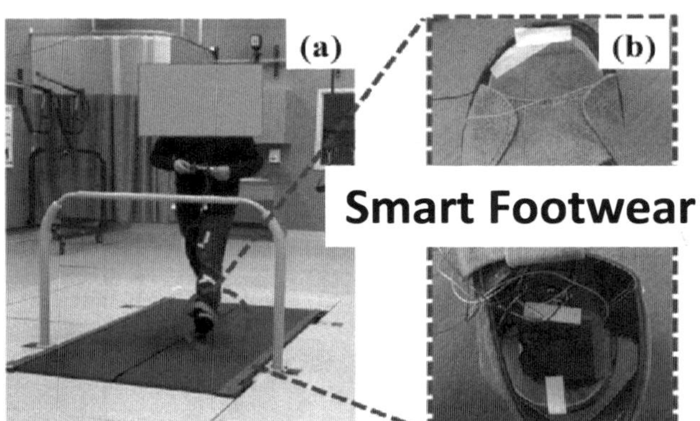

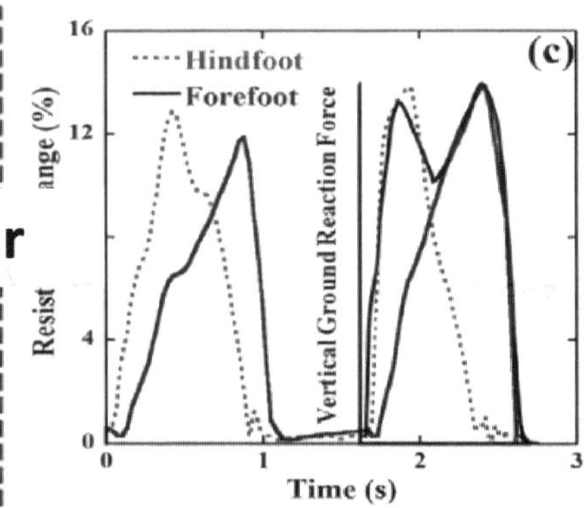

Pharmaflux

Yaling Liu, Chris Uhl, Yuyuan, Zhou

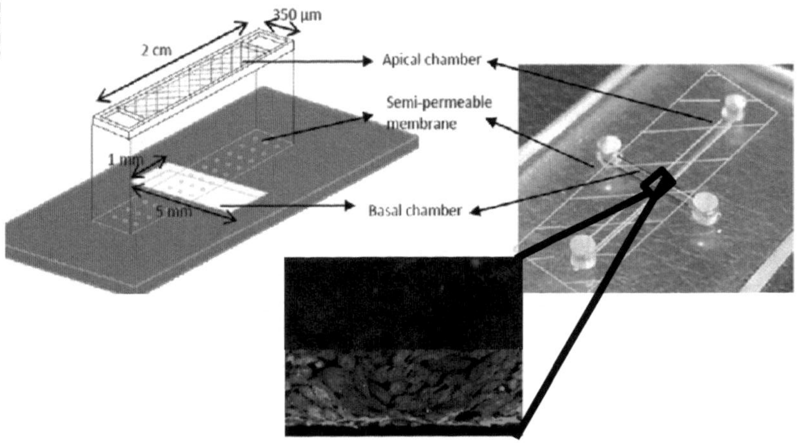

Microfluidic device for High through-put cell-based analysis of early stage candidates.

Bench-scale system for mimicking blood vessels in vitro

Improved imaging and data collection in small flexible platform.

Fast tumor spheroid growth

Allows for discrete use of multiple cell lines within single device

Problem and Relevance

Drug development is Incredibly expensive (~$1-$1.5B) to push single candidate to market.
Time consuming (~10-15 years) process.
Poorly predictive testing.

End User and/or Persona

Directors of Pre-Clinical Pharmacology & Senior Investigators

Solution & Business Model

Providing predictive drug performance under disease-state conditions as a means of improving customer's competitive advantages in the market.

FAST TUMOR SPHEROID GROWTH IN MICROFLUIDIC DEVICE

Yaling Liu[1,2*], Chris Uhl[2], Yuyuan Zhou[2]
[1]Department of Mechanical Engineering and Mechanics
[2]Department of Bioengineering
Lehigh University, Bethlehem, PA

INTRODUCTION

Patient derived organoids have emerged as robust preclinical models for screening anti-cancer therapeutics. Current 2D culturing methods do not provide physiological responses to therapeutics, therefore 3D models are being developed to better reproduce physiological responses. 3D culturing however often requires large initial cell populations and one week to one month to grow tumors ready for therapeutic testing. As a solution a 3D culturing system has been developed capable of producing physiologically relevant tumors in an expedited fashion while only requiring a small number of initial cancer cells. .

METHODS

Device Fabrication

Bi-layer microfluidic platforms were produced following similar techniques outlined in previous microfluidic works from the Liu group45–49. The devices produced for expedited tumor spheroid growth contained a semi-permeable polycarbonate membrane which functioned to separate the apical and basal halves of the microfluidics. The circular chamber where the cancer cells were grown measured 3mm in diameter and 1.5mm in height. The straight sections of channel in the microfluidic devices measured 600μm in height and width. The use of the semi-permeable membrane within the system functioned to prevent the cancer cells introduced into the system from being washed downstream and out of the channel, which had an average pore diameter of 800nm. A depiction of the microfluidic setup can be viewed in Fig. 1.

HCT116 Spheroid Growth

Tumors were grown using HCT116 human colorectal cancer cells. The HCT116 cells used for testing were purchased from ATCC and all tests were carried out with cells within 5 passages from the initial frozen stock provided by ATCC. All spheroids utilized in this work were grown from an initial suspensions of 30 cells which were pre-formed into loose tumor spheroids via an 8 hour incubation in a low adhesion round bottom well plate under static conditions. The Ultra-Low Attachment Multiple Well Plates were manufactured by Corning that feature a covalently bound hydrogel layer that effectively inhibits cellular attachment. No additional coatings were required to facilitate spheroid formation in the round bottom wells. A hemocytometer was used to verify the concentration of cells in a stock suspension from which appropriate volumes of the suspension were separated to ensure that exactly 30 cells were introduced into each well. Due to the small number of cells used in each test, visual confirmation was performed for each well to ensure that 30 cells were present. Well containing greater or fewer than 30 cells were discarded and not used for growth testing. The pre-formed spheroids were then collected and flown into their respective microfluidic devices. Once inside the microfluidic devices, the semi-permeable membrane functioned to capture the pre-spheroids and hold them in place over the duration of the experimentation. The same process of an 8 hour incubation in a low adhesion well plate to pre-form spheroids was utilized for the static based tests. After pre-forming the initial static test case spheroids, Matrigel was added to the well plate in order to further facilitate spheroid growth over the course of 120 hours. Tumor spheroids grown under flow conditions were established as noted in the leukemia growth section at specific flow rates of $3.4X10^3$μL/hr and $3.4X10^4$μL/hr. Nutrient consumption for the HCT116 cells was also visualized and estimated as described for the leukemia cancer cells. Likewise, the HCT116 cells were maintained and grown under the same standard culturing conditions of 37°C and 5% CO2 in an incubator as the leukemia cells.

Cancer Growth Measurements

Cancer growth was monitored daily over the course of five days with brightfield and fluorescence imaging. Data collection for spheroid growth was accomplished utilizing CellTracker™ Stain (10μM red, Thermo Fisher Scientific) and confocal microscopy (Nikon C2+, Apo 4X). Images of each cancer growth set-up were captured every 12 hours over the course of 5 days. All images were processed with FIGI (ImageJ) in order to measure HCT116 tumor sizes and cell counts for leukemia tests.

Statistical Analysis

Statistical analysis of all obtained results were run utilizing IBM's SPSS statistical software package (IBM Corp.). All of the figures have significant differences indicated above elements within the plots. One way ANOVA tests were run for each data set with confidence levels of 95% held throughout all plots. All analyses were carried out under conditions of Tukey equal variances assumed, along with tests of homogeneity of variance further verified by both Brown-Forsythe and Welch analyses.

RESULTS

On-chip cell culture and Imaging

Adherent cancer cells (HCT116) grown within the devices were first aggregated together under static culturing conditions within low adhesion well-plates. The cells were added to the wells and given 8 hours to adhere into loose spheroids before being introduced into the microfluidic devices. HCT116 cancer cells were successfully cultured within the microfluidic devices. Fig. 3b depicts representative images of HCT116 cells grown when subjected to a flow rate of $3.4X10^4$ μL/hr over the course of 120 hours. Both representative figures show overall increases in cell counts as time progresses as is indicated by the presented growth data in Fig.2a

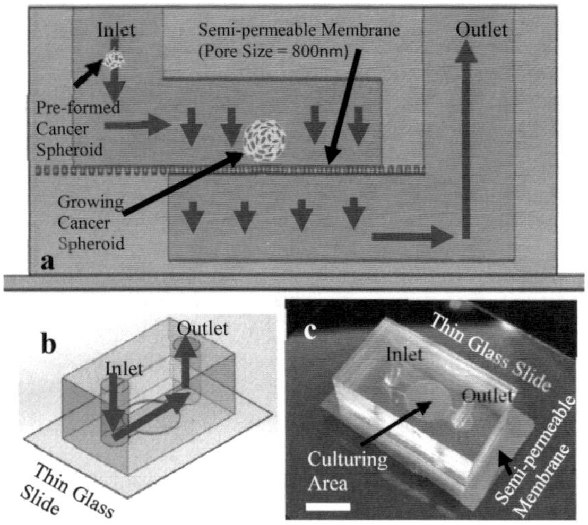

Figure. 1 Microfluidic device schema depicting device structure and cancer cell capture function utilizing semi-permeable membrane. Flow direction indicated by red arrows. (a), Device schema cross-section. (b), Device schema overview. (c), Image of sample microfluidic device utilized in experimental testing (white scale bar is 3mm long).

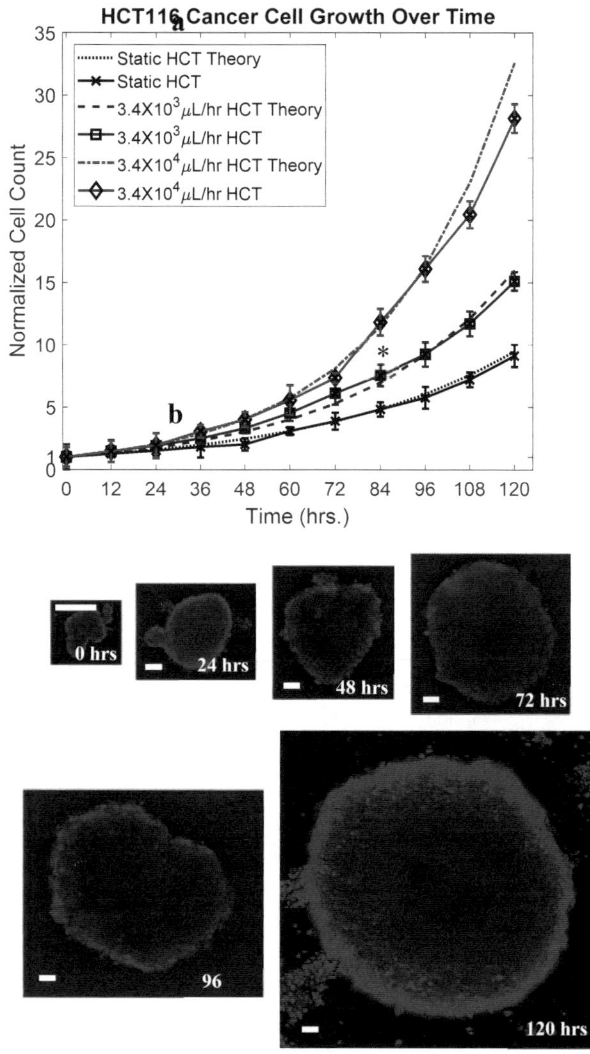

Figure 2. HCT116 cell growth over time based on flow conditions (30 cell initial count). a, Normalized HCT116 cell count over time for various flow conditions and theoretical model predictions. b, Normalized nutrient availability within microfluidic system over time for various flow conditions. Equilibrium point indicates when nutrient availability within the system can no longer satisfy the requirements of the entire cancer cell population. c, Representative fluorescence images of HCT116 tumor spheroid growth over the course of 120 hours under a flow rate of 3.4X104 μL/hr (scale bars are 60μm). Statistical significance indicated by "*" between flow based tests and static conditions and "**" between flow based tests at p≤0.05.

Cancer Growth Measurements and Nutrient Availability

Successful culturing of cancer cell lines was accomplished, as demonstrated by the growth of human colorectal (HCT116) in Fig. 2a. Growth of the cancer populations within the devices was measured directly through confocal imaging with CellTracker™ red stain. Through the application of various flow rates within the microfluidic devices, the rate of growth for the cancer cell populations was controlled. The use of convective flow allowed for a maximum of 4.76 times faster cancer cell growth when compared to the static growth cases. When compared to the static test cases, introducing $3.4X10^3$μL/hr nutrient flow into the systems containing HCT116 results in an overall increase in the cell population by a factor of 2.47 at the end of 120 hours. Even more impressive was the observed increase in HCT116 growth when supplied with a nutrient flow of $3.4X10^4$μL/hr, which resulted in a 4.76 factor increase in the cell population.

REFERENCES

1. A. Thomas, D. Ou-Yang, L. Lorentz, V. Muzkantov, Y. Liu, "Biomimetic Channel Modeling Local Vascular Dynamics of Pro-inflammatory Endothelial Changes", *Biomicrofluidics*, 10, 014101, 2016
2. C. Uhl, V. R. Muzykantov, Y. Liu, "Biomimetic Microfluidic Platform for the Study of Transient Endothelial Monolayer Confluency and Therapeutic Transport under Mimicked Disease Conditions", Biomicrofluidics, 12, 014101, 2018
3. W. Shi, J. Kwon, Y. Huang, J. Tan, C. Uhl, R. He, C. Zhou, Y. Liu, "Facile Tumor Spheroids Formation in Large Quantity with Controllable Size and High Uniformity", Scientific Report, 68372, 2018

ACKNOWLEDGEMENTS

This work was supported by National Institutes of Health (NIH) grants R01HL131750, National Science Foundation (NSF) grants PFI:AIR-TT 1701136, DMS 1516236, and the Pennsylvania Infrastructure Technology Alliance (PITA) program.

Auditory Assistive Laboratory Application for Blind and Visually Impaired Students

Haoyang Chen, Aida Kupa, Junyu Lu, Pushpita Rahman

Our technology enables blind and visually impaired (BVI) students to get real time feedback from color change reactions. We want to stimulate their interest in science, and increase their confidence and participation in STEM field.

Problem and Relevance

- BVI students are discouraged from interacting with lab instrument and reactions involve color change

End Users (in the US)

- 47 schools for the blind (from tsbvi.edu)
- 706,400 BVI students (<20 years old, from National Federation of the Blind)

Solution

Delivery: iPhone application* & custom holder

1. Application identifies each instrument and their locations on the lab table using custom trained neural network
2. Users select the instrument to work with
3. Application analyze color information in selected instrument from real time video and output in audio frequencies
4. Users can "see" the color change by hearing

* Current prototype is developed on MATLAB, and will be transferred to Apple platform

Auditory Assistive Laboratory Application for Blind and Visually Impaired Students

Haoyang Chen, Aida Kupa, Junyu Lu, Pushpita Rahman
School of Biomedical Engineering and Health Systems, Drexel University.

Introduction: In 2015, 83.1% of legally blind students were attending non-specialized schools, limiting their access to learning[1]. Science classes can be difficult for these students because they are heavily dependent on visual introduction[2]. In addition, students can be discouraged from touching the lab equipment or manipulating chemicals[3]. In labs, color changes are present in many reactions and the inability of these students to visualize this information can discourage them[2]. There are currently a few devices on the market to address this issue; however, either it cannot be adapted for lab use or does not provide real-time feedback. Thus, the objective is to develop a customized visual-substitution application for live detection of position and color for laboratory use.

Methods: In our solution, color information is converted to audio frequencies, substituting visual information to auditory information. Video is acquired via a user's cell phone positioned on a custom holder above the lab bench, and our custom application acquires and processes live video

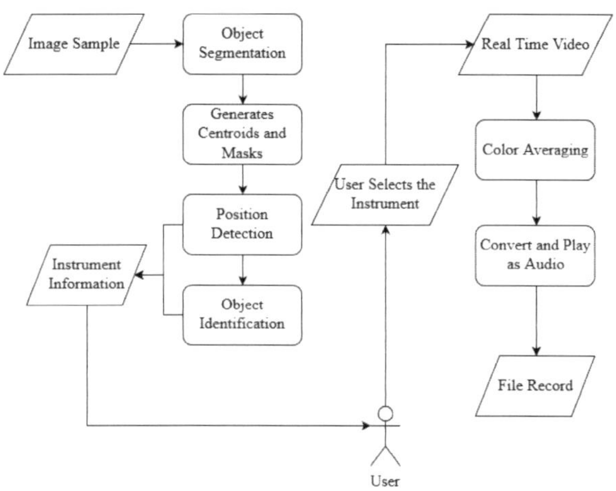

Figure 1: Software Flow Chart (text box and group)

data as described by the flow chart in Figure 1. Object segmentation is accomplished in the first frame of the video. It finds the centroid and detects the area of each instrument on the bench. Centroids in x-y pixel coordinate are converted to the polar system to pinpoint the location of each instrument with respect to the user's body frame. The custom trained GoogleNet neural network identifies each segmented lab instrument. The information of location and identification of each object is played back to the users in speech. Then users select the instrument for color detection. A color detection algorithm calculates the average color within the mask of the selected instrument and this averaged color is converted and outputted as an audio frequency. The color and object identification data are then stored into a .xlsx file for record.

Result: Currently, the application is tested and able to read more than 60 image samples, find their centroids of the instrument in and create masks on the region of interest to segment the instrument out. The customized GoogleNet was trained with 1000 pictures each for beakers, test tubes, cylinders, and flasks and has a final accuracy of 94.3%. The application is tested and can successfully recognize the lab instruments listed above and convert the centroids to the positions represented by 9 o'clock to 3 o'clock (directions with respect to the user's body frame). The software on MATLAB is tested on more than 60 images and can compute the averaged HSV values to obtain the color information in the object of interest and save the in a .xlsx file. The team is currently working on the conversion of the color data to audio frequencies from 100 Hz to 760 Hz for auditory output. The current version of the application is designed only for precipitation labs because of the opacity characteristic.

Discussion: The current prototype is built on MATLAB, and can successfully conduct functions including object segmentation, position detection, object identification, color averaging, and file recording on an image. As stated, the fully functional application will be accomplished in MATLAB by March 2019, and then transferred to a phone application. The custom holder will also be developed by the time which the application is shipped in the App Store. Although the current prototype is constrained to only work in precipitation labs, future revisions of the application are scheduled to be compatible with various color changing labs. The application will be sold with the holder at an affordable price, and ideally, this application and device would be applied in science classes in the schools for the blind to increase their participation in hands-on laboratory subjects. With the help of the application, the students who are blind can "see" the color change in a chemical reaction in real-time, which is a different experience than any other device on the market. We hope this unique experience can stimulate these students' interest in science labs, help them build confidence in conducting lab-related researches, and further increase the population of the blind in the STEM field.

References:

1. Arne D. et al. Home. 2018
2. Mehmet S. et all. US-China Education Review. 2009
3. Cary S. Braille Monitor. 2012

Acknowledgements:
Dr. Catherine Von Reyn, Drexel University
Overbrook School for the Blind

Validating Hydrogel Microencapsulated Insulin Secreting Cells for Wound Healing

Shreya Soni, Zeiny Aubdoollah, Eddy Iturbide, Ronke Olabisi, PhD
Department of Biomedical Engineering, Rutgers University, Piscataway, NJ, USA, 08854

Introduction:

Previous work demonstrated an accelerated *in vitro* and *in vivo* wound healing response when insulinoma cells were used to deliver insulin to scratches in keratinocyte monolayers and chronic diabetic excise wounds in mice, respectively.[1] When combined with mesenchymal stem cells (MSCs), the response was amplified (healing in 14 vs. 35+ days).[2] To isolate the role of insulin in the response and exclude any potential contribution of an unknown cancer moiety released by the insulinoma cells, the effect on wound healing of multiple lines derived from the same insulinoma was examined.

Methods:

Rat insulinoma (ATCC) RIN-5F, which secretes only insulin and RIN-14B, which secretes only somatostatin were propagated in RPMI-1640 medium (ATCC) supplemented with 10% fetal bovine serum (FBS) and 1% w/v penicillin-streptomycin (pen-strep) in a tissue culture incubator. Cells were encapsulated into polyethylene glycol diacrylate (PEGDA) hydrogels sheets by photopolymerizing cells suspended in a precursor solution formed by combining 0.1 g/mL 10 kDa PEGDA (10% w/v; Laysan Bio, Inc.) with (1.5% v/v) triethanolamine/HEPES buffered saline (pH 7.4), 37mM 1-vinyl-2-pyrrolidinone, 0.1 mM eosin Y. The prepolymer solution was combined with cells (1×10^4 cells/μL), pipetted into 1 cm^2 custom made molds and exposed to white light for 20 seconds to achieve 400 μm thick cell-laden hydrogel sheets containing 2 x 10^6 cells per 400 μm thick hydrogel sheets. Hydrogel sheets were maintained in a 37°C / 5% CO_2 humidified incubator in complete culture media. Conditioned media (CM) containing insulin and/or MSC factors released from the cell-laden hydrogel sheets were collected on days 1, 7 and 21. Scratch assays were conducted with human keratinocytes (HaCaT; Addex Bio, San Diego, CA) as previously described.1 Briefly, HaCaT cells were propagated in DMEM, supplemented with 10% FBS, penstrep in tissue culture incubators. For scratch assays, 180,000 HaCaT cells/well were seeded onto 24 well plates (Greiner Bio-one; Monroe, NC). Cells were cultured for 48 hours to form a confluent monolayer at which time a single scratch of 415 μm ± 73 μm was produced in the center of the well with a 10 μL pipette tip. The wells were washed twice with Dulbecco's Phosphate Buffered Saline (DPBS) to remove cell debris. HaCaT cells were then stimulated with conditioned media containing insulin. CM was sampled from positive control monolayers on day 1 or cell-laden hydrogels on days 1, 7 and 21. Media from empty hydrogels was used as negative controls. Insulin levels were assessed with ELISA. HaCaT cell migration across the scratch was imaged at 0, 24, 48, 72 and 96 hours using phase contrast microscopy and analyzed on NIH ImageJ software.

Results:

For matched concentrations (high = 2×10^6 cells per hydrogel sheet), ELISA results demonstrated decreased insulin release from RIN-14B cells than from the original RIN-m cells at Day 1, greater insulin at Day 7, and equivalent insulin at Day 21 (Figure 1). RIN-5F released the lowest insulin levels at all time points. The original low RIN-m concentration (0.5×10^6 cells per hydrogel sheet) had the highest consistent insulin levels over time. Keratinocytes stimulated with conditioned media from RIN-14B cells migrated faster than those stimulated with RIN-5F cells.

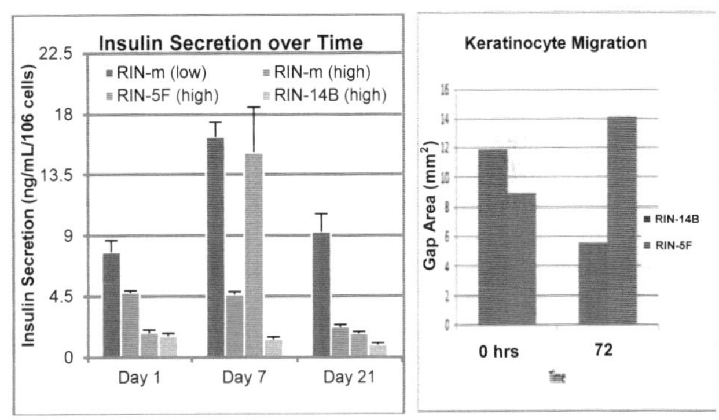

Figure 1. Left: ELISA from original data (RIN-m) compared to RIN-5F and RIN-14B cells. Right: Scratch areas were measured over time to assess keratinocyte migration.

Conclusions:

The decreased insulin levels observed in the RIN-14B cells are consistent with previous data where RIN-m cells encapsulated at higher concentrations produced reduced levels of insulin. These studies will be repeated at low concentrations (0.5×10^6 cells per hydrogel sheet). The accelerated scratch closures observed when keratinocytes were treated with RIN-14B cells, but not with RIN-5F supports our hypothesis that insulin from these cells is responsible for the accelerated response rather than an unknown promotor common to cancer cells.

References:

1. Aijaz A, et al. *Tissue Engineering Part A* 21.21-22 (2015): 2723-732.
2. Aijaz A, et al. "A dual-cell therapy for chronic wounds," *The Society for Biomaterials Annual Meeting*, Minneapolis, MN, Apr 5-8, 2017

Acknowledgements: This work was supported in part by a Rutgers Aresty Undergraduate Research Award and an NSF CAREER Award (CBET-1752079).

Fluid flow rate dictates the efficacy of low intensity anti- vascular ultrasound therapy in a microfluidic model

Brandon DeOre[†], Peter Galie[†], Chandra Sehgal[*]

[†]Rowan University, Department of Biomedical Engineering; [*]University of Pennsylvania, Department of Radiology

Introduction: The tumor vascular bed exhibits heterogeneity in both its structure and the volumetric blood flow rate it transports. An immature, leaky endothelium and elevated interstitial fluid pressure both act to reduce the velocity of blood flow within the vessels [1]. This variation in blood flow rate impedes effective drug delivery of anticancer therapeutics. Low intensity ultrasound therapy is an attractive option for targeting vasculature within the tumor microenvironment. Microbubbles injected into the circulation act as transducers that transform acoustic energy into thermal energy, creating a localized heating effect that restricts the damage to areas treated with insonation [2]. This study uses an *in vitro* vascular model to interrogate the effects of flow rate and sonication intensity on the efficacy of ultrasound treatment. Specifically, this approach yields insight into the effects of insonation on endothelial barrier integrity and cytoskeletal dynamics in real-time, which is not possible to ascertain in animal models. **Methods:** Microvascular Endothelial cells (MVECs) at p4-p6 were injected into a cylindrical void of 180 m diameter within a collagen type I hydrogel at densities of 7.5M/ml to 10M/ml. The hydrogel was supported by a PDMS microfluidic device. The cells were exposed to 0.7dyn/cm^2 of fluid shear for two days forming a confluent, 3D monolayer. Vessels were then injected with 10 L of microbubble solution in 2mL of PBS. Insonation was applied for one minute at either 0.5 W/cm^2 or 1w/cm^2. MVECS transfected with RFP-LifeAct were used for live imaging to track morphological changes. 70-kDa dextran was used to test the barrier permeability before and after Insonation. Vessels were stained to visualize VE-Cadherin and actin after sonification. Two Samples were used to analyze cell-cell junctions using Transmission electron microscopy. **Results:** In vitro vessels consisting of LifeAct-transduced microvascular endothelial cells (MVECs) were exposed to one minute of insonation with either 0.5 or 1 W/cm^2. The vessels were perfused with PBS containing microbubbles at three different flow rates: 10, 50, or 100 µL/min. Experiments were conducted to determine the effect of varying ultrasound intensity and perfusion rate on the morphology of the endothelial cells lining the lumen of the vessel. Cells were traced along the junctions at each ten-second interval of the one-minute tests. The results showed that the magnitude of flow rate dictated the cellular response to intensities (Fig 1A) The live-cell experiments also indicated the presence of actin depolymerization at certain locations along the cell-cell junctions. Permeability tests were conducted to validate the changes observed in endothelial morphology and actin.. The highest perfusion rate applied, 100 µL/min, yielded the lowest permeability values measured for both sonication intensities (Fig. 1B). In contrast, the lowest perfusion rate, 10 µL/min, increased the permeability of vessels exposed to both levels of ultrasound intensity. At the 50 µL/min flow rate, the ultrasound intensity was a significant factor in the measured permeability. In order to assess changes to the cell-cell junction complex VE-cadherin was labele. Vessel's exposed to 100 µL/min exhibited VE-cadherin localized mostly to the junctions. In contrast, vessels exposed to 50 and 10 µL/min showed extensive positive labeling for VE-cadherin within the cell body. Lastly, TEM was used to observe structural differences in the endothelial cell-cell junctions. Two channels were processed for imaging to represent the least and most damaging combinations of intensity and flow rate: 0.5 W/cm^2 and 100 µL/min (Fig. Ci) and 1 W/cm^2 and 10 µL/min (Fig. Cii). The junction of the first vessel appears largely intact along the entirety of its length. In contrast, the other exhibits a large disruption for the majority of its length.

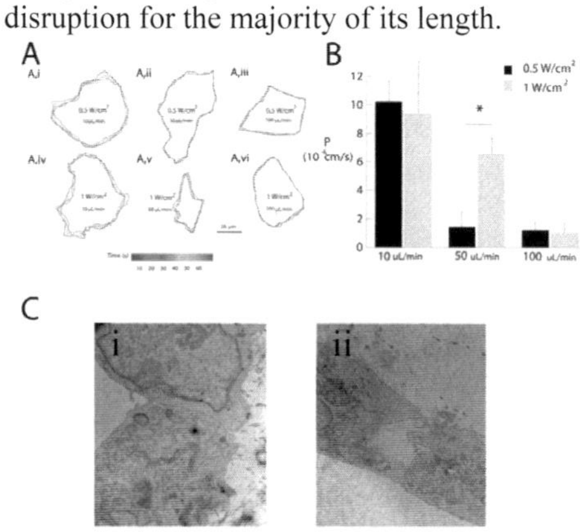

Figure 1. A) The effect of varying flow rate on area reduction is shown for the 0.5 W/cm2 A representative cell was chosen from each condition, with the perimeter traces overlaid from time 0 to 60s. * denotes p < 0 .05 B) B) Permeability of vessels exposed to varying levels of ultrasound intensity and microbubble perfusion rates C) TEM imaging for 0.5 W/cm2 and 100 µL/min (i) and 1 W/cm2 and 10 µL/min (ii) testing conditions. **Conclusions:** The results of this study verify that the rate of blood flow modulates the efficacy of low intensity anti-vascular ultrasound therapy. Unlike anti-vascular approaches that rely on delivering biochemical stimulation to the tumor environment, ultrasound therapy provides a direct means to target vessels with fewer side effects. Its primary strength over other interventional strategies to exert localized heating is the ability of ultrasound probes to measure blood flow. Given that the blood flow rate is a crucial parameter dictating the efficacy of the ultrasound therapy, the probe can potentially measure the blood flow prior to applying sonication. Using the probe as both a diagnostic and therapeutic tool creates a multifunctional strategy to disrupt vasculature within the tumor microenvironment. **Acknowledgements:** This work was supported by NIH grants R01 CA204446 (C.S.) and R01 EB022612 (C.S.) and a Scientist Development Grant from the American Heart Association (P.G.)

References:

1. Jain, R.K., *Normalization of tumor vasculature: an emerging concept in antiangiogenic therapy.* Science, 2005. **307**(5706): p. 58-62.
2. Wood, A.K. and C.M. Sehgal, *A review of low-intensity ultrasound for cancer therapy.* Ultrasound Med Biol, 2015. **41**(4): p. 905-28.

Spatial Organization of Biochemical and Physical Properties in 3D-Printed Scaffolds

Anne Behre[1], Paula Camacho[1], Hafiz Busari[2], Kelly Seims[2], Peter Schwarzenberg[3], Colin Bader[3], Hannah L. Dailey[3], Lesley W. Chow[1,2]
[1]Department of Bioengineering, [2]Department of Material Science and Engineering, [3]Department of Mechanical Engineering and Mechanics, Lehigh University, Bethlehem, PA, USA

Introduction: Tissues are not homogeneous, and their function is closely linked to the specific organization of multiple components. For example, the osteochondral (OC) interface between bone and cartilage contains a transition consisting of gradients in biochemical and physical properties that are critical for normal biomechanical function. Fabricating a biomaterial that recreates these gradients is challenging, which limits ability to engineer native-like, functional tissues. Solvent-cast direct-write (SC-DW) three-dimensional (3D) printing uses computer-controlled deposition of polymer-based ink solutions to form 3D constructs and expands the library of printable materials currently available for melt-based 3D printing. During extrusion, SC-DW inks undergo a liquid-to-solid phase change as solvent evaporates, allowing the printed fiber maintain shape.[1,2]

The Chow Lab is exploiting this technique to 3D print inks containing poly(caprolactone) (PCL), a biodegradable polymer commonly used in biomedical applications[3], and peptide-modified PCL (peptide-PCL) conjugates to fabricate peptide-functionalized scaffolds in a single step.[4] Here, two different peptide-PCL conjugates, HAbind-PCL and E3-PCL, which have been shown to promote human mesenchymal stem cell (hMSC) differentiation into chondrogenesis[5] or osteogenesis[6], respectively, were used to demonstrate multi-peptide organization within a single scaffold. The print pattern was also modified during fabrication to create a variety of scaffold architectures. This platform thus enables us to simultaneously and independently control both the biochemical and physical organization within a continuous construct.

Methods: All inks were prepared with unmodified PCL (MW 80K) dissolved at 37% w/v in hexafluoroisopropanol (HFIP). For peptide-functionalized scaffolds, HAbind-PCL and E3-PCL conjugates were added to the PCL ink at 1, 3, or 6 mg/mL or 6, 12, 18 mg/mL, respectively. Inks were 3D printed using a blunt-tip 32-gauge needle (100 μm inner diameter) and layer-by-layer deposition to construct 14-layer scaffolds with varying porosity and spatial presentation of peptides. Print speed was varied from 0.2 to 0.4 mm/sec to optimize fiber shape in each layer. Scaffold architectures were modified by printing with 120 or 260 μm fiber spacing or 90° (orthogonal) to 30° (non-orthogonal) fiber orientations. Dual porosity scaffolds were fabricated by changing the print pattern in different regions. Dual-peptide scaffolds were printed using two different printer heads.

To confirm peptide presentation and compare peptide concentration, constructs were fluorescently labeled using fluorescein-HA and/or amino-Cy3 to specifically label HAbind and E3 peptides, respectively. Fluorescence intensity was quantified using ImageJ. Scaffold morphology was characterized using scanning electron microscopy (SEM), and its mechanical properties were measured using microindentation. For dual-peptide scaffolds, a cryostat was used to section the scaffolds to visualize peptides across the scaffold thickness.

Results: We successfully 3D printed scaffolds with independent control both biochemical and physical organization. Peptide concentration on the scaffold surface was increased by increasing ink conjugate concentration before printing. Fluorescence intensity measurements showed statistically significant differences across most groups. In parallel, we demonstrated ability to change scaffold architecture by varying fiber spacing (120 μm or 260 μm) or changing fiber orientation (orthogonal vs. non-orthogonal). The fiber spacing was also changed during fabrication to generate dual porosity scaffolds (Fig. 1A). Finally, we 3D printed with both peptide-PCL conjugates to control multi-peptide organization within the same construct (Fig. 1B).

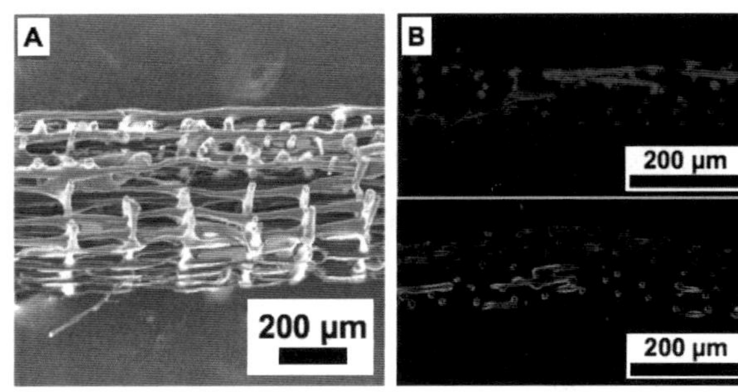

Figure 1. (A) SEM image of the cross-section of a 14-layer dual porosity PCL scaffold where seven layers were printed with 120 μm fiber spacing while the remaining layers were printed with 260 μm fiber spacing. (B) Fluorescent images of cross-sections of the same 14-layer dual peptide-functionalized scaffold, with 260 μm fiber spacing. HAbind and E3 peptides are respectively labeled with fluorescein-HA (**green**) and amino-Cy3 (**red**), confirming peptides are located in discrete spatial regions.

Conclusions: Scaffold architecture and peptide functionalization were independently and simultaneously controlled within a single construct. This platform thus enables spatial control of both biochemical and biophysical cues to optimize scaffold properties to guide functional regeneration of complex tissues, such as the bone-cartilage (osteochondral) interface.

References: [1] SZ Guo +, Langmuir 2014; [2] SZ Guo +, Small 2013. [3] MA Woodruff, WH Dietmar, Prog Polym Sci 2010; [4] LW Chow +, Adv Healthc Mater 2014. [5] PA Parmar +, Adv Healthc Mater 2016; [6] OJ Karaman +, Tissue Eng Regen Med 2016.

Effects of Robotic Exoskeleton Gait Training on an Adolescent with Chronic Brain Injury

Kiran K. Karunakaran [a,b,c], Naphtaly Ehrenberg [a,b], JenFu Cheng [b], and Karen J. Nolan [a,b]
Kessler Foundation, Human Performance and Engineering Lab, 1199 Pleasant Valley Way, West Orange 07052 [a]
Children's Specialized Hospital, Research Department, 150 New Providence Road, Mountainside, NJ 07092 [b]
New Jersey Institute of Technology, Department of Biomedical Engineering, University Heights, Newark, NJ 07102 [c]

Introduction: Children and young adults with brain injury (BI), such as cerebral palsy, anoxic brain injury and stroke, can present with unilateral motor deficits that affect motor control and coordination resulting in decreased walking speed, increased asymmetry, and almost double the support time, leading to pathological ambulation[1]. Motor recovery involves neurorehabilitation through intensive physical therapy. Neurorehabilitation is based on the theory that the human brain is capable of self-reorganization (i.e.plasticity) due to continuous, consistent, repeated practice of motor activity [2]. In the chronic phase of recovery, individuals often have residual motor deficits leading to inability to walk or employing compensatory mechanism [3].
Lower extremity robotic exoskeletons (RE) deliver comprehensive motor and balance training. It provides the user with intensive, consistent, goal-directed repetition of movement and also balance and stability [4]. An RE can provide increased repetitions in a consistent, controlled training environment, which is ideal to induce neuroplasticity [5] and may lead to improved functional recovery in children and young adults with chronic BI.
The objective of the study is to evaluate the effects of 4 weeks of RE gait training on an adolescent with chronic BI using neuromechanical outcomes.

Methods: A participant with hemiplegia due to BI (17 years of age, > 6 months post-injury) with no additional orthopedic, neuromuscular, or severe neurological pathologies (unrelated to their BI) that would interfere with their ability to walk. The participant walked in the RE for an average of 50 minutes, three days per week for 4 weeks. Participant also participated in two assessment sessions (baseline and follow up after 4 weeks of RE training). During each session the participant walked (approximately 6, 10 m walks) with and without the RE. A licensed PT administered all RE gait training. Data collection and analysis included neuromechanical (temporal and spatial parameters) outcomes.

Preliminary Results: The results from one participant with BI is shown in Figure 1. Overall stance time and swing time was longer in the RE as compared to without RE at baseline and follow up, indicating a slower gait speed in RE. However, the step length and spatial symmetry were higher with RE compared to without RE at baseline and at follow up. Furthermore, increased swing and stance time was observed at follow up compared to baseline with RE. Without RE, at follow up (therapeutic effect), there was a slight increase in step length and an increase in swing and stance time. These preliminary results suggest an improvement in neuromechanical outcomes after 4 weeks of RE gait training. Furthermore, the participant performed bilaterally spatial symmetrical gait at baseline and follow up with and without RE.

Discussion and Conclusion: The results from this case study indicate that RE provided the participant with BI a

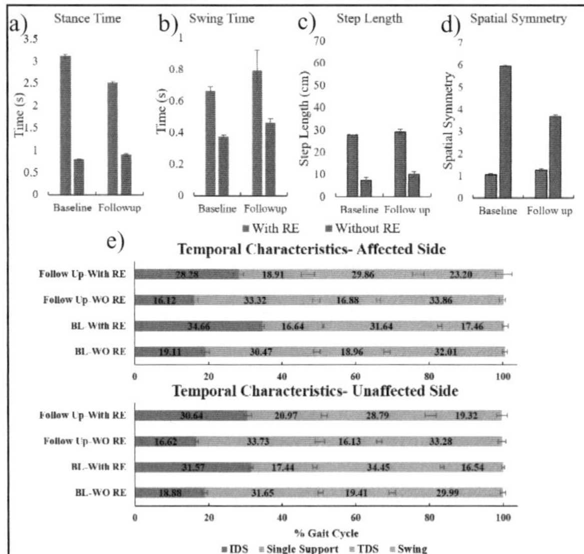

Figure 1. Mean ± standard error of a) stance time b) swing time c) step length d) spatial symmetry e) Temporal Characteristics of the affected and unaffected side for a participant with BI

highly consistent bilaterally symmetrical training environment throughout therapy with increased step length. Furthermore, the increased step length and spatial symmetry at follow up without RE indicates improvement in bilateral symmetry. These preliminary results from this case study suggest that an improvement in neuromechanical outcomes after 4 weeks of RE gait training can be achieved in adolescents with chronic BI. However, these results require further validation with larger sample size.

References: 1) G. Williams, M. E. Morris, A. Schache, and P. R. McCrory, "Incidence of Gait Abnormalities After Traumatic Brain Injury," Arch. Phys. Med. Rehabil., 2009.
2) H. Igo Krebs, N. Hogan, M. L. Aisen, and B. T. Volpe, "Robot-aided neurorehabilitation," IEEE Trans. Rehabil. Eng., vol. 6, no. 1, pp. 75–87, 1998.
3) D. C. Kerrigan, E. P. Frates, S. Rogan, and P. O. Riley, "Hip hiking and circumduction: Quantitative definitions," Am. J. Phys. Med. Rehabil., 2000.
4) A. M. Dollar, H. Herr, "Lower Extremity Exoskeletons and Active Orthoses:Challenges and State-of-the-Art," IEEE Trans. Robot., vol. 24, no. 1, pp. 144–158, 2008.
5) D. J. Reinkensmeyer, J. L. Emken, and S. C. Cramer, "Robotics, Motor Learning, and Neurologic Recovery," Annu. Rev. Biomed. Eng., 2004.

Acknowledgements: Research supported by New Jersey Health Foundation PC5-18, Children's Specialized Hospital, and the Reitman Foundation.

Modeling and Evaluation of an Admittance Controlled Hand Exoskeleton for Neuromuscular Rehabilitation

Xianlian Alex Zhou, Ashley Mont, Sergei Adamovich.
Department of Biomedical Engineering, New Jersey Institute of Technology, Newark, New Jersey, USA.

Introduction: Stroke, one of the leading causes of adult disability, affects approximately 800, 000 individuals each year in the United States. Nearly 80% of stoke survivors suffer hemiparesis of the upper arm and thus impaired hand function, which is integral to most activities of daily living. It is well established that highly repetitive training can aid in the recovery of motor function of the hand however this can be labor intensive for the providing physical therapist in addition to the cost. In the past decade, more robotic hand rehabilitation systems have been introduced to help patients recover hand function through assistance during repetitive training of the hand [1-3]. We developed a motor-based hand exoskeleton for rehabilitation of individual fingers. This exoskeleton consists of a base equipped with a servo motor, an index finger component and a thumb component. In order to provide assistance based on user intent, the exoskeleton's control system was designed to generate suitable actuation torques at the proper timing. The goal of this study is to model the exoskeleton interacting with a neuromuscular hand model to evaluate the effectiveness of an admittance control algorithm on providing different levels of assistance.

Methods: The exoskeleton was modeled as an articulated rigid body with three parts, whose mass and inertia properties were either measured or computed based on material and part geometry. Each part has a gear of equal size, with the base gear driving the index gear that in turn is connected to the thumb gear. For assembly with a generic hand musculoskeletal model (Figure 1), the base part was fixed on the back of the hand, the index part was linked to the index finger through a ring attached to a tri-axial force sensor, and the thumb part is linked to the thumb through a ring. Four extrinsic finger muscles were considered including: flexor digitorum superficialis (FDS), flexor digitorum profundus (FDP), extensor digitorum communis (EDC), and extensor indicis (EI).

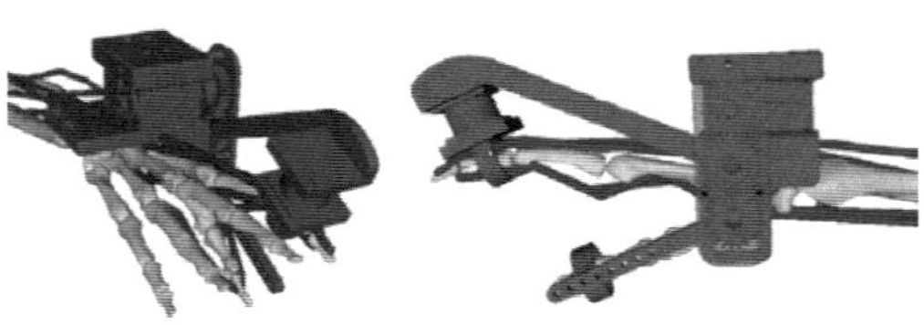

Figure 1. The assembled hand and exoskeleton.

The admittance control framework integrates the motion of a single mass intially placed at the center of the index ring and moved by the force applied by the index finger. The desired angle (θ_d) is obtained from an inverse kinematics computation and the motor is commanded to achieve this position. This is controlled by a proportional derivative (PD) controller which prescribes a joint torque (τ) to the index part joint:

$$\tau = k_p(\theta_d - \theta) - k_d\dot{\theta}$$

where k_p and k_d are tunable parameters, θ and $\dot{\theta}$ are the current angle and angular velocity, respectively.

Four simulations were conducted including a passive mode and three active modes with different control parameters: $k_p = 1, 2, 6$ ($k_d = 0$ for all). In all simulations, the flexion-extension motion of the index finger was prescribed to a trajectory of closing and opening within 2 seconds.

Results: The interaction forces between the index finger and the ring are shown in Figure 2. The passive mode has the largest force close to 4.2N, while the three active cases have the maximums around 3.2 N, 2.5 N, and 1.6N, respectively. The $k_p = 6$ condition has the smallest interaction force while producing the largest actuation torque (Figure 2), clearly indicating substantial motion assistance. If k_p is increased to 10, the interaction force will raise again due to overshoot. The muscle activations of the flexors and extensors are shown in Figure 3. The flexors are active mostly during the finger closing phase while the extensors are active during the opening phase. The activation level of extensors is much higher than that of the flexors except for the third active case.

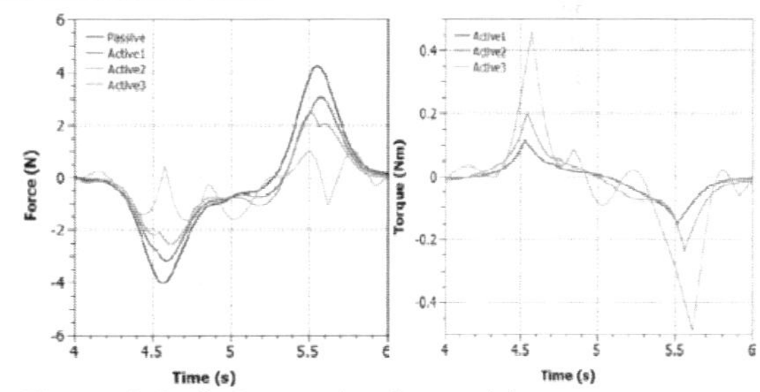

Figure 2. Left: Interaction force; right: motor torques.

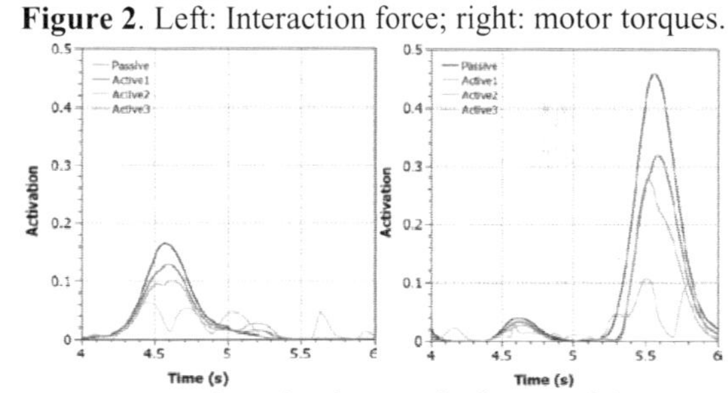

Figure 3. Muscle activations. Left: flexors; right: extensors.

Conclusions: Through modeling of a hand exoskeleton and its interaction with a hand musculoskeletal model, we were able to evaluate the effectiveness of an admittance control method. The results demonstrated that the assistance provided by the motor reduces the muscle activation significantly as a result of reduced interaction forces. An experimental study is being conducted to calibrate model parameters and validate simulation predictions.

References:

1. Adamovich SV, et al. JNER, 2009, 6:28
2. Heo P. et al. IJPEM, 2012;13(5):807-824
3. Thielbar, KO, et al. JNER, 2014, 11:171

Acknowledgements:
NIDILRR Grant #90RE5021-01-00

7B-2

Detection of DNA Mutations in Extracellular Vesicles Derived from Patients with Malignant Ground Glass Opacities

Yi Wen, Komal K. Abhange, and Yuan Wan
Department of Biomedical Engineering, The Pq Laboratory of Micro/Nano BiomeDx, Binghamton University-SUNY, Binghamton, NY 13850 USA

Introduction: Ground glass opacities (GGO) is a hazy opacity that does not obscure the underlying branchial structure or pulmonary vessels (1). It can be observed in both benign and malignant conditions, which cannot always be efficiently and timely differentiated with medical imaging through radiological traits. Although tissue-biopsy can suggest appropriate management guidelines, it is not frequently performed due to multiplicity of GGO, localization difficulties, and sample bias (2,3). Hence, there is a need to develop a new approach to overcome this challenge. Extracellular vesicles (EV) are cell-released lipid-bilayer-enclosed vesicles, containing a tissue specific signature wherein a variety of proteins and nucleic acids are selectively packaged (4). Their roles in tumorigenesis, metastasis, therapeutic resistance, and liquid biopsy have recently been demonstrated. In our previous study, we demonstrated mutations in tissue-DNA of patients with non-small cell lung cancer (NSCLC) can be identified in EV-DNA derived from plasma. We further found in early-stage NSCLC EV-DNA is superior to circulating free DNA (cfDNA) in mutation detection. Due to protection of the lipid envelop from degradation by DNase, EV-DNA and harbored mutations can be well preserved, making it more attractive than highly degraded cfDNA, especially in early-stage cancer when only few tumor-derived DNA is available. Therefore, analysis of EV-DNA derived from patients with GGO might assist in differential diagnosis of GGO and management. In this cohort study, we used next generation sequencing to detect DNA mutations harbored in EV-DNA and GGO tissue-DNA from 10 patients, respectively, and further analyzed their concordance.

Methods: Three-layer heparin and two-layer polyDDA were alternately coated on 1-μm polystyrene microspheres, forming HPMs, through electrostatic force. The various characterization of HPMs, optimization of operation, EV isolation efficiency with HPMs, and EV recovery efficiency were determined. All HPMs were cryopreserved in DI water at -20 C till use. Patients' peripheral blood samples were centrifuged at 2500× g for 15 min followed by filtration with 0.22-μm filter to remove cell debris and microvesicles. Then, 1-ml plasma sample was incubated with 125-μg HPMs at room temperature at 4 C for 1 h. The HPMs were collected by brief centrifugation, and subsequently EV-DNA was extracted. In the other group, cfDNA from 10-ml plasma was extracted. GGO tissue samples were collected after resection followed by DNA extraction. All DNA samples with minimal 30 ng were prepared for next generation sequencing. Totally 2863 hotpots in 65 relevant genes were analyzed.

Results: The five-layer surface coating with heparin/polyDDA forms a dense layer with ~50 nm thickness on microspheres. The HPMs can be preserved at both 4 C and -20 C for at least 6 months. The non-specific absorption assay determined that only ~0.6% albumin and ~2% dsDNA can bind to HPMs, demonstrating high purity of EVs isolated with HPMs. 125-μg HPMs can isolation ~81% EVs from spike-in plasma samples. Moreover, ~72% of isolated EVs can be further detached from HPM surface using 1 M saline within 5 min by dissembling of heparin/polyDDA layers. The released intact EVs can be directly lysed for downstream molecular analyses. Next, we used HPMs to isolate EVs from 10 clinical samples followed by EV-DNA extraction and sequencing. Meanwhile, cfDNA and tissue-DNA were also extracted for sequencing. The total mutations were determined first (Fig 1, left). The mutation concordance between EV-DNA and tissue DNA is 34.7%±14.9%. In comparison, the mutation concordance between cfDNA and tissue-DNA is 34.1%±12.1% (Fig 1. Right). There is no significant difference between two concordance (p>0.05, two tailed t-test).

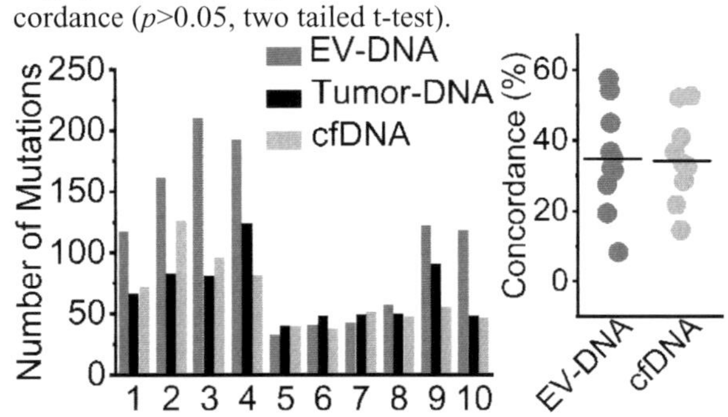

Figure 1. Comparison of mutations in EV-DNA, cfDNA and tissue-DNA. Left: total mutations identified in three types of DNA. Right: Concordance of EV-DNA/tissue-DNA and cfDNA/tissue-DNA, respectively.

Conclusions: HPMs can efficiently isolate EVs from plasma sample. The release of EVs from HPM surface further facilitate downstream molecular analyses. The cohort study demonstrated the EV-DNA derived from 1-ml plasma and cfDNA derived from 10-ml plasma potentially can be used to probe the somatic mutations in GGO.

References:

1. Hansell DM. et al. Radiology. 2008;246;697-722
2. Zhang Y. et al. Lung Cancer. 2014;84;236-41
3. Chae HD. et al. Radiology. 2014;273;285-93
4. Colombo M. et al. Annu Rev Cell Dev Biol. 2014;30;255-89

Acknowledgements: Y.W. thanks the Binghamton University Startup Funds 910252-35.

Superparamagnetic Iron Oxide Nanocrystal Clusters for Cancer Therapy

Qingbo Zhang[1], Zhen Xiao[1], Linlin Zhang[2], Sheng Tong[2], Gang Bao[2], Vicki L. Colvin[1,3]

1. Department of Chemistry, Brown University.
2. Department of Bioengineering, Rice University
3. Center for Biomedical Engineering, School of Engineering, Brown University.

Introduction: Superparamagnetic iron oxide nanocrystals have many potential applications in cancer diagnosis and treatment, such as magnetic hyperthermia, magnetic resonance imaging, magnetic targeting, et .al. The performance of iron oxide nanoparticles, unfortunately, are not satisfactory for many of these applications, especially in the clinical trials. In this work, we report superparamagnetic iron oxide nanocrystal clusters for the application in magnetic hyperthermia and magnetic targeting of cells.

Methods: The clusters are synthesized through a hydrothermal method. In a typical synthesis, $FeCl_3$16H_2O, Poly(acrylic acid), urea, and water are added into 20 mL ethylene glycol under vigorous stirring. The solution is subsequently transferred to a 40 mL autoclave and heated at 200 °C for 6 hours. The as-synthesized iron oxide clusters were magnetically purified for application.
The magnetic heating efficiencies of clusters were measured using an EASYHEAT induction heating system. 1 mL of aqueous solution of clusters was added to a 2 mL of cryovial and placed in the center of the coil. The temperature increase in the solution was recorded in real time with a fiber-optic temperature probe.
The U2OS cells were used for the magnetic targeting experiment. The clusters were loaded into the cells by incubating the clusters and cells overnight in a glass chamber. The chamber was placed on a magnetic letter and then shake for 30 minutes using a mechanical shaker.

Results: The iron oxide nanocrystal cluster synthesized in this work is a hierarchical structure composed of many small primary particles (Figure 1A). While the size of the cluster is much bigger than individual particles, they show little coercivity and remanence (Figure 1 B). Therefore, these clusters are stable in biological media and useful for medical applications.
The magnetic nanocrystal clusters can convert electromagnetic energy into heat in alternating magnetic field. This feature can be used to kill tumor cells. The heating efficiency of clusters was measured at magnetic field of around 80 Gauss, which is relevant to clinical application. Our results show that the heating efficiency of clusters is much higher than individual particles. The heating efficiency of clusters is dependent on cluster size (Figure C & D). The specific absorption rate can reach around 150 W/g iron oxide (Figure 1D), which is 5 times higher than individual particles.
Magnetic targeting of cells is useful to diagnose and treat cancer. We use a very magnetic letter with a low surface magnetic field (~50 Gauss) to control the cells. The cells loaded with clusters can be attracted by magnetic field at a very low field (Figure 1 E&F).

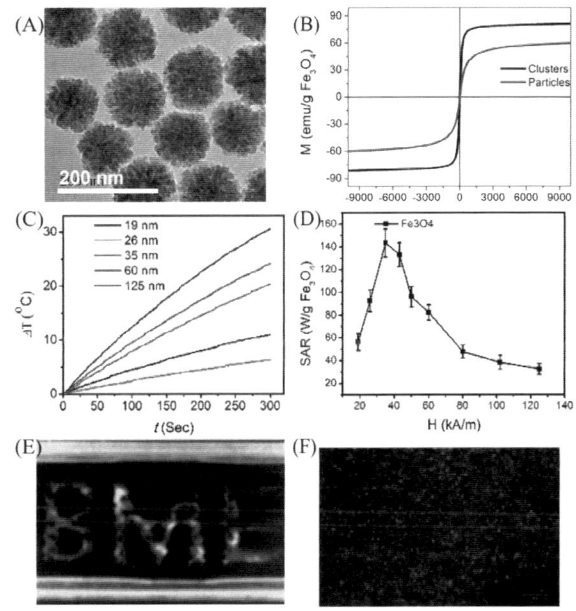

Figure 1. A: TEM image of iron oxide nanocrystal clusters; B: magnetic hysteresis loop of clusters; C: increase of temperature of cluster solution in magnetic hyperthermia; D: size dependent magnetic heating efficiency of clusters; E: fluorescent image of cells loaded with clusters; F: the high magnification fluorescent image of cells

Conclusions: Superparamagnetic iron oxide nanocrystal clusters with different sizes were synthesized for cancer therapy. These clusters display superior performance than individual particles in magnetic hyperthermia and magnetic targeting of cells.

Tunable Hydrogels to Investigate Breast Cancer Dormancy and Metastatic Relapse

Shantanu Pradhan, John H. Slater.
Department of Biomedical Engineering, University of Delaware, Newark, DE 19716.

Statement of Purpose: Metastasis is the leading cause of breast cancer related mortality.[1,2] Targeting metastatic disease is challenging due to the presence of disseminated tumor cells (DTCs) that can lie dormant in secondary organs for long time periods before reverting to an actively proliferative state (metastatic relapse).[3] Pre-emptively targeting dormant DTCs is difficult owing to their quiescent behavior, poorly understood biological characteristics, and limits of clinical detection.[3,4] DTC dormancy can be classified as cellular dormancy (G_0/G_1 arrest) or dormant micrometastasis (balanced proliferation and death). We demonstrate that highly metastatic breast cancer cells (MDA-MB-231s: (231s)) can be induced to adopt a cellular dormancy, dormant micrometastasis, or invasive phenotype by modulating the crosslink density, degradability, stiffness, and adhesivity of the hydrogel they are cultured in. We also present the ability to re-activate dormant cells via dynamically changing matrix adhesivity. This hydrogel platform can be used to induce desired dormant and invasive phenotypes in metastatic breast cancer and may provide useful pre-clinical information for future drug development applications.

Methods: Poly(ethylene glycol) diacrylate (PEG)-based hydrogels containing the proteolytically-degradable peptide GGGPQGIWGQGK (PEG-PQ, 5% w/v), integrin-ligating peptide RGDS (0, 1, 5, 10 mM) and co-monomer n-vinyl pyrrolidone (NVP) (0.0, 4.7, 9.4, 18.7 mM) were used for encapsulation and 3D culture of 231s over 15 days. Hydrogel biochemical (ligand density, degradability) and physical (pore size, elastic modulus) properties were characterized for each formulation. Cell viability, proliferation, early apoptosis, metabolic activity and single cell and cell cluster morphology were quantified as a function of time and matrix properties.

Results: We hypothesized that biochemical and physical properties of the hydrogel could be tuned to control the fate of metastatic breast cancer cells with respect to dormancy or growth. Our results indicate that hydrogel crosslink density and adhesivity can be tuned to induce four different phenotypes (high growth, moderate growth, single cell dormancy, balanced dormancy) quantified by changes in cell viability, viable cell density, proliferation, early apoptosis, metabolic activity, and 3D cell/cluster morphology. A systematic classification was conducted to delineate hydrogel formulations that support these specific phenotypes.

High growth was observed when encapsulated cells displayed high viable cell density and high metabolic activity, with and the cell population contained clusters with filopodial projections (Figs. 1,3). This behavior was induced by the softest, permissive hydrogels that supported adhesion (5-10 mM RGDS) and were highly degradable (0 mM NVP) (Fig. 2).

Single cell dormancy was observed in hydrogels with low viable cell density and low metabolic activity, and the dominant presence of single cells (Figs. 1,3). This behavior was induced by restrictive hydrogels with zero adhesivity (0 mM RGDS), regardless of other properties (Fig. 1). These cells were able to be reactivated through a dynamic increase in matrix adhesivity after 40 days in culture (Fig. 3).

Moderate growth was observed when live cell density from days 0-15 was significantly higher than dead cell density (days 0-15). This phenotype was induced by hydrogels that supported moderate degradability (0-4.7 mM NVP) resulting in attenuated growth and invasion.

Balanced dormancy was observed when live cell density was closely balanced with dead cell density over 15 days in culture. This phenotype was induced by hydrogels with the low degradability (9.4-18.7 mM NVP), regardless of RGDS concentration, resulting in tight cell confinement.

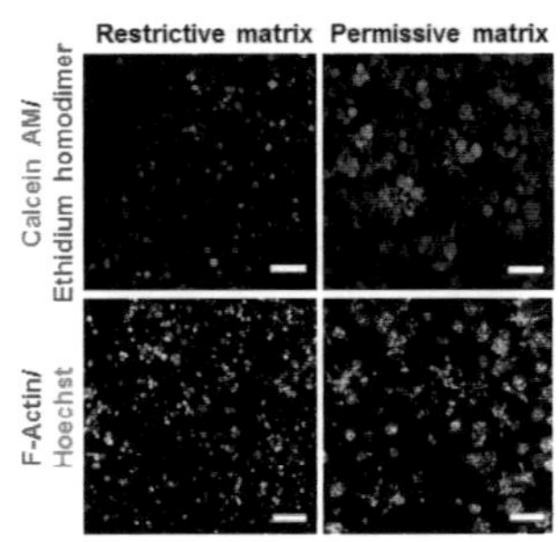

Fig. 1: Viability and morphology in restrictive (0 mM RGDS, 18.7 mM NVP) and permissive (10 mM RGDS, 0.0 mM NVP) hydrogels. SB = 100 μm.

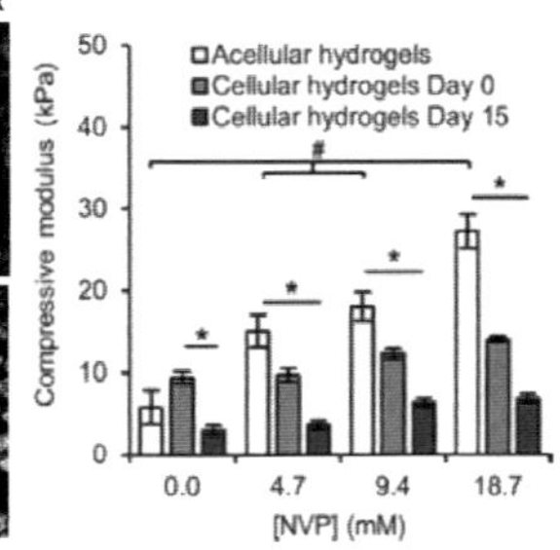

Fig. 2: Elastic modulus from compression tests of acellular and cellular hydrogels (day 0 and 15) with increasing [NVP].

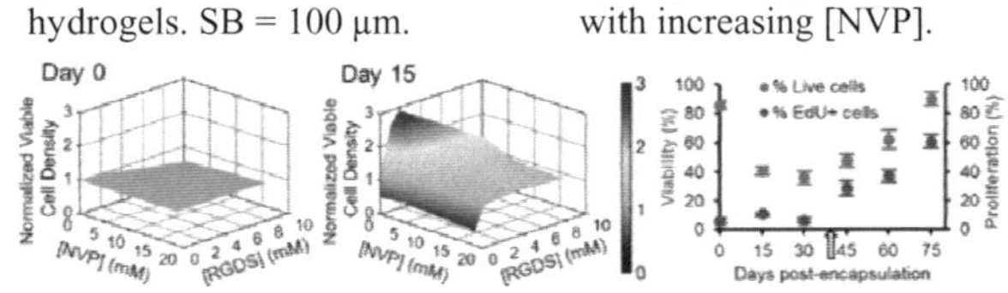

Fig. 3: 3D plots of cell viability as a function of [NVP] and [RGDS] over 15 days. Reactivation of dormant cells by increasing matrix adhesivity on day 40 (arrow).

Conclusions: We developed a hydrogel platform that allows modulation of breast cancer phenotype, invasive growth or various dormant states, via tuning of matrix adhesivity and crosslink density. We also demonstrate the ability to reactivate dormant single cells via a dynamic increase in matrix adhesivity. This platform provides potential use in determining mechanisms underlying cancer dormancy, genomic characterization, and for future drug development applications.

References: 1. Gomis, Mol. Oncol. 2017. 2. Giancotti, Cell 2013. 3. Aguirre-Ghiso, Nat. Med. 2013. 4. Almog, Cancer Lett. 2010.

Spectrally Encoded Compressive Imaging For Improved Resolution in Fiber-Bundle-Based Endoscopy

John P. Dumas[1], Muhammad A. Lodhi[2], Batoul A. Taki[2], Waheed U. Bajwa[2], Mark C. Pierce[1]
[1]Rutgers University, Dept. of Biomedical Engineering, 599 Taylor Road, Piscataway, NJ, USA 08854
[2]Rutgers University, Dept. of Electrical and Computer Engineering, 94 Brett Road, Piscataway, NJ, USA 08854

Introduction: Endoscopic imaging can provide real-time visualization of tissue using coherent fiber optic bundles to transmit images from inside the body. Fiber bundles are appealing because of their narrow diameter and flexibility, but their image quality is limited. Each individual fiber within the bundle transmits only a single pixel of information and the cladding between fibers blocks detail, resulting in images with low pixel density and missing data. Post-processing [1] and image compounding [2, 3] techniques have demonstrated the ability to improve the appearance of images taken through fiber bundles, but most still rely on individual fibers to transmit a single pixel of information. Recently, wavefront shaping techniques have been investigated to resolve multiple points using information delivered through a single multimode optical fiber [4], but the sensitivity to fiber bending has limited these techniques to imaging with rigid fibers [5]. Here, we present a "compressive endoscopy" platform that uses a compressive imaging (CI) framework to resolve multiple points within the diameter of each fiber within a bundle. CI enables high-resolution image reconstruction by collecting multiple observations of a scene through different coded masks. We also adapt a spectral encoding technique [6] to recover sample details that lie behind the inter-fiber cladding.

Methods: We constructed a benchtop platform to image fluorescent samples using a digital micromirror device (DMD) to rapidly switch between coded binary masks. A 470 nm LED with a 429-474 nm bandpass filter illuminates the sample, and emitted fluorescence is captured through the fiber bundle and a 506-594 nm bandpass filter. A "witness" camera is used to capture a ground truth image of the sample before transmission through the fiber bundle. We also place a prism at the distal end of the bundle to encode different spatial locations on the sample with different wavelengths of light. The dispersive action of the prism leads to each wavelength carrying a spatially shifted projection of the sample to the fiber bundle, ensuring that all spatial locations on the sample plane are transmitted through the bundle. A color CCD camera collects images of the sample in the red, green, and blue channels. CI reconstructed images from each channel are combined to display an image of the scene.

Results: The system was first tested for CI without using spectral encoding. We imaged several different fluorescent objects, including an *in vivo* fingertip coated with fluorescent highlighter (Fig. 1). In the ground truth image, the fingerprint ridges are clearly visible. Conventional fiber bundle imaging of the same fingertip could not resolve this detail. The CI approach resolved multiple points within the diameter of each fiber in the bundle, recovering much of the fingerprint detail. Post-processing to interpolate between fibers is used in images (b) and (c) to make the images appear more continuous.

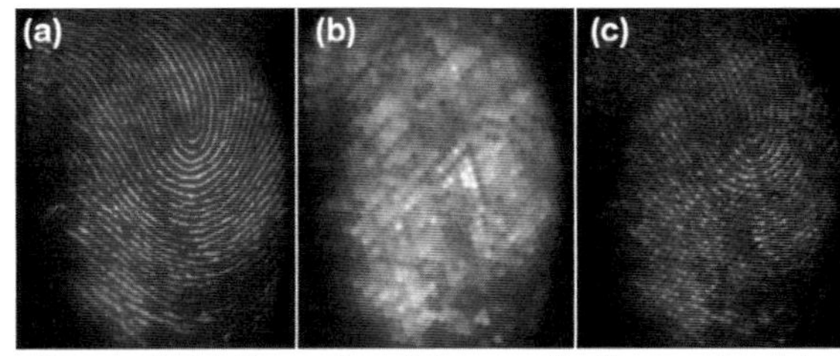

Figure 1. (a) Ground truth image of a fingerprint. (b) Conventional fiber bundle image. (c) Computational reconstruction.

We then imaged a USAF resolution target with our compressive endoscopy platform, additionally using the distal prism for spectral encoding (Figure 2). This allowed us to recover sample detail from regions lying behind the fiber bundle cladding. Compressive imaging alone can resolve multiple features within the diameter of a single fiber that were not resolvable with conventional fiber bundle imaging. The addition of spectral encoding enabled simultaneous recovery of intra- and inter- fiber detail.

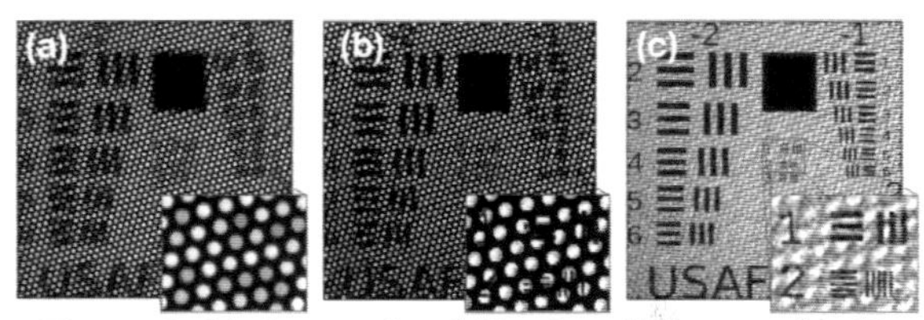

Figure 2. (a) Conventional fiber bundle image. (b) CI with 150 observations. (c) CI with spectral encoding.

Conclusions: Our compressive endoscopy platform produces higher quality images than conventional fiber bundle imaging. The improvement in image quality seen when imaging a fingertip coated for fluorescence are an encouraging initial step toward true *in vivo* imaging. This approach provides a method for improving spatial resolution while maintaining field of view for fiber-bundle-based endoscopy, without increasing the diameter of the bundle itself.

References:

1. A. Shinde and M. V. Matham, J. Med. Imag. Heal. Inform. 4, 203 (2014).
2. S. Park, *et al.*, IEEE Sig. Proc. Mag. 20, 21 (2003).
3. K. Vyas, *et al.*, Biomed. Opt. Express 9, 4649 (2018).
4. O. Tzang, *et al.*, Nat. Photonics 12, 368 (2018).
5. S. Ohayon, *et al.*, Biomed. Opt. Express 9, 1492 (2018).
6. N. Bedard *et al.*, J Biomed Opt 17, 080508 (2012).

Acknowledgements: Funding from the National Science Foundation (NSF) (CCF-1453073, ECCS- 1509260).

Environmentally-Controlled Approaches for Near Infrared Spectroscopic Assessment of Cortical Bone Water

Jack Oswald, Ramyasri Ailavajhala, Nancy Pleshko
Temple University Department of Bioengineering

Introduction: Cortical bone water content is an emerging biomarker for bone quality. Gravimetric analysis is a destructive, 'gold standard' method for evaluating water content in tissues. Vibrational spectroscopic approaches, such as near infrared spectroscopic imaging (NIRSI), are emerging as non-destructive water assessment techniques. Water absorbances at 5184 and 7008 cm^{-1} dominate the near infrared (NIR) frequency region and can overshadow absorbances from other important bone components, including collagen at 4608 and 6688 cm^{-1} [1]. Further, environmental conditions in the laboratory can cause exposed bone samples to absorb or evaporate loosely bound water, imposing limitations on data interpretation and the repeatability of vibrational spectroscopic studies [2]. These challenges motivated the development of environmentally controlled chambers (ECC) for spectral data collection under consistent relative humidity (RH) conditions. Spectral data acquisition in a desiccated environment has proven essential for serial lyophilization studies used to associate water absorbances with individual bone compartments. The assignment of NIR absorbances to bone components can be useful to monitor bone quality in disease states and aging [1]. The present work aims to non-destructively validate NIR absorbances associated with water under controlled environmental conditions.

Methods: The initial ECC (Fig 1a) was assembled from a microscope slide base, acrylic walls and lid, and a glass coverslip viewing window. This model was tested with bovine tibiae (Research 87, Bolyston MA). To collect NIRSI imaging data from larger, clinically relevant human bone samples (NDRI, Philadelphia, PA), a larger ECC (Fig 1B) was constructed from acrylic and two 76.2x76.2x0.5 mm Corning® Gorilla® Glass imaging windows (Corning, NY). Each model utilizes a Bosch BME-280 (Stuttgart, Germany) humidity sensor to monitor RH. Compressed air regulated to 4 psi was fed into the chamber to generate an internal RH of <3%. A high RH environment (>70%) was achieved by aerating a 50 mL conical vile partially filled with room temperature water with a 4 psi pump and feeding the resulting humidified air into the chamber. Each ECC model was validated by monitoring the previously reported 5184 cm^{-1} water absorbance of a bovine bone sample within an hour timespan under high and low RH conditions. The study was repeated after the samples had lyophilized (Martin Christ-Alpha 1-2) for 24 hours. NIRSI data was obtained on a Perkin-Elmer Spotlight 400 imaging system with a mercury cadmium telluride detector (Shelton, CT), with 50 µm pixel size, 64 cm^{-1} spectral resolution from 4000-7800 cm^{-1}, and 32-coadded scans.

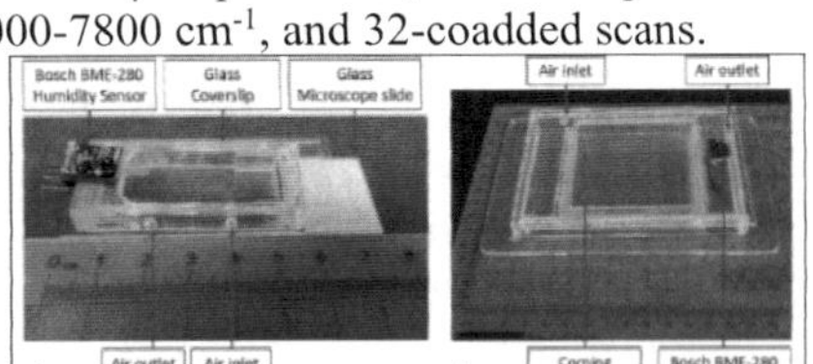

Figure 1: Small (a) and Large (b) ECC models

Results: The previously described ECC chamber setup maintained high or low RH conditions during the entire data collection window. Under each condition, a decrease in 5184 cm^{-1} absorbance is observed after 15 minutes in wet bone samples. Under low RH conditions, the 5184 cm^{-1} absorbance continues to decrease while the peak fluctuates under high RH conditions. After 24 hours the samples were lyophilized for 24 hours, the sample absorbs the environmental moisture under high RH conditions as interpreted from the increase of the 5184 cm^{-1} peak. The 5184 cm^{-1} peaks of the lyophilized same maintains a constant under low RH conditions (Fig 2).

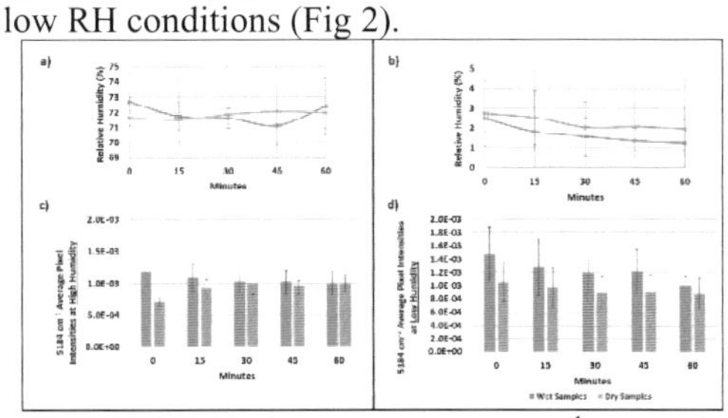

Figure 2: Correlations between 5184 cm^{-1} water absorbance RH inside ECC

The low RH setup was consequentially chosen to evaluate the serial lyophilization of human bone tibiae after the time durations shown in Fig. 3. As predicted, 5184 cm^{-1} and 7008^{-1} spectral peaks associated with water absorbance decrease with lyophilization. The dried samples allow for greater illumination of the collagen peaks at 4608 cm^{-1} and 6688 cm^{-1}.

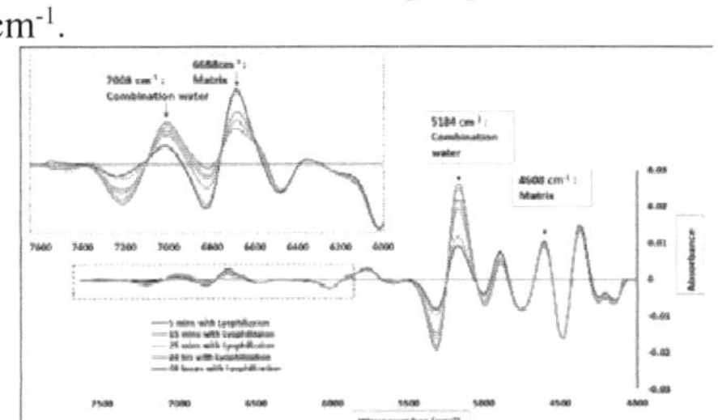

Figure 3: Second derivative Spectra from serial lyophilization of bone tissue

Conclusions: The ECC models developed for the present research reliably facilitated data collection under known environmental conditions. The decrease in spectral absorbances at 5184 cm^{-1} and 7008 cm^{-1} after lyophilization support the previous association of those peaks with loosely and tightly bound water. The lyophilized bone tissue is useful for assignment of spectral absorbances at 4608 cm^{-1} and 6688 cm^{-1} from collagen, though data collection in a desiccated environment is essential for adequate interpretation of the sample data.

References:

1. Rajapakse CS. et al. Bone. 2017;103:(116-124).
2. Dong, D. et al. Proc Natl Acad Sci USA. 2017;114(28):E5486-E5487

Acknowledgements: This study was funded by NIH R21 AR071704

Short-Wave Infrared Spatial Frequency Domain Imaging For Non-Invasive Quantification Of Tissue Hydration

Christine C. Sahyoun[a], John Paul Dumas[a], Yanyu Zhao[b], Matthew B. Applegate[b], Darren M. Roblyer[b], Mark C. Pierce[a]
[a]Rutgers, The State University of New Jersey; [b]Boston University, Boston MA

Introduction: Spatial frequency domain imaging (SFDI) is a non-invasive technique that can quantify tissue chromophore concentrations. SFDI has been implemented using visible and near-infrared wavelengths to provide information on oxy- and deoxy-hemoglobin concentrations for applications including assessment of burn wounds, pressure ulcers, and tumor resections. Further tissue characterization can potentially be achieved using short-wave infrared (SWIR) wavelengths (approx. 1,000 – 1,700 nm), due to distinctive absorption bands of water, lipid, and collagen in this spectral range. Quantification of these tissue components may have clinical significance in relation to such topics as inflammation, obesity, and wound healing. There are currently no reports in the literature that directly perform SFDI in the SWIR range. In this study, we developed an SFDI system using laser sources at 980 nm and 1,550 nm to capture differences in hydration in a biomedically significant context.

Methods: Light from fiber-coupled laser sources at 980 nm and 1,550 nm was collimated and directed onto a digital micromirror device (DMD) after combining beams with a dichroic filter. The DMD was used to project three phase-shifted sinusoidal illumination patterns with spatial frequencies of 0 mm^{-1} (DC) and 0.1 mm^{-1} onto the sample plane. Diffusely reflected light was captured with an InGaAs camera. Orthogonal linear polarizers in the illumination and imaging paths was used to reduce specular reflection. To convert raw images measured in pixel counts to absolute values of diffuse reflectance, the system was calibrated with a 10% Intralipid phantom of known absorption and reduced scattering coefficients at the two wavelengths. Six Spectralon reflectance standards of known diffuse reflectance at planar illumination were then imaged to test the accuracy of this calibration. To determine the system's ability to detect changes in sample hydration, a commercial dish sponge was imaged at different levels of hydration with distilled water; an analytical balance was used to determine water content as a function of sponge mass. Diffuse reflectance, absorption and reduced scattering coefficient maps were calculated using look-up tables generated by Monte Carlo simulations. Finally, an *in vivo* imaging test was performed to observe and quantify different hydration states of the back of the hand (dry, hydrated with water, and hydrated with lotion).

Results: Images of the reflectance standards (2%, 10%, 20%, 40%, 80%, 99%) yielded diffuse reflectance measurements that were in close agreement with the true values, confirming the accuracy of calibration. For the sponge hyrdation experiment, both the 980 nm and 1,550 nm data showed decreasing diffuse reflectance with increasing water content, with the diffuse reflectance values of the 1,550 nm data being overall lower than that of 980 nm. The mean absoprtion coefficient increased with increasing water content at both wavelengths, with greater differences observed between hydration states in the 1,550 nm data compared to 980 nm (Fig. 1).

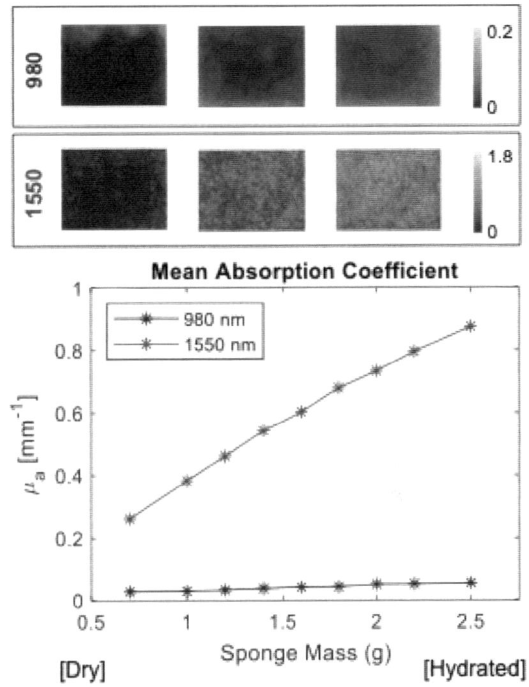

Figure 1. Mean absorption coefficient of the sponge during the transition from a dry to hydrated state for the 980 nm (blue) and 1,550 nm (red) wavelengths.

Mean reduced scattering coefficient values were similar between the two wavelengths and decreased with increasing water content. For the hand hydration experiment, the 980 nm data did not show any difference between the three hydration states. For the 1,550 nm data, decreased diffuse reflectance, increased absorption coefficient and decreased rediced scattering coefficient were observed in the hydrated states.

Conclusions: This study has demonstrated that SFDI can be directly performed in the SWIR wavelength range, and while both the 980 nm and 1,550 nm data can distinguish differences in hydration state, the 1,550 nm data is more sensitive due to stronger water absorption. This was confirmed with the absorption coefficient differences in both the sponge and hand hydration experiments. For all hydration experiments, the same trends were observed with increasing water content: decreased diffuse reflectance, increased absorption coefficient, and decreased reduced scattering. Future work includes optimizing the spatial frequency selection and the wavelength choices for accurate extraction of absolute water concentrations.

Diagnosis of Myocardial infarction in cardiac cine MR images using corr-entropy based endocardium detection and ventricle wall motion analysis

G. Kavitha, M.Muthulakshmi
Department of Electronics Engineering, MIT Campus, Anna University, Chennai, Tamilnadu, India.

Introduction: Cardiovascular diseases (CVDs) account for 31% of global mortality. In coronary heart disease, the blood vessels that supply blood to the heart muscles are blocked. This causes damages to heart muscles due to deprivation of oxygen and leads to irreversible myocardial infarction (MI) [1]. The contractility and morphology of cardiac chambers are important predictors of myocardial infarction [2]. Reliable computation of cardiac wall motion is based on proper positioning of endocardial contours. The left ventricular volume change over a cardiac cycle is an essential parameter for clinical evaluation of various CVDs [3]. Accurate regional cardiac wall motion abnormality due to MI can be detected by sector based LV analysis. The aim of this work is to study regional contractility of LV in MI and normal subjects using magnetic resonance images. The main contribution of this work is analysis of regional abnormality in cardiac wall motion using segmented LV cavity and sector based study using Hausdorff distance (HD) measure.

Methods: This experiment uses 200 cine steady state free precession short-axis MR slices of normal and infarcted subjects from Medical Image Computing & Computer Assisted Intervention (MICCAI) left ventricle segmentation database. Initially the LV blood pool is segmented using corr-entropy based local bias field corrected image fitting (CELBIF) model [4]. The performance of the segmentation algorithm is compared with local bias field corrected image fitting (LBIF) model. The segmentation results are validated with Dice coefficient. The cardiac chamber abnormalities are diagnosed based on LV volume tracking and clinical indices such as EDV, ESV and ejection fraction. Further, the segmented LV is subdivided into six sectors to study the regional wall motion, with the help of Hausdorff distance measure and wall motion score index.

Results: The following parameters are used in the CELBIF method: Gaussian kernel parameter Ã= 10, radius of circle neighborhood Á= 21, time step ”t = 0.1, coefficient of length penalty term Å= 0.002*2552, coefficient of distance penalty term μ = 1 and correntropy criterion ¨ = 170. The CELBIF algorithm (0.92) produces an overall higher mean Dice coefficient compared to the LBIF algorithm (0.89) as shown in fig. 1.

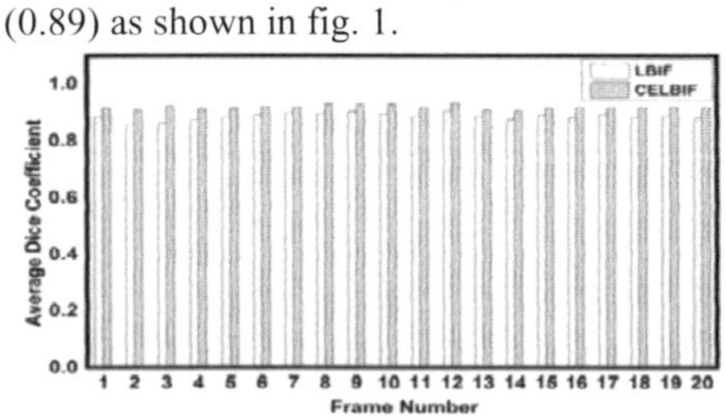

Figure 1. Average Dice coefficient for segmentation of LV in a cardiac cycle

It is observed from fig. 2 that the LV volume of normal subjects ranges from 27 ml to 120 ml. The volume of LV varies from 108 ml to 349 ml for abnormal subjects over the entire cardiac cycle. Fig. 3 depicts that the considered infarcted subjects have a lower ejection fraction (<45%) compared to normal subjects.

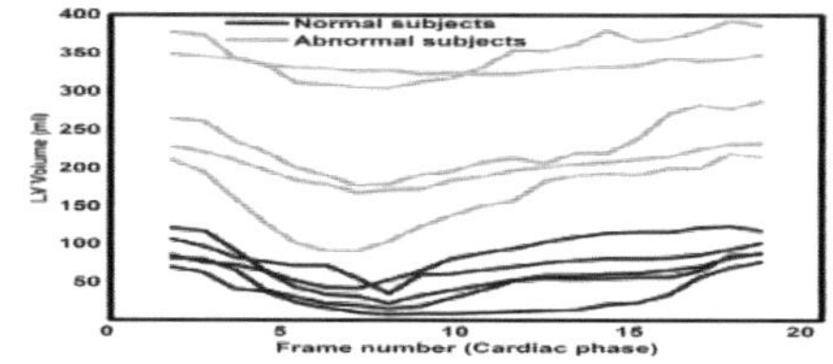

Figure 2. Tracking of LV volume for entire cardiac cycle.

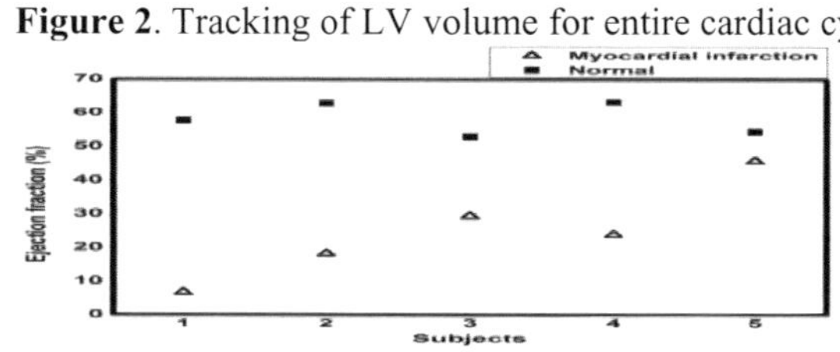

Figure 3. Variation in LV ejection fraction for typical normal and infarcted subjects

The sectorwise wall displacement for normal and abnormal subjects in terms of HD is represented in the form of Bull's eye plot in Fig. 4. The calculated Hausdorff distance is illustrated by different colours. Here, each concentric circle represents a single frame. The 6 sectors are named as follows: A - Anterior, AS - Anteroseptal, IS - Inferoseptal, I - Inferior, IL - Inferolateral and AL - Anterolateral.

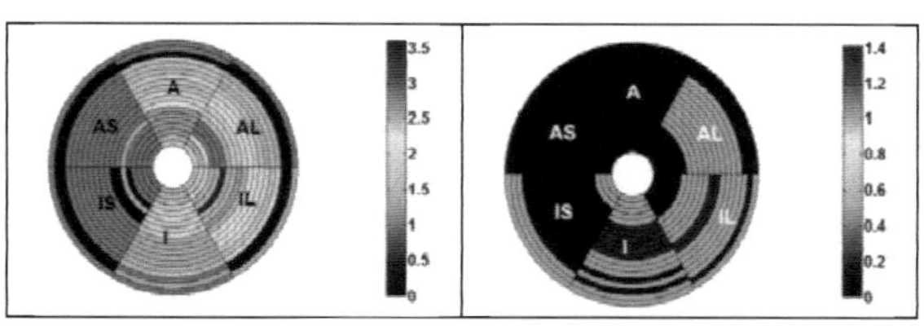

(a) Normal subject (b) Infarcted subject

Figure 4. Bull's eye plot of cardiac wall displacement

Conclusions: The CELBIF algorithm better segments LV endocardium. The infarcted subjects have a higher EDV, ESV and lower EF compared to normal subjects. This indicates enlarged LV cavity, systolic dysfunction and lower percentage of blood pumped out. Further the sector based analysis indicates lower contraction in inferoseptal and anteroseptal myocardium with a low Hausdorff distance value. The future study will be to find discriminative features for diagnosis of MI tissue pattern.

References:

1. Welch TD. et al. Curr Probl Cardiol. 2012;37(7):237-310.
2. Bastarrika G. et al. Eur Radiol. 2008;18(5):939-946.
3. Sani ZA. et al. J Digit Imaging. 2015;28(1):91-98.
4. Zhou S. et al. Neurocomputing. 2017; 234: 216–229.

Title of Abstract: Targeted Tumor Drug Delivery by Magnetic nanoparticles ; An Invitro Study

Negin Farzad BS, Peter Klesczews, Christina Zito Ph.D, Hashini Mohottala Ph.D, Saion Sinha Ph.D
Depts. of 1 Cellular and Molecular Biology, 2 Electrical & Computer Engineering 3 Biomedical Engineering Program
UNIVERSITY OF NEW HAVEN, WEST-HAVEN, CT-06516

Introduction: Breast cancer has continued to be fundamental in field of cancer research because of incidence in the general population. The cells used for this study are MDA-MB-231 cells and they are referred to as a triple negative breast cancer (TNBC) [1]. TNBCs are difficult to treat because there is no receptor overexpression to target [1]. Doxorubicin is a commonly utilized chemotherapeutic agent that affects healthy tissue, it is important to find a method of distribution targeting only the afflicted area. Nanobiotechnology is defined as the biomedical application of nanomaterials [2]. Today, nanobiotechnology, especially involving nanostructured particles such as polymeric macromolecules, has both improved drug delivery to cancer cells and facilitated their eradication [2]. This study aims to produce a biocompatible $MnFe_2O_4$-MNP that will allow for greater targeting of cancerous tissue and tumor sites. The MNP will be coated with polyethylene glycol (PEG) and coupled to Doxorubicin (DOX), then exposed to magnetic field and high frequency electromagnetic field (EFM) exposure. This novel MNP vehicle for chemotherapeutic delivery may provide a delivery system that can be coupled to multiple chemotherapeutic agents and improves their efficacy.

Methods: #The intrinsic MnFe2O4 (from US Research Nanomaterials, Inc (Houston, TX)) magnetic nanoparticles (MNPs) were coated with PEG-6000 (purchased from Sigma Aldrich (St. Louis, MO)) to produce biocompatible and functionalized MNPs. A solution of MNPs in distilled water were purified against 1 mM HCl to eliminate excess salt, then coated with 20 μM sodium oleate in aqueous solution. The MNPs were then incubated in a hybridizer for 1 hr at the room temperature. Then, the MNPs were washed with distilled water and incubated in hybridizer at room temperature in 1 mL of a 20 μM PEG solution for another 1 hr to form PEG-MNPs. Doxorubicin (DOX) (Selleckchem, Houston, TX) as the chemotherapeutic agent was coupled to the MNPs using the SelfMeg Transfection kit from Oz Biosciences (San Diego, CA). Coupling percentage following each step of the polymer coating reactions was determined by UV-vis spectroscopy.The MNP pegylation [3] was determined by AFM(Atomic Force Microscopy). One-way ANOVA and Tukey tests were utilized on Minitab to determine any significance between the mean MNP size after each stage of the polymer coating process. MDA-MB231 breast cancer cells, ATCC (Manassas, VA), were treated with five types of treatment; MNPs, coated with PEG, and coupled with DOX. Every treatment was in three different treatment environments; dc magnetic field (for 90 minutes), high frequency AC (alternative current) exposure (34 Mhz for 10 minutes) and no magnetic field in incubator at 37 ℃ and 5% CO_2. After treating the cells in different levels, the cell proliferation was determined via MTT Assay after 24 h and 48 h was performed.

Results: As shown in Fig 1, in coating process a one-way ANOVA was performed, which found that the mean diameter at each step of the process increased at each step of the coating process (P-value = 0.018). In DOX-MNP treatment was most effective treatment (P-value =0.045) in different levels of treatments. According to the several combinations of DOX and MNP treatments, a two-way ANOVA found a significant difference for the treatment factor at each time point (P-value =0.002). Tukey test did not find a significant difference in cell viability between the treatment groups in dc and ac magnetic field after both 24 and 48 h MTT assay.

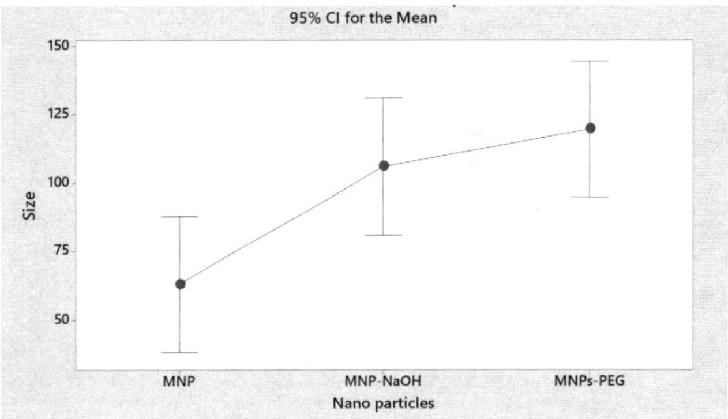

Figure1. Analysis of magnetic nanoparticles in each step of coating process.

Conclusions: Pegylation of MNPs was successful through AFM analysis and the mean PEG-MNP diameter was earned 119.67 nm. DOX was successfully coupled to MNPs and mean coupling efficiency was determined at a rate of 90% (n=3). The most effective treatment was DOX-MNP after 90 min magnetic field exposure and in 10 minutes Ac electromagnetic field.

Further studies are needed, however the DOX with an MNP vehicle in magnetic field was successfully delivered.

References:

1. Holliday, D. and Speirs, V. 2011;13(4):215
2. Kayser, O. Lemke, A. Hernandez-Trejo, N. T. 2005; 6 (1): 3-5.
3. Illes, E, Szekeres, M, Kupcsik, E, Toth, IY, Farkas, K, Jedlovszky-Hajdu, A, Tombacz, E. 2014; 460: 429-440.

Acknowledgements: We would like to thank Dr. Saion Sinha and Dr. Christina Zito for their guidance and advice throughout this project. I would also like to thank the University of New Haven for funding the project.

Design of a standardized *in vitro* model of skeletal muscle regeneration for implantable microthread scaffolds

E. E. Heinle, E. M. Morra, E. M. Mossman, A. M. Paul,
M. E. Carnes, C.F. Whittington, G. D. Pins
Department of Biomedical Engineering, Worcester Polytechnic Institute
100 Institute Road, Worcester, MA 01609

Abstract - Current treatments for Volumetric Muscle Loss (VML) defects limit the restoration of muscle function at the site of the wound, where the primary action and uniaxial growth of the myoblasts is destroyed. Wounded warriors and trauma patients undergo months of painful and expensive treatments, with limited outcomes. Three-dimensional (3D) outgrowth assays model cellular proliferation onto a scaffold that could be implanted into the site of the VML wound. Fibrin microthreads are a polymerization of fibrin and thrombin and provide a provisional matrix to model C2C12 cellular proliferation and muscle tissue regeneration. The goal of this project is to bridge the gap between in vivo animal testing and in vitro assays, by creating a standardized six-well assay that will be used to test cellular outgrowth, migration, and proliferation.

Keywords - volumetric muscle loss (VML), fibrin microthread scaffolds, polydimethylsiloxane (PDMS), in vitro model system

I. INTRODUCTION

Volumetric Muscle Loss (VML) is an injury sustained as a result of battle, trauma, or cancer ablation. During Operation Iraqi Freedom, 82% of soldiers injured on the battlefield suffered a musculoskeletal extremity wound [1]. The average lifetime disability cost for veterans suffering from a VML injury is on average $341,300. Potential treatments include amputation/prosthetics, autologous tissue transfers, and skin flaps [2]. These current treatments do not restore muscle function at the site of the wound and can cause donor site morbidity, decreasing quality of life [3]. Thus, a treatment is needed to restore muscle function and direct myoblast growth at the site of the wound. Fibrin microthread scaffolds allow for uniaxial 3D tissue outgrowth [4]. However, it is crucial that these scaffolds be tested *in vitro*, under multiple conditions (crosslinked fibrin microthreads, growth factors release) to determine which conditions would enhance functional tissue growth *in vivo*. The Pins lab has shown promising research that fibrin microthreads are an ideal 3D scaffold for cellular outgrowth. The current *in vitro* model uses an assembly process that introduces a great deal of user variance and a 25% failure rate when users attach fibrin microthreads to a polydimethylsiloxane (PDMS) frame and stretch them across a raised Thermanox® stage (Figure 1) prior to adding the fibrin gel containing myoblasts [3, 4]. The goal of this project was to design a standardized 3D *in vitro* model system to support, image, and analyze cellular outgrowth on fibrin microthread scaffolds. The model decreases user variability, improves reproducibility, and increases data throughput.

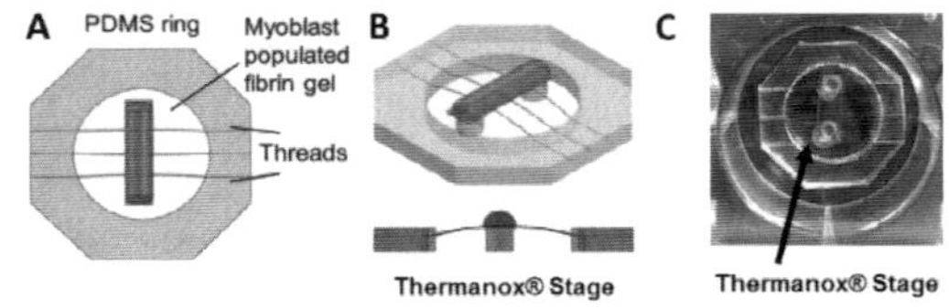

Figure 1*: Current PDMS frame model: A – Top view, B – Thermanox® stage with fibrin microthreads stretched across, C – Ring and stage assembly in a well of a 6-well tissue culture plate. Adapted*

II. METHODS

A. Final Device Design

To make the model more reproducible, a mold was designed to produce identical PDMS frames to support fibrin microthreads. A Solidworks CAD model was designed, and a six-frame mold was machined out of Delrin™ (DuPont) plastic (Figure 2).

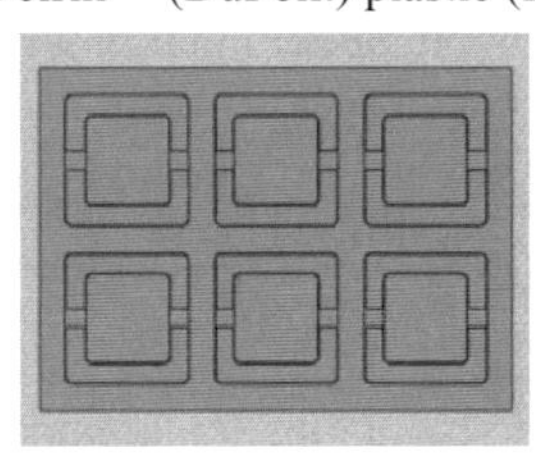

Figure 2: *CAD model of the mold for six PDMS frames to be created simultaneously prior to PDMS curing and frame removal.*

Each frame is shaped as a square with a side length of 24.5 mm to ultimately fit into each well of a 6-well tissue culture plate. The height of each frame is 3.8 mm. The width of each frame is 4 mm. A cutting guide was made out of acrylonitrile butadiene styrene (ABS) using the Ultimaker 3 and combined with three razor blades aligned equidistant apart. The height of the cutting guide is 1.9 mm (Figure 3). The cutting guide is placed around the PDMS frames to ensure that the depths and spacing of the slits created in the PDMS to fixture the fibrin microthreads are consistent.

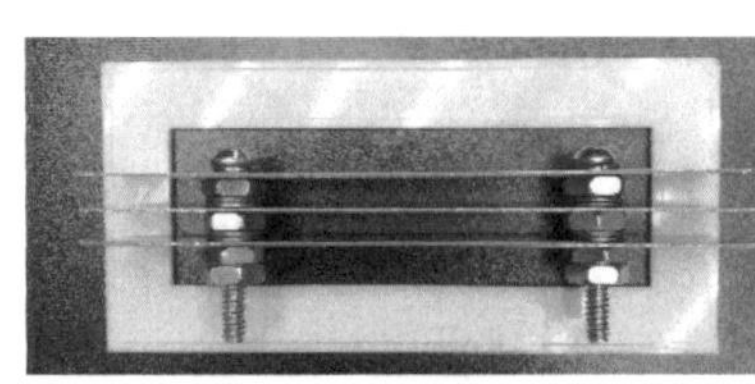

Figure 3: *Cutting tool is placed on top of cutting guide. The device is long enough to create slits in three frames simultaneously.*

To eliminate the need to construct an individual Thermanox® stage for each frame, a groove was created in the mold to create a slot to hold the Thermanox® coverslip, which will support a fibrin gel containing C2C12 cells for cellular outgrowth onto the fibrin microthreads (Figure 4). The groove is 3.5 mm wide and 2.1 mm deep. The mold was also machined with a 3° draft angle, and rounded edges to allow for easy PDMS release for each frame.

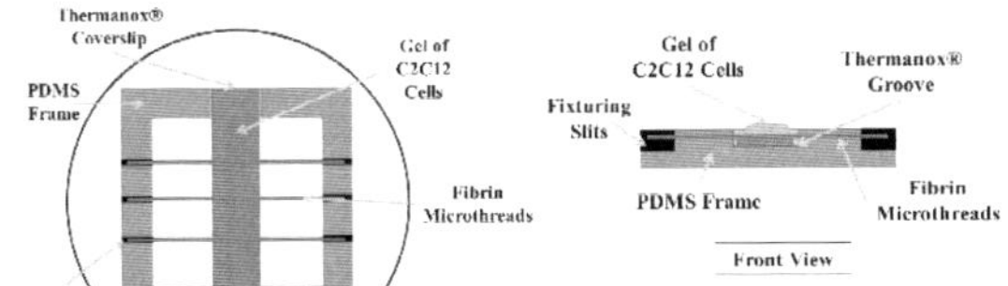

Figure 4: Schematic of the top and front views of the PDMS frame setup with each component.

Bend testing was performed on the PDMS frames to determine the ideal radius of curvature to open the slits for thread placement without causing the frame to fail. PDMS with 1.9mm deep slits was bent until failure, and the optimal bend angle was determined. Based on this information, a PDMS bending tool was created to hold the frame open during thread placement.

B. Assembly Testing

To determine the process flow of the new design versus the existing model in the Pins lab, assembly testing was done to compare the two models. Initial assembly testing included cutting and shaping six PDMS rings, cutting slits into the PDMS rings using a razor blade, placing three fibrin microthreads per ring, and constructing six Thermanox® cell-gel stages. First, the time to assemble the existing model was compared between one experienced user and three inexperienced users. This process was compared to the six PDMS frames that the team removed from the mold. Using the ABS cutting guide and razor cutting tool, razor slits were made into the PDMS frames, and a bending tool was used to open the slits and place three fibrin microthreads into each PDMS frame. The assembly time for each model was collected.

III. RESULTS

It was found that the inexperienced users were 37-42% slower than the experienced user over the entire assembly process (Figure 5). Thread placement appeared to be the step that produced the largest gap in time. Once assembly was compared between the existing model and the new model, differences in the workflow were observed. Overall, the new model system decreased the number of steps to 8 steps from 15 steps for the existing assembly procedure. Collectively, the PDMS 6-well mold, cutting tool and guide, and bending tool decreased the number of assembly steps by 53%. We anticipate that the six identical PDMS frames in a single machined mold

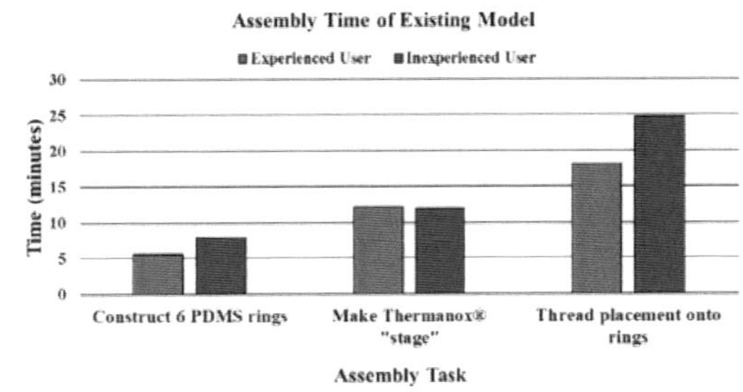

Figure 5: Comparison of three major assembly steps for the existing model performed by an experienced user (n=1) versus inexperienced users (n=3).

decrease variability between users, increases reproducibility, and improves throughput. The cutting tool and guide further standardizes the assembly process by creating equidistant slits in the PDMS frame, three at a time, with uniform depth to fixture the fibrin microthreads. Finally, the bending tool opens the slits to ease fibrin microthread placement by making it easier handle the threads and shortening thread placement time.

IV. CONCLUSIONS AND FUTURE WORK

Initial testing has shown that the new model system supports decreased assembly time with fewer opportunities for user variability. A kit will be created to include two PDMS molds, a razor cutting guide, and bending tool. This kit will be used for 3D scaffold research in the Pins Lab but has potential to be used more broadly. Additional time tests and cell-based studies will also be performed for further validation. With this standardized and reproducible assay system, researchers will be able to accelerate the study of multiple experimental conditions to better identify and predict conditions that will be successful *in vivo*. Ultimately, this will expedite the testing process for fibrin microthread scaffolds that can be used as clinical treatment for a range of VML patients.

V. REFERENCES

[1] D. Devore, T. Walters, R. Christy, C. Rathbone, J. Hsu, D. Baer, and J. Wenke. Military medicine, vol. 176, no. 6, pp. 660-663, Jun. 2011.
[2] B. Corona, J. Rivera, J. Owens, J. Wenke, and C. Rathbone. Journal of Rehabilitation Research & Development, vol. 52, no. 7, PP. 785–792, 2015.
[3] J. Grasman, D. Do, R. Page, and G. Pins. Biomaterials, vol. 72, pp. 49-60, 2015.
[4] J. Grasman, R. Page, and G. Pins. Tissue Engineering. Part A, vol. 23, no. 15-16, pp. 773-783, 2017.

VI. ACKNOWLEDGEMENTS

The team would like to acknowledge Leonard Polizzotto Ph.D., Distinguished Executive in Residence, for his commitment to assisting the team with project development.

P3-3

Inflammatory Secretome of Macrophages Treated with Hemoglobin-Haptoglobin Complexes

Paulina Krzyszczyk[1], Kishan Patel[1], Maurice O'Reggio[1], Kristopher Richardson[2], Martin Yarmush[1], Andre Palmer[2], François Berthiaume[1]

[1]Department of Biomedical Engineering, Rutgers, The State University of New Jersey

[2]Department of Chemical & Biomolecular Engineering, The Ohio State University

Introduction: Macrophages are recruited to sites of injury or inflammation in order to aid in the repair process. In a wound healing context, the first macrophages that are recruited have pro-inflammatory characteristics, and are referred to as *M1-like*. As wounds progress in the healing process, the macrophage population shifts to having predominantly anti-inflammatory characteristics; these are called *M2-like* macrophages. In chronic inflammatory conditions, *e.g.* chronic wound healing, the M1-to-M2 transition is impaired. Macrophages retain pro-inflammatory characteristics, which stall wound healing, or prevent a return to homeostasis in other chronic conditions. Therapeutic approaches aim to attenuate M1-like, and promote M2-like characteristics in macrophages. There is evidence that hemoglobin-haptoglobin (Hb:Hp) protein complexes can lead to these changes [1]. Hemoglobin is often found at sites of injury, as it is released when blood vessels rupture. Haptoglobin (Hp) is an acute phase protein that can bind extracellular hemoglobin (Hb) and activate anti-inflammatory pathways within macrophages. The ultimate goal is to use Hb:Hp complexes in a model of chronic wound healing to promote anti-inflammatory properties in macrophages [2]. The current study aims to examine the *in vitro* cytokine, chemokine and growth-factor secretion profile of macrophages treated with Hb:Hp complexes.

Methods: Human monocytes were isolated from peripheral blood donations via density gradient centrifugation and CD14+ magnetic bead separation. To create M1-like macrophage phenotypes *in vitro*, monocytes were stimulated with 5ng/mL granulocyte-macrophage colony stimulating factor (GM-CSF) and allowed to attach and differentiate over 7 days. Resulting M1-like macrophages were cultured in serum-free media (SFM) and activated with 1μg/mL lipopolysaccharide (LPS) to increase pro-inflammatory protein secretion. Macrophages were also treated with 0.2mg/mL Hb alone, Hp alone, or Hb:Hp complexes for 24 hours, at which point supernatants were collected. Secretion of 27 inflammation-related cytokines, chemokines and growth factors were measured using a Bio-Plex MultiPlex Immunoassay System (BIO-RAD; Hercules, CA). Two-way ANOVA with Fisher's LSD post-hoc analysis was used to identify statistical significance ($p<0.05$).

Results: The secretion of 9 out of 13 M1-associated factors were affected following treatment with Hb:Hp complexes (Figure 1). In several of these factors (*e.g.* interferon-³, IFN-³, tumor necrosis factor-±; TNF-±), Hb alone significantly increased secretion above baseline, whereas Hb:Hp complexes brought levels back to baseline. Other factors, such as MIP-1±, IL-8 and IL-17, were significantly decreased below baseline following Hb:Hp treatment.

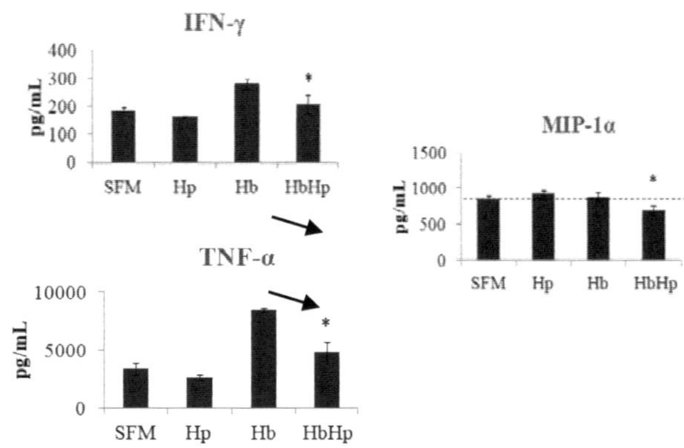

Figure 1. Secretion of IFN-³, TNF-α, and MIP-1α from Hp, Hb, and Hb:Hp-treated M1 macrophages

Conclusions: These results suggest that Hb:Hp complexes aid in attenuating M1-like macrophage characteristics *in vitro*. In the current study, M2-like factors were not increased in response to treatment. This may be due to the fact that the starting macrophage population was stimulated to have strong M1-like characteristics, and promotion of M2 factors was unlikely in this inflammation-challenged system. Future studies will investigate the effect of Hb:Hp complexes on starting macrophage populations primed to have M2-like characteristics. This system will allow us to study the effects of Hb:Hp complexes on a wider range of macrophage populations, which is more biologically-relevant. This will allow for a more comprehensive approach in determining if Hb:Hp can both attenuate M1, and promote/protect M2 characteristics in macrophages.

References:

1. Finn AV, J Am Coll Cardiol. 2012;59(2):166-77.
2. Philippidis, P., Circ Res. 2004;94(1):119-126.

Acknowledgements: The primary author, Paulina Krzyszczyk, was provided a stipend from two fellowship sources over the course of this work: Graduate Assistance in Areas of National Need, DOE and T32GM008339, NIH.

Nanoparticle Containing V domain of sRAGE for Diabetic Chronic Wounds

Hwan June Kang[1], Arielle D'Elia[1], Suneel Kumar[1], Biraja Dash[2], Henry Hsia[2], Martin L. Yarmush[1], Francois Berthiaume[1]
1 Department of Biomedical Engineering, Rutgers University, Piscataway, New Jersey, 2 Department of Surgery (Plastic), Yale School of Medicine, New Haven, Connecticut

Introduction: Diabetic foot ulcers (DFUs) are a common long-term complication of diabetes and cost annually \$9-\$13 billion, affecting 8 million patients in the U.S. Approximately 25% of patients with diabetes develop a DFU during their life time and the cost of treatment per ulcer can be up to \$20,000 [1]. Diabetes is associated with the accumulation of advanced glycation end products (AGEs) that form when proteins and lipids react with reactive sugars during hyperglycemia [2]. The binding of AGE to the receptor for AGEs (RAGE) activates the pro-inflammatory transcription factor, NF-° B, which results in increased expressions of several pro-inflammatory genes including intercellular cell adhesion molecule-1 (ICAM-1) [3]. The soluble form of RAGE (sRAGE) is thought to be a potential therapeutic agent to prevent the AGE-RAGE binding as a competitive inhibitor of RAGE [4]. Recent studies have shown that the V domain of RAGE is the binding site to AGEs involved in the AGE-RAGE interaction [5]. However, exogenous peptide therapies may not be effective due to rapid proteolysis in the wound environment [6]. We developed a nanoparticle system containing fusion proteins of elastin like polypeptides (ELPs) with the V domain of sRAGE (vRAGE). ELPs consist of pentapeptides repeats of Valine-Proline-Glycine-Xaa-Glycine, where Xaa can be any amino acid except proline. ELPs form nanoparticles by aggregating above a transition temperature, protecting peptides from degradation, and solubilizing below the transition temperature [6]. Here, we show that vRAGE-ELP fusion proteins bind to AGEs, thus inhibiting AGE-RAGE binding, as indicated by the decreased level of ICAM-1 *in vitro*.

Methods: The full-length encoding sequence for the V domain of human sRAGE (amino acids 23-121) was fused to the N-terminus of 50 repeats of ELP encoding cassette via linker sequence, $(G4S)_3$. The expression vector containing vRAGE-ELP was transformed into *E. coli*. The bacteria were grown in Terrific Broth and protein expression induced by adding 0.5mM IPTG. Cells were lysed by sonication and the fusion proteins purified by inverse transition cycling. As control, ELP proteins alone were also expressed and purified. Fusion proteins were characterized by SDS-PAGE gel, Western blot and Dynamic Light Scattering. To investigate the bioactivity of fusion proteins, human umbilical vascular endothelial cells (HUVECs) were stimulated with AGEs and treated with vRAGE-ELP nanoparticles. Immunostaining and cell-based ELISA for surface ICAM-1 expression were performed to see whether vRAGE-ELP could suppress this pro-inflammatory gene in AGE-stimulated cells.

Results: The characterization results confirmed the presence of vRAGE-ELP fusion proteins (MW: ~35.8 kD) in the purified products. The radius of nanoparticles at 37°C was 648.3 ± 55 nm at 25 μM. Furthermore, vRAGE-ELPs suppressed the expression of ICAM-1 in AGE-stimulated HUVECs while non-fused ELP proteins alone did not suppress the expression as shown in ELISA experiments (Figure 1) as well as immunostaining (data not shown). TNF-α, which is a pro-inflammatory cytokine, was used as a positive control and N-acetyl cysteine (NAC), which acts as an antioxidant, was used as a negative control.

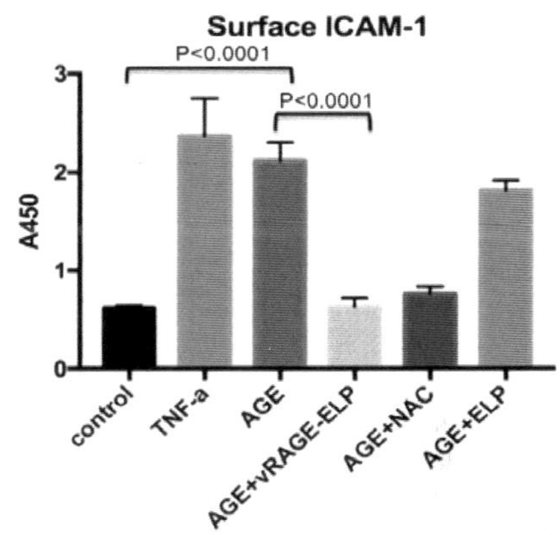

Figure 1. Surface ICAM-1 Expression. TNF-α: 10ng/mL, AGE: 100ug/mL, vRAGE-ELP: 1mg/mL, NAC: 30mM, ELP: 1mg/mL

Conclusions: We have developed and characterized vRAGE-ELP fusion proteins that could form nanoparticles at physiological temperature. These fusion proteins can be easily purified from bacterial culture systems by inverse temperature cycling, thus enabling rapid and inexpensive purification. Our results suggest that vRAGE-ELP could bind to AGEs and inhibit AGE-RAGE bindings in *in vitro* cell culture system. Future studies will investigate the bioactivity of vRAGE-ELP *in vivo*, possibly in diabetic mouse wound model to see whether the fusion protein could enhance the wound healing process.

References:

1. Nussbaum SR. Value Health 2018;21(1):27-32
2. Chen Q. Biochemical and biophysical research communications, (2009) 381(2):192-7.
3. Bierhaus A. Annals of the New York Academy of Sciences, (2005) 1043:676-80.
4. Goova MT. The American Journal of Pathology, (2001) 159(2):513-25.
5. Xue J. Biochemistry, (2014) 53(20):3327-35.
6. Yeboah A. J Control Release, (2016) 232:238-47.

Acknowledgements: We would like to thank Dr. Vikas Nanda for help with Dynamic Light Scattering experiments. This project is supported by a National Institutes of Health Grant (R21EB021570).

P3-5

Engineered Extracellular Vesicles Derived from Human Umbilical Cord Mesenchymal Stem Cells for Skin Rejuvenation

Komal K Abhange[1], Yi Wen[1], and Yuan Wan[1],*

[1]The Pq Laboratory of Micro/Nano BiomeDx, Department of Biomedical Engineering, Thomas J. Watson School of Engineering & Applied Science, Binghamton University, Binghamton, NY 13902; *Email: ywan@binghamton.edu

Introduction: Extracellular vesicles (EVs) are lipid-bilayer-enclosed vesicles released by cells. They are novel intercellular communication mediators in physiological and pathological scenarios. A variety of proteins and nucleic acids can be selectively wrapped within EVs, and delivered to recipient cells. Recent studies demonstrated the use of mesenchymal stem cells derived EVs can promote skin repair, regeneration and rejuvenation [1,3]. Moreover, in contrast to stem cells; EVs when injected into a tissue or organ have no risk of causing an unwanted inflammatory response, tumorigenicity and vascular occlusion [4,5]. They also can be sterilized by filtration and have a longer shelf-life than parental cells. These unique advantages make them be a novel and promising cell-free therapeutic modality for skin treatment [2]. Nevertheless, the slow growth rate of stem cells and the moderate isolation efficiency of prevalent ultracentrifugation (UC) make the massive production of EVs very difficult, and thus it is very difficult to translate stem cell derived EVs into practical applications [2]. Here, we used an extrusion method to produce engineered EVs derived from human umbilical cord mesenchymal stem cells (hucMSCs), and we further demonstrated that these engineered EVs show similar morphology and size distribution in comparison with EVs naturally secreted from cells. Moreover, the engineered EVs also can promote synthesis of fibronectin, collagen and elastin in human dermal fibroblasts (HDFs).

Methods: Fresh umbilical cords were collected from informed, consenting mothers and processed within the optimal period of 6 h following standard procedures. These cells were further passaged for expansion into flasks. The primary HDFs were purchased from ATCC, and routinely cultured. The culture protocols of hucMSCs and HDFs can be found somewhere else. Approximately, $\sim 6.7\times10^7$ hucMSCs were cultured without FBS for 48 h, and the supernatant was collected followed by filtration and ultracentrifugation at 180,000×g at 4 C for 4 h. The pellet was resuspended in 500-µl PBS and cryopreserved at -80 C. In the other group, $\sim 7\times10^6$ hucMSCs suspension in 500-µl PBS was continuously filtered with 10-µm and 5-µm track-etched membrane (Whatman) at 2,000 rpm for 5 times, and each filtration took 5 minutes. The suspension was further centrifuged at 2500×g for 15 min to remove cell debris followed by filtration using Ultra-15 filter (Amicon, 10 kDa). The filtered supernatant was then stored at -80 C. Both samples were characterized with TEM. The amount and size distribution were determined with Nanosight. Subsequently, ~100-µg EVs in respective group were used to educate 1×10^4 HDFs for continuous 3 days. Subsequently, HDFs were harvested and the concentration of soluble fibronectin and soluble collagen in the supernatant was measured using a Human Magnetic Luminex® Screening Assay (R&D Systems, Minneapolis, MN, USA) and SirCol assay kit (Carrickfergus, UK), respectively. Each measurement was repeated five times.

Results: Naturally secreted EVs and engineered EVs show characteristic saucer-like structure under TEM (Fig. 1a-b). Engineered EVs have a diameter of ~130 nm, and the naturally secreted EVs have a diameter ~100 nm. After 3-day education with EVs, the average fibronectin concentration in groups of control, engineered EVs, and natural EVs was 1.65, 2.31, and 2.2 µg/ml, respectively. The collagen concentration in respective group was 2.1, 4.3, and 4.4 µg/ml. The concentration of elastin in respective group was 497, 1236.7 and 1117.6 pg/ml (Fig. 1c-e). There is a significant difference between control group and engineered EV group (p<0.01, two tailed t-test), and between control group and natural EV group (p<0.01, two tailed t-test). On the contrary, there is no significant difference between engineered EV group and natural EV group (p>0.05, two tailed t-test).

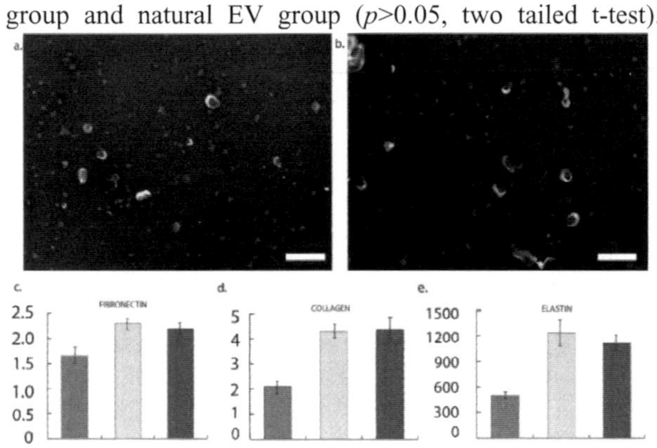

Figure 1. Morphology of EVs and measurement of fibronectin, collagen, and elastin. a-b: TEM images of naturally secreted EVs and engineered EVs, respectively. The scale bar is 500 nm. c-e: concentration of fibronectin, collagen, and elastin released from HDFs without EV treatment (blue), treated with natural EVs (yellow), and engineered EVs (red), respectively.

Conclusions: Engineered EVs derived from hucMSC using extrusion method can be massively produced. Meanwhile, these engineered EVs can promote proliferation of HDFs and increase the synthesis of fibronectin, collagen and elastin. Altogether, engineered EVs prepared by extrusion could be translated into skin treatment in future.

References:

1. Kim, Yoon-Jin, et al. Biochem. Biophys. Res. Commun., 2017;493(2):1102-1108.
2. Phan, J, et al. J Extracell Vesicles., 2018 7:1, 1522236.
3. Basu, Joydeep, et al. Opin Biol Ther., 2016, vol. 16:4,489–506.
4. Arango-Rodriguez, Martha L et al. World J Stem Cells,2015, vol.7,2: 408-17.
5. Ramdasi S, et al. Int J Hematol Oncol Stem Cell Res., 2015 1;9(2):95-103.

Alginate Encapsulation for Bupivacaine Delivery and MSC Co-therapy

Mollie S. Davis[1], Xiomara I. Perez[1], Ileana Marrero-Berrios[1], Charles P. Rabolli[1], Rene S. Schloss[1], Joel Yarmush[2], and Martin L. Yarmush[1]
[1]Rutgers University, Department of Biomedical Engineering, Piscataway NJ
[2]Department of Anesthesiology, New York Presbyterian- Brooklyn Methodist Hospital, Brooklyn, NY

Introduction: Local anesthetics (LA), which reversibly block nerve transmittance, are commonly used to minimize patient discomfort and post-surgical pain. LAs can also be co-administered with cellular therapies to improve regeneration and reduce inflammation. However, LAs have been shown to have negative effects on mesenchymal stromal cells (MSC) in a potency-and-time-dependent manner [1]. To help improve the efficacy of MSC therapies in conjunction with LA, we have developed a sustained release bupivacaine alginate-liposomal construct which enables at least four days of bupivacaine release. However, a co-therapy between MSC and LA would require positional control of the cells. For this reason, our lab utilizes alginate encapsulated MSC (eMSC), which also promote cytokine and chemokine secretion [3]. However, the effect of LA on eMSC has not been evaluated. Interleukin (IL)-6 and prostaglandin E2 (PGE2), two important immunomodulatory molecules, as well as transforming growth factor beta 1 (TGF-^{2} 1), an indicator of tissue regeneration, were studied to determine the immune and regenerative properties of eMSC in the presence of bupivacaine.

Methods Unilamellar liposomes containing bupivacaine were encapsulated in a 2.2% ultrapure alginate hydrogel. MSC were freely cultured or encapsulated in alginate microspheres. The cells were left untreated or dosed with bolus, liposomal, or construct bupivacaine and were untreated or activated with TNF-± and IFN-^{3} to simulate inflammatory conditions. The cells were analyzed at 24hrs, 48hrs, and 96hrs and secretory profiles were assessed. Matlab was used to model the diffusion of bupivacaine from the construct to the eMSC. A one dimensional implicit finite difference model of diffusion was used in conjunction with diffusion data from Maguire et al [2].

Results Encapsulating MSC in alginate protects the cells from LA. eMSC secrete similar levels of IL-6, PGE2, and TGF-^{2} 1 regardless of LA modality. IL-6 levels for free monolayer MSC were similar to eMSC for most of the culture period - For all time points, PGE2 was lower for free MSC and TGF-^{2} 1was higher for free compared to eMSC. This indicates that eMSC in the presence of LA are more immunomodulatory and less regenerative than free MSC. However, for eMSC, the construct increases TGF-^{2} 1 secretions compared to bolus LA conditions. Modeling the eMSC system (Figure 1) suggests that encapsulating the cells decreases LA exposure for all LA modalities to levels consistent with 90% viability [1], which concurs with the secretory results. The model also indicates that higher levels of LA (up to 9.5mM) can be used in the construct in conjunction with eMSC while retaining cell viability and function. *In vitro* experimentation results concur with the expected results from the mathematical model. eMSC dosed with 7mM and 9mM LA-constructs showed improved PGE2 and IL-6 secretion levels compared to other LA modalities at lower levels, and maintained viability consistent with lower LA concentrations.

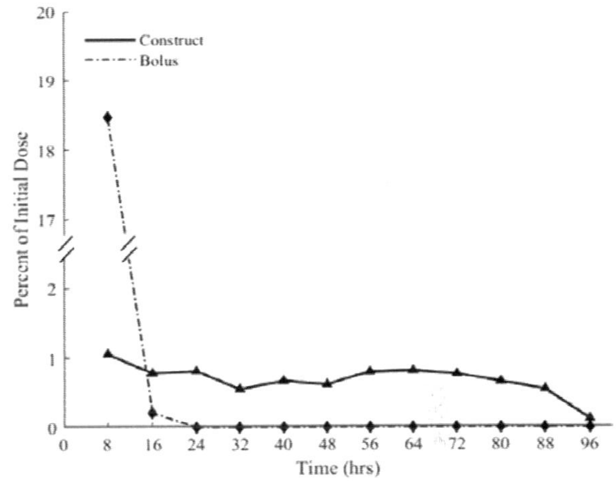

Figure 1: Computational Model of eMSC and LA.

Conclusions: eMSC, but not free MSC, are protected from 1mM LA, regardless of LA modality. eMSC and the construct can likely be used in conjunction to reduce pain as well as improve regeneration. The computational model indicates that a higher concentration of bupivacaine can be used in the LA-construct while retaining eMSC functionality and viability. These higher levels of LA may be more clinically relevant for *in vivo* studies than the typical *in vitro* doses used.

References

1. Gray, A., et al., Nano LIFE, 2015;5;155001-1-13
2. Maguire, T., et al., Int. J. Anes. & Pain Med., 2016;2 (1:3)
3. Barminko, J., et al., Biotechnol Bioeng., 2012;108(11): 2747–2758.

Acknowledgements: Research reported in this abstract was supported by the National Institute of General Medical Sciences of the National Institute of Health under award number T32 GM008339.

P3-7

A Thermoreversible and Photoactive Collagen-Based Scaffold for Tissue Engineering Applications

Yoliem S. Miranda-Alarcón, Dorota Jazwinska, and David I. Shreiber
Department of Biomedical Engineering; Rutgers University, New Brunswick, NJ

Introduction: The bioactivity, structural anisotropy, superior biocompatibility, and versatile chemical composition make collagen a major biomaterial of interest for tissue engineering. Collagen has been used as a building block to develop scaffolds, but without crosslinking, its weak mechanical properties serve as a major impediment. To this end, we have synthesized collagen methacrylamide (CMA) by adding methacrylate groups to lysine residues on type-I collagen, which renders the collagen photoactive; CMA can be rapidly crosslinked with spatial control through exposure to UV light in the presence of a photoinitiatior. Functionalization with methacrylamide groups also imparted a second, serendipitous property that makes CMA a unique material for 3D printing scaffolds. Like collagen, CMA can self-assemble into fibrillar hydrogels at physiological pH and temperature, but unlike collagen, CMA reversibly disassembles into a liquid macromer suspension when temperature is lowered to 4°C. Photocrosslinking after self-assembly eliminates thermoreversibility. Combined, these two properties enable a number of exciting applications, such a 3D printing and cell encapsulation. In this work, we have tested the potential for cell encapsulation of CMA using human mesenchymal stem cells (hMSCs) due to the cell's potential to differentiate into a variety of cells. We hypothesize that methacrylic acid grafted to collagen type-I would afford a tunable matrix for the differentiation of hMSCS into chondrocytes and adipocytes in the absence of growth factors.

Materials and Methods: Cell-laden collagen and CMA hydrogels were prepared by incorporating cells in the buffered hydrogel solutions at pH 7.2. A 0.1% w/v solution of the photoinitiator Irgacure 2959 was added. hMSCs were expanded in culture flasks until 70% confluence and added to the buffered hydrogel suspension at 40,000 cells/mL. A volume of 200mL of cell-gel suspension was incubated at 37°C for 1h to allow for self-assembly and entrapment of the cells within the fibrous hydrogel. Gels were then crosslinked with UV light (365nm, 50 mW/cm^2) for 90s or 120s, which represented the durations required for crosslinking to 95% or 100% of the maximum achievable shear modulus, respectively. After 48h incubation, cell viability of hMSCs entrapped in CMA was assessed using a LIVE/DEAD Viability/cytotoxicity kit and imaged using an Olympus IX81 fluorescence microscope.

Results and Discussion: Preliminary data indicates good cell viability of hMSCs encapsulated in CMA hydrogels. Free radicals from the photoinitiator cause a decrease in cell viability (Figure 1). In the absence of photoinitiator, UV light does not affect cell viability. In the presence of photoinitiator and no UV light exposure there is a reduction in cell viability that can be attributed to cytotoxic effects of Irgacure 2959. This suggests that the additional exposure to UV light did not produce more free radicals. Cell viability for the fully crosslinked gel and the overexposed gel were similar. As such, we can conclude that the limiting factor for the crosslinking of CMA is the photoinitiator concentration. A one-way ANOVA indicates no statistical significance between the conditions.

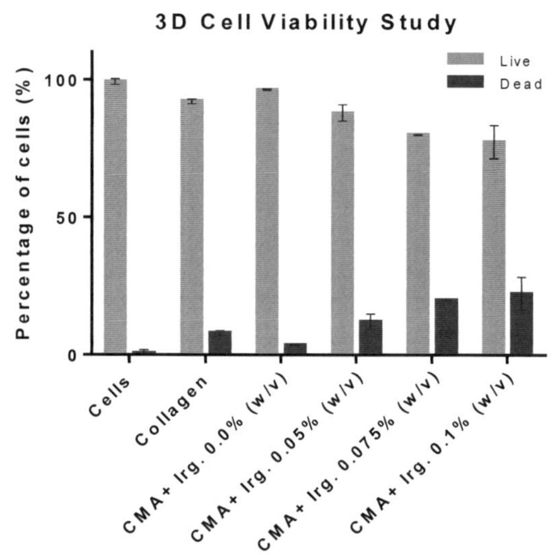

Figure 1. Cell viability of self-assembled CMA gels crosslinked at different UV exposure at 365nm and photoinitiator (Irgacure 2959) concentrations and collagen and cell controls.

Conclusions: In this work, we have assessed the optimization of the conditions for UV-crosslinking to maintain cell viability. Our next step will be to study photoinitiator concentrations to ultimately find the optimal hydrogel stiffness without compromising cell viability. For future work, we plan on leveraging the innate mechanosensitivity of stem cells to develop a matrix with tunable stiffness to differentiate cells.

References:

1. Gaudet ID. Biointerphases. 2012;7(1-9).
2. Williams CG. Biomaterials. 2005; 26(1211-1218).

Acknowledgements: This work is supported through the NIH-Biotechnology Training program (NIH T32 GM008339), the NSF-Graduate Research Fellowship Program (NSF DGE 2433187), and the Rutgers Neu-roengineering Working Group (RU-NEG).

Nanofibrous nerve conduits pre-seeded with bone marrow stromal cells and pre-cultured in bioreactors for peripheral nerve regeneration

Gan Zhou* and Xiaojun Yu*

Department of Biomedical Engineering, Chemistry, and Biological Sciences, Stevens Institute of Technology, Hoboken, New Jersey 07030, USA

Introduction: Peripheral nerve injury (PNI) can cause permanent dysfunction in movement and sensation. Synthesized nerve guide conduit with Schwann Cells (SCs) can help peripheral nerve regeneration, however poor accessibility of SCs and lack of full coverage of seeded cells on the nerve conduit can lead to failure of nerve regeneration across long gaps. In order to overcome these limitations, we have seeded bone marrow stromal cells on a nanofibrous nerve conduit previously developed in our group and cultured it with a rotating bioreactor. We tested the nerve conduit in long nerve defect (15mm) in a rat sciatic nerve model. The survival study last for 12 weeks.

Based on the results from the study, the sciatic functional index, electrophysiologic assessment, and histology analysis showed comparable results between the experimental group and the autograft group, which indicate the possibility of alternative therapy to replace the gold standard treatment(autograft) for peripheral nerve injury.

Methods: Nerve conduits were made from porous Polycaprolactone (PCL) sheets and covered by electrospun aligned PCL nano-fibers, and rolled into conduits with a spiral structure. Bone marrow stromal cells (BMSCs) were seeded onto the nerve conduit and cultured with a rotating bioreactor as long as 7 days before implantation. In the animal study, a 15-mm long nerve defect on the right sciatic nerve was created and bridged with nerve conduits or autograft. The recovery period was 12 weeks, during which the sciatic functional index (SFI) was recorded every four weeks. Upon sacrifice, electrophysiologic studies, pinch test were performed to evaluate the functional recovery. The histology samples were collected and the regenerated axons were quantified.

Results: The results of the sciatic functional index (SFI), pitch test and histology showed a significant difference between BMSCs and BMSCs+Bioreactor group, which indicated a better functional recovery would be achieved with the help of bioreactor.

Autograft group showed better regeneration outcome in histology and gastrocnemius muscle ratio, while no significant difference with BMSCs+bioreactor group in SFI, nerve conducting velocity (NCV) and pitch test. These results showed that with the help of bioreactor, our nerve conduit with the seeded BMSCs could achieve comparable regeneration outcome with autograft. I showed the promising role of the bioreactor in promoting peripheral nerve regeneration.

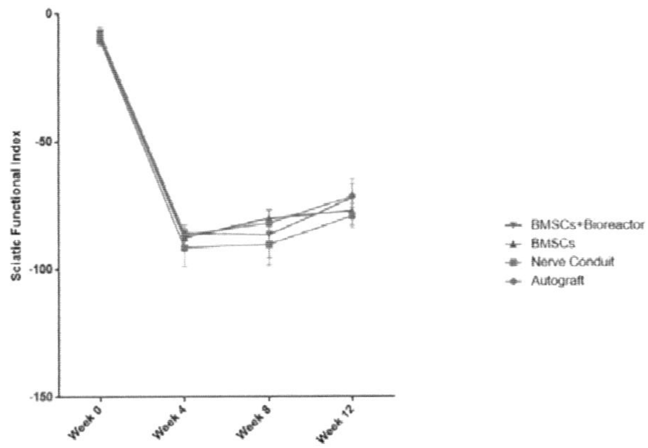

Figure 1. Sciatic Nerve Functional Index

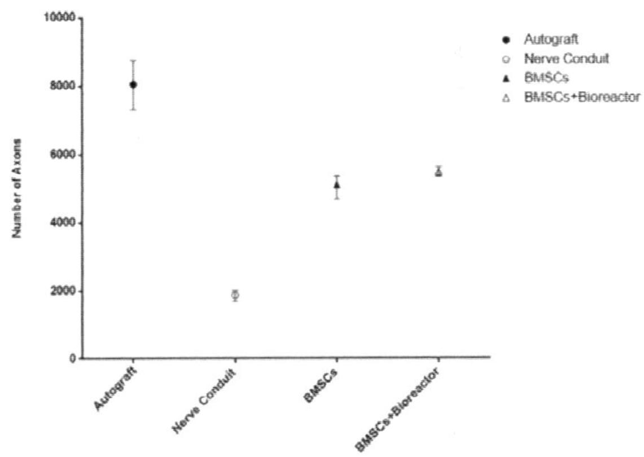

Figure 2. Histology Analysis

Conclusions: In conclusion, the combination of BMSCs seeded nerve conduit and bioreactor showed great promise in peripheral nerve regeneration and could be a potential strategy to replace autograft in treating peripheral nerve injury.

References:

1. Lode, A. J Tissue Eng Regen Med 2008. 2(7):400-407.
2. Gobrecht, P., et al. J Neurosci. 2016. 36(14): 3890-3902

Acknowledgements: This work is supported by the Stevens I&E Doctoral Fellowship Program and Office of the Assistant Secretary of Defense for Health Affairs through the Peer Reviewed Orthopaedic Research Program under Award No. (W81XWH-13-1-0320) and (W81XWH-16-1-0132).

P3-9

Investigating the use of low-cost bioprinting for the fabrication of complex tissue scaffolds

Robert Warren[1], Carolina Leynes[2], Joseph Freeman[1]
[1]Rutgers University, Piscataway, NJ, [2]University of Texas Rio Grande Valley, Edinburg, TX

Introduction: Fabricating multiphase biomaterials using bioprinting is expensive, preventing academic institutions and startup companies from using these materials more frequently in their research. By introducing a low-cost modification to a relatively inexpensive commercial 3D printer, we can lower the cost of entry into the field. In this study we assess the feasibility of converting a thermoplastic 3D printer into a low-cost 3D printer capable of printing thermoplastics and bioinks. We also demonstrate the printer's ability to fabricate a variety of complex tissue scaffolds from commonly used biomaterials such as polyethylene glycol diacrylate (PEGDA) hydrogels and polycaprolactone (PCL) thermoplastic filament.

Methods: We developed a microextrusion based device that modifies a MakerBot Replicator Original Dual and is capable of extruding polymer solutions alongside thermoplastics. A custom Python program was developed to allow for the proper deposition of bioinks. This program converts pre-sliced 3D printing files by manipulating the rotation rates of the bioink depositing stepper motor. Printer efficacy was determined by using ImageJ software to define positional accuracy. The pixel width of printed thermoplastic and hydrogel lines were analyzed and compared to each other (n=4). Mechanical testing was performed by exposing printed scaffolds to a 2%/min tensile test on an Instron 5869 mechanical testing machine and calculating the elastic modulus, yield strength, and ultimate strength (n=5).

Results: We were able to achieve a high level of printing accuracy with the use of an 18-gauge syringe tip. Bioprinting efficacy improved by reducing the size of the tip. We were able to construct a variety of complex polymeric scaffolds including a hydrogel sample sandwiched between two layers of PCL (figure 1a) and PCL fibers imbedded into two layers of hydrogel (figure 1b). Additional scaffold types were also created. Mechanical testing and analysis revealed that the addition of PCL in either fiber or layer form increased the mechanical properties of the scaffolds.

A)

B)

Figure 1. CAD rendering and final product images of two complex scaffolds (red-PCL, gray-hydrogel). A) Hydrogel sandwiched between two layers of PCL, B) PCL fibers sandwiched between two layers of hydrogel.

Conclusions: This low-cost bioprinting modification for commercial 3D printers is a functional alternative to other bioprinting options. We have demonstrated the ability to construct complex polymer scaffolds made of thermoplastic and bioink materials. Printer efficacy can be improved by implementing a pneumatic pressure system to control bioink deposition.

References: [1] Hinton, T.J., Three-dimensional printing of complex biological structures by freeform reversible embedding of suspended hydrogels. Science Advances, 2015. 1(9).

[2] Murphy, S.V. and A. Atala, 3D bioprinting of tissues and organs. Nature Biotechnology, 2014. 32(8): p. 773-85.

Intelligent Rock Climbing Shoes That Provide Haptic and Interactive Feedback to Users with Lower Leg Prosthetics

Lauren Bartlett, Gianna Morrongiello, Meaghan O'Connell, Jiang Wu, PH.D, Ying Sun, PH.D
Department of Electrical, Computer, and Biomedical Engineering University of Rhode Island, Kingston RI, 02882
Correspondence email: yingsun@uri.edu

Introduction: The Providence Veterans Association (VA) Adaptive Sports Program provides veterans with means of rehabilitation through adaptive sports and activities. Active rehabilitation programs help handicapped veterans transition back to civilian life, reducing emotional stress and providing physical benefits. This program helps disabled veterans overcome perceived limitations, create a sense of pride, and challenge themselves by taking reasonable risks. One activity that the VA has looked to for active rehabilitation is rock climbing. Veteran amputees with lower limb prosthetics lack the proprioception required to perform such activities. A climber using a prosthetic cannot feel with their own foot when they are fully stabilized on the rock-climbing wall. For this reason, Jordan Anderson of the Providence VA asked the founding "iClimb" Capstone team of 2017 to create an intelligent pressure sensing rock-climbing shoe for contact detection of lower limb prosthetics. The innovation of this design is to create a cheaper and more lightweight alternative to pre-existing rock-climbing-specific prosthetic systems. The previously designed shoe is embedded with pressure sensors attached to the sole of the shoe that are able to measure the amount of force applied by the climber when the shoe comes in contact with the rock wall. The sensors relay the data based on the user's stability to Android Studio, which in turn produces an audio feedback "beep" via a smartphone Android application. The newly developed system was designed to better meet the specific needs and requirements of veteran climbers. As studies show that a sizable amount of veterans suffer from some type of hearing impairment, it was decided that the previous system would be modified to relay haptic feedback in lieu of auditory feedback [3]. The haptic feedback based on the user's stability is transmitted via vibrational motor discs positioned on a wearable neoprene thigh cuff. Therefore, the climber feels a more normal haptic sensation on the residual limb of the leg when stabilized. This system also eliminates the use of the Android application and the need for the user to carry a smartphone while climbing, enhancing safety and efficiency. The purpose of this design is to provide handicapped veterans with sensory guidance and assurance that they are well stabilized on the wall while rock climbing with a more usable and compact system. Additionally, this design is intended to provide an evaluation of the stability detection algorithm for user safety assurance.

Methods: The block diagram of the system is shown in Figure 1. The pressure sensors in the sole of the shoe are connected to a small printed circuit board (PCB). The PCB, which is attached directly to the shoe, is not a discomfort risk to the user because the prosthetic cannot feel its presence. The voltage data from the pressure sensors are read by a PIC18F4525 microcontroller on the PCB. The microcontroller determines the stability of the climber by evaluating the relationship of the pressure sensors in two parts of the sole. The pressure sensors in the sole are placed at the inner ball of the foot and the middle of the big toe due to the fashion of the foot placement while rock climbing. Climbers use the inside edge of the shoe placed on the edge of the rock climbing hold for the best stability [1]. If the algorithm of the PIC determines that the data from the climber is sufficiently stable, an output pin that is connected to the vibrating motors will be activated. The user will feel the vibrational feedback from the motor disks on his or her thigh only when the safest stability is determined.

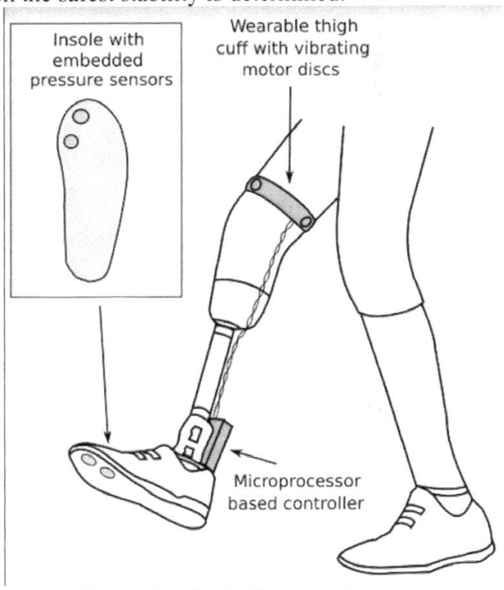

Figure 1. Block diagram of system

Results: The schematic shown in Figure 2 provides the details of the hardware circuit. These implementations make it possible to provide power to the vibrating motor discs and act as an on/off switch as the algorithm detects stability and instability. Software was developed in C to collect the data from the pressure sensors and analyze it with the microcontroller to evaluate the relationship between the two sensor points. When the data falls within the desired relationship based on the stability algorithm, the microcontroller activates the designated output pin that is connected to the motor driver and motor discs. The motor discs in the thigh cuff then vibrate, communicating to the user that they are in a stable position.

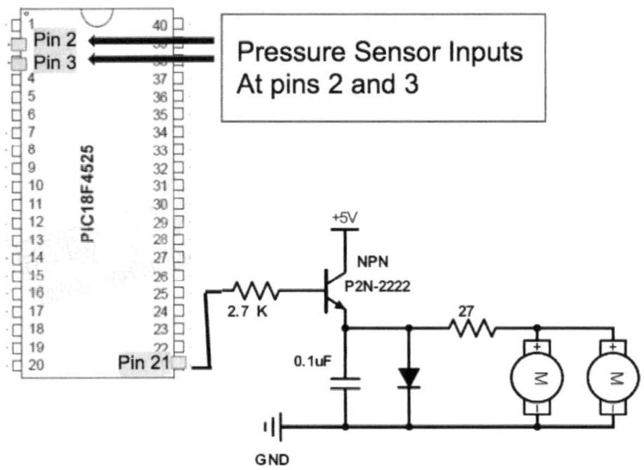

Figure 2. Schematic implementation to run the vibrating motor discs

As stated in the Methods section, two pressure sensors in the sole are placed at the middle of the big toe and the inner ball of the foot. Figure 3 shows an oscilloscope screenshot of a recording of pressure sensor data. The pressure sensor at the middle of the big toe (sensor zero) is represented by the orange line. The pressure sensor at the inner ball of the foot (sensor one) is represented by the blue line. This figure shows that when pressure is applied to the sensor, there is a voltage peak. This is followed by a return to the resting voltage state after pressure is taken away. In the stability algorithm, stability is determined to be sufficient once sensor zero is greater than 40% of sensor one and less than 60% of sensor one. In this recording, pressure was applied to sensor zero (orange) and slowly released at the same time at pressure was slowly applied to sensor 1 (blue). Therefore, the stability "sweet spot" is shown where the two lines overlap each other. Only at this specific time, stabilization is at its highest, and the motors will vibrate.

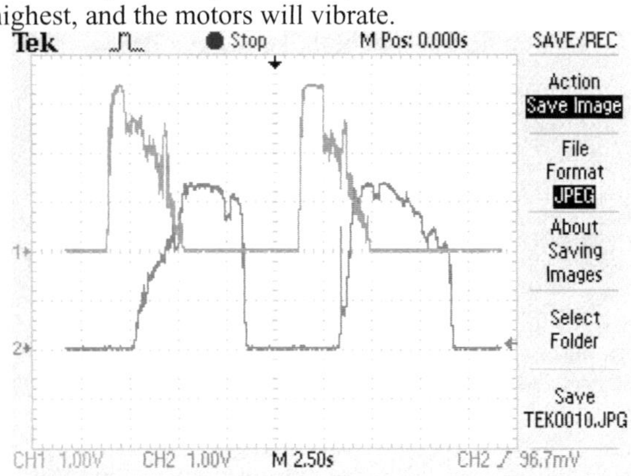

Figure 3. Oscilloscope screenshot of pressure sensor readings

Conclusion: Further testing is being conducted using various climbers to provide qualitative data evaluating the effectiveness of the stability detection algorithm of the system. This experimentation will test the confidence levels of the system's algorithm with subjective data from users, ranging from novice to experienced climbers. The testing collects data of false and true negatives and positives with respect to the users discretion. The main purpose of this experimentation is to ensure the quality of the product in order to protect the safety of the climber. A safe and reliable system will not produce any false verifications of stabilization for the climber. There are hopes of further product development in the future by the preceding students of the University of Rhode Island's biomedical engineering degree program. For example, as the system currently includes long wires connecting the PCB on the shoe to the thigh cuff, further development is necessary in order to construct a completely wireless system. Testing should also be continued using true lower limb prosthetic using veterans as human subjects.

References: [1]. Holden, Jillian, Emma Orton, and Riley Temple. *"Intelligent pressure sensing rock climbing shoe for contact detection of lower limb prosthetics",* 16 April 2018.

[2]. Fecht, Sarah. "New Prosthetic Lets Amputees 'Feel' Leg." *Popular Mechanics*, Popular Mechanics, 15 Feb. 2018, [Online]. Available: www.popularmechanics.com/science/health/a12060/feedback-system-lets-amputees-feel-prosthetic-leg-14821609/.

[3]. "Office of Research & Development." *VA Research on Hearing Loss*, 12 Aug. 2016, [Online]. Available: www.research.va.gov/topics/hearing.cfm.

[4]. R. E. Fan *et al.*, "A Haptic Feedback System for Lower-Limb Prostheses," in *IEEE Transactions on Neural Systems and Rehabilitation Engineering*, vol. 16, no. 3, pp. 270-277, June 2008.

Acknowledgments: The team would like to acknowledge Jordan Anderson of the Providence Veterans Association for her contribution to this design and her advice on how to carry out development and experimentation of the system. The team would also like to acknowledge Dr. John DiCecco and Joseph Maestri for their efforts and aid with software development and problem solving.

Catechins Inhibit Toxin Activity by Changing Secondary Structure of Toxin

Joanne Huang, En Hyung Chang, Angela C. Brown
Lehigh University

Introduction: *Aggregatibacter actinomycetemcomitans* (*A.a.*) is a Gram-negative bacterium that is associated with aggressive forms of periodontal disease. The bacterium releases a leukotoxin (LtxA) that allows the bacteria to evade the immune response of the host by targeting white blood cells. Catechins found in green tea have been found to inhibit the growth of some bacteria, and recent studies are showing evidence of its potential in controlling oral infections. However, the mechanism of this inhibitory activity is not yet known. In these experiments, we studied the interaction between LtxA and catechins using confocal microscopy and circular dichroism (CD) spectroscopy. Six types of catechin were tested: epicatechin (EC), epicatechin gallate (ECg), epigallocatechin gallate (EGCg), gallocatechin gallate (GCg), epigallocatechin (EGC), and catechin (C).

Methods: LtxA activity was measured with cytotoxicity experiments. LtxA was pre-treated with one of the six catechins for 30 minutes and then incubated with THP-1 cells. Cell viability was measured after three hours. To measure the binding between LtxA and cholesterol-containing membranes, giant unilamellar vesicles (GUVs) composed of POPC, 20% cholesterol, and a small amount of NBD-PE (green) were made to model THP-1 cell membranes. GUVs were incubated with LtxA labeled with Alexa Fluor 647 (red) for 30 minutes and then imaged using confocal microscopy. The secondary structural effect of catechins on LtxA was determined by incubating EGCg or C with LtxA for 30 minutes and analyzed using CD spectroscopy.

Results: We found that five out of the six catechins inhibited the activity of LtxA. Galloylated catechins were most influential in preventing the binding between LtxA (red) and giant unilamellar vesicles (green) (Fig. 1A). These same five catechins inhibited the activity of LtxA on THP-1 cells by significantly increasing cell viability when they were pre-treated with LtxA (Fig. 1B). Incubation of LtxA with catechin resulted in a change in the secondary structure of the toxin, preventing LtxA from binding to cholesterol.

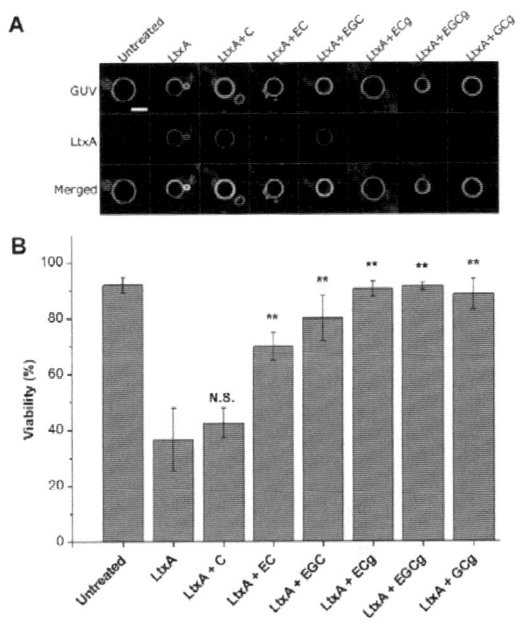

Figure 1. Interaction of LtxA with GUVs and THP-1 cells. A. Confocal microscopy images show the binding of LtxA (red) to GUVs (green) after the LtxA was pre-treated with catechin for 30 minutes. B. LtxA was pre-treated with one of the catechins and then incubated with THP-1 cells for 3 hours. Cell viability was then measured using Trypan blue assay.

Conclusion: Results indicate that catechins have the ability to inhibit the toxic and activity of *A.a.* by changing the secondary structure of its leukotoxin.

References:

1. Kachlany S.C. J Dent Res. 2010; 89(6):561-570.
2. Taylor P.W. et al. Food Sci Technol Bull. 2005;2:71-81.

Acknowledgements: This work was supported by a Clare Boothe Luce fellowship to JH and by the National Institutes of Health (DE022795 and DE025275) and National Science Foundation (1554417).

P3-12

Collagen Type-I Antibacterial Hydrogel for Wound Healing Applications

Dorota Jazwinska, Yoliem S. Miranda-Alarcón, and David I. Shreiber
Department of Biomedical Engineering; Rutgers University, New Brunswick, NJ

Introduction: According to the CDC, 31% of all healthcare associated infections are due to surgical site infections (SSI), and these account for about $5 billion dollars in costs annually.[1] The current treatments for SSIs include high doses of antibiotics, additional surgery, or possible implant removal, all of which happen after the patient has begun showing symptoms of infection. The presently used techniques which utilize antibiotics release it all into the entire body, increasing the chances of antibiotic resistance developing; however, a controlled release mechanism could inhibit the antibiotics ability to spread about the body and concentrate it only in the area under treatment. To this end, we propose the use of collagen type-I, as a biocompatible material with a versatile chemical composition that is able to self-assemble into a hydrogel at physiological conditions, to develop a preventative hydrogel with antibiotic controlled release to stop surgical site infections. We focus on antibiotics from the [2] -lactam family, due to their chemical structure and broad-spectrum neutralizing abilities. We aim to tether the antibiotics to type I collagen and have it presented within the collagen scaffold.

Materials and Methods: After synthesis, collagen materials were purified through size exclusion chromatography followed by concentrating the suspension with a concentrator. Antibiotic conjugated collagen (CarbColl) hydrogels were buffer to a pH of 7.4 at a concentration of 2.5 mg/mL and were further incubated at 37°C for 1 hour to allow for self-assembly. A serial dilution of soluble carbenicillin with a starting concentration of 270 μg/mL, equivalent to the concentration of antibiotic conjugated, was prepared using Luria broth (LB). The antibiotic-LB solution was then added to collagen hydrogels without undergoing conjugation. Similarly, controls of soluble carbenicillin were added to LB solutions. *E.coli* (NEB 5-alpha competent) 10% (v/v) in LB was incubated for 1 hour at 37°C prior to its addition to the gels. *E.coli* was plated on top of the hydrogel and incubated for 72 hours at 37°C. Optical density (OD) measurements of each well were taken at 595nm periodically for 72 hours. To confirm self assembly into a fibrillar hydrogel, the mechanical properties were measured using a rotation rheometer (Kinexus Ultra Rotational Rheometer-Malvern Instruments). A 200uL sample of each collagen suspension was plated on a temperature-controlled stage of a rotational rheometer, and the temperature was raised from 25°C to 37°C.

Results and Discussion: Preliminary data shows a direct dependence of bacterial inhibition on antibiotic concentration for soluble carbenicillin in LB (Figure 1). Similarly, soluble carbenicillin in the presence of collagen hydrogels show an increase in bacterial inhibition. Antibiotic conjugated collagen presented a similar trend to soluble carbenicillin in collagen hydrogels. Collagen controls with no antibiotic showed higher bacterial growth than the LB control. Mechanical testing showed a decrease in the shear modulus of the antibiotic conjugated gels compared to collagen type-I. This can be attributed to the conjugation of antibiotic to the collagen monomers, which compromises the ability of collagen to fully self-assembly.

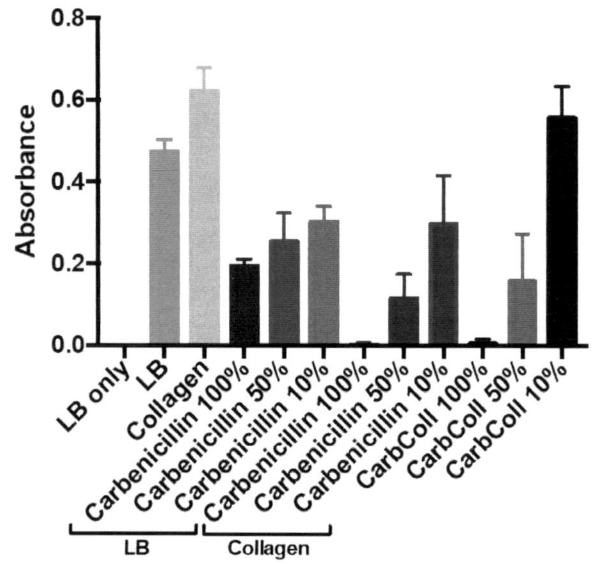

Figure 1. Optical density measurements at 72 hours of bacterial incubation with CarbColl hydrogels and corresponding controls.

Conclusions: In this work we have tested the abilities of antibiotic conjugated collagen hydrogels to inhibit *E. coli* growth. We also showed that the presence of collagen enhanced the antibacterial properties of the antibiotic under study. We wish to further study the antibiotic conjugated to collagen's ability to neutralize different bacterial strains.

References:

1. World Health Organization, 2016
2. Berríos-Torres SI. JAMA Surg. 2017;152(784-791)

Acknowledgements: This work is supported through the NIH-Biotechnology Training program (NIH T32 GM008339)

Effect of Muscle Cell Differentiation on AAV Transduction Efficiency

Kedar Patel, Alexandra Burr, Dr. Biju Parekkadan
Rutgers University- New Brunswick, Department of Biomedical Engineering

Introduction: Adeno-associated virus is a promising vehicle for gene therapy. For example, AAV has been explored for treatment of hemophilia with certain clotting factors through liver transduction and recessive disorders in the retina by encoding for the absent proteins. However, the field has struggled with developing a universal method for gene delivery through viral vectors. Host immune response and low transduction efficiency of host cells have delayed progress for effective therapeutics. An early target for stable delivery was post-mitotic muscle cells, due to longevity and stability. Myocytes provide a promising target for systemic release to the circulation. The purpose of this study is to increase the efficiency of AAV transduction in C2C12 (mouse myoblast cell line). Differences in transduction of muscle cells in 3 different states provide an avenue of increasing transduction efficiency: myoblasts, activated myoblasts and differentiated myotubes. Studies have shown that activated myoblasts are more apt to take up viral DNA and express the transgene. To increase activation, stimulants have been used. Theoretically, if the proportion of activated cells are increased through external stimulation, and they have stronger transduction rates, then a more robust method of transducing muscle cells can be developed. This paper demonstrates the difference in transduction rates during the muscle cell differentiation cycle, as a platform to test the effect of agents on stimulation and transduction in the future.

Methods: C2C12 cells were grown in 24-well plates. 3 different groups were cultured: myoblasts, activated myoblasts and differentiated myotubes. Myoblasts were seeded at 6.25% confluency and grown for 3 days in DMEM with 10% FBS. Activated myoblasts were seeded at 12.5% confluency and grown for 3 days in DMEM with 10% FBS. Myotubes were seeded at 50% confluency and grown in DMEM with 2% FBS. On Day 4, all cell groups were transduced with AAV2-CMV-GFP at dilutions of 1:10, 1:50, and 1:100. After 72 hours, the cells were analyzed using flow cytometry with a BD FACSCanto II (BD Biosciences, San Jose CA), against positive and negative controls for transduction efficiency. Additionally, cells were stained against the satellite cell marker, Pax7 with goat anti-human Pax and anti-goat AlexaFluor567 (Invitrogen, Eugene OR) and imaged with fluorescent microscopy. In addition to just confirming cell state, satellite cell markers would allow for an objective measure when manipulating cell state through additives.

Results: The transduction % was determined for each group. At the highest titer, activated myoblasts show the highest efficiency with 18.4% and myoblasts had the least with 8.68% (Fig1A). As seen in Fig 1B, the image on the left has greater Pax7 (red) expression than more confluent cells, indicating myoblast properties. Conversely, the myotubes express significantly less Pax7 which aligns with visible myotube formation and differentiation.

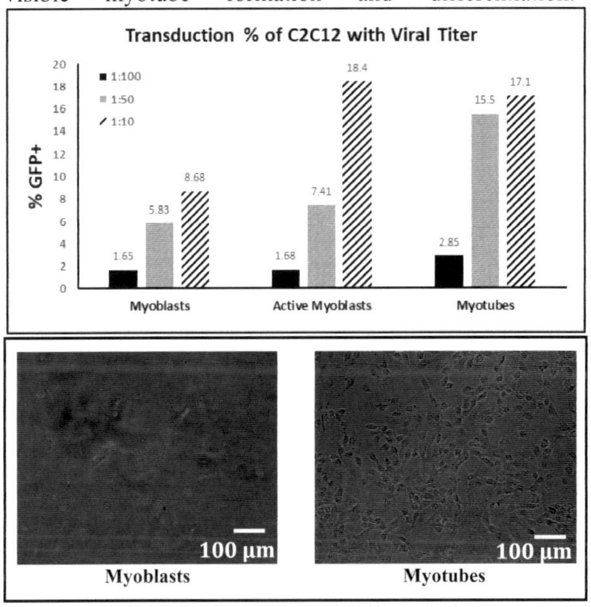

Figure 1: **(A)** Effect of C2C12 differentiation state on transduction efficiency with AAV-GFP using 3 dilutions of viral particles. **(B)** Microscopy images of anti-Pax7-AlexaFluor567 staining of C2C12 cells pre- and post-differentiation.

Conclusions: In showing that Pax7 expression is higher in myoblast cells and transduction efficiency is highest in activated myoblasts, potential for increasing efficiency of transduction using stimulants to activate myoblasts could boost efficiency and longevity. An increase percentage of activated myoblasts, as a subset of the entire cell population that would be targeted with an injection based delivery mechanism, should lead to higher transduction. This hypothesis can be further supported through experiment tracking biomarkers Pax7 and a satellite cell activation marker, MyoD, to confirm different cell states and assess the impact of additives on efficiency and longevity of transduction. Improvements to this project include reruns to conduct tests of significant and quantitative measures of Pax7 staining, through flow cytometry. The next step is to track the impact of additives on cells, through analysis of activation markers and transduction efficiency.

References:

1. Anguela X, High K. AnnuRev Medicine. 2019;70: 273-288
2. Yin H. et al. PhysiolRev. 2013;93(1):23-67
3. Willkomm L. et al. Stem Cell Res. 2014;12(3):742-753

P3-14

Significant Glycosaminoglycan Deposition in the Aorta of a Murine Model of Hutchinson-Gilford Progeria Syndrome Associates with Microstructural Alterations and Compromised Biomechanical Properties

Yuki Kawamura[1], Kristin Zimmerman[2], Demetrios T. Braddock[2], Jay D. Humphrey[1,3], Sae-Il Murtada[1]
[1]Department of Biomedical Engineering, Yale University, New Haven, CT, USA
[2]Department of Pathology, Yale School of Medicine, New Haven, CT, USA
[3]Vascular Biology and Therapeutics Program, Yale School of Medicine, New Haven, CT, USA

Introduction: Hutchinson-Gilford Progeria Syndrome (HGPS) is an ultra-rare premature aging disease resulting from a point mutation in exon 11 of the Lamin A gene (*LMNA*), which leads to aberrant splicing and consequent accumulation of a truncated protein called progerin[1]. HGPS results in death from cardiovascular events between 7 to 20 years of age[2], and vascular stiffening appears to be a pervasive mechanism underlying the vascular pathology associated with HGPS[3]. In this study, we investigate the microstructural basis of the severe vascular phenotype seen in HGPS by histopathological analysis of a knock-in $Lmna^{G609G/G609G}$ mouse model with the equivalent C>T point mutation.

Methods: Aortic samples were mounted and tested on a custom computer-controlled biaxial testing device for active and passive mechanical studies[4]. The samples were fixed after *in vitro* mechanical testing in 10% formalin, then stored in 70% ethanol at 4°C until embedded in paraffin and sectioned at 5μm. Sections were stained for glycosaminoglycan using Movat's Pentachrome stain, elastin using the Verhoff-Van Gieson stain, cytoplasm using Masson's Trichrome stain, and for collagen fiber undulation analysis using the Picrosirius Red stain. Images of the aorta were obtained using an Olympus BX51 microscope at 20x magnification and were analyzed using a custom MATLAB program that calculates constituent area density through background subtraction and pixel-based thresholding[5].

Results: End-stage (140 days old) HGPS aortas exhibited severe loss of elastic energy storage, a drastic increase in structural stiffness, and almost complete loss of contractility in comparison to both age-matched wild-type aortas and 100-days-old HGPS aortas. Histological analysis revealed significant reduction in medial thickness but a significant increase in adventitial thickness in HGPS mice between 100 days and 140 days. Medial glycosaminoglycan (GAG) cross-sectional area density in HGPS aortas increased five-fold or more in both thoracic and abdominal regions and was markedly different, whilst elastin and collagen densities displayed changes comparable to those in murine models of natural and pathological aging. In particular, immunohistochemistry revealed a significant increase in chondroitin sulfated GAGs in the medial layer. GAG deposition also increased drastically in end-stage HGPS aortic adventitia and contributed to its thickening. Overall vessel thickening was not the only ramification of increased glycosaminoglycan deposition; collagen fiber undulation exhibited a strong ($R^2 = 0.6089$) negative correlation with GAG area density, suggesting altered fiber properties as a ramification of GAG deposition and a potential explanation for severely compromised biomechanical phenotypes seen in end-stage HGPS aortas.

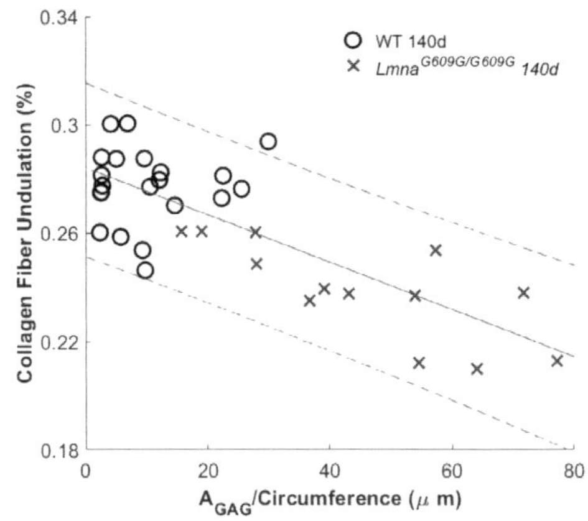

Figure 1. Scatterplot showing negative linear correlation between collagen fiber undulation and GAG area normalized over circumference. Dotted line denotes the 95% confidence interval.

Conclusions: Drastically increased glycosaminoglycan deposition in HGPS aortas associates not only with vessel wall thickening but also with microstructural reorganization and altered properties. Comparatively modest alterations in elastin and collagen area density in HGPS aortas between 100 days and 140 days of age mirror alterations in regular aging and underscore GAG deposition as a manifestation peculiar to premature aging in HGPS. GAG-mediated wall thickening contributes to increased structural stiffness of the aorta and is associated with straightening of collagen fibers, which may be the cause of compromised functional properties of end-stage HGPS aortas. Further studies of GAG accumulation in the aortic wall could aid to better understand the dramatic vascular pathology of HGPS aortas.

References:

1. Eriksson M. et al. Nature. 2003;423:293-298
2. Olive M. et al. Areterioscler Thromb Vasc Biol. 2010;30:2301-2309
3. Gerhard-Herman M. et al. Hypertension. 2011;59:92-97
4. Gleason, R. L. et al, J. Biomech. Eng., 2003;126(6):787–795
5. Bersi et al., Int J Adv Eng Sci Appl Math. 2013;4(4):228-240.

Acknowledgements:
The project was funded in part by the NIH grant HL105297. YK was supported by a research grant from the Masason Foundation.

Project Title: Kinetic Characterization of Nitrite Reduction to NO by the Molybdopterin Enzyme mARC2

Eric Cecco, Jesus Tejero, Mark T. Gladwin, and Courtney Sparacino-Watkins
Pittsburgh Heart, Lung, Blood and Vascular Medicine Institute, School of Medicine, University of Pittsburgh

Introduction: Nitrite is a nitric oxide (NO) precursor that is endogenously produced in human tissues. Endothelial dysfunction is a condition involved in cardiovascular disease that is characterized by a decreased production of NO in the vascular endothelium. New therapeutic strategies using nitrite can enhance NO levels and potentially treat endothelial dysfunction. However, the mechanisms involved in NO production from nitrite are incompletely defined and the Enzyme(s) responsible for biological activation of nitrite vary in each tissue.

Our lab identified a novel human nitrite-dependent NO synthase, the mitochondrial amidoxime reducing component (mARC) enzymes [1]. Human mARC-1 and mARC-2 are molybdenum (Mo)-dependent oxidoreductases that function as part of a three-enzyme metabolon with cytochrome b5 (CYB5) and cytochrome b5 reductase (CYB5R). However, we did not report kinetic values for human mARC-2 metabolon nitrite reductase activity. We hypothesize that the mARC-2 enzyme metabolon would exhibit similar kinetic characteristics to that of previously characterized mammalian molybdenum-dependent nitrite reductases. To test this hypothesis, we used isolated recombinant human mARC-2, CYB5, and CYB5R enzymes and NO-production rate measurements to determine (1) for affinity for nitrite, and (2) effect of pH.

Materials and Methods: Recombinant human mARC-2, CYB5, and CYB5R were isolated and NO formation rates were measured using a Nitric Oxide Analyzer (NOA, Sievers), as previously reported [1]. The mARC2/CYB5/CYB5 metabolon (Fig. 1*A*) was incubated in 50 mM Bis-Tris buffer with NADH at 37 °C with 1 mM NADH at a range of nitrite concentrations (10 pM to100 μM) (Fig. 1*B*) pH (5.9, 6.4, 6.9, 7.4, and 7.9) (Fig. 1*C*). The nitrite data was fit with Michaelis-Menten equation to determine the affinity for nitrite (Km). Heat denatured mARC-2 enzyme was used as a negative control and human mARC-1 as a positive control.

Results and Discussion: First we analyzed the mARC2/CYB5/CYB5 metabolon's affinity for nitrite using excess NADH as a reducing agent (Fig. 1A). We found that the NO-formation rates increased with increasing nitrite levels (Fig. 1*B*). Fitting the data generated a Km of 12.0 ± 2.6 mM NO_2^-, comparable to the previously reported established human nitrite-dependent NO synthases (e.g., xanthine oxidase [22.9 mM]; aldehyde oxidase [2.7 mM]; sulfite oxidase [80 mM]; and mARC1 [9.5 mM]) [2].

Next, we evaluated the pH dependence. Our lab and others have observed that the rate of nitrite reduction to NO is enhanced under acidic pH levels. Thus, we also measured the effect of pH on nitrite reduction to NO with the mARC2/CYB5/CYB5 metabolon. We found that NO-production was elevated at low pH (Fig 1C). While the mARC2 metabolon doesn't have the same magnitude of change as heme-based nitrite reductases (slope of 1.0), our data is in line with the molybdenum enzymes (slope ~ 0.7).

Conclusions: This study characterized the kinetics of the human mARC2 enzyme metabolon catalyzed nitrite reduction to NO. The reduction of nitrite to NO by the mARC2/CYB5/CYB5 metabolon has similar reactivity to previously reported molybdenum enzymes and is enhanced under acidic conditions, in line with mammalian nitrite-dependent NO synthase.

References: [1] Sparacino-Watkins, *et al., Nitrite reductase and NO synthase activity of the mitochondrial molybdopterin enzymes mARC1 and mARC2.* JBC, 2014.

[2] Maia, Moura. *Nitrite reduction by molybdoenzymes: a new class of nitric oxide-forming nitrite reductases.* JBIC, 2015.

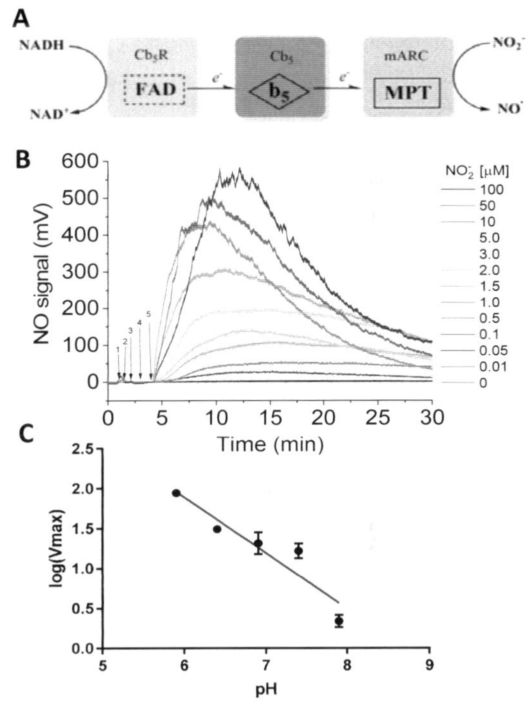

Figure 1 A. Diagram of the three-enzyme metabolon comprised of Cb5R, Cb5, and mARC (1 or 2). NADH oxidation is coupled to nitrite reduction to NO. Redox cofactors in each enzyme are displayed; flavin adenine dinucleotide (FAD), *b5* type heme *(b5)*, and molybdopterin (MPT). **B.** NO-formation by the mARC2 metabolon. Numbers represent the sequential addition of (1) nitrite, (2) 1mM NADH, (3) 4μM CYB5, (4) 1μM CYB5R, and (5) mARC2 (4.5-6μM). **C.** Quantified data of log of maximum rate of NO production at increasing pH levels, plotted with linear trend line

P3-16

Methods for Encapsulating Cells into Hydrogel Sheets for Wound Healing

Eddy Iturbide, Zeiny Aubdoollah, Shreya Soni, Ronke Olabisi, PhD
Department of Biomedical Engineering, Rutgers University, Piscataway, NJ, USA, 08854

Introduction: Hydrogels are cross-linked polymer networks that are widely used in tissue engineering due to their specific characteristics, such as hydrophilicity, biocompatibility, and tunable mechanical properties that are similar to a variety of soft tissues. Alginate is most commonly used to form cell-encapsulating hydrogels. Such hydrogels shield encapsulating cells from immune attack. Alginate can itself be immunogenic,[1] therefore we encapsulated cells within synthetic polymer hydrogels formed using polyethylene glycol diacrylate (PEGDA). Building on prior research whereby RIN-m insulinoma cells were encapsulated into sheets to deliver insulin for wound healing, we encapsulated additional insulinoma cell lines at 2 cell densities to optimize their secretion profiles from encapsulating hydrogels.

Methods: Rat insulinoma RIN-5F, which secretes only insulin and RIN-14B, which secretes only somatostatin were propagated in RPMI-1640 medium supplemented with 10% fetal bovine serum (FBS) and 1% w/v penicillin-streptomycin (pen-strep) in a tissue culture incubator. Cells were encapsulated into PEGDA hydrogel sheets by photopolymerization of precursor solution. Precursor solution was formed by combining 0.1 g/mL 10 kDa PEGDA (10% w/v) with (1.5% v/v) triethanolamine /HEPES buffered saline, 37mM 1-vinyl-2-pyrrolidinone, 0.1 mM eosin Y. The precursor solution was combined with cells at two different cell densities (1×10^4 or 4×10^4 cells/μL), pipetted into 1 cm^2 custom made molds and exposed to white light for 20 seconds to achieve 400 μm thick cell-laden hydrogel sheets containing $0.5x10^6$ or $2x10^6$ per 400 μm thick hydrogel sheets (Fig 1).

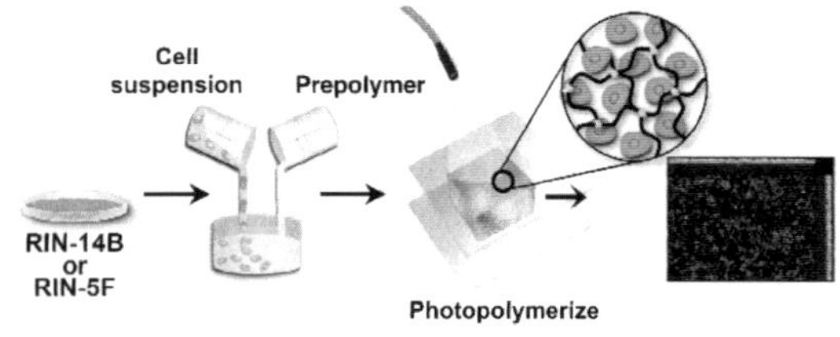

Figure 1. RIN insulinoma cells were expanded, harvested, combined with prepolymer solutions, photo-polymerized in masks, and kept in culture for 21 days.

Hydrogel sheets were maintained in a 37°C / 5% CO2 humidified incubator in complete culture media. Conditioned media (CM) containing insulin and/or MSC factors released from the cell-laden hydrogel sheets were collected on days 1, 7 and 21. HaCaT cells were propagated in DMEM, supplemented with 10% FBS (v/v) and 1% pen-strep (v/v), seeded onto 24 well plates at $1.8x10^5$ cells/well in tissue culture incubators. Scratch assays were conducted with human keratinocytes as previously described with conditioned media containing insulin.[2] Media from empty hydrogels was used as negative controls. Insulin levels were assessed with ELISA. HaCaT cell migration across the scratch was imaged at 0, 24, 48, 72- and 96-hours using phase contrast microscopy and analyzed on NIH ImageJ software.

Results: ELISA results showed that the maximum insulin released from RIN-5F hydrogel sheets was 15.2±3.3 ng/mL/10^6 cells at $2x10^6$ cells per sheet, which compared well to RIN-m encapsulated at $0.5x10^6$ cells per sheet, which released 16.3±1.1 ng/mL/10^6 cells. At $2x10^6$ cells per sheet, RIN-m had released only 4.6±0.2 ng/mL/10^6 cells. The maximum insulin released from RIN-14B was 1.6±0.2 ng/mL/10^6 cells. Scratch assays showed scratches closing faster when exposed to conditioned media from RIN-5F cells (Fig 2).

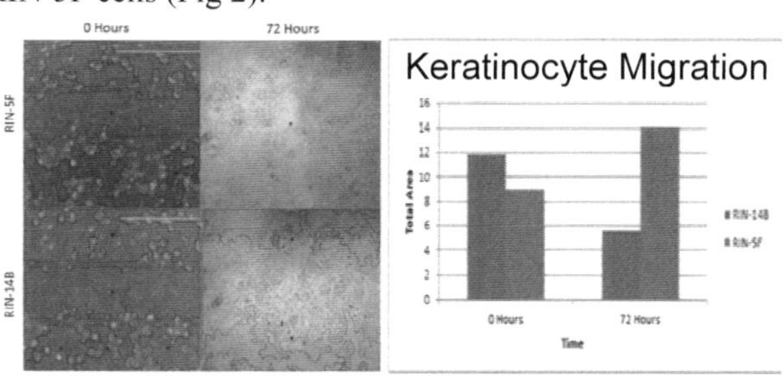

Figure 2. Images: Scratches at 0 and 72 hours for RIN-5F and RIN-14B cells with analyzed area from the ImageJ software. Graph: Keratinocyte migration induced by RIN-14B and RIN-5F cells.

Conclusions: ELISA results demonstrated insulin could exit the hydrogel sheets while scratch assays demonstrated that the insulin exiting was bioactive. The study here assessed cell function rather than viability, as our priority is insulin release. Hydrogel encapsulation did not impede insulin release and function.

References:

1. Ménard M, et al. *JBMR B: Appl Biomat*. 2010 May; 93(2):333-40.
2. Aijaz A, et al. *Tissue Engineering Part A* 21.21-22 (2015): 2723-732.

Acknowledgements: This work was supported in part by a Rutgers Aresty Undergraduate Research Award and an NSF CAREER Award (CBET-1752079).

Development of a New Approach to Characterize Hearing Loss Using Overlaid Spoken Words

Mohammed Janoudi, Liam Stamp, Justin Jewell, Jiang Wu, Ph.D., Ying Sun, Ph.D.
Biomedical Engineering Program, Department of Electrical, Computer and Biomedical Engineering
University of Rhode Island, Kingston, RI 02881, USA. Correspondence email: yingsun@uri.edu

I. Significance

Bilateral speech frequency hearing loss is a problem that affects 11.2% of the US population aged from 50-59 and drastically increases to 24.7% for individuals aged from 60-69 [1]. Noise related hearing loss is the leading cause in the workplace. Hearing loss is typically most prevalent on the higher frequency end.

Modern hearing tests use methods such as a tuning fork or audiometer test to produce tones. The method described in this study uses spoken words to test specific frequencies for hearing loss. This approach was based on the viral audio illusion 'Laurel - Yanny'. The audio clip observed nearly half of the listeners hearing 'Laurel' and the other half hearing 'Yanny'. This study strives to create a new method of testing for hearing using spoken words.

The issue in most tone tests today is that they do not give any meaningful information with relation to communication. Human hearing commonly ranges from 20 - 20,000 Hz [2], however, most humans cannot make sounds outside of about 9 KHz [3]. A tone test will test many frequencies between 20 and 20,000 Hz. This test could inform someone if they cannot hear specific sounds in nature that have similar frequencies to the tones used, however, it is very tough to decipher how this will impair one's ability to communicate audibly. The test described in this study aims to use spoken words to test within the spoken frequency spectrum. This will help determine whether a person will have trouble hearing specific words because of the frequency they must be spoken at. In this project, an Android application was created to test four different audio clips comprised of seven different words between the frequencies of 20 and 8,000 Hz. These audio clips are words overlaid on top of other words. Through internal testing, it is believed that these clips will sound like one word to a healthy ear and will sound like a different word to an unhealthy ear. Table 1 shows the words that are currently being used. More words are in the process of being created for more frequencies.

WORD A	WORD B	TEST FREQUENCY
MOM	SOX	1000
SHOE	WHO	3000
STAY	FRAY	5000
MOM	TOP	7000

Table 1, Words that will be used

II. Design Description

A. Word Choice

Words with four or less letters were selected because they usually only have a few consonants. This prevented sudden frequency changes from stress marks that divide consonants allowing for tighter frequency ranges. Consonants are sounds in the English language that help create syllables in words. They offer breaks that are important for the spoken English language. Words with less consonants often lead to words with less syllables which leads to less changes in frequency.

Words that include constants such as s, f, and x generally use a wide range of frequency. This is because those consonants require a higher frequency sound, and when combined with vowels, can create a wide sound frequency. The words needed for this test required a narrow frequency range.

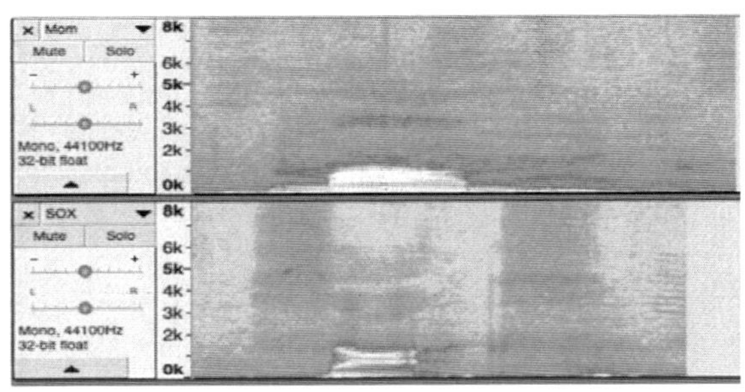

Fig 1. Spectrogram of two overlaid words, "Mom", and "Socks"

An example of a word and its overlay that was used is "mom" and "socks". In this example, "mom" is a low frequency word, only having a frequency range from about 0 to 1,000 Hz. "Socks" is a word with a higher frequency range that is outside the 0 to 1,000 range that "mom" is in, allowing the filtering of the lower frequency while still retaining the higher frequency. When these two words were overlaid on top of each other and filtered at 1,000 Hz, the "m" sound at both the beginning and end of "mom" would disappear, allowing the sound of "s" and "cks" to dominate.

B. Digital Filtering Using Matlab

Taking the words from the original recording to the final test required signal processing tools which were available in Matlab. The first step was to make sure that both words for the test were the same length. This was done by trimming the unnecessary signal from before and after the word to remove noise and comparing the number of samples. If both discrete time sample values were within 100 samples then the filtering process would begin. Each word is put into a bandpass filter to make the frequency range as narrow as possible while still being able to understand the word. Then the words are overlaid and the power of each adjusted until the target frequency is the only audible word to a healthy ear. To check the audio test a bandpass filter is used to eliminate the target frequency and to ensure the other sample is audible.

Fig. 2. Flowchart of audio recording to final product.

C. *Application Design*

The Android application we have created aims to be as simple and user-friendly as possible. An Android application was chosen for the delivery of the test for convenience of use, anybody with a smartphone can have access to this hearing test. The app itself does not do any of the audio filtering, it is just a medium to present audio clips to the user and provide immediate feedback.

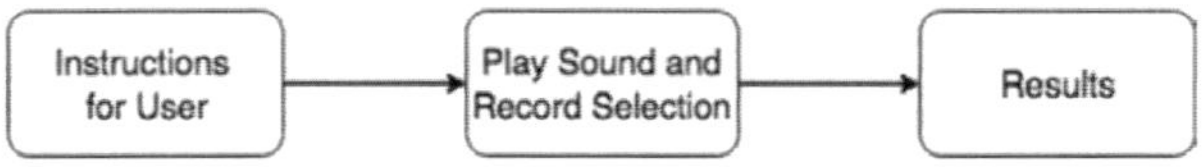

Fig. 3. Android application flowchart

Figure 3 outlines the process of using the designed application. After this process is repeated for every sound clip, the results are compiled and displayed graphically to the user. The graphical representation spans the frequency range we are testing at, with a green circle indicating a correct choice, and a red circle indicating an incorrect choice, as displayed in Figure 4.

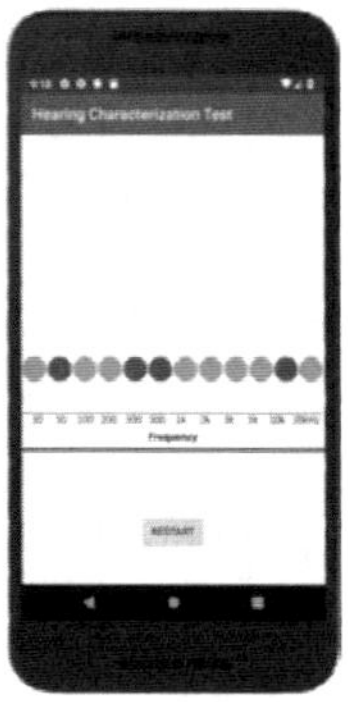

Fig. 4. Android application result screen

The results screen provides immediate feedback to the user. This application is paired with a second, control application that has been created from the same template, with audio clips utilizing pure tone audiometry. These tones are at the same frequencies as the test words, and the results from the control study will be compared to the overlaid word study to determine a relationship between affinity for hearing at specific frequencies, and the ability to distinguish overlaid words.

III. Design Criteria

The proposed design of this product had convenience as its top priority. Many hearing tests currently used can only effectively be done within a clinical setting. This application aims to solve that problem present within this industry.

Another priority was having an easy-to-use program. This needed to be something that anyone can pick up, use and get immediate feedback. With a results screen that shows exactly where mistakes were made, the user knows better of how their hearing loss is affecting their ability to communicate.

IV. Results

The results produced ten audio test spanning up to 8,000 Hz. The immediate proof of concept came from running each test through a filter to imitate what someone with a hearing deficiency would hear. These worked as intended, as filtering out the target frequency resulted in hearing the wrong word. Unofficial tests have been done on various students within the Biomedical engineering program and the clips seem to stay consistent with the expected results. Someone with a diagnosed hearing deficiency is yet to be tested. Many of those tests are on hold until an official IRB has been approved.

V. Future Implications

This project contributes a new approach to characterizing hearing loss. The ability to have this technology within an Android app makes it easy to access. Now a hearing test can be done anywhere there is an Android device and headphones. There is room for improvement too with making the word tests more exact to a frequency and creating more words to test. This test also removes the abstract noises and has the user select exactly what they are hearing.

REFERENCES

[1] National Health and Nutrition Examination Survey, 2011 - 2012. analysis reported in JAMA Otolaryngology- head & neck surgery, December 2016. https://www.NIDCD.NIH.gov

[2] Rosen, Stuart (2011). Signals and Systems for Speech and Hearing (2nd ed.). BRILL. p. 163. For auditory signals and human listeners, the accepted range is 20Hz to 20kHz, the limits of human hearing

[3] Hall, Jeff. "High Frequency Hearing Loss." ZipHearing Blog, 7 Oct. 2018, www.ziphearing.com/blog/high-frequency-hearing-loss/.

[4] D. Pressnitzer*, J. Graves, C. Chambers, V. de Gardelle, and P. Egre Laboratoire des Syste` mes Perceptifs, Departement d' etudes cognitives, ENS, PSL University, CNRS, Paris, France *Correspondence:daniel.pressnitzer@ens.fr https://doi.org/10.1016/j.cub.2018.06.002

P3-18

Monitoring coral health via allosteric transcription factor biosensor technology

Kevin Lorch, Melanie Martinsen, Pulkit Mittal, Rachel Petherbridge
Boston University, College of Engineering

Introduction: Rising levels of air and ocean surface temperatures due to climate change have led to increased coral bleaching events, severely impacting coral communities around the globe. Bleaching occurs when coral is under stress for prolonged periods of time, and, unfortunately, the state of coral stress prior to bleaching is currently undetectable. A monitoring capability for early stress detection is the first step towards mitigating future bleaching events. This project leverages genetically encoded biosensors derived from bacterial allosteric transcription factors that can detect dimethylsulfoniopropionate (DMSP) - an organosulfur compound released by coral undergoing stress - in real time, and provide a fluorescent signal proportional to the concentration of DMSP. Future deployment of these sensing devices has the potential to halt further ocean acidification, thus preserving the biodiversity of our aquatic ecosystems.

Methods: Current methods of measuring DMSP require spectroscopy[1] and are therefore incapable of providing real-time stress measurements in the field; thus, we measured DMSP concentrations indirectly through detection of its metabolites. Our approach was biomimetic, leveraging the mechanism used by a species of marine bacteria, *Rhodobacter sphaeroides*, to metabolize and consequently sense DMSP in their environment. *R. sphaeroides* catalyzes the metabolism of DMSP with a DMSP-lyase, DddW[2], and one of the resulting metabolites, acrylate, regulates the bacterium's gene transcription by binding to AcuR, an allosteric transcription factor. We used both DddW and AcuR from *R. sphaeroides* to serve as the basis of the DMSP sensor. To observe the molecular interactions of our sensor, we employed a technique developed in the Galagan Lab at BU where allosteric transcription factors, such as AcuR, are attached via polyhistidine-tags to quantum dots (QD); using QD technology, we monitored in real-time the fraction of QD-AcuR complexes bound to AcuR's DNA binding sequence, acuO, versus our inducer, acrylate. Fluorescence Resonance Energy Transfer (FRET) was analyzed to transduce QD-AcuR binding events to a light-based signal, providing an indirect measure of DMSP. To characterize our sensor, we created theoretical dose response curves (acrylate v. FRET sensor output) using Matlab's ode45 software tool (Mathworks, Natick, Massachusetts). These were created from modelling the binding kinetics of the biosensor.

Results: Using Biolayer Interferometry (ForteBio Octet), we verified binding of AcuR to acuO and determined the dissociation constant, K_D, of AcuR to acuO, to be*4*. × ⁻ *M*. A Fluorescence Lifetime Spectrometer (Edinburgh Instruments LifeSpec II) was used to empirically measure the FRET output of the molecular sensor (QD, AcuR, and acuO) in the presence and absence of acrylate. Normalizing the sensor FRET output to be 1 in the absence of acrylate, the sensor output in the presence of 100 nM acrylate was found to be 0.47. As a negative control, scrambled DNA was tested and yielded an output of 0, indicating no non-specific binding of scrambled DNA to AcuR. To validate our experimental results, we modeled the sensor output as a function of acrylate concentration.

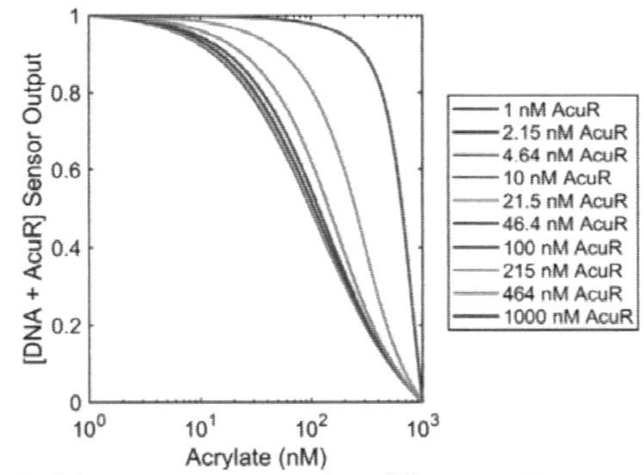

Theoretical dose response curves given different starting concentrations of AcuR. Initial DNA concentration is set to 464nM.

Numerous starting values of DNA and AcuR yielded similar dose response curves with equivalent linear ranges and IC50's, given the dissociation constant of AcuR to acuo specified previously. We also mathematically explored sensor response to different dissociation constants and found that lower dissociation constants led to greater variation of dose response curves over varying initial values of AcuR. Such a change in dissociation constant could be accomplished by altering the binding pocket of AcuR to acuo, or AcuR to acrylate. Using our mathematical model, we found a variety of potential sensor set ups capable of sensing acrylate, and therefore DMSP, across concentration ranges stretching from single nanomolars to micromolars.

Conclusions: By empirically verifying our sensor's ability to detect acrylate, and thus DMSP, as well as modelling the molecular binding to recapitulate experimental data, we envision our sensor has great potential to be employed as a tool for real-time detection of coral stress. As a next step, we plan to integrate this biosensing mechanism into a portable, waterproof, and cost-effective device to be used for rapid sampling in the field during reef surveys. Such a device would fill the need for an affordable, point-of-sampling sensor capable of quantifying DMSP levels quickly and without the need for additional lab equipment. Enabling early detection of coral stress would aid in preventing bleaching events, or at least mitigating their impacts, as local stressors such as pollutants or erosion could be removed upon detection of coral stress, improving chances of reef recovery. Coral sensing could prevent potential economic, environmental and pharmaceutical damages from a loss in reefs and associated biodiversity.

References:

1 Tapiolas DM et al., *J. Exp. Marine Bio.* 2013;443:85-89.
2 Todd JD et al., *The Isme Journal.* 2011;6:223

Establishing a Three-Dimensional Collagen Based Cell Culture of Human Monocytes

Blair Landon[1], Yun Jiao[1], Laura Simpson[1], Jennie B. Leach[1], Gregory L. Szeto[1,2].
[1]Department of Chemical, Biochemical, and Environmental Engineering, University of Maryland, Baltimore County, Baltimore, MD, USA. [2]Marlene and Stewart Greenebaum Comprehensive Cancer Center, University of Maryland School of Medicine, Baltimore, Maryland, USA.

Introduction: Cells in the human body live in tissues, three-dimensional (3D) environments densely packed with cells and proteins. The physical features of this environment affect the biology of cells, changing how they sense and respond to conditions like trauma or infection. Typically, cells are tested *in vitro* on two-dimensional (2D) plastic surfaces or as liquid suspensions. Recent studies show that 3D cultures cause significant differences in cell structure and function compared to 2D culture, and cells in 3D more closely mimic *in vivo* phenotypes[1,2]. Here, we characterize how human monocytes (U937 cell line) change when grown in 3D compared to 2D. In this study, we vary conditions in a collagen-based 3D cell culture model to determine their effects on monocyte growth and cell viability. We also compare culture models by measuring activation levels of the monocytes after incubation and stimulation with LPS. Our long-term goal is to use these 3D cultures to dissect the immune-related risk factors for diseases such as Alzheimer's.

Methods: A 40 μl collagen gel was used as our 3D culture model. The gel consisted of type I collagen, DMEM, sodium bicarbonate, U937 cells, a human monocyte cell line, and water. Gel components were mixed together using a pipette and placed on circular cover slides inside multi-well plates. Mixed gels were incubated for 30 minutes at 37°C to polymerize. Five hundred microliters of 1x DMEM was added per well.

After incubation for 24, 48, or 72 hr, gels were dissolved using 4 mg/ml collagenase type I in HBSS for 10 minutes at 37°C. Solutions were centrifuged at 300 *g* for 5 minutes to pellet cells. Viability of recovered cells was determined using trypan blue exclusion and a Countess II FL automated cell counter.

Results: We expected cell viability would be positively influenced by increasing collagen concentration and initial cell seeding density. Collagen was tested at 1 mg/ml and 2 mg/ml concentrations, while initial cell seeding density was varied from 100,000 to 500,000 cells per 40 μl gel. Our data showed that cell viability depended on both collagen concentration and initial seeding density. Viable cell recovery at 48 hr was significantly higher for 2 mg/ml gels seeded with 250,000 cells (p=0.006) or 500,000 cells (p=0.038) when compared to 1 mg/ml gels seeded at the same densities. Cell recovery at 24 hr was also significantly higher for gels seeded with 500,000 cells (p=0.03; Figure 1A). Cell growth inside the gels was normalized to a 2D control with matched seeding density. We observe a significant decrease in cell growth at 24 hr in 1 mg/ml gels at 500,000 cell seeding density compared to that of the 2D control (p<0.019). There were also significantly fewer cells at 48 hr in the 1 mg/ml and 2 mg/ml gels compared to 2D control at a seeding density of 100,000 cells (Figure 1B, p<0.001).

To determine if collagen concentration alone affected cell viability, we kept the seeding density at 250,000 cells and varied collagen concentration from 1 mg/ml to 3 mg/ml. Gels were incubated for 24, 48, or 72 hr. There were no significant differences in cell viability between different collagen concentrations. Growth curves at each collagen concentration were fit to a non-linear exponential equation. The rate constant and plateau of each curve were not statistically different from each other (p>0.05). Altogether, our results demonstrate that U937 human monocyte viability in 3D culture was dependent on time and independent of collagen concentration.

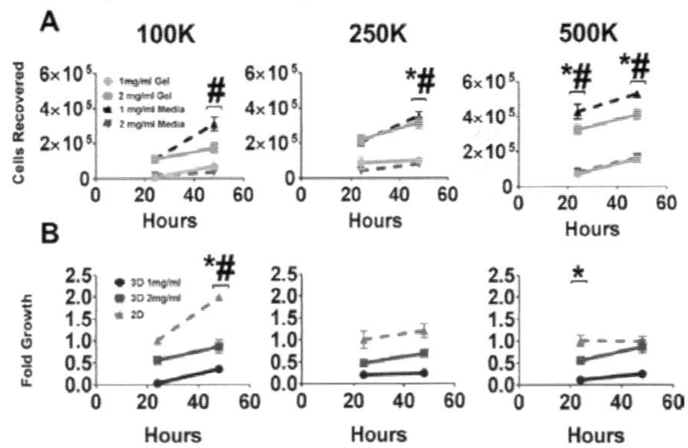

Figure 1. (A) Cell recovery from different initial cell seeding densities and collagen concentrations. # denotes difference between the two media measurements. * denotes difference between the two gel concentrations. (B) Cell growth in 3D cultures normalized to 2D controls. * denotes significant difference between 1 mg/ml and 2D. # denotes significant difference between 2 mg/ml and 2D.

Conclusions: Our data demonstrate that collagen concentration alone has no effect on cell viability of encapsulated U937 cells. This suggests that collagen is favorable for 3D cultures because it does not decrease cell viability. There is more cell growth in 2D compared to 3D cultures . In future studies, we will test the hypothesis that 3D cultures of monocytes are better mimics of *in vivo* cytokine production in response to disease-related tissue inflammation.

References:

1. Baharvand H. Int J Dev Biol 2006;50:645-652
2. Edmondson, J.J. Assay Drug Devl Technol. 2014;12:207-2818

Acknowledgements: This study was supported by NIH/NIGMS MARC U*STAR T34 HHS 00001 National Research Service Award to UMBC

The Automation of the OculoMotor Assessment Tool

Nabih Armout*, Christopher Morris*, Taiga Akiyama*, Anthony Chirayath*
John Vito d'Antonio-Bertagnolli, M.S., Chang Yaramothu, Ph.D., Tara L. Alvarez, Ph.D.
New Jersey Institute of Technology, Newark, NJ; *Equal Contribution

Over 62,000 concussions occur in high school sports every year [1]. Many of these injuries go undiagnosed, but there have been significant correlations between the presence of concussions and oculomotor deficiencies [2]. To confirm these past correlations in data, we have created a device to establish a normative data set to facilitate the diagnosis of concussion. The device will work with the OculoMotor Assessment Tool (OMAT) to objectively count the maximum number of eye movements the subject makes between two fixed points in a fixed duration of time.

I. Significance

Oculomotor dysfunctions associated with traumatic brain injuries can cause accelerated fatigue of the eye in performing the OculoMotor Assessment. The OMAT is a tool that provides visual targets in specific locations as demonstrated in Figure 1. Current studies suggest that an individual affected by mild concussions will fatigue much faster and thus have a lower number of saccades. During the Assessment, the clinician performs two tasks, counting the number of eye movements and keeping track of time. The requirement of the clinician to perform these two tasks simultaneously can lead to error. This device will remove the need for the clinician to track both the time and number of movements as the device will do both.

Figure 1: OculoMotor Assessment Tool (OMAT)

II. Design Description

The device is a pair of goggles that will use infrared light coupled with cameras capable of recording in the infrared range to track the movement of the eyes during the test. The device will work in conjunction with the OMAT to ensure that the targets the patient must look at will be consistent in every test. The hardware components include a custom 3D printed mount that attaches to the goggles, two Raspberry Pis, two infrared cameras, one Raspberry Pi display, and four infrared-emitting diodes (IREDs). This design can be seen in figure 2. The patient will wear the goggles, and the OMAT will extend away from the patient's face in the same manner seen in Figure 1. The patient will alternate looking at the left target and the right target as many times as possible in sixty seconds. The test is then repeated with the OMAT rotated 90 degrees. The custom-built software will calculate the number of movements completed and display that number. Two IREDs are mounted onto each side of the rims of the goggles with 3D printed mounts; each pair will illuminate the two corners of each eye. One camera is positioned in between each pair of IREDs, also using the 3D printed mount.

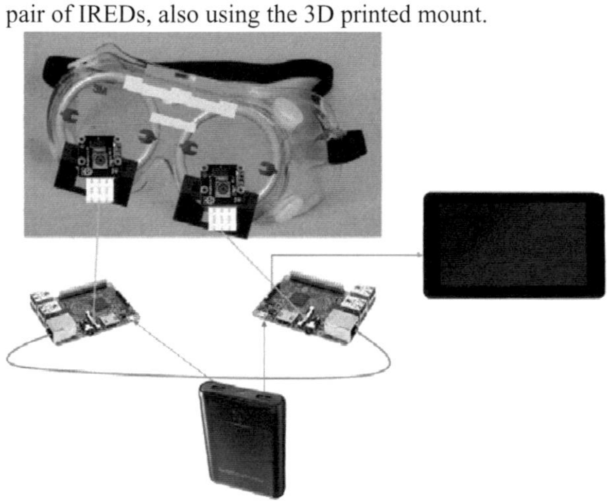

Figure 2: System Diagram

Software requirements ensure that the camera can track saccades as the participant performs the OMAT protocol. The eye tracking algorithm operates using a Raspberry Pi for each eye. It receives data from each camera the follows the flow outlined in Figure 3; specifically, 1) capture eye image, 2) convert image to grayscale, and 3) threshold image to binary, where the pupil is the darkest spot on the image, and a correct thresholding value will result in the areas defined as one being the location of the pupil, and zero being the location of the rest of the image.

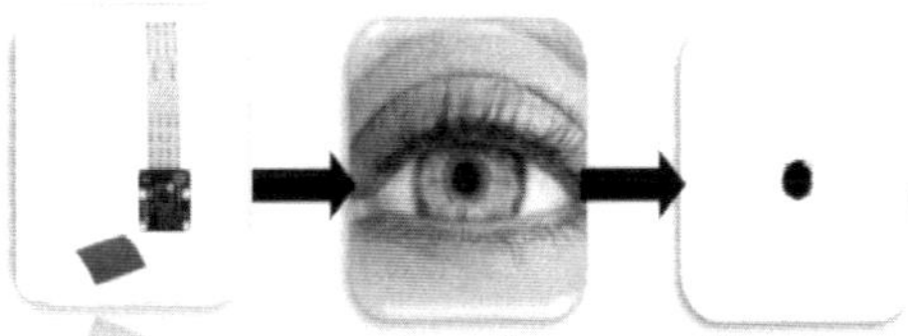

Figure 3: Image Conversion

Using this image, the centroid of the portion of the image that is the pupil is calculated and saved to an array. Before the test is run, a calibration occurs, where the location of the pupil when looking at both targets and when looking at the center between the two targets is calculated in the same manner that the pupil location is calculated during the test. Once the eye tracking is complete, the code checks when the eye location met the criteria to record a saccade. A saccade is defined as one movement from one fixed point to a second fixed point. The number of saccades is calculated by measuring how many times the pupil location went from the area defined as one target to the area defined as the other target whilst also passing through the center. The number of saccades completed during each 30 second test is displayed on the screen.

III. Design Criteria

The device's success is measured by a correct output of the number of saccades completed in a certain time interval. The needs of the project include accurately tracking eye movements while working in accordance with the OMAT protocol. The device should also be portable, inexpensive, and reliable. The design for the project was selected due to its strength in cost and portability. Utilizing Raspberry Pi-s instead of laptop computers reduce the cost while also increasing the portability of the system. Constraints of the project include the output of the power supply and the infrared LEDs in order to optimally illuminate the eye for tracking while remaining within safe limits for use with the eye. The constraints of the 3D printing material depend on the model and precision of the 3D printer utilized.

IV. Results

Structural weaknesses were identified through iterations of 3D printed prototypes. Test results revealed that the configuration of the goggle clip attachment and the IRED mounts easily broke. These issues were addressed in further iterations of the design.

According to ANSI Z136, the safety limit of IREDs on a human eye is 10 mW/cm^2. The IREDs of this instrument will be at a maximum of 1.2 mW/cm^2. The IREDs will be tested individually to ensure safety. Initial testing of the IREDs were measured to be at 4.5 mW/cm^2. This value is under the limitation of the standard, but higher than the limit set by needs in the project and other commercial products. Adjustments need to be made on the power supply as well as in the circuit.

V. Future Implications

Final iterations of the hardware and software are still in development. After achieving functionality, the project will begin trials to collect baseline data and ensure that the device functions as intended. Once the device is operational, it will be used to collect data on subject both with and without concussions. If successful, the device can be used in many situations outside of laboratory settings.

References

[1] Sports-Related Concussion and Head Injury Fact Sheet and Parent/Guardian Acknowledgement Form. (n.d.). Retrieved February 7, 2019, from https://www.nj.gov/education/aps/cccs/chpe/concussions/fact.pdf

[2] Ocular motor assessment in concussion: Current status and future directions Ventura, Rachel E. et al.
Journal of the Neurological Sciences , Volume 361 , 79 - 86

Enhanced properties of liquid infused paper for bacteria handling and point-of-care diagnostics

Emily LeClair,[1] Daniel Regan,[1,2] Caitlin Howell[1,2]
[2]Graduate School of Biomedical Science and Engineering, University of Maine, Orono, Maine

Introduction: Paper-based methods of bacterial detection are a dependable option for point-of-care (POC) diagnostics. Current approaches to POC microbial diagnostics are often hindered by low sample concentration and adhesion challenges, due to the tendency of these organisms to adhere to most surfaces. Nonetheless, issues such as the rise of antibiotic resistance increase the urgency for innovative POC microbial diagnostic systems. One solution may be to use slippery surfaces inspired by *Nepenthes* the pitcher plant, which uses an immobilized layer of water to cause insects to slide down into its cup for digestion (**Figure 1A**, left).[1] Recent work has shown that these layers can be reproduced on simple polymers in the lab (**Figure 1A**, right).[2,3] Recent work in our labs has suggested that these surfaces can also be made from low-cost polymers coated onto paper, leading to the design of cost-effective material for use in POC bacterial diagnostic tests,[4] while also leveraging Maine's paper resources. This project explores the use of bio-inspired 'slippery surfaces' on paper substrates toward the creation of a low-cost POC microbial diagnostic system.

Methods: Single 50 µL droplets of water containing blue dye were placed on two types of paper that were either untreated (control) or coated with a polymer and treated to have a liquid layer. Images were taken from directly above and to the side to observe droplet shape. The adhesion of green-fluorescent-protein (GFP)-expressing *Esherechia coli* to the surfaces was tested by running a 50 µL droplet containing ~1×10^8 cells/mL down a either a control surface (polymer coating alone) or a surface which had a liquid layer. The samples were then examined using several fluorescent microcopy images stitched together to provide a wide-field image of the sample surface.

Results: Droplets placed on the untreated paper surfaces were absorbed into the paper. After creation of a liquid layer on the surface, however, the droplets maintained their shapes, as seen in **Figure 1B**. In **Figure 1C**, the results of the test of bacterial adhesion are shown. On the control surfaces, path of residual droplets remained where the initial droplet had travelled across the surface. Further inspection also showed individual bacterial cells between the residual droplets. On the surface treated with a liquid layer, however, no observable bacteria were left behind.

Conclusions: We have shown how creating a slippery liquid layer on paper can reduce the adhesion of droplets and bacteria on surfaces. The liquid-infused surfaces were shown to decrease absorption into the substrate as well as inhibit bacterial adhesion. These results suggest that liquid-infused paper could help create cost-effective POC diagnostics for microorganisms by making bacterial samples easier to handle and transferring all bacterial cells to the desired location. Future work will include tests on other microbial species and investigations into more POC-specific droplet handling techniques.

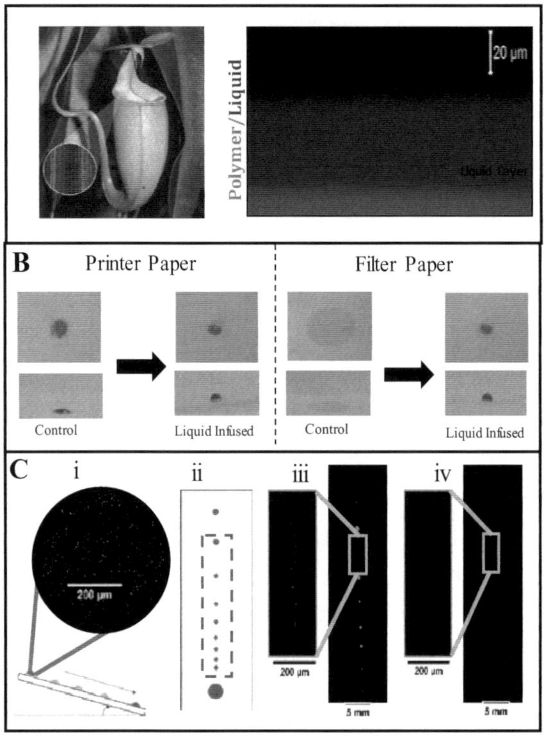

Figure 1: (A) The surface chemistry of a pitcher plant (left)[1] serves as the biological inspiration for lab-made 'slippery' liquid surfaces (right).[2] **(B)** Effect of liquid-infused substrate on droplet properties. **(C)** Microscopy from residual bacteria test. (i) Schematic of the sliding droplet with an image of the initial bacterial solution. (i) Schematic of where on the sample the microscopy images were taken. (iii) The control surface after droplet sliding. (iv) The liquid surface after droplet sliding.

References:

1. Bohn, H. et al. .*PNAS* 101.39 (2004): 14138-14143.
2. Sotiri, I., et al. *Biointerphases* 13.6 (2018): 06D401.
3. Kovalenko, et al. *Advanced Healthcare Materials* 6.15 (2017): 1600948.
4. Regan, D. et al.. *Curr. Op. Col & Int. Sci.* (2019). In press.

Acknowledgements: This work was supported by the University of Maine Vice President for Research and the Graduate School of Biomedical Science and Engineering.

Rapid Assessment of Cellular Activity of Growing Tissue Engineered Constructs

Sabrina Zouaghi, Mugdha Padalkar, PhD, William Querido, PhD, Rutvin Kyada, Shital Kandel, MS, Nancy Pleshko, PhD
Institution: Department of Bioengineering, Temple University, Philadelphia, PA

Introduction: The key to achieving successful implantation of the engineered tissue may be the ability to select constructs with optimal properties. In particular, assessment of cellular activity and metabolic function is essential to verify if a construct is still viable, and hence has the potential for successful tissue regeneration. Current approaches for assessment of metabolic function consist of biochemical analysis that measure the concentrations of lactate and glucose in the cell media, as active cells metabolize glucose and produce lactate over time. AlamarBlue assay can also be used to evaluate cell viability. However, these assays are time consuming and require external reagents. An alternative approach for assessment of metabolic function is attenuated total reflection (ATR), a non-destructive and simple method that does not require any sample preparation. Lactate and glucose have distinct absorbances in the infrared at 1122 cm^{-1} and 1573 cm^{-1} for lactate, and 1079 cm^{-1} and 1361 cm^{-1} for glucose (1). We hypothesize that ATR can be used to evaluate different levels of cellular activity in tissue engineered constructs by detecting a difference in glucose and lactate absorbances in the cell culture media.

Methods: Human osteoblasts hFOB (ATCC) were seeded in 1% methacrylated hyaluronic acid hydrogels with a cell density of 60 million cells/mL in 1 mL of serum free medium initially containing 4.5 g/L of glucose. The constructs were plated in a 24-well plate with 1,2,4,6 construct per well (N=3). **ATR data collection:** spectral data collection was done using Thermo Scientific Nicolet iS5 FT-IR spectrometer with diamond ATR. Spectra were collected from 80 µl of sample media at 8 cm^{-1} resolution and 32 co-add scan in the 800-1800 cm^{-1} frequency range. **AlamarBlue cell metabolic assay:** Cell metabolic activity was assessed using a standard alamarBlue cell viability fluorescence assay (Bio-Rad). Additionally, a compositional assay for glucose (Cayman Chemical) were performed to determine the concentrations of the glucose and lactate in the samples. **Glucose and lactate standards:** a glucose and a lactate assay (Cayman Chemical) were performed to establish calibration curves, followed by spectral collection. **Spectral data analysis:** The 2nd derivative of the spectra was obtained using Unscrambler X 10.4 software (Camo). The spectra were inverted to show positive peaks. The inverted 2nd derivative peak heights specific to lactate and glucose reflective of the glucose and lactate concentrations were calculated. **Statistical analysis:** Correlations between lactate and glucose peak heights of the inverted 2nd derivative, alamarBlue assay and compositional assay results were investigated.

Results: ATR spectra of the media with constructs had visible glucose at 1079 cm^{-1} and 1361 cm^{-1} and glucose peaks at 1573 cm^{-1} and 1122 cm^{-1}. The main differences between the blank media and the constructs media are the appearance of lactate peaks and the reduction in glucose peaks in the construct media (Figure 1). As the number of constructs per well increased the lactate peak heights, 1573 cm^{-1} and 1122 cm^{-1}, increased and the glucose peak heights, 1079 cm^{-1} and 1361 cm^{-1}, decreased, indicating that the greater the number of constructs consume more glucose and produce more lactate (Figure 2). The inverted 2nd derivative peak heights of lactate, 1573 cm^{-1} and 1122 cm^{-1}, correlate positively with alamarBlue references, cell metabolic activity indicator, ($r=0.94$, $p < .00001$ $r=0.88$, $p= 0.00016$). Further, the inverted 2nd derivative glucose peak heights correlates negatively with the alamarBlue reference ($r=-0.81$, $p=001404$, $r= -0.96$, $p < .00001$). The glucose peaks at 1079 and 1361 cm^{-1} also showed strong positive correlations ($r=0.86$, $p=0.\ 000333$ $r=0.98$, $p < .00001$ respectively) with the glucose concentration determined by the compositional glucose assay.

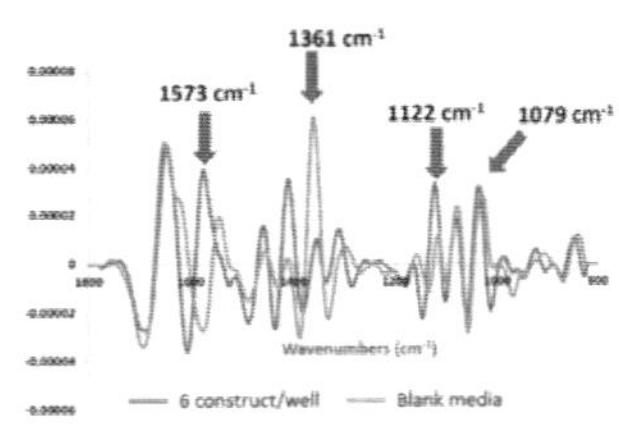

Figure 1. Inverted 2nd derivative spectra of cell media with 6 constructs/ well and cell media without constructs

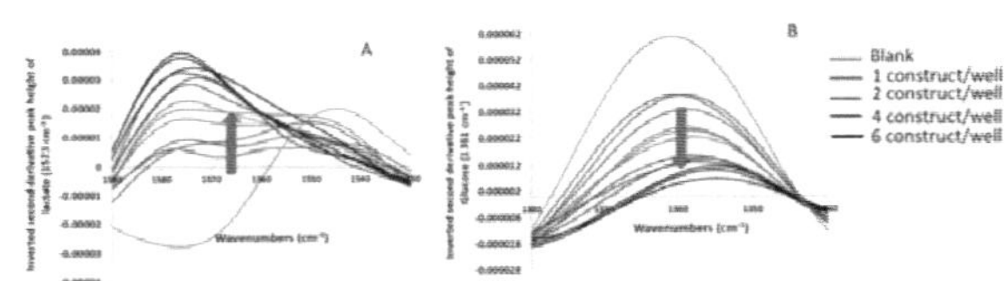

Figure 2. (A) Increase in lactate peak (1573 cm^{-1}) with increase in constructs/well and (B) Decrease in glucose peak (1361 cm^{-1})

Conclusions:

ATR spectroscopic analysis was able to detect changes in the glucose and lactate concentrations in the cell media from engineered constructs with different amounts of cell activity. The glucose and lactate concentrations reflect metabolic function of the growing cells. The strong correlations between inverted 2nd derivative peak heights of lactate peaks and glucose peaks with the alamarBlue reference showed that this approach can be used as a surrogate for assessment of cell activity of growing constructs. Additional advantages of this technique include the requirement for obtaining a very small amount of media for the assay, which will not interfere with construct growth. Moreover, data collection by ATR is very rapid and takes less than one minute per sample. Future work will consist of growing engineering construct for 6 weeks and using ATR spectroscopy to monitor their growth.

References:

1. M. Rhiel, P. Ducommun, I. Bolzonella, I. Marison, U. von Stockar, *Biotechnology and Bioengineering,* Vol. 77, pp. 174-85, January 2002

Acknowledgements: This study was funded by NIH R01 AR056145.

Modulation of Covalently Crosslinked 3D Collagen Hydrogels Regulate Metabolic and Fibrotic Gene Expression

Nikolas Di Caprio, Evangelia Bellas, Ph.D.
Department of Bioengineering, Temple University, Philadelphia, PA

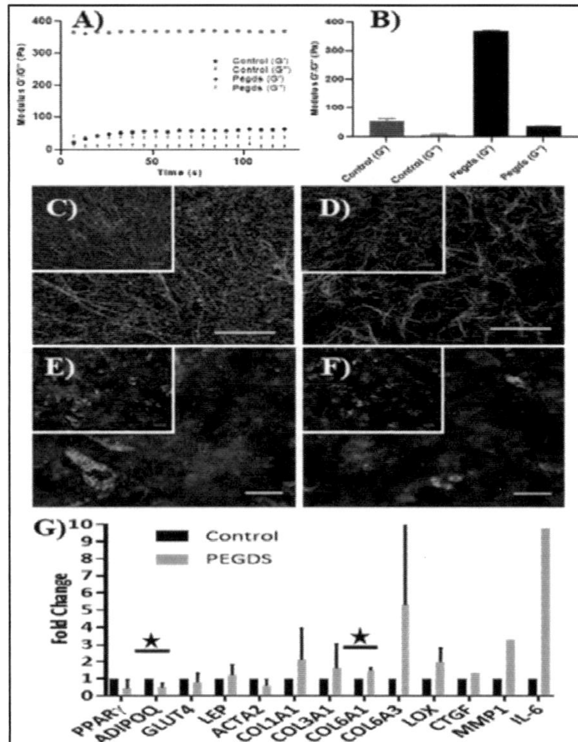

Figure 1. Rheological time sweep analysis (A&B). Reflective microscopy analysis of control (C) and pegds:collagen (D) hydrogels. Confocal microscopy analysis of adipocytes in control (C) and pegds:collagen (D) hydrogels (Red=actin, Blue=nucleus, Green=lipid).Scale Bar: 50μm. Adipocyte gene expression fold change after 5 days in control collagen 3D culture and 0.33% PEGDS:Collagen 3D culture. N=4, except CTGF, MMP1, IL-6 N=1. (G) Student t-test were performed for each gene of interest. 1 star: p-value<0.05.

Introduction: Obesity is a prevalent disease that affects more than 1 in 3 U.S. adults and increases the risk of major comorbidities, such as type II diabetes, cardiovascular disease, stroke and some cancers.[1,2] In obesity, adipocytes grow hypertrophically (size) at a rate where vascularization becomes insufficient thus causing an hypoxic tissue environment. Hypoxia results in Hypoxia Inducible Factor 1α (HIF1α) nuclear localization, where it upregulates pro-fibrotic and pro-inflammatory factors, resulting in stiffening of the extracellular matrix (ECM).[3] The impact of this ECM stiffness on adipocyte dysfunction remains poorly understood. In this study, we aim to investigate effects of stiffness on adipocytes encapsulated within healthy and diseased Type I 3D collagen hydrogels to better understand its role in adipose tissue fibrosis.

Methods: 2mg/mL collagen type I (Corning) was utilized to encapsulate and culture human adipocytes (8 million/mL) in 3D. Collagen hydrogels were placed in incubator and gelled at 37°C for 30 minutes. Collagen hydrogels were tuned to different stiffnesses through varying the mass ratio (0 and 0.33) of the covalent crosslinker, polyethylene glycol disuccinimidyl ester (pegds) to collagen. Time sweep rheological mechanical testing (AR2000, TA Instruments) was performed on post-gelled hydrogels (w/out adipocytes) at 0.1% strain with a 1 Hz frequency for 2 minutes. Reflective image analysis (Olympus Fluorview) was performed at 488nm on post-gelled hydrogels (w/out adipocytes) to visualize collagen architecture (i.e. fiber formation and porosity). Confocal image analysis (Olympus Fluorview) was performed after 5 days in culture. 3D hydrogels were fixed and stained for lipid accumulation (Bodipy), F-actin cytoskeleton (Phalloidin) and the nucleus (Hoechst). QPCR gene expression analysis (Eppendorf Mastercycler) was performed on day 5 to observe changes in adipocyte metabolism and ECM related gene activity.

Results and Discussion: Time sweep rheological testing resulted in control collagen hydrogels having a storage modulus (G') = 58.4 Pa and a loss modulus (G") = 9.3 Pa. Pegds:collagen (0.33) hydrogels resulted in a G' = 368.1 Pa and G" = 34.7 Pa. **(A&B)** Addition of pegds, resulted in a 6-fold difference in G' and 4-fold difference in G". Visualization of collagen architecture through reflective microscopy revealed increased fiber formation and porosity in 0.33% pegds:collagen compared to control collagen hydrogels. **(C&D)** Confocal microscopy analysis portrayed adipocytes encapsulated within pegds:collagen hydrogels contained more cytoskeletal tension (via actin stress fibers) and less lipid droplet accumulation in contrast to the control. **(E&F)** When adipocytes were encapsulated in 0.33% pegds:collagen 3D hydrogels for 5 days, qPCR results revealed statistically significant upregulation in extracellular gene expression (COL6A1) and a downregulation in adipocyte and metabolism gene expression (ADIPOQ) compared to the control. **(G)**

Conclusion: Based on these results, we have shown that modulation of ECM stiffness independently of collagen concentration can regulate adipocyte metabolic and fibrotic gene expression towards dysfunction. Our future work entails quantifying collagen architecture and performing diffusion based experiments to verify ECM stiffness can regulate adipocyte gene expression towards dysfunction independently of ECM architecture.

References:

1. Carroll MD. et al. JAMA. 2016; 315(21): 284-2291
2. Martinez JA. et al. Nat R. 2017; 3: 17034
3. Kos K. et al. Meta. 2017; 78: 118-127

Acknowledgements: We acknowledge funding support from Dr. Bellas' startup funds (Temple University) and the NIH NIDDK Diabetic Complications Consortium DK07616 and DK115255 grants (to EB). We also acknowledge the Burdick Lab (UPenn) for help with mechanical testing.

The Design of an Assisted Intravenous Transfer System for Patient Ambulation

Scott Selig, Lillian Margolis, Laura Parra, Jiang Wu PhD, Ying Sun PhD, Jordan Anderson
University of Rhode Island, Kingston, RI, USA

Significance: Immobility in hospitalized patients has been linked to a functional decline and complications affecting the respiratory, cardiovascular, gastrointestinal, integumentary, musculoskeletal, and renal systems. For surgical patients, early ambulation is the most significant factor in preventing complications [1]. As a result, health care clinicians must constantly help patients move from beds or chairs to ambulatory systems during their stays at the hospitals. Nurses will typically spend around ten minutes individually transferring all of the equipment on an intravenous (IV) pole to an ambulation device, such as a walker, which is time consuming and can result in injury. There are products on the market that have attempted to minimize this problem, such as the TAD™ IV Transfer System by Nexxspan [6] and the Nezzie by Blickman [4]; a single Nezzie can cost up to $5,000 per device. These alternatives are expensive and nurses still have to manually transfer the medical devices from one of these systems to a normal IV pole to be able to share the equipment throughout the hospital. The Assisted IV Transfer System, proposed in this paper, is an affordable solution, as it can readily adapt to the standard IV poles and walkers that are already found in hospitals. This system will aid in an effective transfer of various medical equipment from an IV pole to a walker, while maintaining the safety of the patients and the hospital staff.

Design Description: The Assisted IV Transfer System for patient ambulation is composed of three main parts -- IV pole, Transfer System Device (TSD) and an Adapted Walker, as shown in *Figure 1*.

IV Pole TSD Adapted Walker

Figure 1. This diagram shows the three integral components of the IV transfer system.

A standard IV pole carries the medical equipment, while a patient is stationary. While a patient is ambulating, a nurse is typically found along side the patient pushing the IV pole. Due to the inherent instability of the pole, the medical equipment is prone to falling over, making it unsafe for the patient. In order to decrease the risk of valuable medical equipment falling over and increase patient safety along with ease-of-mobility, an Adapted Walker with a female portion of an IV Pole was designed to accommodate medical equipment being transferred on the male portion of the IV pole, using the TSD.

The most integral portion of the proposed system, with the sole purpose to decrease the risk associated with patient ambulation, is the TSD. This part of the system is composed of a hydraulic jack, a steel pole, a swing arm, and a specialized IV pole clamp. It assists the transfer of the medical equipment, which was originally attached to a standard IV pole and brings it to equipment more apt for patient ambulation. The specialized IV pole clamp is a product designed to hold poles and clamp them tight, like the one that was used in the US Patent number 5,699,988 [1].

Design Criteria: The proposed system will facilitate the transfer of medical devices from an IV pole to an adapted walker. The TSD was designed, constructed completely from steel, and simulated with Autodesk to produce a computer-aided design (CAD). By simulating the anticipated loads, multiple iterations of the transfer system were modeled determining the distribution of weight and safety of the system hinting towards the most optimal design. In *Figure 2*, two simulations show contrasting results of different designs during the prototype stage, the figure on the right represents the most current design. Also, notable, the unique supporting arm relocated the point of highest stress-strain to a low and safer area, reducing the chance that the right-angle joint will fail under load. The stress strain analysis complies with the International Electrotechnical Commission (IEC) 60812:2006 Standard, the analysis techniques for system reliability - procedure for failure mode and effects analysis (FMEA), to ensure the safety of the healthcare professionals who will be using the TSD.

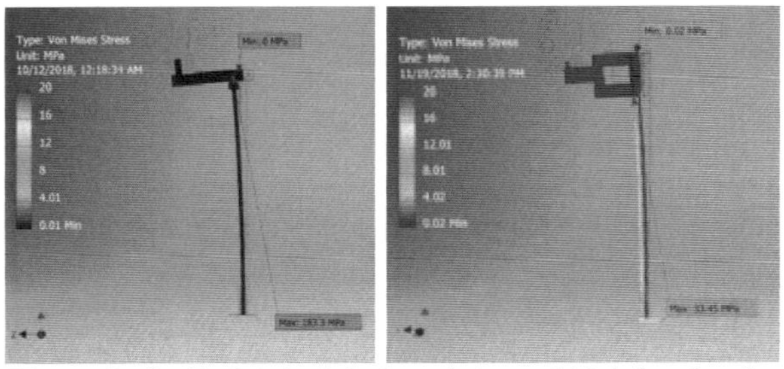

Figure 2. (Left) Original designed showed high levels of stress-strain at a critical design area. (Right) With additional struct design, the TSD had reduced levels of stress-strain at lower point on design. These stress-strain simulations aided early prototyping efforts to minimize possibility of design failure.

Under IEC 14971:2007, the application of risk management to medical devices, a tilt test minimized the risk of the Adapted Walker toppling over by finding the correct placement of the medical equipment on the IV pole. The tilt test will be further explained in the result section. When designing the system, human factors were taken into consideration including: 1. If the IV pole is not extended fully before it is clamped to the TSD the male connector will not clear the female end of the IV pole, 2. If the threaded-twist clamp on the IV pole is not secure the pole may slide down during the transfer or ambulating, and 3. The healthcare professionals may expose themselves to the inherent risk of hitting their head on the TSD while bending down to raise the hydraulic jack.

In order to ensure maximum safety for the patient, the following criteria were set by the team: 1. The bottle jack will be fixed to the ground to maximize stability and 2. The system should only be used in a hospital environment with patients of Category 4 functional ambulation. These patients depend on the supervision of health care clinicians to maintain an adequate level of safety [3]. Further, this design stems from standard IV poles and should not exceed a maximum weight of 25 lbs. (11.34 kg) on the upper half of the IV pole.

Results: Two tests were performed: an average transfer time and a tilt test, to verify the systems' practicality and safety, respectively.

For the average transfer time test, ten new users performed a complete transfer of 14.4 lbs. (6.53 kg) from the IV pole to a walker using the transfer system. The weight on the IV pole simulated the weight of an Alaris ® PC Unit, 2 controllers, and 2 IV bags. The results yielded an average transfer time of 2 minutes and 26 seconds, *Figure 3,* significantly less than the average 10 minutes it takes for a nurse to individually transfer each piece of equipment.

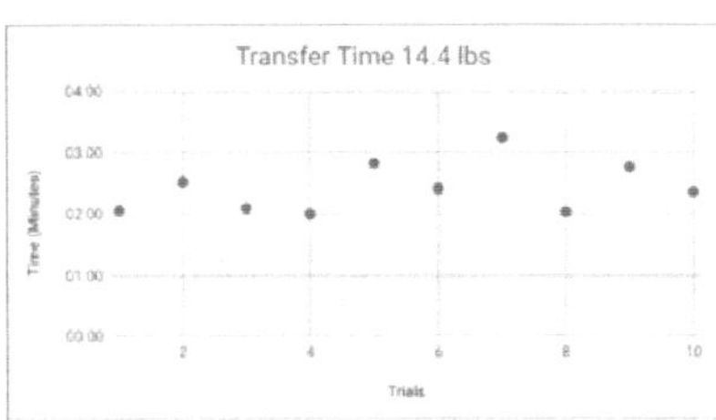

Figure 3. Transfer time of simulated load of 14.4 lbs. (6.53 kg), from IV pole to adapted walker for ten randomly selected participants.

The tilt test was performed to analyze the safest position on the adapted walker for the medical devices during patient ambulation. Our team placed a simulated load of 13 lbs. (5.90 kg) on the right side and then the left side of the pole at different heights - 18 inches, 24 inches, 30 inches, and 36 inches off the ground (45.72 cm, 60.96 cm, 76.2 cm, and 91.44 cm), as shown in *Figure 4*, and conducted controlled tilt testing using a gyrometer for accurate angle measurements. The results indicated that weight mounted on the inside (left side) of the pole proved to be the most stable. Additionally, as the height of the load increased, the tilt angle also increased due to unique design of the adapted walker and its offset frontal wheels.

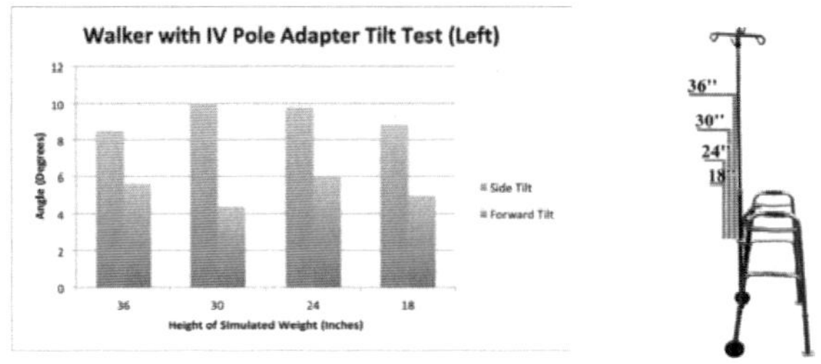

Figure 4. (Left) The average angle before adapted walker toppled due to simulated load of 13 lbs. (5.90 kg). The x-axis represents the vertical height of the weight on the pole. (Right) Marked locations show the different heights for the different distribution of weight on the adapted walker during testing.

Future Implications: The Assisted IV Transfer System provides an efficient and safe way for nurses to transfer medical devices from an IV pole to an ambulation system, while being an affordable and novel solution due to its adaptability to the standard IV pole and walker that can be found in hospitals. It is important to note that this system is just a prototype and there is more work to be done before it can be implemented in a hospital. However, due to the need of a product like this on the market, future commercialization of this product is essential. Some future modifications to investigate and possibly implement are, an automatic jack, an ergonomic foot pedal, a cross bar design on the walker to increase stability [5], and a magnetic locking collar to avoid screwing and unscrewing the IV pole apart. There have also been considerations of including a counter weight on the walker to increase stability, but the trade off is that it will increase the weight that the patient will have to lift when using the walker. These modifications will increase the ease of use and human factors of the system.

References:

[1] Boettger, C. H., & Hawks, B. J., Jr. (1997). U.S. Patent No. 5699988. Washington, DC: U.S. Patent and Trademark Office.

[2] Doyle, Glynda Reese, and Jodie Anita McCutcheon. Clinical Procedures for Safer Patient Care. BC Open Textbook Project, 2015.

[3] Holden M, et al (1986). Gait assessment for neurologically impaired patients: Standards for outcome assessment. Physical Therapy, 66, 1530-1539.

[4] "Meet Nezzie." Nezzie Patient Ambulation by Blickman | Meet Nezzie, www.patientambulation.com/nezzie.

[5] Olson, K. L. (1981). U.S. Patent No. 4251044. Washington, DC: U.S. Patent and Trademark Office.

[6]"Patient Handling and IV Solutions." Nexxspan, nexxspan.com/iv-transfer-system/.

Assistance Calling Device for Advanced Stage Amyotrophic Lateral Sclerosis Patients

John Furey, Andres Gutierrez, Chadwick Dotson-Jones
Drexel University School of Biomedical Engineering, Science & Health System

Introduction: Amyotrophic Lateral Sclerosis (ALS) is a progressive neurodegenerative disease in which upper and lower motor neurons are affected at all levels of the spinal cord. Symptoms may begin in the limb or bulbar muscles, ultimately resulting in loss of functional speech and/or movement. This leads to constant caregiver assistance and thus a call device to accommodate progressed ALS patients is needed. Certain existing solutions require capabilities which are not functionally possible in progressed ALS patients such as force and limb movement to activate the device.

This design must meet the following requirements in order to be successful: (1) Activation force of at least 0.20N (2) Alarm range of at least 20 meters (3) False Positive rate less than 5% (4) 5 volts voltage source (5) Immediate feedback call indicator and (6) Internal device supervision for contact with receiver and battery source. This design has three constraints: (1) Must not further restrict finger movement (2) Must function between all fingers (3) Function throughout patients disease progression.Our design is a Force-Sensitive activated assistance calling device. It will consist of a wearable strap with the FSR (Force Sensitive Resistor), a transmitter, and a pager. This will allow for minimal movement/force required to activate the device compared to current devices. While the transmitter will be attached to a wheelchair and relay to the caretaker's pager upon activation of FSR.

Methods: *Materials:* The device will be activated by a force-sensitive resistor. The resisted voltage will be direct into an ATmega328p microcontroller to determine whether a call is legitimate. If so, a signal will be transmitted to a receiving device using a 433MHz radio frequency transmitter. The device will be powered using 2000mAh lithium ion battery and can be recharged safely using a Adafruit PowerBoost Charger. In the receiving device, a RF receiver, paired to the RF transmitter, will receive a signal and activated an LED on the device. A digital caliper was used to verify the average finger length and width. A set of weights was also used to confirm the accuracy of the FSR.

Population: Twenty-five (25) healthy subjects' index finger length and width were measured in millimeters (14 males, 11 females) using a Carbon Fiber Composites Digital Caliper. Lateral pinch force in pounds was recorded from sixteen (16) patients suffering from neurodegenerative diseases and converted to grams. The lateral pinch force was obtained from a de-identified database.

Verification Tests: The first test involved measuring the lengths and widths of female and male dominant index fingers using calipers. Next the FSR was tested to meet the 20g requirement needed to activate the sensor. The data will be collected live through Serial Plotter in the Arduino IDE, as differents weights, starting at 10g, are added onto the sensor till it reaches 20g. The RF transmitter and receiver signal will be tested at various distance until 20 meters is reach to ensure that it meets the requirement. An immediate feedback test will be conducted by simply testing that the LED light functions when the FSR is activated. We will be testing four scenarios, the first that the LED light up from a light activation force while set at a low threshold. The second scenario had the activation force set at a higher threshold, so hard continuous pressure was tested on the LED. The last two scenarios will test for false positives, so the activation force threshold will be set a low level but hard continuous pressure will placed on the FSR and vice versa for a high threshold. If the LED lights up, the test failed. Lastly, the Arduino will be programed to detect a loss of contact from the receiver if it occurs for more than 90 seconds. In the case that it does, an alarm will alert the caregiver to attend to the issue.

Results: Average finger length and width were found to be 77.66 +/- 8.32mm and 17.44 +/- 2.13mm, respectively. Finger width is most important for this design and was found to be 18.28+/-2.35 for males and 16.38+/-1.24 for females. Average lateral pinch force for right and left hands was found to be 9.03+/-6.98lbs (4.09+/-3.17kg) and 9.38+/-6.39lbs (4.25+/-2.9kg), respectively. For males the right and left lateral pinch forces were 11+/-9.04lbs (4.99+/-4.1kg) and 11.07+/-7.05 (5.02+/-3.19kg) and females were 7.5+/-4.91lbs (3.4+/-2.23kg) and 8.06+/-5.91 (3.65+/-2.68kg). After running 18 trials in each of the four tests explained previously to test the LED immediate feedback indicator the device passed all tests.

Conclusions: Verifications tests for (1) range of alarm transmission and (2) if the device can detect loss of contact with the receiver for more than 90 seconds must still be completed. Verification testing to ensure device meets military specifications will not be addressed in this design due to time constraints. Further studies would need to be conducted to minimize the false positive of the device. Additionally, advances in the detection algorithm can be added by incorporating a sequence recognition or specific duration of FSR activation.

References:

1. Logroscino, G., Piccininni, M., Marin, B., Nichols, E.A., et al. (2018); 1083-1097. doi: 10.1016/s1474-4422(18)30404-6
2. Chiò, A et al. (2013): 118–130. PMC.

Acknowledgements: Alisa Brownlee of the ALS Association (ALSA), the stakeholder in this project, Dr. Joseph Sarver, our Biomedical Engineering Faculty advisor and the MDA/ALS Center of Hope Clinic at Temple University Hospital..

P3-26

Glucose Sensitivity in Engineered Adipose Tissue Model of Type 2 Diabetes

Alexandra Abruzzo, Michael Struss, B.S., Evangelia Bellas, Ph.D.
Department of Bioengineering, Temple University, Philadelphia, PA

Introduction: Over 30 million adults in the United States have Type 2 Diabetes (T2D) (Diabetes.org). T2D is directly linked to how adipose tissue (AT), primarily made up of adipocytes, metabolizes glucose in the body. Adipocytes that are chronically overexposed to glucose become insulin resistant. When this happens, the signal, in which insulin is actively bound to its receptor, no longer activates adipocytes to uptake the excess glucose. We hypothesize that as the glucose levels in the environment around the adipocytes increase, the cells will become insulin resistant, resulting in several biological and environmental changes. Not only will this provide a better understanding of how the glucose pathway works within the cell but also it will indicate how metabolic diseases, such as T2D, can be reversed.

Methods: In this study, human adult mesenchymal stem cells were differentiated into adipocytes and then encapsulated within a 3D collagen matrix to mimic the physiological environment of human AT while allowing the cells to remodel the microenvironment. We compared two groups: a normal glucose (1000mg/L), NG, group, which served as the control, and a high glucose (4500mg/L), HG, group, while the insulin levels remained the same. After 48 hours, the groups were sacrificed for further analysis. qPCR (ThermoFisher Scientific, Philadelphia, PA) was utilized to examine the gene expression for TNFα, IRS1, and GLUT4 between groups. Additionally, lipolysis assays (Sigma-Aldrich, Allentown, PA) were performed to measure the glycerol secreted by the adipocytes, along with a glucose uptake assay using 2-NBDG, a fluorescent glucose analog, and standard imaging. In the glucose uptake assay, glucose molecule (2-NBDG) localization around the cell nuclei (Hoechst 33342) was examined.

Results: Results from 2-NBDG imaging, lipolysis assays, and qPCR are presented in Figure 1. In the NG condition (Fig.1a), glucose molecules are localized at and adjacent to the cell nuclei. In the HG condition (Fig.1b), the nuclei and glucose molecules are not co-localized, where the glucose signal is dispersed throughout the cell. The results of the lipolysis assay (Fig.2d) show that in HG, lipolysis is decreased. The data obtained between HG and NG lipolysis are statistically significant, with a p-value of 0.0343. Lastly, the genes of interest (Fig.2a-c) reveal that in HG, GLUT4 is downregulated, while IRS1 and TNFα are upregulated when compared to the NG group. The present studies are still ongoing.

Conclusion: It was anticipated that glucose would be localized to the cell nuclei in the NG cells as opposed to the HG cells. Since HG cells have been overexposed

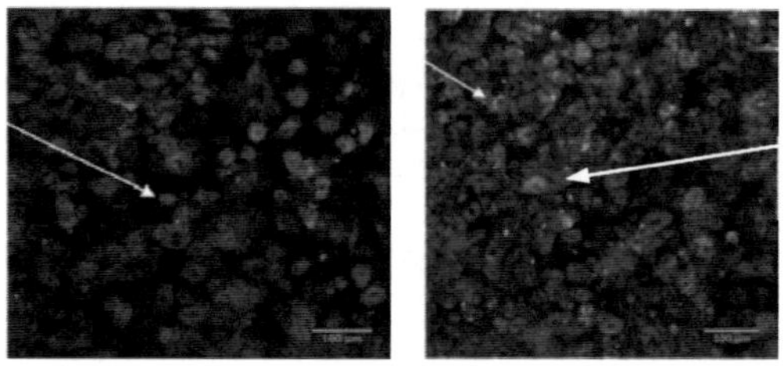

Figure 1. ***(a)*** *Glucose molecules(magenta) are localized to cell nuclei(green) in the NG cells in the image located at the top left.* ***(b)*** *However, glucose molecules surround nuclei less in the HG image, located at the top right, and nuclei are more clustered.*

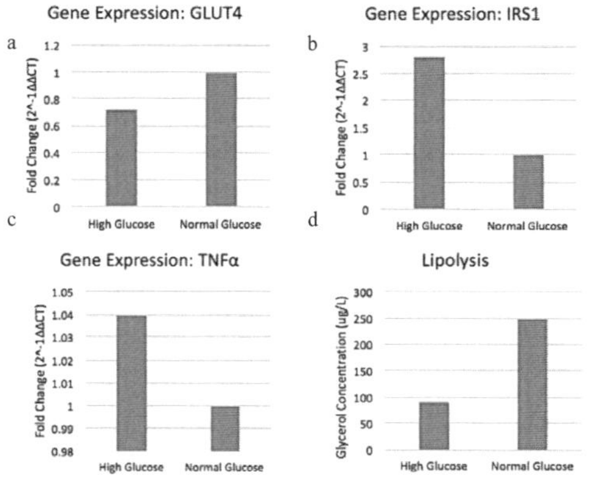

Figure 2. ***(a)*** *GLUT4 gene expression and* ***(d)*** *lipolysis were downregulated in HG cells. Conversely,* ***(b)*** *IRS1 and* ***(c)*** *TNFα was upregulated in HG cells.*

to glucose, they have stopped responding to insulin as a trigger to uptake the surrounding glucose. The 2-NBDG images support this claim, as the NG cells have more glucose localization. Additionally, it was predicted that in insulin resistant cells, lipolysis would be downregulated. Since insulin triggers the cell to utilize glucose, in turn secreting glycerol, the HG cells would secrete less glycerol. The lipolysis data above is representative of this hypothesized trend. The results of the lipolysis assay and 2-NBDG also suggest that exposing cells to high glucose concentrations for 48 hours can, indeed, begin to stimulate insulin resistance in cells. Further, the HG cells do not respond to GLUT4 as well as the NG cells, and IRS1 and TNFα are highly expressed in HG cells, which is likely due to insulin resistance. In the future, protein expression could be studied, seeing as phosphorylation of proteins, such as AKT, are other critical players in the glucose uptake pathway.

Acknowledgments: We acknowledge funding support from Dr. Bellas' startup funds (Temple University) and the NIH NIDDK Diabetic Complications Consortium DK07616 and DK115255 grants (to EB).

The Effect of Macrophage Infiltration on Adipose Tissue Inflammation

Eleanor Caston, Evangelia Bellas, Ph.D.
Department of Bioengineering, College of Engineering, Temple University, Philadelphia, PA

Introduction: Obesity is an epidemic that can lead to stroke, type 2 diabetes, cardiovascular disease and certain types of cancer and affects over 93.3 million US adults (CDC.gov). Adipose tissue (AT) macrophages play a role in AT inflammation and the development of insulin resistance, both of which are conditions linked to obesity. We hypothesize that AT inflammation occurs with increased macrophage infiltration due to crosstalk between the macrophages and adipocytes, the main cell type in AT, that leads to the release of inflammatory factors.

Methods: To study the relationship between macrophages and adipocytes, a physiologically relevant in vitro model was used. Human mesenchymal stem cells were grown to confluency and differentiated for 7 days using adipogenic media. Adipocytes and THP-1 macrophages were then co-cultured using a transwell system with both a 2D and 3D environment for the adipocytes. The 3D environment used a 2 mg/ml type I collagen gel to encapsulate the adipocytes in an environment which allowed for the rounded cell shape typically found in healthy states. The cross-talk was observed in a 1:0, 1:1, and 0:1 ratio of adipocytes to macrophages at a density of 8 million cells per mL of collagen. After 48 hours, adipocytes and macrophages were harvested separately and sacrificed for end-point analysis.

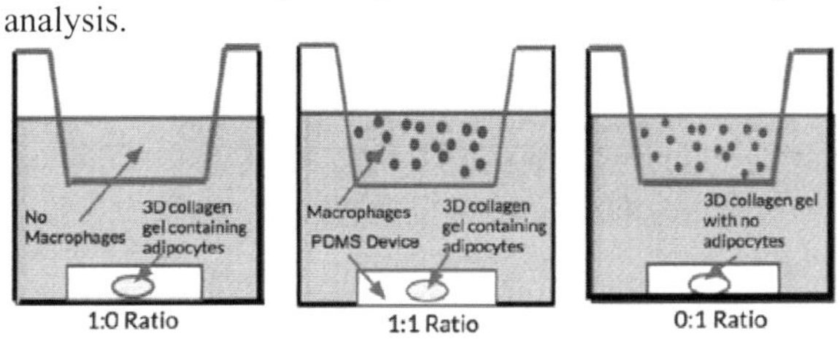

Figure 1. Experimental Setup using Transwell Inserts. Ratio is Adipocytes:Macrophages.

Results: QPCR was performed for genes relevant to adipocytes, macrophages and inflammation, including adiponectin (ADIPOQ), leptin (LEP), peroxisome proliferator activated receptor gamma (PPARG), C-C motif chemokine ligand 2 (MCP-1), matrix metallopeptidase 9 (MMP-9), and tumor necrosis factor alpha (TNFa). MCP-1, a pro inflammatory monocyte chemoattractant, was upregulated in the 3D 1:1 adipocyte group (Fig. 2A). This trend was also observed in the 2D 1:1 adipocyte group, but with a weaker signal, when comparing to the 1:0 adipocyte 2D control. A similar upregulation trend was found with the 3D 1:1 adipocyte and the 3D 1:1 macrophage group with the MMP-9 gene, which encodes for the degradation of type IV and V collagens (Fig. 2A). Ongoing studies are underway for both the 2D and 3D experiments for confocal microscopy of cell organization, including lipid droplets (BODIPY) and cell nuclei (Hoechst 33342). Media samples collected at 24 and 48 hours have been banked for biochemical assays.

(A)

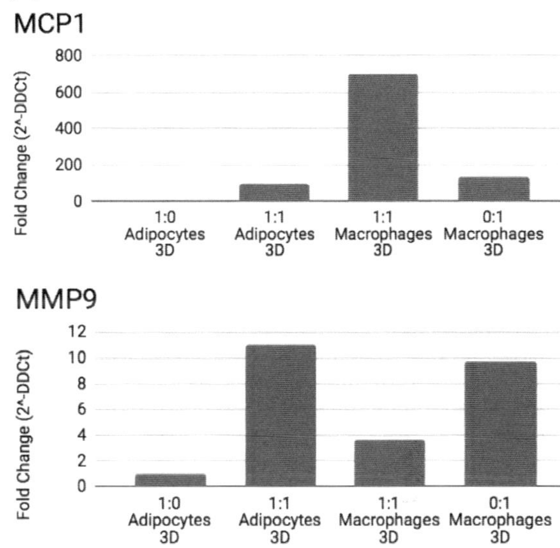

(B)

Group	APIDOQ	LEPTIN	MCP-1	MMP-9	PPARG	TNFa
3D 1:1 Adipocytes	0.994	6.6125	96.919	10.968	2.2696	1.8468
3D 1:1 Macrophages	0.32761	0.42930	697.01	3.6642	4.6926	106.97
3D 0:1 Macrophages	0.064921	0.27921	129.85	9.7282	1.2888	369.43

Figure 2A, 2B: Adipocyte and macrophage gene expression fold change normalized to 3D 1:0 Adipocytes after 48 hours in collagen 3D culture.

Conclusions: The relevant 3D model used demonstrated the ability of adipocytes and macrophages to cross-talk through soluble factors secreted in co-culture. The upregulation of proinflammatory genes, such as MCP-1 and MMP-9, demonstrates that there is a relationship between macrophage infiltration and inflammation and remodeling in adipocytes, which can contribute to a diseased state. Gaining a deeper understanding of the cross-talk between macrophages and adipocytes that leads to AT inflammation will assist in solving the widespread issue of obesity.

References:

1. Permana P.A et al. Biochem Biophys Res Commun. 2005;341:507-514.

Acknowledgements: We acknowledge funding support from Dr. Bellas' startup funds (Temple University) and the NIH NIDDK Diabetic Complications Consortium DK07616 and DK115255 grants (to EB).

The Effect of Vascularization on Adipose Tissue Glucose Metabolism

Jennifer Hammel, Michael Struss, Golnaz Anvari, M.S., Evangelia Bellas, Ph.D.
Department of Bioengineering, Temple University, Philadelphia, PA

Introduction: Adipose tissue regulates the entire body's metabolism. In Type 2 Diabetes (T2D), the body is unable to utilize insulin properly, causing a range of complications due to high blood sugar. These complications can include heart disease, kidney disease, nerve damage, and an inability to heal. According to the CDC, 30.3 million Americans are afflicted and Type 2 accounts for 90-95% of all diabetes cases, with 87.5% of adults diagnosed with diabetes falling under the category of overweight or obese (cdc.gov). Though a surplus of fat tissue can have poor consequences, well-vascularized fat tissue is necessary for the human body to properly metabolize nutrients. Inadequate vascularization leads to tissue inflammation, which can prevent the necessary exchange of insulin and other key glucose uptake molecules. To understand how blood vessels contribute to the health of fat tissue and its ability to use insulin, we performed a direct co-culture of adipocytes and endothelial cells at different ratios in 3D Type I collagen gels.

Materials and Methods: Human mesenchymal stem cells were differentiated into adipocytes and human umbilical vein endothelial cells (HUVECs) were used for vascularization. Ratios tested include 9:1 adipocytes to HUVECs, 1:1, and 1:3, with adipocyte density kept constant at 4 million/mL of 2 mg/mL type I collagen gel. Media was banked at days 3, 5, and 8. At day 8, some gels are sacrificed for end-point analysis. A subset of gels is live-treated with 2-NBDG to evaluate glucose uptake or forskolin to induce lipolysis. Genes involved in the insulin signaling pathway, inflammation, and glucose transport are evaluated via qPCR. Genes surveyed include adiponectin, glucose transporter 4 (GLUT4), interleukin 6 (IL-6), peroxisome proliferator-activated receptor gamma (PPARG), vascular endothelial growth factor (VEGF), and leptin. Glycerol concentration in basal spent media as well as induced media are evaluated via lipolysis assay. Confocal microscopy is used to visualize lipid droplets (Bodipy), cell nuclei (Hoechst 33342), and endothelial cells (CD31). The size and area of lipid droplets and glucose uptake are quantified via ImageJ and vascular organization is manually quantified in terms of branches and nodes. Statistical analysis was performed utilizing ANOVA and students' t-test with an alpha value of 0.05.

Results and Discussion: Adiponectin is significantly downregulated in co-cultures compared to the control. PPARG is found to be downregulated in ratios 1:1 and 1:3 (Fig 1B). Leptin, VEGF, GLUT4, and IL6 were found to be statistically the same as the control (not shown). Confocal microscopy confirms the presence of endothelial cells and early network formation (Fig 1A). Preliminary lipolysis and glucose uptake results suggest that there is no difference between groups (not shown). Preliminary endothelial cell network results suggest that the 1:1 ratio has the most organization (not shown). Preliminary lipid droplet analysis suggests that the 1:3 ratio has the least lipid droplet formation (not shown).

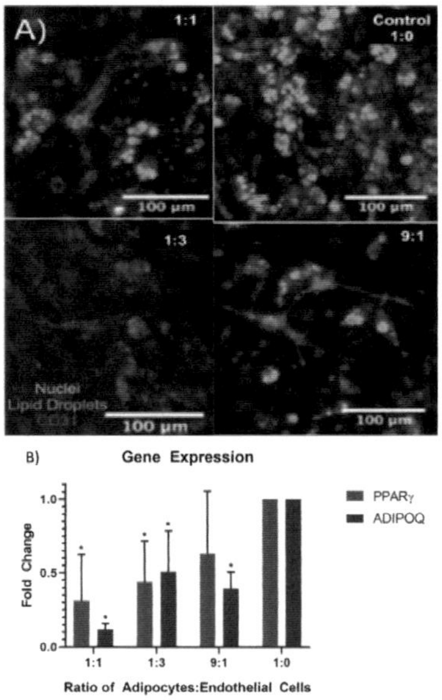

Figure 1.

Conclusions: By cultivating a better understanding of the cellular cross-talk between adipocytes and endothelial cells, we can better understand the role adipose vasculature plays in metabolic disorders such as obesity and type 2 diabetes. Future directions include increasing the n value for newer analysis methods and evaluating protein expression for GLUT1, GLUT4, VEGF, and adiponectin.

Acknowledgements: This study was supported by the MARC-U*Star Program, NIH T34 GM 087239 (JH), the NIDDK Diabetic Complications Consortium (www.diacomp.org), grants DK076169 and DK115255 (EB).

Exploring New Parameters of The K562 Cytotoxicity Assay

Riya Patel, Biju Parekkadan, PhD.
Rutgers University School of Engineering

Introduction: Approximately 200,000 people in the US alone are diagnosed with a form of leukemia, lymphoma or myeloma each year.[1] There is a growing field of cell therapies that are emerging to combat these forms of blood cancers. One specific type of cell that is being used in various cell therapies is known as the Natural Killer (NK) cell. These cells are lymphocytes that are part of the innate immune system that serve to destroy virally infected cells. Due to NK cells natural ability to recognize and lyse tumor cells containing various receptor types, they show great potential as a therapeutic to fight against various blood cancers. The standard method of testing NK cell cytotoxicity is using a killing assay against the myelogenous leukemia cell line K562. This assay is currently only testing the cytotoxicity using set limited ranges. However for purposes of understanding the system itself, we feel that more parameters of this assay have not been manipulated and studied to its full potential. We hypothesize that taking a systems engineering approach to identify and manipulate some of the key variables of the system, we can better utilize this assay to allow us to see various factors that could affect the NK cells interactions with blood cancer cells as well as their viability and persistency *in vivo*.

Materials and Methods:

CFSE and Ethidium Homodimer (EtHD) Staining: K562 cells were stained using CellTrace CFSE proliferation kit by adding 2.5 uM CFSE stain per well in a 96 well containing $2x10^5$ cells. Tube was covered in foil for five minutes at RT, centrifuged and rinsed with PBS twice before plating in RPMI complete media. EtHD stain is used at end of assay, and 2 uM EtHD is added to cells, which are incubated in the dark at 4°C for 7 minutes.

K562 Cytotoxicity Assay: The K562 killing assay was performed as described previously, adapted to a 96 well screening format.[2] Briefly, CFSE-labeled K562 target cells were cocultured with effector NK cells at varying ratios. Flow cytometry compensation controls of non-stained, CFSE only stained, Ethidium Homodimer (EtHD) only, and both were used. Cells incubated for the standard five hours at 37°C. The percentage of K562 cell death was analyzed by flow cytometry analysis.

Statistical Analysis: All data were presented using the mean +/- SD. Experimental groups were run in triplicate. Multiple groups were compared using a one-way ANOVA.

Results: Initial experiments were run to determine a baseline killing curve for NK effector cells at ratios of 3.12 and 6.25 cells per target K562 cell. The importance of this data is to develop standard death ratios to which key parameters can later be varied and more properly understood. Figure 1 represents the results of both ratios compared to the groups of K562 cells with no effector cells which showed near minimal cell death as expected.

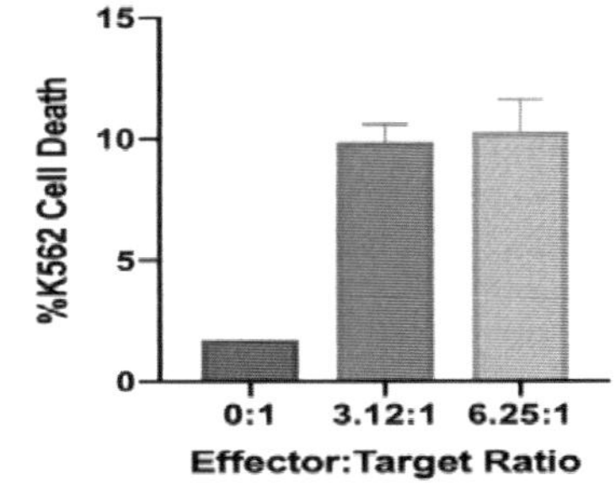

Figure 1. K562 cell death by effector:target ratio.

Conclusions: Our initial results gave us significant data to now be able to assess how different system parameters have an effect on NK cell killing capacity, time of co culture (extended past the five hours) subpopulations of proliferating cells which appear over a span of days, and being able to deduce key interactions between the NK and K562 cells that other groups have not yet investigated. These are outlined in Table 1 below.

Table 1. Changeable Parameters of K562 Killing Assay

Parameters	Study
Time of co-culture	Varying incubation time for effector and target cells
Looking at subpopulations of proliferating cells	Allow effector cells to proliferate and see if second generation of cells are more potent
Imaging NK/leukemia interactions	Visualizing interactions of effector and target cells in coculture end to end

The importance of changing parameters for the K562 cytotoxicity assay is so we can test various factors that could potentially affect NK cells and their interactions with K562 cells, as well as potentially many other cells. From these new approaches being conducted on this assay, we are able to learn various new uncharted territory about NK cells and possibly unlocking a new potential to use them in cell therapies.

References:

1. LaRussa.. (2015, March 03). Facts and Statistics. Leukemia and Lymphoma Society.
2. Abcam (2012). ab133073–7-AAD/CFSE Cell-Mediated Cytotoxicity Assay Kit.

Angiogenic response to IL-4 eluting coatings in mesh tissue explants

Hannah Geisler[1], Alexis Nolfi[1], Aimon Iftikhar[1], Bryan Brown, PhD[1,2]
1: Bioengineering, University of Pittsburgh; 2: McGowan Institute for Regenerative Medicine

Introduction: Pelvic organ prolapse is characterized as the descent of the pelvic organs due to diminishing support. Susceptibility to developing prolapse increases as a woman ages, as well as with other associated factors such as weight, prior hysterectomy, and vaginal birthing. Surgical repair to compromised tissue is possible, but usually fails within a few years, requiring additional surgeries. While the use of native tissue for restoration of support is ideal, native tissue lacks the mechanical durability present in synthetic implants. Polypropylene mesh (PPM) implants remain the most promising approach for sustainable repair. However, studies have shown that the PPM implant can behave like an internal chronic wound with long-term inflammation, particularly associated with pro-inflammatory macrophage populations. Prior investigations in our laboratory have addressed this controversy via implementation of a cytokine-delivery system used to induce an early shift in macrophage polarization at the host-implant interface, enhancing implant integration [1]. As macrophage-implant interactions have been identified as a determining factor of downstream outcome, early shifts to anti-inflammatory phenotypes serve to mitigate acute host-immune response. The present study serves to quantify vasculature, an important trademark of healing tissue and implant integration, present at the host-implant interface via immunolabeling of the CD31 marker for endothelial cells to investigate effects of the cytokine-delivery system on angiogenesis. As previous studies have confirmed the role of anti-inflammatory macrophages in angiogenesis, particularly as an essential aid in stabilizing fresh vasculature otherwise prone to regression, expected results include increased vasculature present in IL-4 eluting groups.

Methods: Young C57Bl/6J female mice (8-10 weeks) were used for subcutaneous abdominal implantation of one of three types of polypropylene mesh (n=3-5): unmodified "pristine" PPM, polymer-coated PPM, and IL-4 releasing PPM. PPM implants and surrounding tissue complex were harvested at 7 and 14-day time points and fixed for histological studies. Paraffin embedded tissue sections were immunolabeled for the CD31 marker for endothelial cells using 4% diaminobenzene-peroxidase (DAB) substrate solution and Hematoxylin QS counterstain. Samples were imaged using a 20x field and processed using a colour deconvolution macro on image analysis software, ImageJ. Two trained investigators blinded to study groups counted the number of blood vessels present within each 20x field. Results of the two individual counts were averaged. Significance was calculated using two-way ANOVA with pairwise post-hoc comparisons and Bonferroni adjusted alpha level with $p < 0.01$.

Results: Results are represented as the number of blood vessels present at the host-implant interface normalized to area. No significant differences were detected in vasculature present at the interface between the three treatment groups at 7 days (Figure 1). However, all three treatment groups did present differences in vasculature present at 14 days; the unmodified pristine PPM contained the fewest blood vessels while the IL-4 eluting PPM contained the highest number of vessels. Additionally, there was a significant increase in vasculature present at the interface in the IL-4 eluting mesh between the 7-day and 14-day time points.

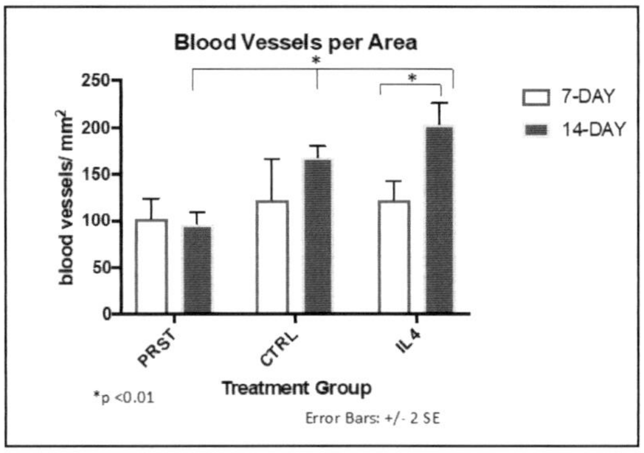

Figure 1. Number of blood vessels present at host-implant interface at 7 days and 14 days post-implantation.

Conclusions: Modified cytokine-releasing PPM implants increased vasculature present at the host-implant interface after 14 days, suggesting enhanced angiogenic activity. As the promotion of blood vessel formation correlates to the delivery of vital oxygen and nutrients to tissue and removal of detrimental wastes, findings of the present study further reinstate that IL-4 eluting PPM implants display enhanced integration into the host immune response. As previously stated, cytokine elution shifts early polarization of macrophage phenotype toward an anti-inflammatory phenotype. Results suggest that this polarization provides some sort of influence over the later acute vasculature, possibly due to macrophage-induced recruitment of pericytes or enhanced stabilization of vasculature prone to regression via recruitment of mesenchymal stem cells. These results display strong evidence for the implementation of similar cytokine-releasing systems in other applications of soft-tissue reconstruction and overall improved integration of non-degradable biomaterials.

References: [1] Hachim D et. al. *Biomaterials.* 2017. 112: 95-107.

Acknowledgements: University of Pittsburgh's Swanson School of Engineering

Gait Analysis Through Pressure Sensing

Stephanie Speaker, Patrick Riolo, Bailey Felix, Emerson Iannone, Miguel Gomez, Gabriela Angel
Syracuse University Engineering in collaboration with Apollo Biomedical

Introduction: The goal of this project is to design and build a wearable device to analyze foot strike mechanics for athletes. The device will slide into the patient's shoe and fit tightly against the foot, producing a pressure map and real-time data including velocity and acceleration. This project was a joint effort with Apollo Biomedical, a startup company of recent Syracuse University graduates. The motivation behind the project was provided by an aspiring physical therapist, Nick Clayton, who works with athletes at Villanova University as a physical trainer. He sees injuries such as Achilles tendonitis, plantar fasciitis, shin splints, IT-band syndrome, and stress fractures, all of which can be easily treated if diagnosed prior to the occurrence of a more serious injury. Most of these injuries stem from improper gait. By analyzing gait, abnormalities can be addressed, leading to longer and healthier athletic careers. Of the current technology options, none are suitable to properly analyze gait.

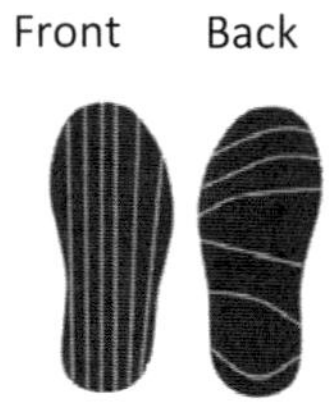

Figure 1: Layout of copper wire

One of the most important aspects of gait analysis is attainment of a pressure map as accurate pressure measurement is essential for effective diagnosis and evaluation. There are eight major locations on the foot where pressure measurement is most important and which give a general sense of the patient's stance. Measuring pressure at additional points will however, produce a more accurate profile. Bluetooth capability is required to ease the transfer of data from the device to the test administrator's computer or phone as opposed to other options such as Wi-Fi. Using Bluetooth allows for data to be transferred regardless of access to a Wi-Fi network. This immediate access to data allows for a timely diagnosis, and eventually tips on preventative measures. The insole has a thin design, enabling it to be easily inserted and removed from the person's shoes while still being comfortable for the patient. It will be reusable upon patient visits or between individuals who share the same shoe size. The goal is to have multiple insoles with different shoe sizes and two microcontrollers (one for each foot) that will be able to plug into the correct size insoles for the athletes. The device also needs to be rechargeable in order to maximize its use. The power source would be located in the microcontroller black box.

The device must be easy to use and affordable to appeal to a wide demographic. It is catered towards athletes but has potential to go beyond that. The universal and user-friendly aspects of this device will set it apart from other devices. Many existing products have features that make the patient feel that he or she is being examined. With this device, patients should receive more accurate readings and seamless evaluations, filling the void in the market and speeding up the rehabilitation process.

Methods: Materials needed for this project were: shoe inserts, copper wiring, ribbon wiring, copper tape, pressure fabric, and an Arduino Uno.

The first step was creating a prototype. This was crafted using women's shoe inserts which sewed to a piece of pressure fabric lined with copper wire on both front and back. See **Figure 1** for the outline of where the wire was placed. The wires were soldered to ribbon wiring in order to transmit signals when plugged into an Arduino Uno. A 3D printed box was made to hold the Arduino Uno and protect it from wear and tear and weather treatments. The next step was creating an Arduino code which would transmit when pressure was placed at certain points on the copper wire. The cross section of the front and back copper wires are the points that were measured. A MATLAB code was created to map the pressure and show the intensity of the pressure.

The next part was to switch the MATLAB and Arduino code to work off of Bluetooth, this would make transfer of data much easier and accessible in even the most remote areas. This part of the code is still in production.

Results: The prototype was tested for comfort by making sure that the copper wiring did not cause any pain. It was concluded that the copper wiring was thin enough that it did not cause any discomfort. As the prototype is still in the works, the accuracy of the data transmitted is currently being tested. The goal is to work with current athletes who have leg/foot injuries and those that do not to test how the gait analysis differs. **Figure 2** shows the initial findings based on the first trial run of the code. The trail run however was based off of a Wi-Fi module and we are currently in the process of switching to a Bluetooth module. We are hoping a newer, more condensed code will yield more accurate results.

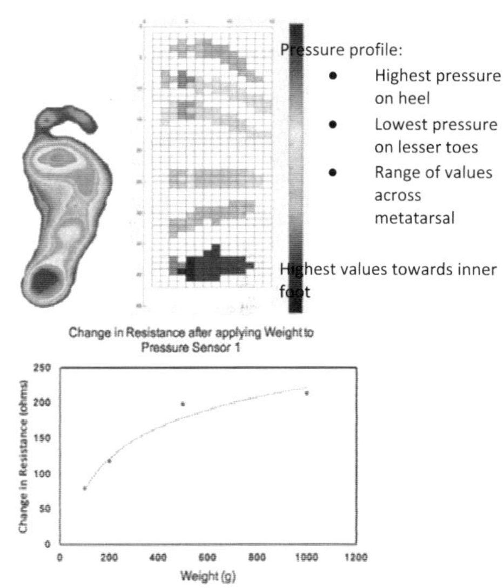

Figure 2: Results based on first run of code

Acknowledgements: Kristen Casella, Devon Foley, Sarah Willsey. These individuals previously worked on this project and crafted the original MATLAB and Arduino code. Douglas Yung, our coordinator. As well as Appollo Biomedical, our clients.

P3-32

Deep Tissue Injury (DTI) Detector

Eric John[1], Amanda Joyce[1], Meenakshi Patel[1], Michael Neidrauer, PhD[1], Richard Huneke, DVM[3], Michael Weingarten, MD[3], Wei-Heng Shih, PhD[2], Wan Shih, PhD[1]

1. School of Biomedical Engineering, Science and Health Systems, Drexel University, Philadelphia, PA
2. College of Engineering, Drexel University, Philadelphia, PA
3. College of Medicine, Drexel University, Philadelphia, PA

Introduction: Deep Tissue Injury (DTI) is a type of pressure ulcer developed next to bony prominences that progresses outward towards the skin. [1] Therefore, DTI can only be identified once it reaches the surface of the skin. There are a few current solutions to detect DTI, but all fall short to be the gold standard. Because of the high cost of treatment, in 2008 the Center for Medicare and Medicaid Services (CMS) stopped reimbursing hospitals for DTI acquired in the hospitals. [2] This incentivizes medical professionals to diagnose DTI prior to hospital admission. Our approach is to use a set of Piezoelectric Fingers (PEF) of varying depth resolutions to detect changes in elastic modulus between the skin surface and the sacral bone. Piezoelectric effect is a unique material property in piezoelectric ceramics, such as lead zirconate titanate (PZT). When electric voltage is applied to a piezoelectric material, this induces mechanical deflection and causes the material to bend. The effect is also reversible in the sense that electric voltage is induced when the material is deflected. DTI tissue has shown to have a higher elastic modulus than healthy tissue in numerous previous studies. [3] Using a set of four PEF probes with varying depth resolutions, changes in elastic modulus can be detected indicating the existence of a DTI and all three dimensions of the DTI can be measured.

Methods: The PEF probe is an intellectual property of Drexel University and Dr. Wan Shih (US8549933B2) and is made of two lead zirconate titanate (PZT) layers sandwiching one stainless steel layer as shown in Figure 1, and wires are soldered to each layer.

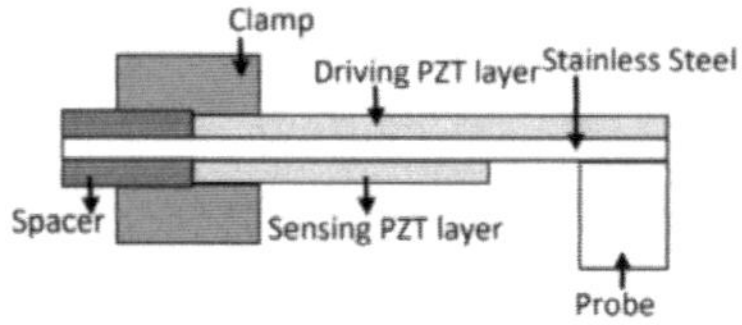

Figure 1. Schematic of PEF probe

When electrical current is applied to the driving PZT layer while the probe is on top of target tissue, the driving PZT layer deflects downward and causes the probe to push onto thee tissue. This resultant deflection then induces voltage in the sensing PZT layer, which is measured as voltage. By tracking the change in voltage measurement with and without tissue, elastic modulus of target tissue can be calculated. Another property of PEF probe is that the measurable depth is double the width of the probe. If the PEF has 4 mm width, the sensor can measure up to 8 mm beneath the skin surface. A set of at least 4 probes can be used together to detect elastic modulus changes at various depths below the skin surface. Once the PEF is created, it must be tested for loss factor on both the driving and sensing layer. While a loss factor below 2% is ideal, PEFs with loss factor's below 5% are deemed acceptable.

In order to test whether the probe can detect difference in stiffness, five measurements at different voltages were taken probe without gelatin, on gelatin and on clay embedded in gelatin. If the probe is functional, the applied and induced voltages would create a linear line. The elastic modulus is determined by using induced voltage values as shown below.

Results: The probe tested is a functional size 4 mm probe, as shown in Figure 2.

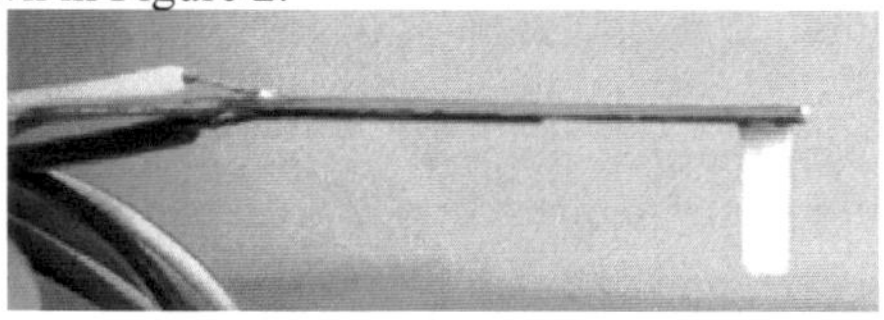

Figure 2. Size 4 PEF Probe

The results in Figure 3 showed that the induced voltages increased when more voltage was applied indicating the measurements were stable.

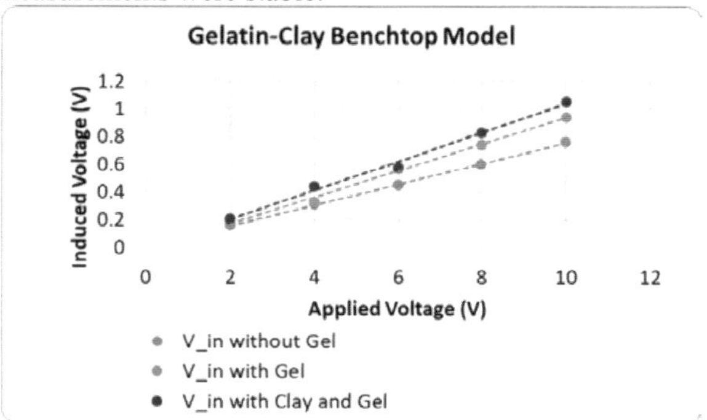

Figure 3. Measurements on Gelatin-Clay Model

Conclusions: It can be concluded from the results that the probe is functional and capable of measuring changes in elastic modulus. More data will be collected using simulated benchtop model mimicking measurements over the sacral bone. This will also be accompanied by an animal study involving Sprague-Dawley rats to prove the device can detect DTI in vitro.

References:

1. Stekelenburg A. et al. Archives of PM&R. 2008;89(7):1410-1413
2. Sullivan R. Ostomy Wound Management.
3. Yamada H. et al. MBEC. 2017;55(1):79-88

Acknowledgements: We want to acknowledge School of Biomedical Engineering, Science and Health Systems, and Close School of Entrepreneurship at Drexel University for funding the study.

An acoustic bioreactor for high throughput chemotherapeutic screening on perfused three dimensional tumors

Matthew Carey, Kayla Van Buren, Ben Ross-Johnsrud*, Bart Lipkens* and Anthony English
Western New England University, Department of Biomedical Engineering,
1215 Wilbraham Rd. Springfield MA
*FloDesign Sonics Inc., 380 Main St., Wilbraham MA 01095

Introduction Current oncological pharmaceutical testing protocols include chemotherapeutic applications to two dimensional cell cultures and animal tumor models. Human tumors *in vivo*, however, often grow in three dimensional clusters and their response to chemotherapeutic drugs can be significantly different from that predicted by animal models. The discrepancies in current methods render them unable to identify many failing drug candidates early in the development process [1]. Current methods for cultivating three dimensional tumor clusters, such as hanging drop plates and non-adherent conical tubes, are a challenge to perfuse. The development of perfused bioreactors capable of cultivating human tumors in three dimensions, therefore, is of great interest.

The goal of this research is to develop a bioreactor for cultivating perfused three dimensional tumors. To overcome the current limitations of hanging drop plates and non-adherent surface cultivation techniques, a novel acoustophoretic method for cultivating cancer cell clusters is proposed. The ability to suspend cancer cell clusters using an acoustic field is examined using finite element simulations. A perfused bioreactor is then designed using computer aided design software.

Methods: Acoustic pressure field simulations were preformed using finite element software (COMSOL Inc. v5.4) based on a 2 MHz ceramic transducer (APC International), and a fused silica reflector (University Wafer 100mm DSP 500μm U01-120920). The results of these simulations were used to design a bioreactor prototype capable of perfusing three dimensional tumors held in an acoustic field using computer aided design software (Dassault Systèmes SolidWorks). A transducer cooling chamber was included to maintain a constant temperature. A rapid prototyper (Objet Eden500V) rendered a three dimensional prototype of the design (Objet FulCure 720). A laser cutter (Trotek Speedy 100) cut fused silica wafers into the appropriate sizes for attachment to the bioreactor using epoxy (Loctite AA 3311), which was cured with a UV light source. A function generator (Agilent 33210A) drove the transducer with a $10V_{p\text{-}p}$ sinusoid. The bioreactor was perfused using a syringe pump (New Era Syringe Pump NE-1000).

Results: After several design iterations, finite element simulations demonstrated the feasibility to suspend cell clusters using acoustic fields. Figure 1 shows a representative simulation of an acoustic pressure field from a 2 MHz transducer. Pressure nodes and peaks, indicated in blue and red respectively, provide potential sites for cell aggregation.

The design, shown in Figure 2 combines a cell chamber, cooling chamber, diffusers for media and coolant exchange, a transducer housing, and a water bath for additional heat exchange. Interlocking connections prevent leakage and a thin mylar film, transparent to acoustic waves, separates the cooling and cell chambers.

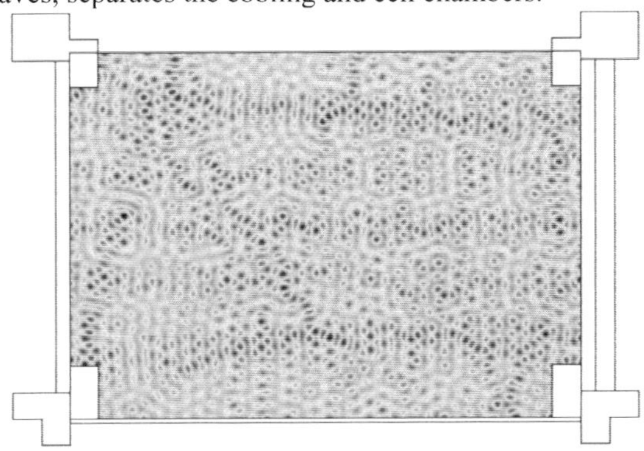

Figure 1. Acoustic pressure field cross section produced by a 2 MHz transducer. Acoustic pressure nodes capable of suspending tumor cell clusters are present. Nodes and peaks, in blue and red respectively, in the pressure fields provide potential trapping points for cell clusters.

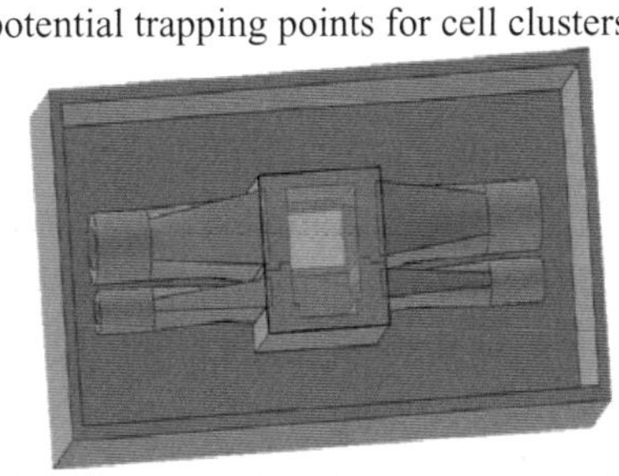

Figure 2. Final concept for a three compartment bioreactor in a cooling bath. The large upper chamber contains the cells and the reflector. The transducer is housed at the very bottom and the small cooling chamber is above it. Compartments are separated with mylar film. (Opaque portions of the water bath top and affixing screws were excluded for clarity of drawing details)

Conclusions: A perfused acoustophoretic bioreactor has been successfully designed based on the results of finite element simulations. The bioreactor consisted of three separate chambers; one for cell culturing, cooling, and for a transducer housing. The design was compatible with a microscope stage, addressed the need for transducer and reflector cooling, and supplied a system to perfuse and test cells.

.References:

[1] G. Dane, K. Roy, R. A. Milton, S. Lichtner, R. Brindle, and L. Pack, Achieving large-scale, cost-effective, reproducible manufacturing of high-quality cells, A technology roadmap to 2025, National Cell Manufacturing Consortium, 2016.

Acknowledgements: Grant support from the Massachusetts Life Sciences Center is gratefully acknowledged. Technical support provided by Lisa Murray during rapid prototyping is also gratefully acknowledged.

Neonatal Real-Time Photoplethysmogram Heart Rate Monitoring System During Resuscitation

Luke McConnaghy[1], Alexander Gray[1], Brandon Williams[1], Ying Sun PhD Gregory Welch M.D.
Biomedical Engineering Program, Department of Electrical, Computer and Biomedical Engineering
University of Rhode Island, Kingston, RI 02881, USA. Correspondence email: yingsun@uri.edu

Significance: About 10% of neonatal patients require some form of resuscitation immediately following birth in order to regulate their heart rate and breathing. Major changes in a baby's external environment can lead to complications in a neonate's circulatory and respiratory physiology. A baby's heart rate helps doctors determine the proper form of resuscitation to prevent any complications following birth.

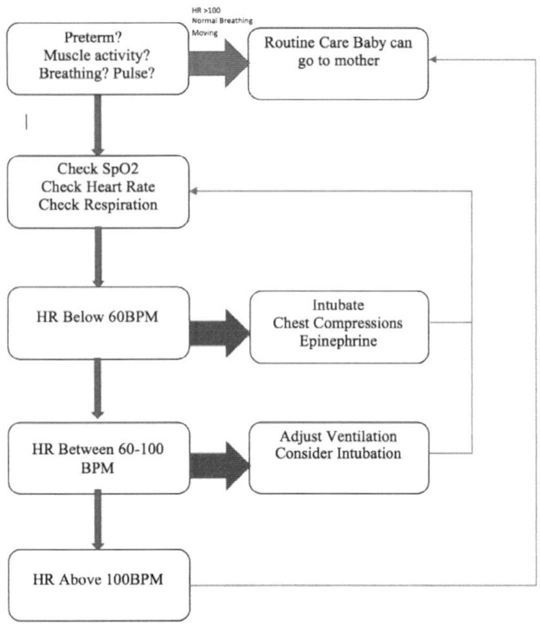

Figure 1. Flow Chart demonstrating general procedure for neonatal resuscitation.

Currently, there is not a universally accepted method for taking accurate real-time heart rate for neonatal patients. One method requires the use of electrocardiogram (ECG) leads on the patient's chest which makes it complicated to perform chest compressions properly, and studies have shown inaccuracies of ECG for neonatal patients. Another method requires the palpation the umbilical cord to count beats for 15 seconds to estimate heart rate. Our goal is to design a portable, low-cost, lightweight photoplethysmogram (PPG) system which can detect and display real time heart rate of neonatal patients immediately after birth. This is a proof of concept system to demonstrate the feasibility of taking real time PPG heart rate measurements on the forehead. This type of system has a high potential for commercial use for hospitals.

Design Description: We have designed a PPG-based heart rate monitoring system to measure heart rate on the forehead. PPG is a low-cost optical technique which detects blood volume changes in microvascular tissue underneath the skin. Utilizing a Gravity heart rate sensor, we can obtain the PPG signal from the supratrochlear and supraorbital arteries on the forehead.

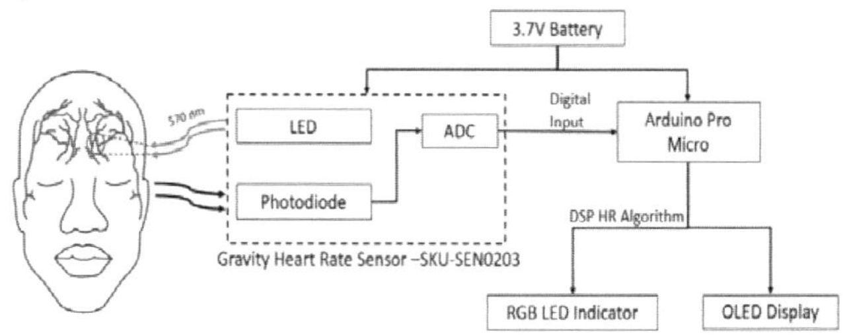

Figure 2. Arterial anatomy of the human forehead and block diagram for prototype neonatal heart rate monitor.

To calculate the subject's heart rate, we have programmed an Arduino Pro Micro to read and process the data sent from the Gravity heart rate sensor. The heart rate detection algorithm determines the time between peak threshold values to calculate beats per minute value. We utilize an array of heart rate values to average three previous values to filter out any inconsistencies in our measurements. The resulting heart rate is shown on an OLED display. A light emitting diode (LED) indicator is included to easily identify the type of resuscitation to be performed by health professionals. Our compact portable system is powered by a 3.7 V, 150mAh, Lithium Polymer cell and fit on a 1.6" x 1.3" perforated circuit board. The device will consist of two layers. The superficial layer will contain the Arduino, LED, OLED display, battery and power switch. The layer which interfaces with the skin contains the Gravity heart rate sensor surrounded by 3M hydrocolloid medical tape.

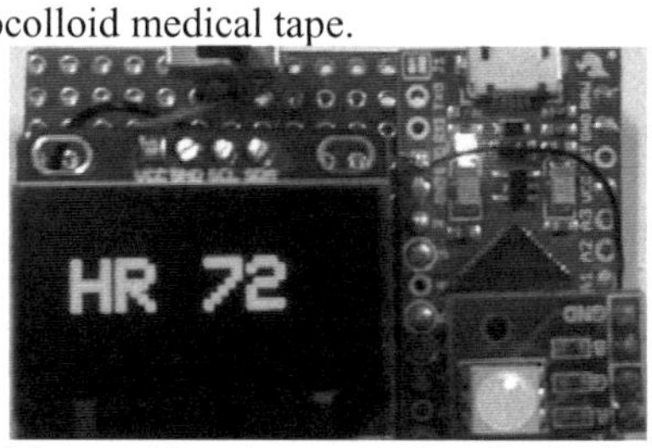

Figure 3. Neonatal heart rate monitor displaying heart rate of team members from the forehead

The tape is a non-harming and moisture resistant adhesive to help to maintain pressure for the sensor on the neonate's forehead. The end goal is to give clinicians the ability to wipe the baby's forehead clean apply the sensor to the forehead, and start recording heart rate measurements.

Design Criteria: The design of this PPG based system requires a small and adhesive device to be able to adhere to the curvature of a neonate's forehead immediately following birth. Our device is limited by the size of the neonate's forehead (approx. 2" x 2"). In order to take accurate measurements, the sensor surface must be flush with the skin requiring the use of a strong and infant safe adhesive. The adhesive portion of the sensor needs to be made of a material that will not cause harm to the baby during this process. It is also beneficial to have an adhesive that can maintain a strong hold, but also be removable for replacement back on the forehead in case the sensor is not placed over the artery. The time sensitive nature for resuscitation requires prompt heart rate determination to limit negative outcomes for a baby in tachycardia or bradycardia. Overall, this type of application requires a compact and robust device that can take accurate readings efficiently.

Results: The device was tested in comparison to a beat-to-beat ECG heart rate monitor and the differences between beat-to-beat values over a 20-data point period is shown in Figure 4. We have demonstrated our device can display an accurate heart rate after a ~5 second delay for initialization. This accounts for a 2 second delay before any heart rate is detected and an additional 3 seconds for the heart rate value to normalize through an averaging method.

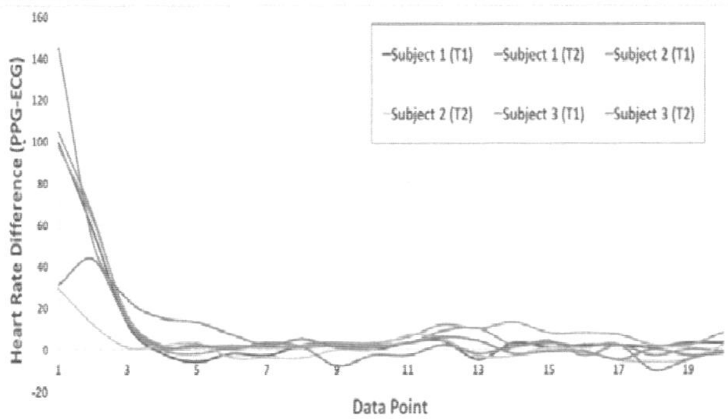

Figure 4. Graph displays the beat-to-beat difference between ECG and PPG heart rate measurements (PPG-ECG) for a 20 second period using a ECG built on a breadboard.

Results for resting heart rate of three adult subjects yielded a consistent reading (within ±13bpm of ECG value) for all trials conducted. This relatively high difference can be accounted for by both the inconsistencies from the ECG as well as the variability of the device. The above results demonstrate the ability to take quick measurements, within 5 seconds, after placing the sensor on the forehead. The device does not yet demonstrate an accuracy low enough for further human trials therefore further testing shall be conducted. Our device has a quicker setup time when compared to the time it takes to apply the ECG leads on the patient. Overall, our device is capable of taking quick and accurate heart rate measurements when compared to conventional methods such as ECG and other medical grade devices.

Future Implications: The ability to quickly and accurately measure the heart rate of a newborn is vital for ensuring a healthy birth outcome. Current clinical procedures such as manually feeling the pulse through the umbilical cord and ECG are severely limited in accuracy and speed. Our PPG-based heart rate monitoring system can provide an accurate and fast calculation of heart rate on the forehead of an adult which eliminates any obstructions to performing resuscitation. Preliminary studies have shown that our device can obtain accurate heart rate measurements within a difference of 13 bpm compared to the ECG less than 5 seconds after being switched on. The simple comparison test between the ECG and PPG monitor expose some of the accuracy limitations within our system. However, our device has a quicker setup time when compared to the time it takes to apply the ECG leads on the patient. In addition to displaying the beat-to-beat heart rate, an LED is used to indicate the correct clinical response to be performed by medical professionals. Due to the low cost associated with fabricating our system, the device can be utilized as a disposable for one-time use in a hospital setting. Further studies will be conducted with more participants to determine the clinical accuracy in comparison to medical grade ECG. Experiments conducted under strenuous exercise could provide insight into how our system might operate in the typically higher resting heart rate range (HR > 100 bpm) of neonates. In the future, a flexible design will be implemented to increase the comfortability and decrease the thickness of the wearable device.

References:

1. Fishman CE. "...delivery room resuscitation". 2018
2. Phillipos E, "Heart Rate Assessment" 2016 109:130-138
3. Grubb, M R, et al. "Forehead Reflectance Photoplethysmography" vol. 35, no. 5, 2014, pp. 881–893
4. Vali, Payam, et al. "Neonatal Resuscitation: Evolving Strategies." vol. 1, no. 1, 2015,
5. A. Johansson, et al. "Monitoring of heart and respiratory rates". 1999 Dec; 15(7-8): 461–467

Acknowledgements: We would like to acknowledge Dr. Gregory Welch for proposing the idea of this project to provide a proof of concept device for neonatal monitoring. We would also like to acknowledge Tanya Wang for her continued technical expertise and support throughout the development of this project. We would like to say thank you to all involved.

A 2-Photon Approach to Imaging the Spatial Relationship of Collagen and Nerve Fibers in MMP-13 Inhibited and Uninhibited Subcutaneous White Adipose Tissue

Mitchell Harling, Karissa Tilbury, William P. Breeding
University of Maine Chemical and Biomedical Engineering Department

Introduction: While neuropathy is easily identifiable, it is not well known how innervation mechanisms in obese tissue are affected, therefore investigation into the architecture of fibrotic adipose tissue is a necessary endeavor to uncover valuable diagnostic and mechanistic information. Studies have demonstrated that the fibrotic nature of obese adipose tissue results in excess collagen deposition [1], which is hypothesized to cause a disruption in tissue innervation, directly resulting in an array of neuropathic related issues and abnormalities [2]. Little structurally specific information is known about fibrotic adipose tissue, as common confocal techniques lack adequate image resolution in complex biological systems. Two-photon microscopy overcomes the limitations of traditional confocal microscopy by minimizing excessive scattering and unwanted absorption with higher wavelength excitation, ultimately allowing for deep, high-resolution imaging. In addition, second harmonic generation (SHG), an energy-conserving two-photon process, allows for the label free imaging of non-centrosymmetric structures, such as collagen [3]. Here we present a two-photon approach to image innervation and structural collagen components in thick adipose tissue pads (75-400 μm), collecting fluorescence signal from Alexa 594-labeled nerves and SHG signal from collagen.
Previous work by the UMaine Tilbury and Townsend groups have demonstrated alterations in the innervation of obese and aging subcutaneous white adipose tissues (ScWAT) [4]. Matrix metallopeptidase 13 (MMP-13) encodes a collagenase that has a significant role in the breakdown and restructuring of the extracellular and collagen matrices, its role will be explored via colocalization of nerves and surrounding collagen in MMP-13 inhibited and uninhibited ScWAT. Colocalization analysis will be performed using the ImageJ plugin, JACoP, and a recently adapted astrophysics technique, the Metric Space Technique (MST). We expect these imaging tools and analyses to provide insight into MMP-13 enzymatic activity and potential collagen deposition contributions to alterations in innervation and thus, neuropathy.

Methods: Mice were fed four diets; normal diet (CV), normal diet with MMP-13 inhibitor (CD), high fat diet (HV), and high fat diet with MMP-13 inhibitor (HFD). ScWAT tissue sections were prepared and immunostained with a protein gene product 9.5 antibody (PGP9.5) (primary) followed by an Alexa 594 antibody (secondary). 3 sets of images were collected from 3 distinct locations in each tissue section with a home-built two-photon microscope modified from an Olympus FV300 system. The Alexa 594 labeled nerve fibers were imaged using an excitation wavelength of 805 nm and a 582/64 nm bandpass filter (Semrock, Rochester, New York). The collagen was imaged using SHG with an excitation wavelength of 890 nm and a 448/20 nm bandpass filter (Semrock, Rochester, New York). Image analysis was performed using the ImageJ JACoP plugin for thresholding and to obtain the Pearson's correlation coefficient (PCC). Manual image thresholding was performed to optimize JACoP PCC values.

Results: After thresholding in JACoP, the average PCC for each mouse type was determined from representative samples of each set.

Table 1. The PCCs determined using JACoP from representative samples of each mouse type.

Mouse Type	CV	CD	HV	HFD
PCC	0.546	0.4684	0.515	0.499

A PCC near +1 or -1 represents strong colocalization, or anti-colocalization respectively, while a PCC of 0 represents no colocalization. These values indicate a similar degree of moderate colocalization across the mouse types.

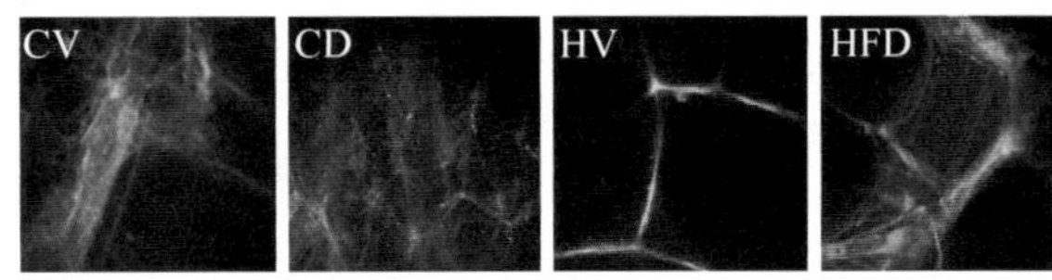

Figure 1. Representative merged nerve (green) and collagen (red) images of ScWAT mouse sets, from left to right: CV, CD, HV, and HFD.

The images above represent the variability in collagen and nerve colocalization across mouse types, and the merged image reveals object overlap.

Conclusions: While the JACoP plugin appears to yield no significant difference in the colocalization of nerve fibers and collagen, qualitative comparison of images, such as those displayed in Figure 1, present high variability and contrast in collagen and nerve structures. Several factors contributing to sample variability, such as density of tissue components, thickness of fat pads, noise, and scattering, will need further consideration before conclusions based on statistically significant results are made. A more robust method of image analysis is actively in development to overcome these factors, which includes the use of the MST as a readout for; 1) distribution of density, 2) distribution of volume (area), 3) distribution of the number of connected components, and 4) the distribution of these components' average filament indices (a morphology indicator). We anticipate that the combination of our current colocalization analysis and future MST analysis will reveal critical information about MMP-13 activity, potential collagen deposition contributions to alterations in innervation, and neuropathy mechanisms in obesity.

References:

1. Wynn TA. J. Pathol. 2008;214(2):199-210.
2. Blaszkiewicz M. bioRxiv; 2018
3. Mostaço-Guidolin, L. IJMS. 2017;18(8):1772.
4. Tilbury K. SPIE 2019; 10882(42).

Acknowledgements: We would like to thank Dr. Kristy Townsend and her lab group for entrusting us with imaging and analysis of their samples, and Dr. Andre Khalil for his ongoing optimization of his self-developed colocalization/MST ImageJ code.

Design of a Bone Morphogenetic Protein Based Carapace Repair Device

Sylvia Baeyens[1], E. McKenzie Cummins[1], Matthew Figucia[1], Daniel Kanter[1], Aislinn Keane[2], Christopher Wilson[2]
[1]Boston University, [2]Bioventus LLC

Introduction: Worldwide, approximately two-thirds of all turtle populations are facing serious population declines.[1] Traumatic fractures of the turtle shell, or carapace, are a large threat to these vulnerable populations.[2] Current treatment techniques require multiple surgeries over several weeks, which are stressful to the animal and diminish long term success once released back into the wild. [3] Unpublished case studies in which the osteogenic cytokine bone morphogenetic protein 2 (BMP-2) was repeatedly administered (in solution) to severe carapace fractures suggest that this FDA-approved drug could dramatically improve care for animals sustaining these injuries. A carrier that retains BMP-2 at the implant site would enable a single administration, thereby minimizing procedures and stress on the animals. The objective of this project was to design an injectable carrier with a high affinity for BMP. One key challenge in this project is identifying carrier compositions compatible with passage through a fine gauge needle, which is necessary for delivering the material to the thin subcarapacial tissue in the wound bed of a fracture.

Methods: To leverage existing FDA-approved biomaterials, type I collagen was selected as a base flowable material. Because type I collagen has a low affinity for BMPs, ceramic particles (~5μm mean diameter) composed of stoichiometric hydroxyapatite (HA) were added. HA has a relatively high affinity for BMP family molecules, similar to the mineral in native bone[4]. Each carrier composition's ease of injectability was evaluated by observing flow through, and relative effort required to draw and eject through an 18, 27, and 30 gauge needle. Mechanical characterization of the composite materials was performed using continuous flow rheometry to measure viscosity over multiple shear rates. The concentration of type I collagen (recombinant human, expressed in Nicotiana tabacum) was constant at 3 mg/mL, and the HA particles were added at 0-100 mg/mL. The measurement geometry was a 20mm 2.019° cone and plate. The shear rate ranged from 0.1 to 10 1/s with 10 data points per decade being obtained for each sample over a duration of 60 seconds.

Results: An injectability test on the 100mg/mL, 10mg/mL, 3 mg/mL, and the 0 mg/mL samples determined that each carrier could be easily drawn/ejected through the 18G and 27G needle; however, discrepancies occurred with the 30G needle. Only the 3 mg/mL and 0mg/mL HA formulations could be pushed and pulled through the 30G needle, but we qualitatively described the process as difficult. Our flow test results indicated that the 30 mg/mL HA formulation had the greatest viscosity at the lowest and highest shear rates. The 10 mg/mL HA formulation had the highest viscosity at a shear rate of 1, showing on average, higher particle densities tend to have greater measures of viscosity.

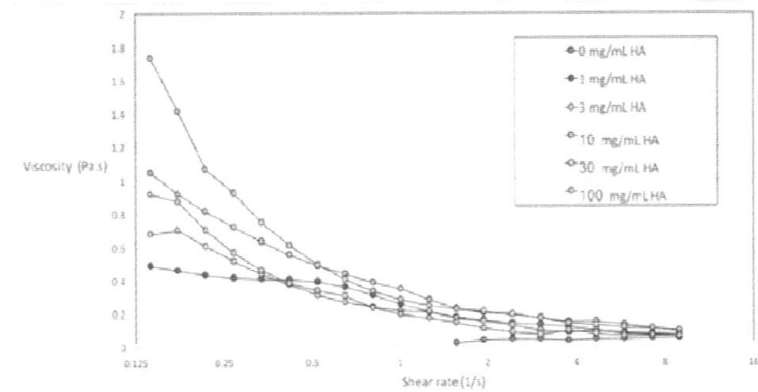

Figure 1. Shear rate-dependent viscosities of a type I collagen carrier with different concentrations of HA particles.

Conclusions: These preliminary data indicate that the composite slurries can be administered through fine gauge needles. In unpublished animal studies performed by our group, a 27G needle was found to be appropriate for injecting materials within the interstitium of subcarapacial soft tissue. Qualitatively, the ease with which these slurries can pass through a needle appears to be inversely related to the HA content. Rheological measurements indicate that the investigated formulations are shear-thinning. This suggests that in clinical use, the formulation will be easy to push through a needle due to the force applied to it, but will become more viscous again at the wound site, and will not flow away as easily. Additionally, preliminary experiments demonstrated that formulations with higher particle densities have greater measures of viscosity, which is consistent with the qualitative evaluations in the injectability tests. Ongoing work aims to identify the threshold concentration of HA for injectability, quantify the BMP-binding affinity of these materials, and ensure repeatability of preparing the slurries. In addition, lead candidate formulations will be assessed for feasibility and ease of use by veterinary staff from the New England Aquarium.

References:

1. Wallace BP. et al. Glob Cons Mar Turt. 2011; 6(9).
2. Bentivegna F. et al. Car Tre Sea Turt. 1993; 61(6).
3. Fleming G. Cli Tec: Sh Rep. 2008; 17(4):246-258.
4. Rohanizadeh R, Chung K. Hydr Car BMP. 2011; 37(6):659-672.

Monitoring Self-Contained Breathing Apparatus Performance through Temperature and Physiological Recordings

Jake Duerwald, Thomas Totillo, David Cleary; Advisor Eugene Chabot
Department of Electrical, Computer and Biomedical Engineering
University of Rhode Island, Kingston, RI 02881, USA. Correspondence email: jake_duerwald@my.uri.edu

I. Significance

Firefighters face many hazardous situations on a daily basis, requiring their equipment to withstand such scenarios. The Self-Contained Breathing Apparatus (SCBA) is considered the weakest component of a firefighter's safety equipment which faces potential failure at extreme temperatures. The polycarbonate lens of the SBCA, begins to soften at 300°F, and melts at 600°F. These temperatures are common in modern structural fires, especially when a flashover occurs [1]. At around 600° Fahrenheit, the gases in a room can ignite, causing a flashover which can raise temperatures upwards of 1100° F in a matter of seconds. This can cause mask degradation in the form of: bubbling, crazing, microfractures and/or complete deformation. Breaks in the mask can lead to direct skin burns, respiratory damage, and in multiple recorded cases, death.

Currently, there is no system in the marketplace available to predict the probability of Self-Contained Breathing Apparatus failure. A device that records the frequency in which an SCBA has been exposed to extreme heat could provide insight into when a failure is imminent. In addition, a Photoplethysmogram integrated into the SCBA could aid in determining failure based off physiological abnormalities caused by inhaling toxic gases. The device also provides a visual means for active personnel to avoid high temperature environments. Using the combination of these components, the device aims to provide a novel solution to mask degradation injuries otherwise not addressed by current marketplace technologies.

II. Design Description

This device contains two major components: the Temperature Indicator Sensor and the Photoplethysmogram.

Temperature Indicator Sensor: The Temperature Indicator Sensor records and displays ambient air temperature as a series of three ranges. This will inform the fire department about at-risk equipment, as well as providing active personnel a tool to avoid high temperatures. The circuit utilizes a thermistor-based resistive network. Through designed comparators, the difference in voltage will signal a transition into the defined temperature range. The activated ranges will illuminate the appropriate LEDs as well as send the results to the Android device for display. The LED array and circuitry fits under the ventilator of the SCBA, with the thermistor itself resting on the external lens of the mask.

Photoplethysmogram: Using a pulse oximeter manufactured by PulseSensor, and a PIC microcontroller, a heart rate can be read. Processing is achieved through this PIC microcontroller, where data returned from the pulse oximeter can be filtered to return an accurate reading. Testing has found that the optimal location for a reflective PPG is on the protuberance of the eyebrow[3]. For integration into the SCBA, the location slightly above this has proved to be functional. The data is sent to a remote Android device via Bluetooth where an Android application displays the information for interpretation by a monitoring officer.

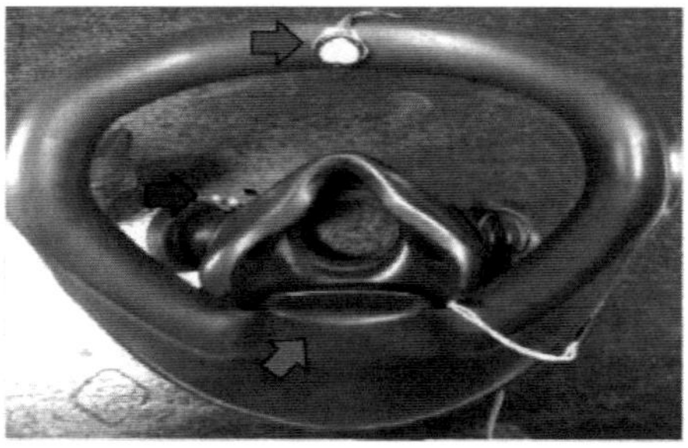

Figure 1: Integration into a SCBA: Red Arrow: PPG; Blue Arrow 3 LED Display; Green Arrow: Housing Unit for Temperature Indicator Sensor circuit.

III. Design Criteria

During the design process, there were multiple constraints that had to be overcome. A main limitation was the lack of space within the SCBA while in use. The design had to remain comfortable for personnel, including placement of device circuitry as well as maintaining a clear field of vision. In relation to size, the Temperature Indicator Sensor and PPG circuitry themselves had to be optimized for practical use of the device. A housing unit was designed to fit securely around these sensors to provide protection from both mechanical and thermal damage. Through trial and error, the location under the ventilator of the SCBA proved to allow for the most room for the device components. The current prototype utilizes a perforated circuit board to display the functionality of the Temperature Indicator Sensor. The technological components for the temperature sensor were set to achieve specific temperature ranges shown in Figure 2 below. The components of the PPG and temperature sensor are working conjunctively on a breadboard as well as on separate perforated circuit boards, displaying remote capability.

LED Temperature Range	Temperature °C (°F)	Resistance (kΩ)
Green	> 25°C (77°F)	10
Yellow	> 33°C (91°F)	8.2
Red	> 63°C (145°F)	2.2

Figure 2: Temperature Indicator Sensor ranges with corresponding colored LED and resistance.

IV. Results

Temperature Indicator Sensor: The current prototype is set to lower temperatures to allow for easier testing and demonstration. For qualitative testing the thermistor was exposed to a butane lighter with a flame approximately 2 inches away for 8 seconds. The results for activation time as follows: Green; 1.42 s; Yellow: 2.09 s; Red: 4.01 s. These times were well below the thermistor heat time constant of 10 seconds. Deactivation time results were considerably slower, averaging 14.5 s due to the ambient air remaining at the higher temperatures. This would not cause double recordings for entering a specific range, nor would it cause the user to enter a higher temperature environment. The sensor functioned as designed both outside and while installed into the SCBA.

Photoplethysmogram: The signal produced from the pulse oximeter placed at the center of the forehead gave clear and consistent results when sitting still with the mask on. Movement of the head and other motion artifacts, however, did cause signal disturbances. The circuit has been placed onto a perforated circuit board to make it easier to integrate into the mask. The PPG circuit currently resides on the outside of the SCBA. The independent PPG circuit is operational, including the transmission to the Android device.

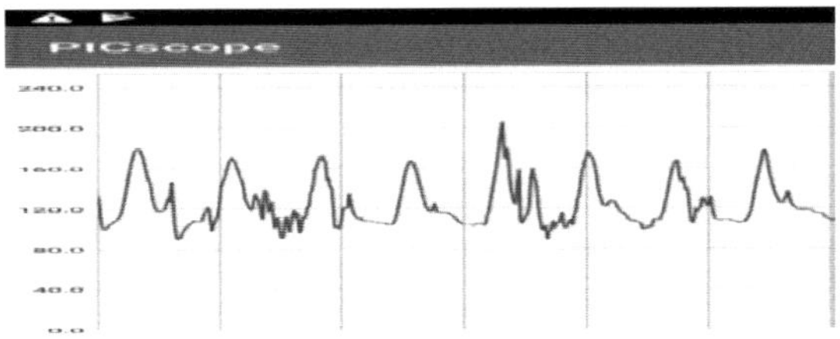

Figure 3: PPG signal received from the forehead brim of the SCBA displayed on the Android applications.

V. Future Implications

The overall purpose of the device is to test for degradation of the SCBA after it experiences different amounts of heat exposure. The current prototype has reached many of the desired goals, however, it would benefit from additional alterations. Currently, the PPG sensor resides outside of the mask leaving it exposed to extreme conditions as well as physical damage. In the future, the PPG will be integrated into the mask with its own housing unit and optimally on the same board as the temperature indication circuit. Currently the sensors are both powered by separate, non-rechargeable battery packs. Possible alternatives could be: rechargeable batteries, wireless charging docks, or Thermo-electric batteries. After making these optimizations, our design will be ready for active duty testing.

The project has been designed as a possible method for determining the probability of an SCBA failure. Although the components of the device have worked independently, no tests have been performed to determine its effectiveness in predicting failure. Future research and testing with these sensors needs to be completed to have a full understanding of the practicality of the device. Although the Android application is capable of displaying the temperature ranges experienced by the mask in real time, it currently does not have a method of data storage. Adding a means of data recording is required for determining the effectiveness of this device as a way to predict SCBA failure.

The commercialization for this design can be marketed towards independent fire departments as a retrofittable unit for current inventory, or as a device for SCBA manufacturers to include in their current products. When purchasing the entire unit independently, the technology would be easily integratable into any existing SCBA. Alternatively, manufacturers can include this technology to enhance SCBA safety. This will provide an avenue for additional purchases to replace high risk of failure SCBAs. This device, when completed and tested thoroughly, will provide an alternative technique for identifying at-risk of failure SCBAs.

References

[1] National Fire Protection Association. "SCBA facepiece lenses may undergo thermal degradation when exposed to intense heat" NFPA Alert Notice, July 2, 2012

[2] A. Mensch, G. Braga, N. Bryner. "Fire Exposures of Firefighter Self-Contained Breathing Apparatus Facepiece Lenses" NIST Technical Note 1724, November 2011

[3] Toshiyo Tamura, Yuka Maeda "Wearable Photoplethysmographic Sensors - Past and Present," Electronics pg. 282-302, MDPI, 2014

[4] Hibbs, James D, and John W Brodhecker. "System and Method for Identifying Unsafe Temperature Conditions." *United States Patent*, Fireeye Development Inc., 9 July 2002

Acknowledgements: Special thank you to Kingston Fire Chief Nate Barrington for bringing the issue of mask failure to our attention and providing a SCBA for prototyping.

Neuron Emulator with Realistic Action Potentials Voltage-Clampable via an Electrophysiological Instrument

Pascaline Uwase, Fatima Issa, Huafen Guo, Samuel Ng, Ying Sun PhD
University of Rhode Island, Kingston, RI, USA

Significance:
A neuron emulator is an electronic device designed to replicate the response of a neuron to electrical stimuli, including the passive and active electrical properties [3]. The NE provides a convenient way to test the functionality of an electrophysiological instrument such as the voltage clamp or patch clamp [1]. The RC circuit reflects on the passive property while the firing of action potentials represents the active property [2]. The signal generated by this device is intended to include both the inward and outward currents to simulate the potassium and sodium channels found in the cell membranes of neurons. The first-generation design of the NE was based on the utilization of a voltage source to create an adequate action potential waveform. However, this method generated a signal that could not be cancelled out by an external feedback current. The second generation of the NE design introduced the application of a capacitance source instead of a voltage source [4]. Pre-charged capacitors were switched into the circuit at appropriate times through analog switches. However, due to the rapid rising time of the depolarization phase, the resulting action potentials were too fast to be clamped entirely by a standard voltage clamp amplifier. Thus, this project aims to provide a more realistic action potential waveform that has a rise time on the order of 1 ms using four switches and three RC circuits. The fourth switch built at the end of the three RC circuits right before the output source wire is hypothesized to slow down the signal before it is amplified thus increasing the time constant of the depolarization phase.

Design Description:
The methodology approach involved in the designation of this project includes the use of Multisim to stimulate a certain portion of the circuit for accuracy, the connection of the circuitry on the breadboard via Mplab and the testing of the circuit using Oscilloscope to generate signals.
Figure 1 shows the schematics of the NE under the control of a microprocessor (PIC18F4525). The inward sodium current and the outward potassium current are represented by the two pre-charged capacitors C_{in} and C_{out}, respectively. Also, the resting potential represented by the parallel connection of Rm and Cm. At the bottom right of figure 1, it shows the pattern of four switches (S0, S1, S2, S3). As four switches operate simultaneously, it will produce the output Vm. Based on Vm, the repolarization of the signal occurs when S0 and S3 are off, and S1 and S2 are on. When the switches S0 and S2 are off while S1 and S3 are on, the output signal depolarizes. The implementation of the PIC microprocessor is necessary to control the four switches [3]. This version of the NE uses Multisim software as a guideline to help in the designing process and stimulation of the circuit portion.

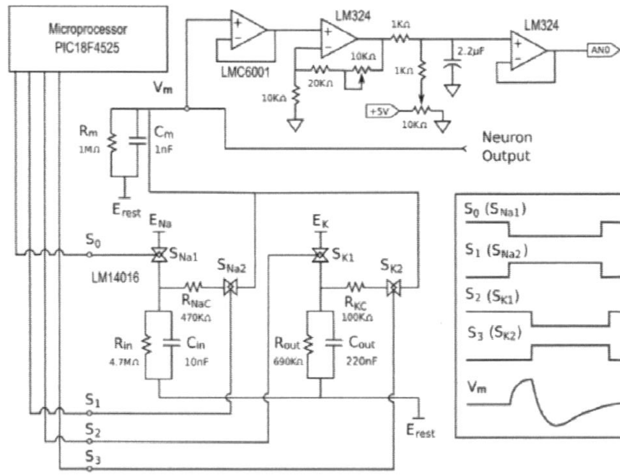

Figure 1. Analog circuit of the neuron emulator controlled by a microprocessor with an inserted panel showing the switching signals and the intended action potential waveform.

Design Criteria:
In addition to the previous NE project, which included pre-charged capacitors and three switches, this current project added a fourth switch at the end of the RC circuit which helps slow down the rising time of Action potential as it propagates.

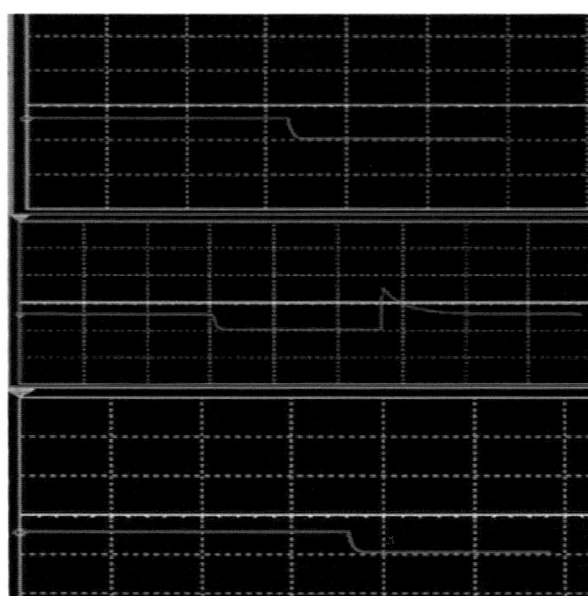

Figure 2. Different Stages of Action Potential Signal (top is part one, the middle is part two, the bottom is part three) of the signal generated by Multisim.

Results:

The usage of Multisim determined that 470kΩ for R_{Na} resistor slow down the rise time (depolarization) which the previous designs were not able to achieve. However, the main issue of the current version of the NE is that the hyperpolarization phase of the output signal shrunk causing an upward vertical shift of the resting state half-way as compare to its ideal position of the resting state. The modification of several other values of components such as the following: R_m = 2.2MΩ, C_m = 2.2nF, R_{out} = 1MΩ, R_{Kc} = 220KΩ, E_{rest} = 2.2KΩ, E_K = 2.7KΩ, E_{Na} = 2.2KΩ, E_k = 56KΩ on the circuitry aided in generating more desirable and stable action potential shape. The simulated and the actual waveforms are not the same, due to the lack of exact representation of certain components such as the microprocessor and the analog switches in Multisim. Figure 2 shows the stages of action potential signal generated as a result of the use of Multisim. The first part of the signal, representing the depolarization of an action potential, corresponds to the S_1 and S_3 open and S_2 and S_4 closed. The second part of the signal, representing the repolarization phase of the action potential, corresponds to when the S_2 and S_3 are open while S_1 and S_4 are closed. The final phase corresponds with the hyperpolarization phase of the signal which occurs when the S_2 and S_4 are open while the S_1 and S_3 are closed. The bottom right of figure 1 shows the pattern of four switches, and the combination of them will produce the output V_m. Figure 3 is the comparison of action potential signals between the oscilloscope and the Multisim model. The idea is that the signal from the oscilloscope should match with the display from Multisim. The utilization of Multisim to estimate the optimal value of certain components are not efficient due to the lack of certain components that are crucial in creating the same RC circuit as the breadboard circuitry. Such components include the PIC microprocessor and the analog switches.

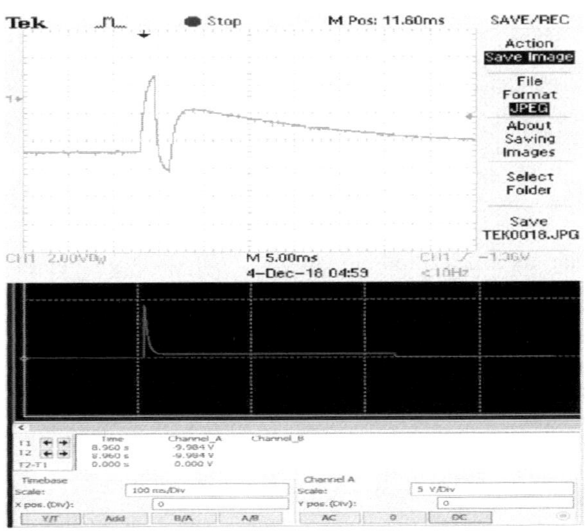

Figures 3. Action potential signals from the NE circuit (top) and Multisim (bottom).

Future Implications:

This project has successfully improved a neuron emulator circuit with the rise time of the action potential close to 1 ms. For the current model of the NE, the main issue is the high vertical shift of the hyperpolarization of the signal. Prioritization will include providing improvements to the circuitry such as possibly adding an extra switch. Further improvements will include an evaluation of the interaction between the neuron emulator and a voltage clamp amplifier. This project serves to provide a better understanding of the relationship between the circuitry and the electrical activities within a neuron cell membrane. The innovation of the Hodgkin and Huxley model introduced the connection between the electrical properties in neurons and circuitry [5]. The NE will serve as a foundation in assisting researchers to test future neurophysiological devices by giving experimental data about what causes an action potential in neurons, how slow or faster these action potential affects other neurons that rely on it to perform a specific task. This breakthrough can help to determine or diagnose patients who might have an underlying disease with their neurons. The materials used to design this device such as resistors, capacitors, PIC18F4525, etc. are in accordance to the RoHS guidelines that prohibit the production of materials with hazardous substances and ethical standards issued by IEEE Code of Ethics that provides ethical standard practices and techniques relating to engineering products in research and industry. The use of the medical instrument called Oscilloscope gave us an understanding of what is happening in the electronic circuit while debugging the design to efficiently identifying errors in circuitry. This project instrument follows the IEC 60812: 2006 and IEC 61010-1:2001 guideline techniques for detecting failure modes, and offering safety requirement regarding measuring, controlling, and analyzing of data from electrical equipment. The NE will provide a more cost-efficient way to test such devices. The NE not only simulates the electrical activities of a neuron but also draws a connection between activities in a neuron and electronic circuitry. This NE used the cheapest and lightest materials to build this circuit which is efficiently designed to achieve a desirable and workable circuitry that favors the economy. Current device has some room for improvement; however, future projects will provide a more accurate waveform in comparison to an action potential of a real live neuron.

References:

1. Rieger R. et al. J Med Biol Eng. 2016;36(4):563–572.
2. Sun Y, R.Rieger, U.S.Patent 8,990,137 B2, 2015.
3. Al Zowelei M. et al, Instrumentation of Measuring Cellular Capacitance via Signal Processing Tool. 2018
4. Hoffman K, Tran V, Blandin J, Wu J, Sun Y, 42nd NEBEC, SUNY Binghamton,Vestal, NY, 2016
5. Hodgkin A. L, Huxley A. F, Membrane Current. J. Physiol. 117(4):500-544,1952

Monitoring the Patency of Pediatric Tracheostomy Patients' Airways using the TrachAlert Device

Kenneth Poser, Matthew Iobst, Franzisca Komar, Julianna Ricci, Yanique Spigner, Sally Shady[1], and Frank Castello[2]
[1]Stevens Institute of Technology
[2]Rutgers-Robert Wood Johnson Medical School

Introduction: Every year, over 100,000 tracheostomies are performed in the United States due to an upper airway obstruction, spinal cord injury, inability to clear secretion or when long term mechanical ventilation is required. Approximately 48% of these patients are affected by a complication with the tracheostomy tube. The most common complications observed are obstructed airways caused by secretions, as well as displacement of the tube. Both may lead to oxygen deprivation and potentially irreversible brain damage or death. Pediatric patients are unable to communicate their suffocation so a device is needed to alert caregivers before the damage becomes permanent. Current solutions such as pulse oximetry and ECG systems are targeted only towards in-hospital patients to monitor oxygen deprivation. These devices, however, are expensive, nonportable, and rely on signs of hypoxia to alarm. Once patients have been weaned off the ventilation system, or are sent to a long-term care facility, they do not have an accurate monitoring system for patency and displacement. TrachAlert will offer a novel and portable way of directly monitoring airflow through a tracheostomy tube for outpatients in order to improve their quality of life.

Methods: The TrachAlert device continuously monitors tracheostomy patients for indications of airway obstruction using a dual sensor system. Breath rate, breath sounds and effort in breathing were measured by the sensors in order to indicate the presence of an obstruction. The sensors are wired to a LabVIEW DAQ and a power source, which are contained in a separate housing unit. Proprietary code was developed based upon volumetric flow rate detection by the sensors. When the device detects a variation from "normal" readings for a prolonged period (>20 seconds), it will sound an alarm to alert the caregiver that there is an obstruction or cessation of breathing.

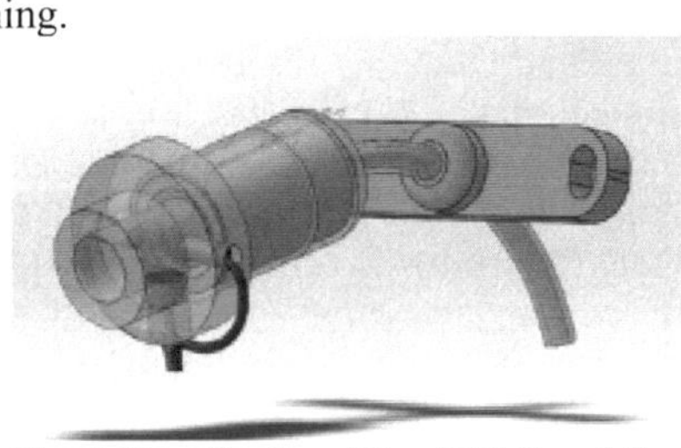

Figure 1: TrachAlert Device without DAQ and Power source

Testing was completed to further analyze the accuracy and efficacy of the device system. The test setup included an air pump simulating breaths into the posterior end of the TrachAlert device, a variable gate valve, as well as the TrachAlert device. Obstruction of the airway was simulated using the gate valve at no obstruction, 50% partial obstruction (representing hypopnea), and 100% fully obstructed (representing complete obstructive apnea). A 6-12 month old child's breath rate was simulated by manual activation of the air pump at a rate of 30 breaths per minute. Thirty trials for each valve position were conducted and the respective results were analyzed.

Results:

Table 1: Breath Volume Results

Position of valve (Percentage closed)	Sensor Trial Results (ml / sec)
0%	31
50%	15
100%	0

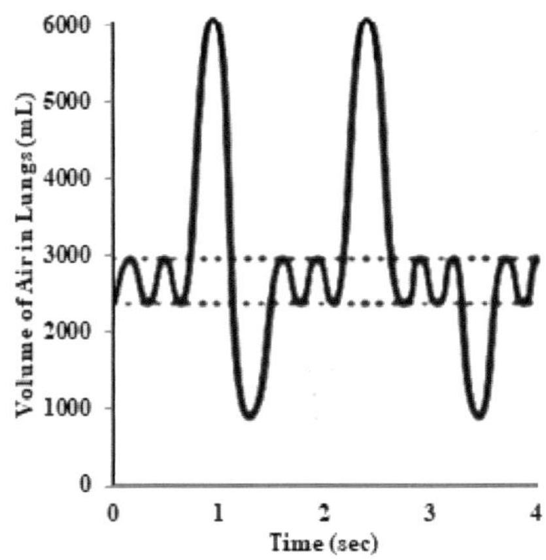

Figure 2: Unobstructed Breathing Pattern and Volumetric Flow for Pediatric Patients Aged 6-12 Months

Conclusions: The TrachAlert device is effective and reliable in monitoring airflow through a tracheostomy tube. The sensors measured a difference in the volumetric airflow levels compared to both 50% obstruction and full obstruction levels. Results were confirmed to be statistically significant based on a p-value d 0.05. This proved the device's ability to accurately detect the presence of an obstruction and alert the caregiver before hypoxia is reached. Further testing is necessary to determine a relationship between the measured response and the size of the obstruction in the tracheostomy tube.

References:

1. G. Manfro, L. Dias, and T.P. Farias, "Tracheostomy Complications." *Tracheostomy*, pp. 307-319, January 2018.
2. K. Heimann, et al. "Basic values for heart and respiratory rates during different sleep stages in healthy infants" *Biomedizinische Technik/Biomedical Engineering,* vol. 58, no. 1, pp. 27-34, January 2013.

Acknowledgements:

We thank Dr. Hazelwood, Val DeAngelo and Deep Parikh of Stevens Institute of Technology.

Monitoring Balance Board Pressure during Ankle Rehabilitation

Meghan Keenan, Aiden Keene, Tessa Arsenault, Eugene Chabot, Craig Simpson
University of Rhode Island, 45 Upper College Road, Kingston RI, 02881
Comprehensive Physical Therapy, North Smithfield, RI 02896

Significance: Ankle injuries are among the most common of all sports injuries. The topic we aim to understand and improve is the rehabilitation process of the ankle after injury. The result of this project is a device designed to quantify rehabilitation of proprioception which would augment typical techniques focused on muscle and range of motion rehabilitation. The intelligent wobble board can be used for any person with an ankle injury to determine their normal ankle function, as well as when they should be cleared for their personal level of activity. This balance board design is hypothesized to assist in proprioception training involving motor control, more specifically the neuromuscular feedback mechanism that is interrupted with injury [2]. With a balance board being the prefered rehabilitation device used by physical therapists, an intelligent balance board with easy to track user interface and pressure sensing capabilities is developed to evaluate the efficacy for reducing the recurrence of ankle injuries. Expanding upon the work presented by Kelley Magill, Ryan Buckley and Thomas Jancura, [1] this design integrates pressure sensing capabilities. With it the board gains the ability to sense compensation being made with respect to foot pressure due to the injury. According to Kim H et al. chronic ankle injury causes a lack of ankle flexibility resulting in compensation by different related muscles [3]. Thus proprioception paired with pressure sensing would be essential to fully rehabilitate the ankle. The board allows for physical therapists to better assess when a person has regained their normal abilities with quantitative results. A human study has been designed to determine variations between an individual's left and right foot. The intent of the study is to contrast differences in proprioceptive ability to determine what is normal and use this information as a landmark for ankle rehabilitation. As of right now physical therapists cannot rehabilitate a patient with a baseline in mind unless the person had been tested before injury. With our balance board, the left and right ankles can be used to better determine when a person has returned to their normal ankle functioning in both proprioception and the ability to apply force. Gathering data to determine the average in left and right ankle abilities would allow for a more accurate determination of when a person is fully recovered, with some numerical data to back the assessment.

Design Description: Our procedure assesses a person's ability to follow verbal instructions. These would include the instruction to move their ankle in given directions while seated with the knee bent at 90 degrees and one foot centered on the board. Keeping the person in a seated position maintains safety minimizes change of further injury. This also focuses all movements on the foot and ankle, rather than further compensation and balance through the legs and hips. During the assessment, pressure will be simultaneously monitored to determine a person's their ability to apply force at different angles. Key landmarks, such as directing the board at 12 o'clock, using plantar flexion of the foot, will be assessed by recording sensor measurements from the accelerometer, gyroscope, magnetometer, and pressure sensors within the board. Following directions and successfully meeting these landmarks prove the patient's proprioceptive and strength abilities. The force the ankle exerts on the board is collected by 3 pressure sensors mounted under a 3D printed foot placement to accurately gather information during the assessment.

Design Criteria: A foot placement board is mounted to the top of the balance board with color coded sections indicating the center of the board for a variety of foot sizes, seen in figures 1 and 2. Below this plexiglass mount is a 3D printed triangle with three cylindrical legs. The foot placement and triangle mount together are less than 2 inches from the board surface, keeping a relatively low profile. Each of the legs sits atop a pressure sensor, directing all pressure applied by the patient directly to the sensors for accurate measurements. Software was designed to record pressure during these movements at a rate of 200Hz. This is used in part to assess if a person is overcompensating with force, or the opposite, unable to apply pressure. The three pressure sensors are housed in three cup like 3D printed structures that guide the legs directly into place. The sensors are wired through the board to our hardware housing found within the half sphere at the bottom of our balance board. The hardware is contained within the half sphere to protect both the person using the board, and the hardware from damage. The hardware contains bluetooth to wirelessly connect to our software program, maintaining further safety and keeping a condensed, wireless design.

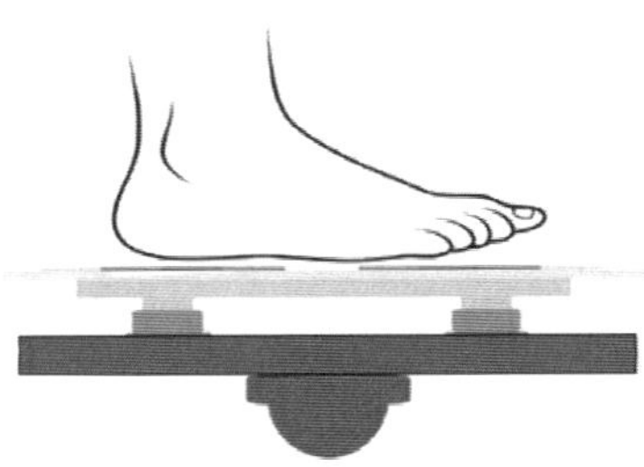

Figure 1: Side view of complete balance board structure

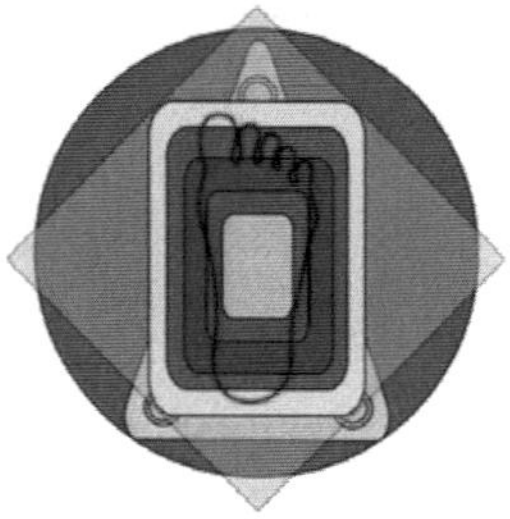

Figure 2: Bird's eye view of balance board

Results: The device is powered by a 5800mAh battery pack that can operate the device for around 8 hours. The battery pack allows for the board to be used without a large power cord, which would restrict the movement of the board. Materials purchased for the pressure sensing capabilities of the wobble board were less than $70. These purchases include three FlexiForce A401 sensors and a 3D printed foot platform. These sensors have a weight range of 0-25lb. This should allow us to account for any range of pressure applied to the board from a sitting position. There are 3 pressure sensors mounted on the board with the 3D printed triangle on top to evenly distribute the pressure between each for all size feet. The circuit used for the pressure sensors is a voltage divider circuit with a 10 and 100 ohm resistor. This circuit was chosen because it prevented over 5V from going into the arduino and damaging it. The circuit board also consists of a bluetooth modem, arduino and gyroscope/accelerometer. The diagram below depicts all the components. The pressure sensors are labeled as A, the arduino is B, the bluetooth modem is F, the accelerometer is C and the battery is D. The arduino reads the information coming from the pressure sensors and accelerometer and sends it via bluetooth to a computer for analyzation. The bluetooth modem allows for the board to be wireless. Any wires connected from the board to a computer would inhibit the range of motion of the board and effect the accuracy of the results. The board is completely constructed and ready for testing. A procedure has been developed and is awaiting IRB approval.

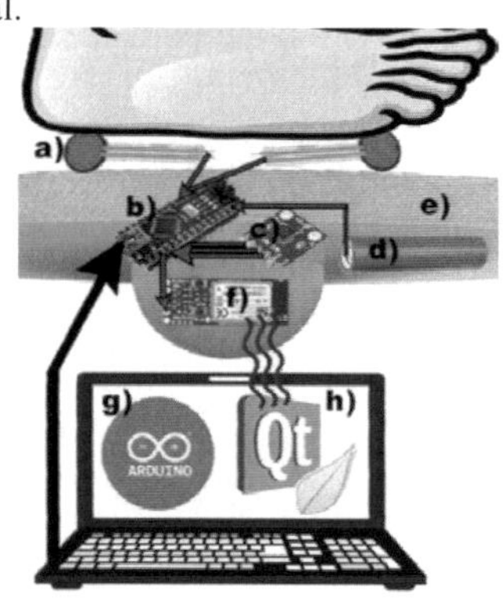

Figure 3: Balance Board Hardware and Software

Future Implications: The continued development of the balance board is to provide a cost effective method for rehabilitation. Determining the average results of a healthy ankle are hypothesized to be extremely beneficial to helping physical therapists in assessing when a patient is back to full health. The pressure sensors will allow monitoring of the amount of pressure a patient can exert into the board from a seated position. The importance of the pressure sensing capabilities is in understanding joints regain their mobility before they regain the strength to endure the pressure the body puts on the ankle. This is made even more important for athletes that will be exerting higher levels of pressure on their ankle due to running, jumping, kicking, etc. A patient may regain their normal ankle proprioception before they regain the strength in their joints and ligaments within their ankle. If the patient were to be cleared based off of their ankle proprioception, the chance of recurring injury is very high, since the patient may try exerting pressure greater than the strength of the ankle. Having data to determine the average difference between left and right ankles can further simplify the rehabilitation process, allowing a physical therapist to determine if the ankle is weak, or reasonably within range of the healthy ankle. Our balance board would allow for accurately determining when a patient is both strong and flexible enough to return to their personal level of activity. Ultimately, the balance board results in reduced medical costs, reduced recovery time, and minimized chance of re-injury.

References:

[1] T.Jancura, K.Magill, R.Buckley, C. Simpson, & E. Chabot, "Characterizing Ankle Proprioception with Embedded Sensor Balance Board," in Northeast Bioengineering Conference (NEBEC), 2017 43nd Annual, 2016, pp. 1-2.

[2] McGuine, T.A. & Keene, J. S. (2006). The effect of a balance training program on the risk of ankle sprains in high school athletes. The American journal of sports medicine, 34(7), 1103-1111.

[3] Kim, Hyunsoo, et al. "Kinetic Compensations due to Chronic Ankle Instability during Landing and Jumping." *Medicine & Science in Sports & Exercise*, 2017, p. 1., doi:10.1249/mss.0000000000001442.

In-theater Concussion Assessment through Reaction Time Evaluation Utilizing an Android Application

Christian Chaneski, Duncan Gandley, Gabrielle McCormack, Caroline Shusta, Magdalena Slonski
Stevens Institute of Technology, Hoboken NJ 07030 USA

Introduction: Mild Traumatic Brain Injuries (mTBIs) are difficult to detect effectively in-theater for the United States military. With over 383k cases reported since the year 2000, it is one of the most common military injuries.[1] It is very difficult to diagnose due to varying symptom presentation such as dizziness, fatigue, and inhibited reaction time. An mTBI self-assessment application was developed for use on an Android device as it is the operating system used for ATAK/phone devices in the military. The application was designed to measure the severity of common mTBI symptoms of active soldiers in remote, in-theater mission environments, with a specific focus on reaction time. This can potentially reduce risk of injury and death of military personnel, mission failure, and subsequently reduce over $234 million[2] in medical costs over a given year.

Methods: Android Studio and open source libraries including OpenCV were used to create the application, and more specifically the reaction time feature. The reaction test consists of images appearing on the screen in random time intervals, with the user being prompted to tap on the image as soon as they recognize its appearance. Function and scoring of the test can be seen in Figure 1.

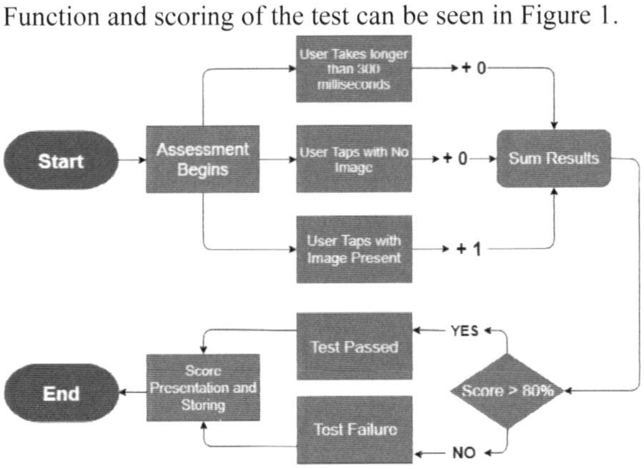

Figure 1: Reaction time flow chart and scoring calculation.

For in-theater assessment, the score would be compared to a baseline. A computer simulation was used to test the reaction time feature and see if the app would produce an accurate score in relation to the user input. The team prepared a specific 'script' of 20 inputs per 7 result scenarios (0%, 20%, 40%, 50%, 60%, 80% 100%) for the test system that included correctly and incorrectly timed inputs. These result scenarios were collected in triplicate. Correct inputs were set to be recorded within 300 milliseconds of an image appearance, which is a value verified as healthy neurocognitive recovery.[3] Incorrect inputs were set to be recorded in the 300 millisecond empty 'space' between image disappearance and the next image appearance or to skip an image appearance signifying a missing input. The app's results were then compared to determine the test's accuracy and effectiveness.

Results: After running the simulations, the scores were recorded and matched against the theoretical scores. Figure 2 shows the plotted comparison of predicted reaction to the app measured reaction.

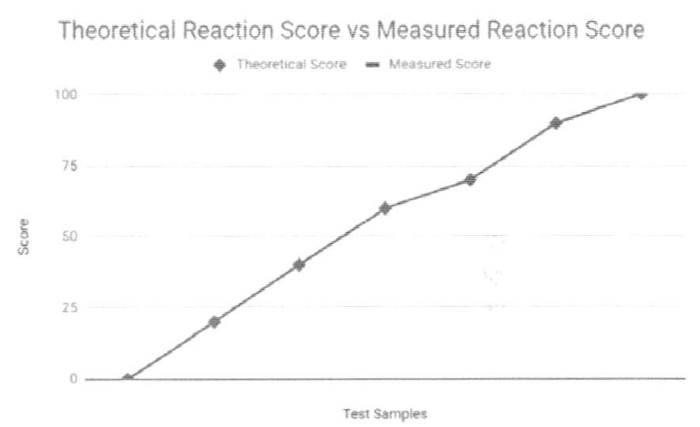

Figure 2: The projected correlation of the predetermined scores and the scores calculated by the app. The app's scores match the predetermined scores.

Conclusions: Because of the accurate alignment of the predetermined scores to that of the scores calculated by the app, the reaction time scoring was validated with a correlation ratio of 1:1. Furthermore, the reaction time scores can be used to compare to the baseline for future in-theater tests. For future development, more features/symptom trackers would be added to the app to help further identify possible mTBI symptoms.

References:

[1] "DoD Worldwide Numbers for TBI." *Dvbic.dcoe.mil*, Defense Health Agency, 23 Jan. 2017.

[2] "Health Care for Veterans: Traumatic Brain Injury." *EveryCRSReport.com*, Congressional Research Service, 9 Mar. 2015.

[3] Norris, Jacob N., et al. "ANAM4 TBI Reaction Time-Based Tests Have Prognostic Utility for Acute Concussion." Military Medicine, vol. 178, no. 7, Oxford University Press, July 2013, pp. 767–74.

Acknowledgements: The team would like to thank the following people: Capt. David R. Whittaker, U.S. Navy Bureau of Medicine and Surgery; Dr. Vikki Hazelwood, Stevens Institute of Technology; Dr. Gregg Vesonder, Stevens Institute of Technology; and SERC Marketplace.

P3-42

Micro-catheters for Targeted Fluid Delivery in Neural Tissues

Michael Fabiunke[1], Jeffrey D. Zahn[1]
[1]Department of Biomedical Engineering, Rutgers University, Piscataway, NJ

Introduction: Convection aided delivery can be used to treat Parkinson's and many types of central nervous system cancers [1]. In the future, it could also be used to treat Alzheimer's. Current methods of convection aided delivery, such as multi-tipped and hollow fiber catheters, treat a large tissue volume which can cause chemical damage from drug side effects as well as mechanical damage to the tissue surrounding the tumors. In light of these challenges, we have developed a microfabricated multi-lumen catheter to aid in convection enhanced drug delivery to the central nervous system, which is able to accurately treat a targeted tissue using a flexible polymer, perylene, to reduce mechanical damage to the brain during insertion while providing a precise delivery volume. The multi-lumen design is also capable of delivering multiple drug types at the same time while also allowing for removal of trapped bubbles and other blockages that are often found in catheters.

Methods: The fabrication of the micro-catheter starts with SU-8 mold fabrication via photolithography. Then polydimethylsiloxane (PDMS) was cast on the mold to create the delivery channel. After sealing the channel with another flat PDMS, Parylene C was deposited to conformally coat the interior wall of the delivery channel. Wet etching was then adopted to remove the PDMS, exposing the Parylene C delivery channel (Fig.1a). The catheters were connected to syringe filled with fluid to confirm its fluidic intactness and functionality.

Results: A series of micro-catheters with its width ranging from (200 μm to 400μm) were successfully fabricated. The microstructures designed to support the channel to increase its robustness and prevent blockage were defined very well in the final catheters. The fluidic test by inserting catheters into a 0.9% agarose gel to mimic the brain tissue showed that no blockage or channel collapsing occurred. (Fig.1b)

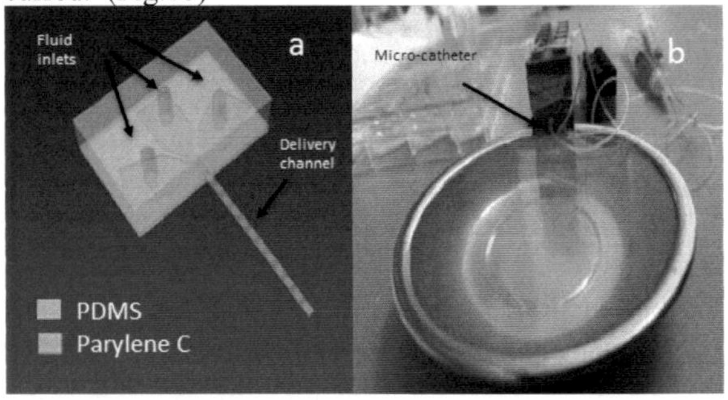

Figure 1. Microcathter with three inlets (a) Schematic illustration of the structure, and there are microstructures in the delivery channel as channel support (b) Fluidic delivery test in 0.9%agarose gel

Conclusions: At this time the catheters can be used to perfuse multiple liquids via syringes. However, sometimes the pressure generated by hand pumped syringes often leads breakage in the device. This breakage can be avoided by using osmotic pumps which will better simulate the drug delivery process in the brain tissue. In the future, we will use it to check the precise delivery volume of the devices.

References:

1. Ndesendo, Valence MK. "Convection-Enhanced Delivery of Neurotherapeutics." *Advances in Neurotherapeutic Delivery Technologies*, Semantic Scholar, 2015.

Acknowledgements: This project is conducted in the BioMEMS and Microfluidic laboratory of Rutgers University

Body Powered Distal Interphalangeal Finger Prosthetic

Ricardo Garcia, Christian Pignataro, Giovanna Nolan, Madison Taylor, Ashley Mont, MS, Sergei Adamovich, PhD*
Department of Biomedical Engineering, New Jersey Institute of Technology (NJIT), Newark, New Jersey, USA

Introduction: Finger/fingertip injuries are increasingly common around the world. Nearly five million injuries of this sort occur each year that result in emergency room visits [1] and they remain the most common type of amputation. Hand injuries account for 33% of injuries in the workplace. Many times, these injuries result in the loss of a fraction of the appendage. A significant number of those affected will suffer with the restricted use of a partial digit, resulting in loss of basic hand function. This can be detrimental to normal work, as well as normal interaction with the environment. In addition to this, the finger prosthetics currently on the market are extremely costly making it difficult for those suffering from this injury to obtain. A large population of those impacted are children and laborers. Children need multiple variations to their prosthetic as they grow, and recreation can lead to potentially damage to the device. In addition to this, many laborers require basic hand function to perform their job and need a reproducible option in the event of damage and ware during occupation. Therefore, affordable and reproducible finger prosthetics that restore function to the amputee is an ever-increasing need. This, coupled with the lack of prosthetics available to those with amputations distal to the proximal interphalangeal (PIP) joint, leads to extremely limited access. In this case study, we developed a methodology for producing a low-cost, high performance, finger prosthetic with a novel artificial distal interphalangeal joint system.

Recently, there have been significant strides in the technology that allows for custom prosthetics as advancements in 3D printing have resulted in a cheaper and faster form of production. The DeXter addresses affordability by utilizing this largely available 3D printing methodology while maintaining significant structural integrity with the use of carbon fiber and nylon. The design is body powered in that it uses the force generated by the remaining appendage to create substantial mechanical motion in the device, mimicking normal finger flexion and extension.

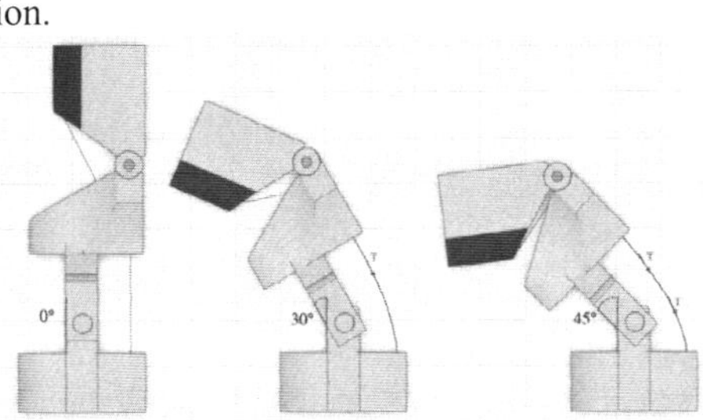

Figure 1: Flexion of the device as remaining middle phalanx bends [0°, 30°, and 45°] at the proximal interphalangeal joint.

The device uses flexion of the PIP joint at the interface with the midsection to create tension in the cable (Figure 1). This tension compresses the legs of the torsion spring together and forces the device to bend at the positioned joints. When the appendage is relaxed, the tension in the cable is released, causing the embedded torsion spring to return to its natural position at 90 degrees.

Methods: Exclusion criteria included amputation proximal to the PIP joint, previous skin/medical conditions resulting in allergy to materials used, and other concurrent injury that would prevent normal use of hand function. Subject chosen is a twenty-four-year-old male with an amputation distal to the PIP joint on the right index finger. The remaining middle phalanx maintains strength and sixty degrees of rotation. The inner dimension (where the residual limb interfaces with the prosthetic) was obtained using a Sense 3D scan of the patient's limb and modeling the prosthetic around the limb in Creo 5.0. The device is composed of three sections (ring, midsection, and fingertip). All sections are 3D printed on a Markedforged Mark Two printer with Markforged Onyx [Black] material with the option to add continuous carbon fiber. The 90 Degree, 0.105" OD Left-Hand Torsion Spring is used to mechanically operate the prosthetic in conjunction with ultra-high molecular weight polyurethane cable [rated at 100lbs] and the strength in the remaining appendage.

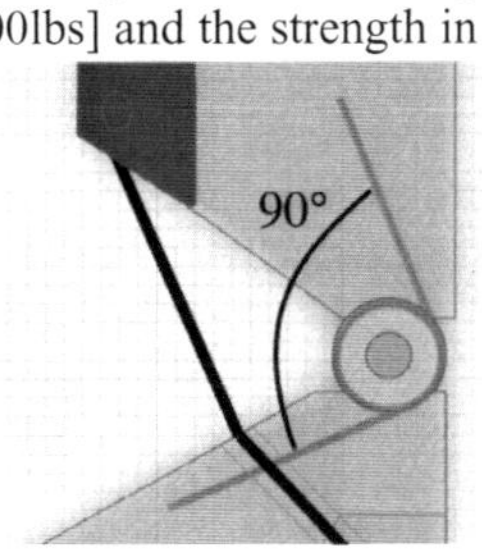

Figure 2: Torsion Spring Location

Results: A functional prototype of the DeXter was created for the subject using the torsion spring with dimensions catered to the subject's residual limb. The prosthetic is composed of three 3D printed components. The ring secures the prosthetic to the residual limb and the midsection acts to connect the ring and fingertip components together. The sizing for the fingertip and midsection components were designed based off of the measurements from the subject's left index finger as a basis of normal finger size. The torsion spring is housed in the hinge connecting the midsection fingertip. The spring constant allows for the subject to compress the spring for prolonged periods with little fatigue. The cable used is able to maintain prolonged use with a high strength, yet soft enough to not irritate the skin when in contact. The prototype has successfully achieved mechanical properties needed to mimic the human finger.

Conclusion: This device shall be tested for functionality and thus, improve the quality of life for the individual impacted with this injury, while allowing for an affordable option when compared to those currently on the market.

References:

1. Sindhu, K., et al., Injury. 2017. 48(12): 2643-2649.

Acknowledgements:
Joel Schesser, PhD
Project funded by NIDILRR Grant #90RE5021-01-00

P3-44

Otto-mobile: The Modular Wheel Chair for Children

Samantha Abate, Maria Catalane, Arianna Lee, Alina Zdebska
Syracuse University: College of Engineering and Computer Science

Introduction: The ability to obtain a power wheelchair can be empowering to those who may have mobility issues. For children, it allows them to have some autonomy and gives them a sense of independence. The Otto-mobile is a store-bought children's motorized vehicle that was altered to provide an alternative to traditional expensive pediatric power wheelchairs. A major cause for development the Otto-mobile was to create an adaptable wheelchair capable of being altered for children with various disabilities. Particularly in a developmental daycare setting there is a combination of children with different skill sets and capabilities, to be able to easily modify a wheelchair would provide a teacher with the ability to educating a larger range of children. In addition to the modularity of the device, the ability for it to last for extended period of time on a single charge is necessary for teaching and assessment purposes. In order to accomplish this, the Otto-mobile is lightweight, highly modular, has a long-lasting battery, and is easy to use.

Methods: Provided with a prototyped version of the Otto-mobile, all of the previous hardware was removed and examined for viable components. The old pre-programmed microchips were replaced with an Arduino microcontroller, dual motor shield, SPDT relay shield and RC Transmitter. These new components inside the car were connected using industrial grade wiring through a process of soldering. The microcontroller was then programmed to collect data from the RC transmitter and interpret it to produce directional and speed related data to be transmitted out to the motors and linear actuator. Next, the controller was stripped down to only the analog joystick. From there, the joystick, an Arduino microcontroller, RC transmitter and battery pack were fitted into a casing to contain the entire device for easy movability. Again, the parts were soldered together, and the microcontroller was programmed to collect data from the analog joystick and relay it to the RC transmitter to be sent to the car's processing unit for interpretation. Finally, to account for power consumption needed to run the product for extended period of time with the weight of a child and the powering of the various components, an additional 12V battery was added to the car. This battery was added in parallel to double the current rather than voltage available in the system.

Results: The adjustments made to the Otto-mobile have shown beneficial impacts on the modularity of the vehicle. Microchips are pre-programmed to only complete one function. Although this allows for faster processing, the trade-off for modularity with a microcontroller outweighed the speed loss. Now anytime a user wants to add a new feature or modify it for a different disability need, they can simply modify the code and add the new equipment rather than needing to rewire and replace various hardware components to account for the addition. In terms of the code, it is broken down into 4 parts for the car: dual motor shield control, relay shield control, data collection from joystick, and interpretation of joystick data. Since the code itself if very module, the addition of another feature is clear and easy. As well, in our modification of making the old joystick wireless, we have already begun increasing the various ways in which the car can be used. In this case, a child whether right-handed or left-handed can now drive the car. Finally, by increasing the current supply in the car, it nearly doubles the run-time for a single charge. This took the car

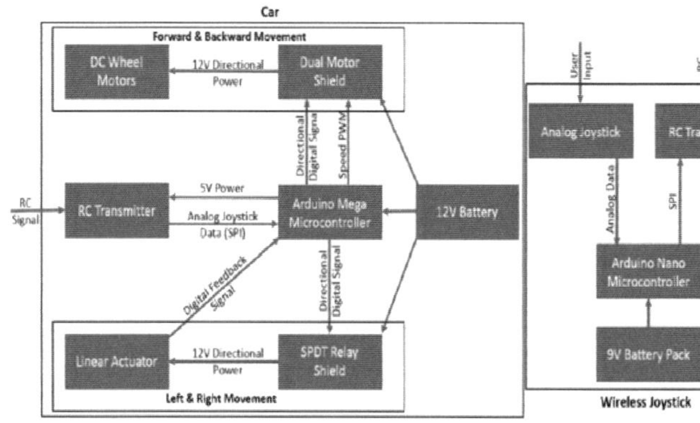

Figure 1. Circuit Schematic

from running continuously for only 15 minutes to being able to last for over 30 minutes. Thus, it allows for a more reliable and realistic experience for the children. In addition, it allows for a teacher or assessor to run more tests and evaluate the children's capabilities. Overall, the desired results of the car's ability were reached.

Conclusions: Overall, the improvements to the Otto-mobile are already allowing for a simple children's ride-on car to act a supplemental adaptable pediatric wheelchair. Further studies with a developmental day-care will help pinpoint any other basic functionality modification that may be needed. Some predicted changes will be the code for faster relay and reaction timing to ensure accuracy of a child's actions and a replacement of the current motors as they are limited in speed control. As well, it may include the addition of a head array for alternative steering for children with disabilities that prevents them from using the joystick.

References: 1. U.S. Department of Health & Human Service. 2. D. Rodgers, M.D., M.P.H., F.A.A.P. "The Evaluation of a Child with Developmental Delay"
3. National Dissemination Center for Children with Disabilities.

Acknowledgements: Professor Pun To Yung, Department of Bioengineering, Syracuse University;
Pruthvi Kilaru, Syracuse Bioengineering Graduate; The Jowonio School, Syracuse, New York

Inter-user Variability Analysis for Manual Bone Surface Segmentation from Ultrasound Data

Hridayi Patel, Ilker Hacihaliloglu
Department of Biomedical Engineering
Rutgers University

Introduction: One of the main steps of any ultrasound (US)-guided computer assisted orthopaedic surgery system is the automatic segmentation of bone surfaces from intra-operatively acquired US data. In order to determine the accuracy of the segmentation, automatic segmentation is usually compared to manual segmentations done by expert US technicians. Bone surface response appears several millimeters in thickness in the US data which can cause both an inter- and intra- user variability when performing manual segmentations [1]. Characterizing these errors can help determine new validation metrics to compare the segmentation data and define its accuracy. The application of these new error metrics can help make the process of manual US segmentations more universal and allow for a more precise image analysis tool. The objective of this study is to examine different methods of quantifying the amount of error that occurs between inter-user manual US segmentations.

Methods: 358 in vivo scans, from radius and femur bones, were collected after obtaining approval from the Rutgers University institutional review board (IRB). Four different datasets, obtained from 4 subjects, were segmented by two observers. Observer 1 had more than 15 years of experience segmenting bone US datasets and is considered the expert in this study. Observer 2 had two years of experience segmenting bone surfaces from US data. The first method of error quantification was the average Euclidean distance calculated between the two segmented surfaces. Euclidean distance was measured by considering Observer 1 segmentations as the ground truth. The next method of quantifying error was the STAPLE algorithm [2]. The STAPLE algorithm was used to obtain sensitivity (p) and specificity (q) values to determine the error and validity between the observers.

Results: The average Euclidean distance between the two manual segmentations was 0.20 mm with the highest error going to 0.56 mm and the lowest error being 0.0071 mm. Figure 1 displays box and whisker plots for the sensitivity and specificity results obtained from the STAPLE algorithm. Investigating the plots, we can see that the STAPLE method correctly identifies the expert as having superior performance to Observer 2.

Conclusions: Manual expert segmentation of bone surfaces is an error prone procedure due to the various US imaging artifacts and bone surface response having a thickness of several millimeters in the acquired data. During this study we have investigated the inter-user variability between two observers with varying expertise level. To the best of our knowledge this is the first study investigating the inter-user variability for segmenting bone surfaces from US data. Although the expertise level between the two observers was significant, the sensitivity and specificity mean values did not show a significant difference. This could be due to the fact that the segmented bone surfaces, radius and femur, did not have a complex shape. However, the mean Euclidean distance error of 0.20 mm indicates a significant difference between the two segmented surfaces. This indicates that during validation of the automatic bone segmentation methods, intra-user variability should be taken into account.

Future work will include the extension of the database to include more scans obtained from various bone surfaces and investigation of intra-user variability analysis. Finally, we would like to announce that our group, together with various international collaboration across multiple research centers, is creating an open database for US bone segmentation consisting of several thousand US images and corresponding bone surface segmentations and invite the community to contact if they are interested to be a part of this initiative.

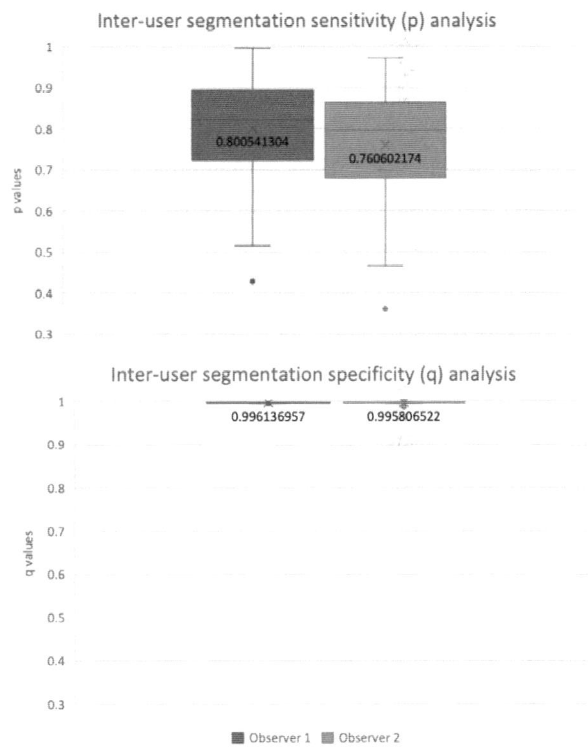

Figure 1. Box and whisker plots showing the sensitivity (p) and specificity (q) results of the segmentations performed by two observers with varying level of expertise.

References

1. Hacihaliloglu I. Technology. 2017; 5(02):74-80.
2. Warfield SK et al. IEEE Trans Med Imaging. 2004;23(7):903-21.

P3-46

Induced Heart Rate Variability Index Derived from the Valsalva Maneuver and Sudden Standing Up

Abdulrahman Alsasa, Nathan Guillemette, Dylan Young, Ying Sun
Biomedical Engineering, University of Rhode Island, Kingston, RI, USA. [sun@uri.edu]

Significance: Coronary Heart Disease (CHD) is a buildup of plaque that leads to atherosclerosis, which commonly leads to heart attack and heart failure. Although cardiac disease is a leading cause of death in the United States, most symptoms are silent, especially those of CHD. It is imperative to form new testing methods that can detect CHD before a cardiac event occurs, and recent research into Heart Rate Variability (HRV) indicates that it can be used to predict heart health. However, there have been no attempts to uncover whether this correlation is due to the preload or the afterload. This paper describes a novel method of isolating preload HRV via the Valsalva and Sudden Standing Up Maneuvers in an attempt to relate it to CHD, thereby forming a new and noninvasive early detection method for CHD patients.

Design Description:

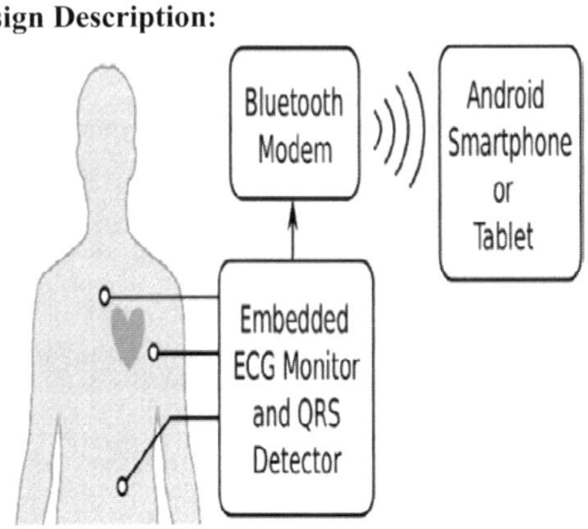

Figure 1: Data collection required each subject to affix electrodes as shown above. Their heart rate information was processed and transmitted via Bluetooth to an Android tablet.

Heart rate variability (HRV) describes the time variation between R-R intervals, the peaks of each heartbeat. Over the last two decades, Heart Rate Variability (HRV) has been linked to Coronary Heart Disease (CHD), and our team aims to leverage this relationship to devise an efficient, noninvasive testing system for CHD. Our team believes that since the greatest strain is put on the heart during left ventricular end-diastolic pressure (LVEDP, commonly referred to as preload), this is also when HRV will be most effective in identifying CHD. In order to isolate the preload, we chose to utilize the Valsalva and Sudden Standing up maneuvers. We chose these because while the Valsalva applies intrathoracic pressure that induces changes in the preload and afterload of a cardiac contraction, the Sudden Standing Up induces a drop in blood pressure that only affects the afterload. We believe that by normalizing the HRV values between these two maneuvers we can efficiently build an index of preload HRV data that can be used to find more specific correlations between CHD and HRV. To facilitate this, we developed a quick and accurate method to collect HRV information from human subjects. While there are several effective methods that could be used to determine a patient's HRV, our team implemented an electrocardiogram (ECG)-centered approach due to its high accuracy. By measuring the time between the R-R peak intervals recorded by the MOBD algorithm, we received dependable HRV data points that were used to create the aforementioned index. To accommodate this, we equipped a working electrocardiogram (ECG) with Bluetooth to transmit heart rate information during each maneuver to an Android app.

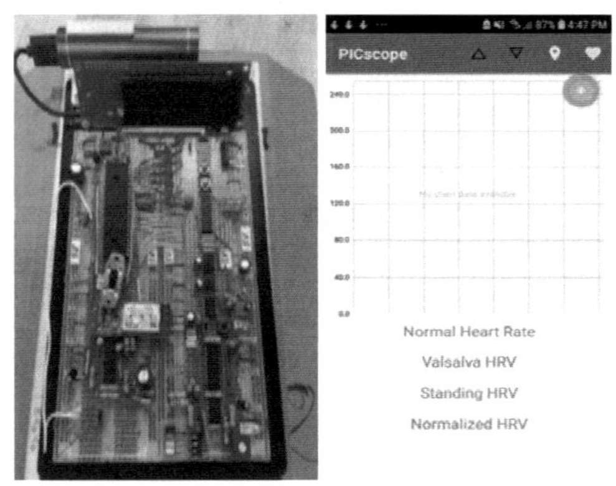

Figure 2: Integrated ECG breadboard and Android application.

Within the app, the data was then stored, converted to HRV data, and normalized. After our human trials, a baseline "healthy" preload HRV value will have been established that can be used to identify correlations between preload HRV and CHD.

Design Criteria: Our team used a PIC18F4525 microprocessor outfitted with the MOBD algorithm as the base of our ECG, allowing us to detect each subject's heart rate. By implementing a Bluetooth RN-42 Modem into our hardware we were then able to transmit all live recordings to an Android tablet. Our team also installed a switch system that allowed us to toggle between two modes: Valsalva and Sudden Standing up data collection. The procedure for each trial was preprogrammed into the PIC as follows: 15 second recording under normal conditions, 15 second recording immediately after the Sudden Standing Up maneuver was applied, another 15 seconds of recording under normal conditions, and lastly a 15 second interval recorded while the Valsalva maneuver is applied. After receiving heart rate information from each interval, the Android device converted the information to HRV and, after all data was collected, normalized the Valsalva HRV values with those of the Sudden Standing Up for each subject. The heart rate data was separated into two categories: normal heart rate (HRn) which was the average heart rate under normal conditions, and the tested heart rate (HRt) which was the maximum heart rate observed during each respective maneuver. HRV was then calculated using the following equation: $HRV(\%) = \left(\frac{|HRt-HRn|}{HRn}\right) * 100$. The HRV values were then normalized by dividing the Valsalva by the Sudden Standing Up due to the following reasoning: $\frac{Valsalva\ Maneuver}{Sudden\ Standing\ Up} = \frac{Afterload \& Preload}{Afterload} = Preload.$

Results:
The embedded breadboard was successfully modified to include a functional toggle switch, Bluetooth transmittance, and HRV normalization. The PIC successfully iterates through the 15 second time intervals allotted for each test, and, as can be observed in Figure 2., the Android device is ready to receive transmitted heart rate information. Although we are unable to conduct a full human study until next semester, our team successfully received data from five test subjects.

	Normal HR	SSU HR	HRV (%)	Normal HR2	VSV HR	HRV2 (%)	Normalized HRV (%)
1	69	87	26	60	76	27	1.02
2	87	109	25	74	98	32	1.28
3	65	78	20	66	83	26	1.29
4	71	84	18	67	88	31	1.71
5	73	81	11	72	85	18	1.65

Figure 3: HRV Data for the 5 test subjects.

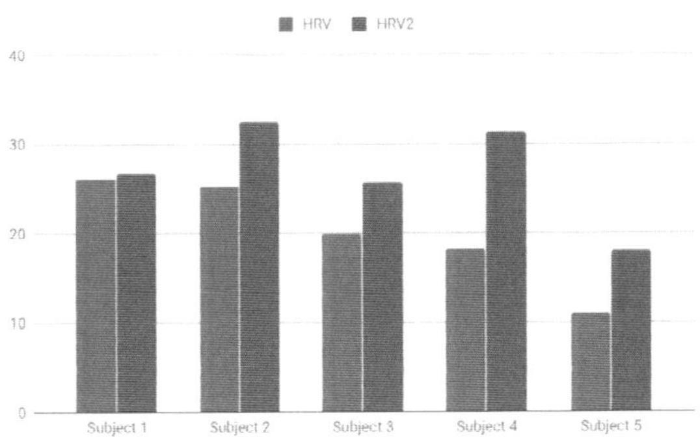

Figure 4: Sudden Standing Up (blue) and Valsalva Maneuver (red) HRV values for each subject.

This information is displayed in Figure 3. In the Sudden Standing Up test, HRV=20% ±6.04% (mean±SD), ranging from 11% to 26%. In the Valsalva Maneuver test, HRV=26.8%±5.54% (mean±SD), ranging from 18% to 31%. The underlying data for these calculations is shown in Figure 4. After normalizing these values for each subject, HRV=1.39%±0.29 (mean±SD), ranging from 1.02% to 1.71%. As expected, the HRV values of each subject were higher during the Valsalva Maneuver.

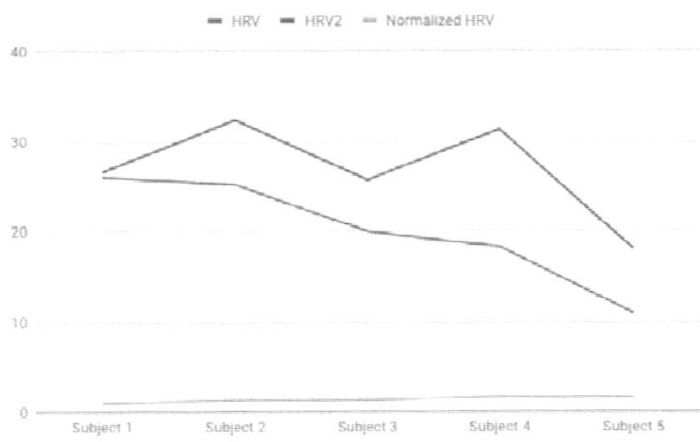

Figure 5: Sudden Standing Up (blue) and Valsalva Maneuver (red) HRV values along with their normalization (yellow).

Future Implications: This project forms the foundation for a future index of preload heart rate variability, which could potentially serve as a useful tool in early identification of CHD. This undertaking produced two important contributions to this goal: an effective method of collecting then normalizing HRV data for our chosen tests, and studying the resultant trends to identify a correlation between preload HRV and coronary heart disease. The embedded ECG device proved to be an efficient collector, and the bandwidth of our Bluetooth chip proved to be sufficient for data transfer. These findings suggest that our device could be a useful tool for medical professionals to quickly gather accurate HRV information from their patients. Although more samples must be collected, it appears that our normalization method effectively clusters healthy subjects together. Both the hardware and android application are fully developed, have been successfully paired, and a practical normalization method has been produced. Our team intends to continue adding to our existing data set in order to create a better foundation for a preload-HRV index.

References:

1. Malik M. Eur Heart J. 1996; 17:381-385
2. Malik M. Br Heart J. 1994; 71:3-6
3. Kleiger R. Annals of Noninv. 2005;10,88-101
4. Porth CJ. PubMed. 1984;13: 507-518
5. Ying S. Biomed Instr & Tech. 1992; 26:477-484

Soothing Patients with Autism in the Emergency Room - BusyBin

Jeff Ung, Craig Hughes, Faith Martin, Daniel Natale.
Stevens Institute of Technology
1 Castle Point Terrace
Hoboken, NJ 07030.

Introduction: To address the needs of individuals with Autism Spectrum Disorder (ASD) and their families in a hospital setting, a "Busy Box" named the 'BusyBin' has been designed for Hackensack Meridian Health. The most recent CDC biennial report stated that 1 in 59 children in the United States have ASD, and the prevalence is increasing as more information is learned about the condition and better screening methods are introduced [1]. Patients with ASD often get anxious in unfamiliar situations and may act out in different ways without something to hold their attention, something most Emergency Room physicians are not trained for. The 'BusyBin' was a portable, multifunctional cube that not only had sensory stimulating features on all sides, but also served as a bin for additional toys that will calm patients. The stimulating features individually focused on the senses of touch, sight, and hearing so that patients could choose to use whichever features they liked. The 'BusyBin' was made to be used in emergency rooms and physicians' offices to soothe, calm, and teach patients with autism. Some toys and activities currently exist already for these situations; however, after garnering feedback from physicians, nurses, and families alike, it was discovered that these toys were seen as childish, boring, and insufficient at easing the tensions of patients. Even if certain toys were geared towards sensory stimulation, they may not have passed joint commission hospital standards for cleanability and sanitation. Both these issues were kept in mind for the development of the bin.

Methods: A 'Busy Box' was created that highlighted different senses for each side of the box. The dimensions of the 'BusyBin' were 18''x 18''x 18'', and fabricated out of HDPE sheets. The first side was named the fidget side as it stimulates touch and provides an outlet for patients to dissipate their nervous energy; the second side contained an "hourglass" that allows patients to observe and be soothed as different density liquids mixed with glitter to make patterns. The third side was named the texture side and provides textured panels that were soothing to the touch; the fourth side was a music side that allows patients to interact with a tube xylophone, and the top lid was a puzzle side that incorporates a sliding number tile puzzle into the top of the box. A model of the 'BusyBin' can be seen in Figure 1. In order to make the 'hourglass' as effective and entertaining as possible, immiscible liquids and glitter were used, along with static mixers, to create visually appealing patterns. The optimal settling velocity was calculated, using the following equation:

$$V_t = \frac{gd^{5}(\rho_p - \rho_m)}{\mu}$$

Equation 1. Settling Velocity Calculations for Glitter in Fluids

Figure 1. BusyBin

Results: The settling velocity of glitter in water was found to be *3.* in/s, and the settling velocity of glitter in oil was found to be *3.* in/s. These settling velocities were slow enough to make continued patterns as the hourglass spins and the differing speeds made the patterns interesting to observe. A similar amount of detail was put into each component ensuring their efficacy and ability to entertain.

Conclusions: The 'BusyBin' was proven to be effective at soothing the tension of patients in the emergency room both by being entertaining and as a stress outlet. This device can be used in the hospital setting to relieve the issues associated with triggers of patients with ASD.

References:

1. Baio J., et al. Prevalence of Autism Spectrum Disorder Among Children Aged 8 Years—Autism and Developmental Disabilities Monitoring Network, 11 Sites, United States, *MMWR* ,2018;67(No. SS-6):1–23. DOI: http://dx.doi.org/10.15585/mmwr.ss6706a1

Acknowledgements: We thank Dr. Randye Huron and Dineen Lesky-Dibari from Hackensack Meridian Health and our advisor Dr. Vikki Hazelwood for her insights and expertise.

Rapid Prototyping to Roll to Roll manufacturing of Microfluidic devices

Chris Toothaker,[1] Bailey Corless,[1] Amber Boutiette[1] Caitlin Howell[1,2]
[1]Department of Chemical and Biomedical Engineering, University of Maine
[2]Graduate School of Biomedical Science and Engineering

Introduction: The Howell lab has been collaborating with SAPPI North America to identify novel applications of their high-throughput roll-to-roll (R2R) manufacturing technology. This technology enables continuous production of high-fidelity pattern replication on the surface of large reams of paper or film. These surfaces can have pattern replication with features from the micro-scale down to the nanoscale. We performed customer discovery and found the Point-of-care (POC) diagnostics market to be of highest interest; an application where personalized medicine allows patients to customize treatments and improve their effectiveness without needing to travel to a care facility or wait extended periods of time for test results [1]. These POC devices often rely on microfluidic components to carry out diagnostic functions, but there are limitations imposed by traditional lithographic manufacturing techniques; requiring expensive cleanrooms [1][2] and complex chemistries for scalable manufacture and functionality. In this project, we evaluated the technology's potential for use in mass-manufacturing microfluidic mixing components among other potential applications.

Methods: The experiments were run by 3D printing microfluidic patterns commonly found in literature. The designs were molded in PDMS (polydimethylsiloxane) which is a silicon based organic polymer. The experimental data gathered was compared to theoretical data gathered through SOLIDWORKS models and COMSOL simulations. Black ink and water were used when running the experimental version of the microfluidic device in the lab to get a stark contrast between the two concentrations. The experimental and theoretical models were then compared using a MATLAB program which took points along the channel and measured the pixel intensity of grey pixels across the channel. The mixing index (MI) was calculated as:

$$MI = 1 - \sqrt{\frac{1}{N}\sum_{i=1}^{N}\left(\frac{I_i - I_{mean}}{I_{mean}}\right)^2}$$

where the I_i is the grey intensity of one point along the channel and I_{mean} is the mean grey intensity across the length of the whole channel.

Results: There was a clear difference between the mixing efficiencies of the two designs tested in theoretical COMSOL models. **Figure 1A** (above) demonstrates a flat mixer with a low mixing index of 0.139. This is in comparison with **Figure 1A** (below), which has diamond protrusions and splits the flow of the fluid into two different channels that merge back together after the protrusion. The diamond mixer had a mixing index of 0.88, which is noticeably higher than the flat channel. **Figure 1B** shows the experimental data of the flat channel (above) and the diamond mixer with protrusions (below). The flat mixer has a lower mixing efficiency than the mixer with the diamond protrusions likely because it is relying on laminar flow which does not promote mixing except through diffusion. The microfluidic mixers, however, will not be long enough to allow for complete mixing through diffusion. When the two mixers were run experimentally, the flat mixer performed better than expected with a mixing index of 0.4458 and the diamond mixer performed worse than expected with a mixing index of 0.4482.

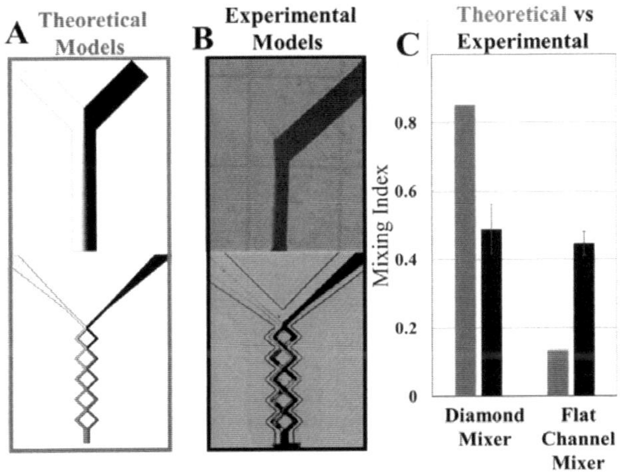

Figure 1. A Results of a theoretical COMSOL analysis of mixing of two fluids (black and clear) a flat channel mixer (above) and a mixer that better promotes mixing (diamond mixer, below). **B.** Images from the experimental analysis of mixing in flat channel and diamond mixers. **C.** Quantitative comparison of the mixing index of the theoretical and experimental designs.

Conclusions: We have developed a method for the rapid fabrication and quantitative evaluation of microfluidic systems using 3D printing to make custom patterns for use in a low-cost roll-to-roll printable, a low-cost process. Although further work on optimizing the more complex systems is required, we expect that this approach has potential to be used for the mass production of micromixers at a much lower cost than is currently possible.

References:

1. Sia S.K. et al. Lab on Chip 2008, 8, 1982-1983
2. Wlodarczyk K.L. et al. Micromachines 2018, 9, 409

Acknowledgements: This work was supported Sappi North America contract DIC Project 3109, as well as a Maine Technology Institute Seed Grant no. SG5725 and a University of Maine System Research Reinvestment Fund MIRTA Accelerator Grant.

Biomarker Predictors of the Development of Alzheimer's Disease

Nathan C. Ackermann, Natalie K. Brunette, Jared A. Klug, Katherine A. Ward
Boston University, Massachusetts General Hospital

Introduction: Alzheimer's disease (AD) is pathologically defined as a buildup of amyloid plaques and tau protein tangles inside and outside nerve cells, respectively. These biomarkers block communication between the nerve cells in the brain and can lead to brain atrophy. Absolute diagnosis is currently determined through autopsy to detect amyloid plaques and tau tangles. Probable AD can be clinically diagnosed through brain imaging after significant and irreversible brain damage has occurred, but diagnosis methods are neither quick nor cost effective. Early diagnosis of AD can lead to early treatment options that have proven to significantly lower the rate of neurodegeneration in the brain. Homocysteine is an amino acid commonly found in blood. Elevated levels of this biomarker have been linked to brain atrophy and the development of heart and blood vessel diseases.[1] Patient blood biomarkers were analyzed from the Alzheimer's Disease Neuroimaging Initiative (ADNI) to determine novel biomarkers that can be used to predict early risks of AD and cognitive decline.

Methods: The dataset *Homocysteine - Results* from ADNI 1 and ADNI GO was downloaded into statistical programming software R using the ADNIMERGE package. This package creates a .csv file containing key patient information (age, diagnosis, etc.) and data from other AD studies (biomarkers, genetics, brain scans) linked through a non-identifying participant roster ID. The data was filtered by diagnosis, retaining cognitively normal (CN) and Alzheimer's individuals, and limited to baseline biomarker levels. The filtered data yielded 228 CN individuals and 188 Alzheimer's individuals with baseline total plasma homocysteine levels. The mean total plasma homocysteine was found for both groups, and compared using a one-sided Welch's t-test in order to determine if the two means are equal.

Results:

	Male	Female	Mean age	APOE4 positive
Cognitively normal	119 (52%)	109 (48%)	75.86	26.3%
Alzheimer's disease	99 (53%)	89 (47%)	75.26	66.0%

Table 1. The demographic distribution of the data between CN and AD individuals, including gender, mean age, and APOE4 status.

The mean averages of total plasma homocysteine for CN and AD were 9.94 and 10.754, respectively. The one-sided Welch's t-test yielded a p-value of 0.004163 at a 99% confidence interval. A second t-test was conducted to determine if homocysteine levels varied with the presence of APOE4. The Welch's t-test yielded a p-value of 0.2454 at a 95% confidence interval. The CN individuals had a mean age of 75.86 and was divided into 119 males and 109 females. 26.3% of CN individuals were APOE4 positive. The AD individuals had a mean age of 75.26 and was divided into 99 males and 89 females. 66.0% of Alzheimer's individuals were APOE4 positive.

Conclusions: Although Alzheimer's is known as a neurodegenerative disease, increasing amounts of evidence show that it can also be linked to vascular disorders. This study demonstrates that patients with diagnosed AD also show higher levels of homocysteine in the blood plasma, suggesting that AD is correlated with vascular health. AD individuals had a higher percentage of patients with the APOE4 allele, a known Alzheimer's risk factor. The second t-test showed that homocysteine levels did not vary with APOE4 allele presence (one or two alleles). Future directions for the homocysteine study include analysis of longitudinal data, specifically how homocysteine levels in the blood change over time as subject diagnostic status changes. The results from this will help determine the role of homocysteine in AD and how it can be targeted for drug therapies in the prevention of the disease. Because this is an initial investigation, the code and methods written for this study will be used in combination with other ADNI data sets to identify additional blood-based biomarkers that may similarly show potential risk for the development of AD, such as tau proteins and amyloid beta. Combining trends from multiple biomarkers would increase the predictive power of a blood-based detection test for AD. Ultimately, a machine learning algorithm will be created based on multiple blood biomarker statistics to predict other subjects' risk of developing AD. This algorithm will be tested for accuracy with longitudinal data. Identification of these biomarkers is essential to the development of early Alzheimer's detection methods, as well as preventative treatments of the disease that minimize the amount of neurodegeneration in individuals.

References:

1. Rajagopalan P. et al. Neuroreport. 2011;22(8):391–395.

Acknowledgements: This project was made possible by the College of Engineering at Boston University, and by Dr. David H. Salat at the Brain Aging and Dementia Lab at Massachusetts General Hospital. All data was obtained from the Alzheimer's Disease Neuroimaging Initiative.

Effects of Substrate Stiffness and Cell-Cell Contact Area on Stem Cell Signaling

Sarah Furman[1], Sebastian Naranjo[1], Kirstene Gultian[1], Abigail Loneker[2], Rebecca G. Wells[2], Sebastián L. Vega[1]
[1]Department of Biomedical Engineering, Rowan University
[2]Department of Medicine, Perelman School of Medicine, University of Pennsylvania

Introduction: Microenvironmental factors, including substrate stiffness and cell-cell interactions regulate stem cell behavior; however, the interplay between outside-in stimuli and stem cell signaling is poorly understood. Notch signaling is one of the most established mediators of short-range signaling between cells, and Yes-associated protein (YAP) signaling has been implicated as a mechanotransductive regulator [1]. To probe the effects of stiffness and cell contact on Notch and YAP signaling, human mesenchymal stem cells (MSCs) were cultured atop polyacrylamide (PA) hydrogels of Low (500 Pa) and High (10 kPa) stiffness. By seeding MSCs sparsely, MSC pairs labeled with antibodies for Notch and YAP were found, allowing for the investigation of how stiffness and physical contact between cells affects cell-cell signaling and mechanotransduction.

Methods: *PA hydrogel formation:* to create hydrogels of varying stiffness (Low, 500 Pa; High 10 kPa), acrylamide, bis-acrylamide, and PBS were mixed at different ratios followed by free radical polymerization as previously described [2]. Fibronectin was covalently bound to PA hydrogel films by treating surfaces with sulfo-SANPAH (0.2 mg/ml solution activated with 10 mW/cm^2 UV light for 10 min) and incubating hydrogels in FN solution (20 µg/ml) for 30 min. *Cellular characterization:* MSCs (Lonza, P3) were cultured atop FN-treated PA hydrogels (3,000 cells/cm^2) for 7 hours and stained with primary (Notch-1 abcam, 1:100, 1 hour; YAP Santa Cruz, 1:200, 1 hour) and secondary (Notch-1 AlexaFluor 488, 1:200, 2 hours; YAP AlexaFluor 594, 1:200, 2 hours) antibodies. To visualize nuclei, MSCs were counterstained with Hoescht (1:5000, 10 mins). *Single cell analysis:* confocal images of Notch and YAP stained samples were used to find cell pairs for analysis. Metrics for cells in MSC pairs (area, nuclear YAP, Notch intensity, cell-cell contact area) were acquired using ImageJ software.

Results: MSC pairs on PA hydrogels with Low and High stiffness were imaged (Notch, green; YAP, red; nuclei, blue) for analysis (Fig. 1a). MSCs on Low stiffness hydrogels exhibited smaller areas (not shown) and nuclear YAP ratios in comparison to MSCs on High stiffness hydrogels (Fig. 1b). To examine the effects of cell-cell signaling on how cells respond to substrate stiffness, cells in MSC pairs with lower and higher total Notch fluorescence were demarcated as "Sending (Send)" and "Receiving (Rec)" cells, respectively. Surprisingly, after separating MSC pairs into Sending and Receiving groups, cell area (not shown) and nuclear YAP of Sending cells was greater than those of Receiving cells (Fig. 1c). This finding was more pronounced in MSC pairs on High stiffness PA hydrogels. To evaluate the effects of cell-cell contact on YAP signaling, nuclear YAP vs. cell-cell contact length scatter plots were generated for Sending and Receiving cells of MSC pairs cultured atop Low and High stiffness PA hydrogels (High stiffness shown in Fig. 1d). Cell-cell contact length did not appear to have a significant impact on YAP signaling.

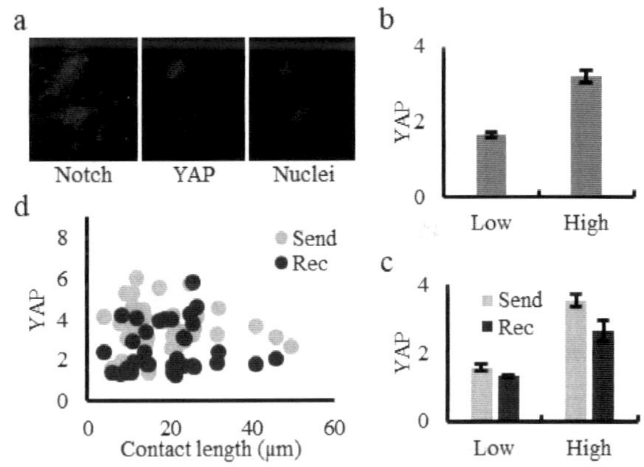

Figure 1. Refer to body text for figure description.

Conclusions: In this study, MSCs were cultured on PA hydrogels of Low and High stiffness to investigate the effects of cell-cell contact and stiffness on MSC signaling. Results from this study showed that cell area and nuclear YAP increase with increasing stiffness. By separating cells based on Notch intensity into Sending and Receiving cells, it was observed that cells with more Notch were smaller and had less nuclear YAP than cells with less Notch. No correlation between nuclear YAP and contact length was observed, and this may be due to the heterogeneity in cell morphology and timing of cell-cell contact. To address these potential caveats, a micropatterning photomask with a constant cell-cell contact area will be incorporated with geometries that induce varying degrees of contractility.

References:

1. Totaro A et al. Trends Cell Biol. 2018;28:560-573.
2. Tse JR and Engler AJ. Curr Protoc Cell Biol. 2010;47:10.16.1-10.16.16.

Acknowledgements: The authors would like to thank Ryan Emenecker, Apurva Limaye, and Kari Miller from the Center for Engineering MechanoBiology (CEMB) for preliminary data and discussion leading to the inception of this study.

P3-51

Creating an Assistive Beach Wheelchair Device

Gabriella Borodyansky, and Jacqueline Chi, Justin Stowe
Stevens Institute of Technology.

Introduction: Taking a trip to the beach has been a significant challenge for individuals who are affected by motor disabilities. Current beach wheelchair models have used inappropriate dimensions and design variations, ultimately limiting the force per area distribution used to move over sandy terrain. Unlike existing beach wheelchairs the product, "Beach Wheels," utilized large balloon wheels to effectively traverse over sand, along with a lightweight aluminum frame, ensuring stability for the user. This novel device can be folded into less than 16 cubic feet, which is the average trunk size of sedans, allowing easy transportation to and from the beach. Beach Wheels had ideal wheel tire pressure as well as optimized surface area in contact with the sand to better propel the user in beach environments. This wheelchair was designed to effectively traverse over sand and has brought the beach back to wheelchair users, and provided a comfortable means of assistance for caretakers.

Methods: This device combined existing beach wheelchair elements to optimize performance in addition to means of transportation, such as polyurethane balloon wheels and a 6061 aluminum framing for a three wheel design. The novel folding option was accomplished by keeping dimensions small and choosing adjustable wye aluminum hinges. The optimal tire dimensions were chosen using push force verification calculations to ensure the design met ISO standard 11228-2[1] for ergonomics (requiring less than 70 lbs of force to push initially). The final rear wheels chosen for the design were a 16.5" polyurethane balloon wheel from Wheeleez Benicia, CA. This design was compliant with ISO standard 7176[2], ensuring stability and a factor of safety that would allow the device to withstand minimum loads.

A push force test was performed on both dry and wet sand to quantify the force required to move the wheelchair in an accurately simulated environment. The governing force opposing the pushing motion was the force of friction. In this test, the fully assembled Beach Wheels prototype was brought to a sandy location and the chair was placed down on a flat, sandy area. A 70 kg weight was securely loaded on the wheelchair seat to simulate the load of a user. The dependent variable was the force required to push the wheelchair. The force of friction formula, seen below in Equation 1 was used.

$$\text{Equation 1: } F_f = \mu F_N$$

A force gauge was secured to the handlebar of the wheelchair with tape to record the required push force. A volunteer then pushed the chair via the force gauge over a 2 meter testing area and recorded the initial push force and sustained push force for 10 trials. The sand was raked back into place after each trial to simulate the most accurate results. The expected results were that the dry sand would be more difficult to push the wheelchair over than wet sand.

Results:

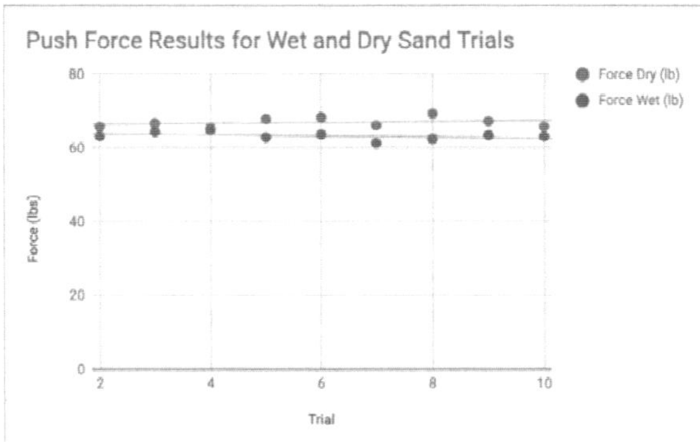

Figure 1. Push Force Results for Wet and Dry Sand Trials

Conclusions:
The results from the push force verification test were successful and can be seen in Figure 1 above. As expected, the average required push force over the wet sand (63.2 lbs) was less than the average push force over the dry sand (67.2 lbs). Overall this test successfully proved that the Beach Wheels device was compliant with the required ISO standard 1122-2 and required less than 70 lbs of force to push on both dry and wet sand. Future studies should include a tiltover test to ensure the device will retain the user in all turning scenarios over both wet and dry sand as well as any hills the user may encounter in sandy terrains.

References:

1. International Standard Organization. 2007. *Wheelchairs* (Standard No. 11228-2) .
2. International Standard Organization. 2014. *Wheelchairs* (Standard No. 7176).

Acknowledgements: The team would like to thank Dr. Vikki Hazelwood at Stevens Institute of Technology for advising, Mr. Edward McCormack and Dr. Herman Morchel at the University of Hackensack Medical Center, and Martin Hardy from Composite Yachts, MD for fabrication.

Development of an Electroactive Tissue Engineering Scaffold

Ruchi Patel, Jeeba Thomas, Madara Dias, Ronke M. Olabisi, PhD
Rutgers, The State University of New Jersey, New Brunswick, NJ, USA.

Introduction:
We previously developed an electroactive scaffold capable of deforming when exposed to a 20 V electrical field.[1] The scaffold was biocompatible and supported cell attachment in the absence of electrical stimulation. Level of scaffold actuation was directly proportional to voltage (5 and 10 V) applied, with no movement at 5 V, minimal movement at 10 V, and robust movement at 20 V or above. All voltages capable of inducing scaffold actuation (20+ V) resulted in cell death of any attached cells. Here we describe strategies to reduce the voltage required for actuation. The stiffness of electroactive hydrogels was reduced to promote actuation under lower voltages capable of supporting cell viability. This was accomplished both by altering the material properties (i.e., increasing polymer molecular weight from 10 to 20 kDa) and by adjusting the aspect ratio of the hydrogel sheets.

Methods:
Two different hydrogel solutions were prepared by dissolving polyethylene glycol diacrylate (PEGDA: 10 kDa or PEGDA: 20 kDa) and acrylic acid in phosphate buffer solution. A photoinitiator solution containing 2,2-Dimethoxy-2-phenylacetophenone in 1-Vinyl-2-pyrrolidinone (300 mg/mL) was prepared and added to the hydrogel solutions (5% v/v) immediately prior to injection into a rectangular glass mold and 2 min exposure UV radiation (365 nm), as previously described.[2] A power supply connected to 2 platinum electrodes was used to stimulate the hydrogels within a petri dish. From pilot data, the aspect ratio influenced the degree of bending in the hydrogel. Therefore, hydrogels of varying aspect ratios were stimulated with 5 V and 10 V and bending angle was determined by imaging. Aspect ratios were varied by holding the width of the hydrogels at 1 mm, while varying hydrogel lengths from 15-25 mm. There were three trials for each experimental group.

Results:
Hydrogels moved with both levels of electrical stimulation (Fig. 1). Higher levels of actuation were most strongly correlated with higher aspect ratios (Fig. 2).

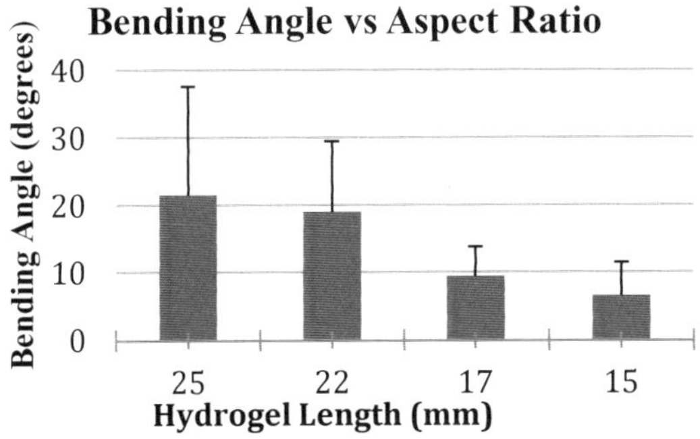

Figure 1. Bending Angle vs. Voltage applied for 20 kDa PEGDA.

When compared to an aspect ratio (length:width) of 15:1, aspect ratios of 25:1 ($p = 0.03446$) and 22:1 ($p = 0.03551$) had a statistical significance . Varying contact time of electrical stimuli and hydrogel sheets also varied actuations.

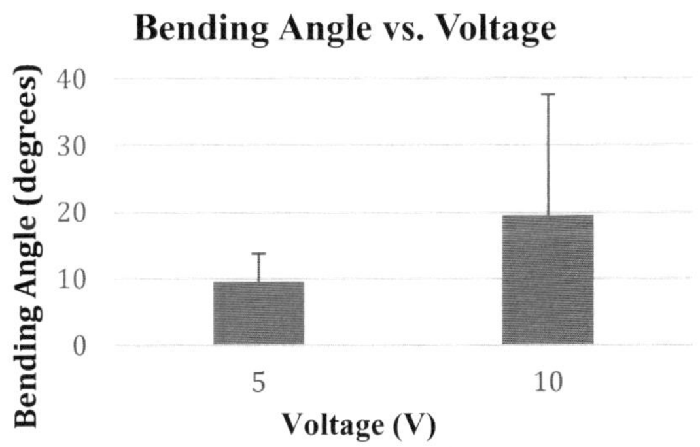

Figure 2. Bending Angle vs. Aspect Ratio of hydrogels stimulated at 5 V.

Conclusions: Originally, actuation was not observed at various voltages (1, 2, 5 and 10 V) when the hydrogel sheets were composed of 10 kDa PEG. Based on these results, the hydrogel sheets were reformulated with 20 kDa PEGDA, in order to decrease their stiffness. Increasing the molecular weight of PEGDA used in the hydrogels successfully resulted in a more compliant electroactive hydrogel. This increased compliance translated into hydrogels that were able of actuating at lower voltages that were not shown to be lethal to cells in concurrent pilot data. Increasing the aspect ratios of the hydrogels permitted further reduction of voltage required to actuate the hydrogels. This work demonstrates the next steps towards a functional electroactive tissue engineering scaffold.

References:
1. Daniel P.B. et al. Characterization and optimization of actuating poly(ethylene glycol) diacrylate/acrylic acid hydrogels as artificial muscles, Polymer, Volume 117, 2017, Pages 331-341
2. Gupta, Kriti. Inducing mechanical stresses in electroactive hydrogels to influence the fate of mesenchymal stem cells. Thesis. Rutgers University-School of Graduate Studies, 2018.

Acknowledgements:
This work was supported in part by a Rutgers Aresty Undergraduate Research Award.

P3-53

Targeted Heating Gadget: A Heating Device for Treatment of Hypothermic Conditions

D. Cleary, S. Alshafie, M. Caba, E. Ferguson, C. Perry
Stevens Institute of Technology.

Introduction: Hypothermia, a condition in which an individual's core body temperature falls below 36°C, is commonly associated with trauma patients. 15% of all trauma patients were reported to have hypothermia in addition to their injuries [1]. Hypothermia increases the risk of developing additional life-threatening complications such as coagulopathy, damage of internal organs, cardiovascular arrest, and death. Current treatment methods used are often invasive and/or cumbersome and require an excessive length of time to normalize an individual's body temperature. The Targeted Heating Gadget (THG) provides a highly-effective method of increasing the body temperature of hypothermic individuals through resistive heating technology. THG is focused on targeting specific locations of the body and operates with a temperature limit of 44°C to comply with safety standards and prevent pain or injury to the individual [2]. This study aimed to demonstrate the effectiveness of THG in surpassing the Standard of Care heating rate value of 1°C/hour [3].

Methods: Blood flow through a blood vessel was modeled using tubing and a metering pump to mimic blood flow at hypothermic conditions. The system was configured with components represented in the process flow diagram, Figure 1.

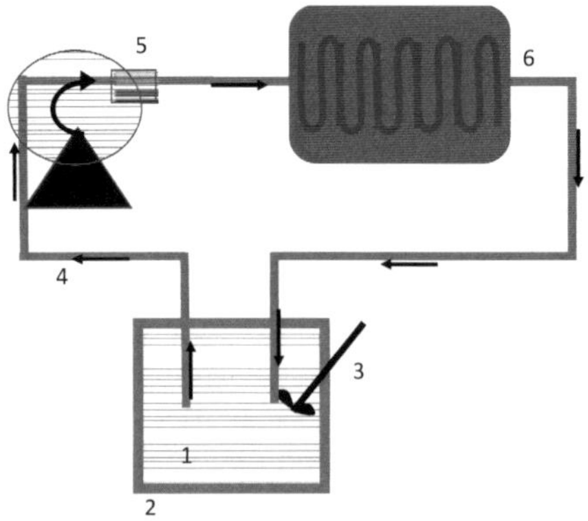

Figure 1. Process Flow Diagram
1. Water Reservoir 2. Reservoir Insulation 3. Mixer 4. Polyurethane Tubing 5. Metering Pump 6. Targeted Heating Gadget

A 1 Liter reservoir of water at 30°C and a metering pump operating at a rate of 40 gph were used to represent blood traveling at a low flow state in the body under severe hypothermic conditions. The reservoir was continuously mixed and fully insulated to prevent any energy loss to the environment. Polyurethane tubing with a ½" inner diameter was used to represent a blood vessel because polyurethane's thermal conductivity was comparable to human tissue. The Targeting Heating Gadget was placed over the polyurethane tubing and applied heat via conduction to the water flowing through the tubing. A temperature sensor was placed in the water reservoir, and the flow system operated uninterrupted for one hour and temperature readings were collected every 5 minutes. The data collected was analyzed to determine if the rate of heating was above the Standard of Care.

Results: The temperature values measured during the experiment are shown in Figure 2 below.

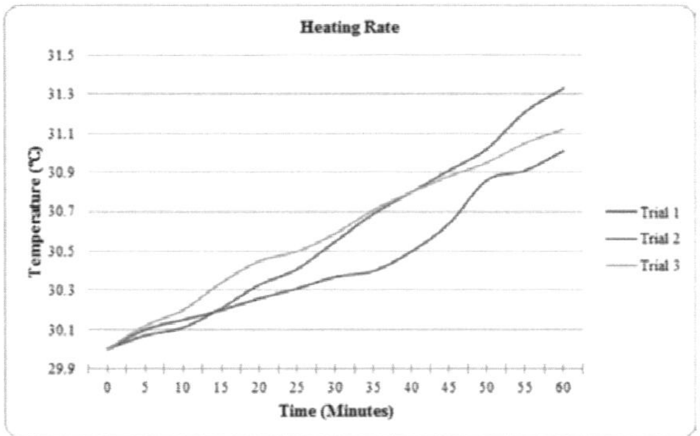

Figure 2. THG Sample Data

Conclusions: A linear trendline was measured for each trial to determine the heating rate. The average heating rate of 1.152°C/hour was determined for this study. Based on this experimental rate of heating, the rate of 1.84°C/hour was calculated for the full device utilized on the body. The study demonstrated an individual patch of the Targeted Heating Gadget may be capable of heating at an effective rate. Upon use of the full system, the device may surpass the standard of care heating rate of 1°C/hour. In a hospital setting, THG may be a noninvasive solution and reduce the length of treatment for hypothermic patients. Future research through clinical trials of the device would be beneficial in demonstrating the effectiveness of targeting specific locations of the body through resistive heating technology.

References:

1. Soreide, Kjetil. Clinical and Translational Aspects of Hypothermia in Major Trauma Patients: From Pathophysiology to Prevention, Prognosis and Potential Preservation Injury. 2014:647-654
2. Ungar, Eugene & Stroud, Kenneth. A New Approach to Defining Human Tough Temperature Standards. 2010
3. Danzl, Daniel. Hypothermia. MSD Manual Profession Version. 2018

Acknowledgements: Thank you to the following individuals in the development of this study: Dr. V. Hazelwood from Stevens Institute of Technology and Dr. K. Lu from Stevens Institute of Technology.

Evaluation of endogenous neural stem cell activity in a mouse model of traumatic brain injury

Rebecca Risman, Jeremy Anderson, and Li Cai
Department of Biomedical Engineering, Rutgers University, Piscataway, NJ

Introduction: Traumatic brain injury (TBI) causes cell death and temporary or permanent loss of cognitive and/or motor function. Approximately fifty thousand people in the United States die annually from TBI due to falls, motor vehicle accidents, and assault. It is known that injury activates neural stem cells (NSCs) which generate mostly astrocytes and oligodendrocytes in the damaged brain. Notch signaling plays an important role in neural development and NSC regulation after injury. The goal of this research is to improve TBI recovery by inducing NSCs after injury to generate mature neural lineages that will help to restore damaged neural circuitry. Using a Notch1CR2-GFP animal model where NSCs are tagged with a reporter green fluorescent protein (GFP), we study Notch signaling in mice after TBI. A closed head injury (CHI) on the mouse was induced by weight-dropping. Lentiviruses carrying neurogenic transcription factors to promote Notch signaling were injected into the mouse brain at the injured site to determine the therapeutic effect on NSC activation and differentiation, with the potential to lead to tissue repair and regeneration after TBI. The brain tissues were analyzed at various time points after injury (i.e., 7 days and 21 days) to capture the early injury response and the lineage adoption. Immunohistochemistry was performed to examine the role of neurogenic transcription factors on neuronal differentiation using markers for neurons (DCX, NeuN), oligodrocytes (Olig2, NG2) and astrocytes (S100β, GFAP). Results with help determine the potential therapeutic effects after TBI.

Methods: Stereotaxic stability during microinjection surgery is performed to deliver the lentivirus into the mouse {Cetin, 2006}. Half of the animals will be subject to a TBI created by dropping a controlled weight on their head using CHI brain injury model. The other half of the animals will be part of the sham group and not receive a brain injury. There will be two subgroups within the injured and sham animals that either receive the lentivirus delivered gene of interest or a blank gene. This allows us to monitor the effect of the genes of interest delivered with and without injury to understand whether the gene induces the same response in the uninjured and injured states.

With the mouse anesthetized using isoflurane and secure ear bars, a nose cone administering isoflurane is used to keep the mouse asleep throughout the procedure and a dose of pain medicine is also administered. The scalp is shaved and cleaned to keep a sterile site. An incision is made using a scalpel along the scalp to have open access to the skull. If the mouse is in the brain injury group, a weight is dropped at 4 cm. For all groups, the tip of the needle was positioned at the coordinates of the hippocampus or cortex where the lentivirus containing the gene of interest of the control are injected into the respective locations. The wound site is stapled and meloxicam is administered.

In vivo experiments will be performed in triplicate by gender to rule out experimental bias and be able to appropriately conduct statistical tests. After surgeries, sectioning, staining, and quantification will be performed.

Results: Beginning at 1 hour after surgery, the animals are observed for behavior using the "Neurobehavioral Severity Scale (NSS-R) for Rodents." This scale is based on a series of testing focusing on different behaviors that could be affected by TBIs. The tests include: observing the animal's balance on a beam, landing from a height of a few centimeters, dragging across a surface, ear reflex, and sound reflex. {Yarnell, 2016}. Analysis will be done to stain for neurons and astrocytes as well as cell death and proliferation. It is expected that when the gene-containing lentivirus is injected, the notch pathway will be amplified and there will be more staining of neurons. Furthermore, the injured animal will have more neurons present than the uninjured animal. Statistical analysis using a two way students' t-test or ANOVA followed by post-hoc analysis (e.g., Tukey) will be performed to find the level of significance (p-value ≤ 0.05). T-tests will be used to compare the uninjured and injured animals as well as lentivirus and blank injections.

Conclusions: Our findings indicate that TBI activates neural stem cells after brain injury, and that these cells have the potential to differentiate into neuronal lineages (e.g., NeuN+ neurons). Further characterization of these cells may help to improve recovery after brain injury. The overall impact of this research project is to expand what is known about the role of lentivirus injections after TBI.

References:

1. Cetin, A., et al. (2006). "Stereotaxic gene delivery in the rodent brain." Nat Protoc 1(6): 3166- 3173.
2. Chapman, H., et al. (2018). "Gsx transcription factors control neuronal versus glial specification in ventricular zone progenitors of the mouse lateral ganglionic eminence." Dev Biol 442(1): 115-126.
3. Gilmore, E. et al. (2010). "Traumatic Brain Injury." AAN Enterprises, Inc.
4. Pei, Z., et al. (2011). "Homeobox genes Gsx1 and Gsx2 differentially regulate telencephalic progenitor maturation." Proc Natl Acad Sci U S A 108(4): 1675-1680.
5. Ran, Q. S., et al. (2015). "Activation of the Notch signaling pathway promotes neurovascular repair after traumatic brain injury." Neural Regen Res 10(8): 1258-1264.
6. Yarnell, A.M., Barry, E.S., Mountney, A., Shear, D., Tortella, F., and Grunberg, N.E. 2016. The revised neurobehavioral severity scale (NSS-R) for rodents. Curr. Protoc. Neurosci. 75:9.52.1-9.52.16. doi: 10.1002/cpns.10

Acknowledgements:
1. New Jersey Space Grant Consortium Academic Year Fellowship 2018-2019
2. Aresty Research Center

P3-55

Designing a high-throughput cell immobilizer for microscopy imaging in a monolayer

Chance Jackson, Tina Ta, JB Lugagne, Nathan Tague, Mary Dunlop.
Boston University

Introduction: Emerging developments in microfluidics has enabled tracking of individual cells over long periods of time under precisely controlled conditions. Single cell imaging allows for the observation and separation of individual cells over time, rather than data that is averaged over a population of cells. This aids in the understanding of stochastic processes of molecules in cells and their impact of cellular dynamics. However, the current microscopy methods for imaging bacteria are not generally amenable to automation for high-throughput screening. New methods using microfluidic devices constructed out of polydimethylsiloxane (PDMS), a transparent, gas-permeable polymer, have shown improvement in single cell imaging. This project aims to create the InstaPad, a device in 96-format well plate format capable of rapidly immobilizing live bacteria into a monolayer for high-throughput imaging across a single, uniform focal plane, that is amenable to multiple imaging modalities. Positive pressurization of a gas-filled membrane will sufficiently create this monolayer of non-adhering cells with minimal perturbation to natural physiology.

Methods: *Developing the InstaPad*: Prototypes consist of a top pressure chamber and a bottom well-plate, which are assembled with screws. The pressure chamber was designed using Solidworks computer-aided design (CAD) software and printed from a Form2 3D printer using V4 resin. There was a hole leading to the pressure chamber to allow tubing to be inserted to serve as air pressure supply. The tube was sealed over with epoxy resin. The printed part had neither a top nor bottom face; a glass slide was sealed to the top face with Loctite 352 UV glue, and a PDMS membrane was sealed to the bottom face. To achieve the seal with the membrane, a groove on the bottom face surrounding the pressure chamber was filled with liquid PDMS resin and cured overnight; this was sealed to PDMS membranes pre-fabricated with a spin coater, using a plasma over. The bottom well-plate was also designed in Solidworks and machined with a computer numerical control (CNC) mill on an aluminum plate. A glass coverslip was sealed to the bottom face of the aluminum sheet with Loctite 352 UV glue. *Testing the InstaPad*: Live E. coli cells were placed into the well-plates and the top and bottom portions of the device were screwed together. A pressure regulator was attached to a normal laboratory air supply source, and attached to the device to achieve inflation of the membrane. Cells were compressed, and imaged under a standard inverted microscope to visually confirm monolayer formation. A propidium iodide viability stain was also conducted on a different sample and imaged under fluorescence to ensure pressurization of a membrane would not cause harm to sample cells.

Results: Figure 1 contains a still image ofE. coli cells taken during microscopy to confirm successful immobilization of cells into a monolayer. While the device was successful in immobilizing cells, there were issues with keeping the cells in focus. We believe that minor fluctuations in pressure, perhaps due to inconsistencies with the pressure regulator, were the source of these issues. The propidium iodide viability stain showed cell growth levels that one would expect in a normal cell culture, validating that the InstaPad could successfully use positive pressurization of a membrane to achieve cellular immobilization.

Figure 1. Image of E. coli cells taken with standard inverted microscope. Issues were encountered with focusing due to inconsistencies in pressure.

Conclusions: The device is capable of rapidly immobilizing live bacterial cells for single cell analysis under a standard inverted microscope, but several improvements must be made before this device can be widely adopted for use in research. A fine-tuned pressure control system is needed to rectify issues with focusing during imaging. Further calculations and experimenting should be done to find the optimal pressure supply for imaging. This iteration of the InstaPad is only compatible with certain imaging modalities due to the use of PDMS as a material for the membrane. The C-H bonds would not be compatible with Selective Raman Microscopy, something our own research group uses often in experiments. A suitable alternative to PDMS with similar material qualities has yet to be identified, but will be necessary moving forward. Once the issues with pressurization are resolved, the cell viability stain should be conducted again; there is some doubt as to the reliability of that experiment given the inconsistency with pressure.

Mechanical Tests on a Polyelectrolyte Complex (PEC)

Hima Tallam, Brandon Mirabile, Rene Schloss and Noshir A. Langrana.
Department of Biomedical Engineering, Rutgers University, Piscataway, NJ, USA.

Introduction: An adhesion is an abnormal union of membranous surfaces, often due to trauma [1]. Our lab focuses on implementing a polyelectrolyte complex (PEC) barrier to prevent adhesions. In the past, Chitosan (Chi) and Polygalacturonic Acid (PgA) based PECs that the lab created were used to determine the mechanical properties of films and the best way to attach them to the peritoneal cavity [2]. Specifically, the feasibility of two methods were tested: suturing the film or sticking it with a fibrin-based glue [3]. Fibrin glue showed more potential for attachment than sutures. Hence, we are now focused on further developing the glue dependent attachment system. In an attempt to strengthen the glue, we decided to double the concentration of the glue's components while maintaining the ratio of the volumes of each component. The following study was developed in order to determine if this change enhances our attachment system.

Methods: The Chi solution is prepared by dissolving 300 mg of Chi in 30 mL of deionized water with 1 mL of 1 M hydrochloric acid added in. The PgA solution is prepared by dissolving 300 mg of PgA in 30 mL of deionized water with 1 mL of 1 M sodium hydroxide added in [4]. Then the Chi solution is added dropwise to PgA solution to create varying concentrations of 80 mL PECs. The solution is sonicated using a Branson Digital Sonifier 450. The mixed solution is poured into a rectangular container and left on a heating plate for 20-24 hours at 37 degrees C. Then the heater is turned off and the film air dries until complete.

The single concentration glue was made by mixing 50 mg of fibrinogen/mL of deionized water solution, 150 units of thrombin/mL of deionized water solution, and 0.1 M calcium chloride. The double concentration glue was made by mixing 100 mg/mL fibrinogen solution, 300 unit/mL thrombin solution, and 0.1 M calcium chloride. For each sample, 27 µL of fibrinogen was placed on the surface of a film, followed by 20 µL of $CaCl_2$, and finally 3 µL of thrombin. Another film was placed on the 50 µL of glue and left to dry. To secure the samples and prevent them from slipping from the grips, duct tape was wrapped over both ends. The thickness, width, and grip separation for each sample were recorded for data analysis. To better emulate the conditions inside the body, the samples were kept moist with PBS before testing.

Results: The slope of the stress versus strain graph provided us with the modulus of elasticity. The modulus of elasticity in both cases varied ranging from 5.8 to 16.63 psi. This large variation can be attributed to the difficulty in controlling the area of glue in the various samples. Furthermore, the maximum stress in single concentration samples ranged from 34.575 to 84.099 psi whereas in double concentration samples, it ranged from 56.820 to 63.651 psi which is not as much variation.

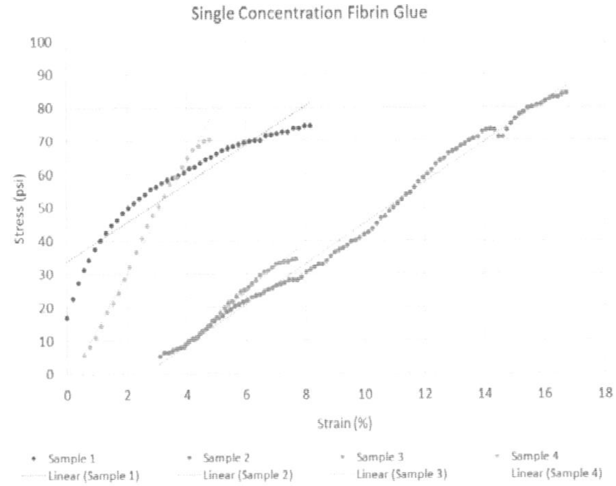

Figure 1: The results of mechanical testing for all samples up to the point of deformation for single concentration

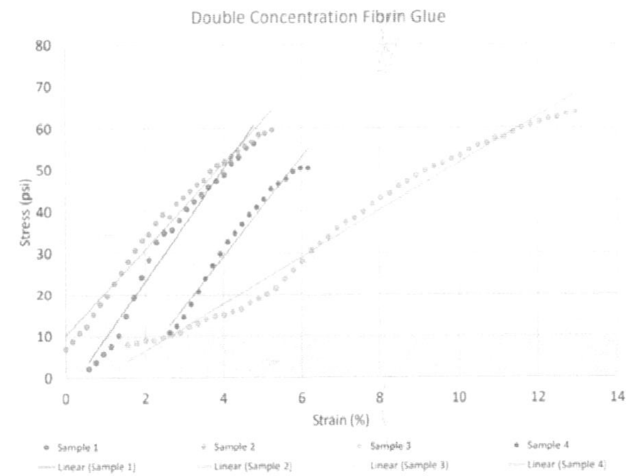

Figure 2: The results of mechanical testing for all samples up to the point of deformation for double concentration.

Conclusions: A t-test was performed to compare the average modulus of elasticity and average maximum stress between the single and double concentration glue samples. The two T-tests had two-tailed p-values of 0.835 and 0.701 respectively for the average modulus of elasticity and average maximum stress, showing no significant differences at a 5% significance level. Overall, both the single and double concentration glue on films resulted in similar behavior. Therefore, either could be used for attachment purposes but it is still more advantageous to use the double concentration glue because it dries quicker.

References:

1. Akhtar, R., et al. Materials Today 14:96-105.
2. Boland, G. M., et al. The Journal of Surgical Research. 132(1):3-12.
3. Radosevich, M., et al. Vox Sanguinis 72(3):133-143.
4. Schatz, C., et al. Langmuir 20:7766–7778.

Mid-Fidelity Simulator Model for Training Cesarean Sections in Kampala, Uganda

Victoria Utria, Rachel Junod, Michelle Krach, Lyndsey Sbarro
Drexel University School of Biomedical Engineering, Science and Health Systems

Significance: In 2015, the maternal mortality rate in Uganda was estimated to be 343 per 100,000 live births, compared to 21.5 per 100,000 in the United States. The Expanding Training in Surgical Skills and Operative Obstetrics program (ETOO) was created by Permanent Secretary Ministry of Health Dr Diana Atwine to improve the clinical skills of current and future OB/GYN clinicians in Uganda, with the overall goal of reducing national maternal and neonatal mortality and morbidity rates. The main component of the program is instructor-led training on physical models. This allows students to practice the techniques of cesarean sections repeatedly, increasing physician competency and reducing likelihood of critical mistakes.

While models exist for training cesarean sections, none meet the complete needs of the program in terms of cost, anatomical accuracy, and accessibility.

Design Description: The objective of this project is to develop an artificial cesarean section simulation model for the ETOO program. Artificial organ and tissue layer analogs will be cast in elastomers of varying elastic moduli and tactile properties to mimic the material properties of their corresponding tissues. In the proposed design, the abdominal wall, uterus and bladder analogs are removable so that they may be utilized individually for part-task training and replaced once the maximum number of incisions are made in them. Replaceable and rotatable components will increase the simulation uses per model while decreasing the price per simulation. The proposed solution accounts for cost effectiveness, transportability, and assembly

The model's ability to replicate the properties of native tissue will be verified through rheometry (ASTM D5099-08) and uniaxial tensile testing (ASTM D412-16).

Design Criteria: Program physicians identified the key procedural methods that the model must support to improve patient outcomes: the Joel Cohen technique, correct manipulation of the bladder, and proper suturing. To meet these requirements the model must accurately represent the anatomical geometry of the uterus, bladder, and abdominal cavity; and the mechanical properties of the uterus, bladder, abdominal muscle, fat and skin. The uterus analog dimensions will be 35x25x20 cm^3 [1], with a wall thickness of 10-20 mm [2], and an elastic modulus between 0.5 and 2.3 MPa [3]. The bladder analog will have a volume of 150 mL, a wall thickness of 13 ± 2.7 mm [4], and an elastic modulus of 1.9 ± 2 MPa [5]. Skin, fat, and muscle analogs will have thicknesses of 1.16 ± 0.26 mm, 13 ± 2.7 mm, and 9.8 ± 1.7 mm respectively [6, 7, 8]. Their elastic moduli will be 4.0 ± 3.8 MPa, 1.6 ± 0.8 kPa, and 22.5 ± 2.6 kPa [9, 10, 11]. Additionally the model must fit in a standard-sized suitcase for transport to Kampala, Uganda. Per stakeholder request, the model must have an overall assembly time of less than five minutes, allow for part-task training, and allow for a minimum of four incisions per tissue analog before replacement.

Results: Material testing is set to begin on February 8, 2019. Elastic moduli accuracy determined by MTS tensile testing and statistical analysis will inform final material selection. Stakeholder feedback and validation testing with Drexel University College of Medicine (DUCOM) residents will determine final material selection.

Future Implications: The model will aid in increasing physician competency at Makerere University in Kampala, Uganda. As the model is used in ETOO trainings and user feedback is reported, the model can be produced in bulk to serve all teaching hospitals in Uganda. Though designed for the ETOO program the final model has global applications. DUCOM physicians have expressed interest in using the model to teach medical students and residents. There is currently no affordable, tissue-representative model on the market-- a need in the developed and developing world.

Future studies can be conducted to determine optimal elastomer, specifically silicone, ratios to effectively mimic the mechanical and material properties human tissue.

Category: Global Health

References: 1. D.M. El-Mowafi. GFMER. 2017. 2. S. Standring. Gray's Anatomy. 2016;1288-1313.e1:(1288-1313). 3. S. J. Manoogian. Stapp Car Crash Journal. 2008; 52:(167-185). 4. N. Kennedy. General Obstetrics. 2015;11:3:(225-232) 5. P. Martins. IUJ. 2011;22:8:(991-995). 6. Eisenbeiss. BJD.

1998;139:3:(462-467). 7. N. Kennedy. General Obstetrics. 2015;123:2:(225-232). 8. M. Ota. AGG. 2012;55:2:(e26-e30). 9. A. Kalra. Journal of Material Science & Engineering. 2016;5:4:(2-7). 10. N. Alkhouli. AJPEndo. 2013;305:12:(E1427-E1435) 11. C. Song. Surg Endosc. 2006;20:6:(987-990).

Disclaimer: The authors are fifth-year undergraduate biomedical engineering students at Drexel University. This abstract submission is for the NEBEC Undergraduate Design Competition under the global health category.

Acknowledgements: The School of Biomedical Engineering, Science and Health Systems provides all Senior Design teams with funding. The team additionally received the 2018 ASTM International Senior Design Project Grant. The team would like to thank its primary stakeholders: Owen Montgomery, MD; Dipak Delvadia, DO; and Greg Alleyne, MD; and its advisors: Michele Marcolongo, PhD; Jamie Dougherty, PhD, and Wan Shih, PhD.

Design of optimization procedure for detection of a breast cancer-specific surface molecule

Senior Design Team: Virginia Tanner, Hayley Roth, Raadiya Qadeer, Priya Gupta
School of Biomedical Engineering, Science, and Health Systems, Drexel University, Philadelphia, PA, USA
Category: Undergraduate Design Competition – Biomedical Imaging

Introduction: Breast cancer is the most common cancer in women. For the treatment of breast cancer, a lumpectomy—a breast conserving surgical technique—is often preferable over a full mastectomy largely for cosmetic reasons. Nearly 25% of women need re-excision surgery following a lumpectomy, which increases the risk of surgical complications and can be a financial burden to the patient. A margin assessment evaluates the surfaces of the resected tumor to determine the presence of cancer cells and doing so intraoperatively would ensure complete removal of the tumor. While intraoperative electric and optical tumor margin analyses exist, a molecular analysis method is preferred due to the higher specificity of the assessment. The Tn antigen found on the surface of 90% of breast cancer cells is a promising target for a molecular-based analysis. The Tn antigen is a carbohydrate surface marker consisting of an N-acetylgalactosamine (which contains three hydroxyl groups and an amide group) attached to a serine or threonine in a mucin (a transmembrane protein) on the cell surface. Due to the complex structure of the Tn antigen, the antibody-antigen binding most likely consists of a combination of hydrogen bonding, van der Waals forces, and hydrophobic interactions which are all affected differently by temperature. The objective of this senior design project is to reduce the incubation time for the antibody-antigen binding and to develop a program to optimize the analysis so that a future intraoperative protocol can be further developed.

Methods: This project is limited by availability of resources, funding and time. In order to develop an optimal protocol, the effects of temperature on the binding kinetics of the Tn antigen and anti-Tn IgM antibody are being investigated. Acrylic substrates were cleaned with a Piranha solution (diluted sulfuric acid and hydrogen peroxide) and then coated with 100 nm chromium and then 100 nm gold using a thermal evaporator. Following 12 MPS treatments, synthetic Tn antigen (Sigma-Aldrich, St. Louis, MO) was bound to the gold plated substrates using an $SM(PEG)_{12}$ linker (ThermoFischer Scientific, Waltham, MA). The substrates were then incubated with anti-Tn antibody [Tn 218] (GeneTex, South San Francisco, CA) and then with 6µm Fluoresbrite® carboxylate fluorescent polystyrene microspheres (Polyscience, Warrington, PA) conjugated to additional synthetic Tn antigen. This process can be seen in Figure 1. This was done at 25°C, 37°C, and 40°C using a water bath. Images of the substrates treated with the fluorescent microspheres will be analyzed using a program. The program quickly analyzes images of fluorescently labeled Tn antigen by evaluating the intensity based on a predetermined threshold value that would indicate the presence of cancer, and relaying the information to the user. Results from the image analysis program will be verified by comparing them to the results of manual analysis ($p>0.5$) through statistical testing. This protocol will then be used with a breast cancer cell line to verify the binding results and analysis method.

Results: It was found that cleaning with the Piranha solution provided a sufficient surface for the chromium and gold coating that resulted in a coating that could withstand further cleaning steps. Testing will continue during the next several weeks. The program will be able to analyze fluorescently labeled cells as well as the fluorescent microspheres used in this study.

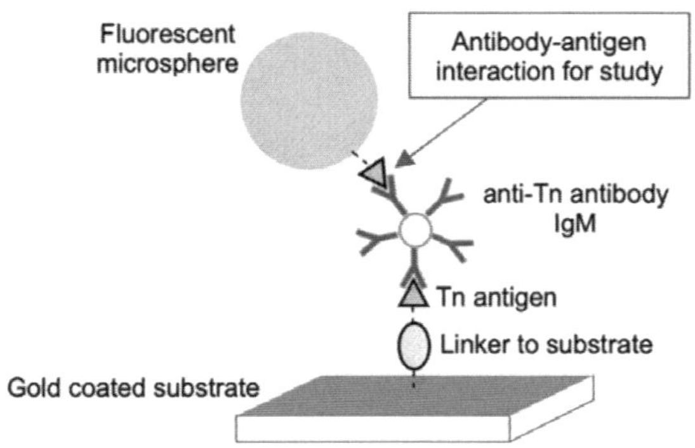

Figure 1: Schematic of process used to study the binding interaction of the Tn antigen and the anti-Tn antibody.

Conclusions: The optimal binding temperature determined in this study will be incorporated into a molecular-based intraoperative margin assessment.

References:

1. Cazet A, Julien S, Bobowski M, Burchell J and Delannoy P. Breast Cancer Res. 2010;12(3):204.
2. Caaveiro J and Tsumoto K, In eLS, John Wiley & Sons, Ltd (Ed.). 2016.
3. Sigurskjold B and Bundle D, J Biol Chem. 1992; 267(12): 8371-8376.

Acknowledgements: The senior design team would like to thank their advisors, Dr. Wan Shih and Dr. Wei-Heng Shih, and mentors, Pawan Rao and Alexandra Jednorski. This project was funded by the School of Biomedical Engineering, Science and Health Systems at Drexel University.

Mapping Functional Networks in the Brain Through Increased Adaptability and Comfort Levels in fNIRS Technology

Kirsten Judge, Stephanie Lemus, Justin Lim, Angelica O'Hara, Samantha Santoni, Elizabeth Tarangelo, Brooke Waldon
Syracuse University Bioengineering in collaboration with MIND Lab

Introduction: The goal of this project is to design and build a wearable framework for fNIRS probes that will accurately and comfortably collect brain activity data from psychology study participants. This project is a joint effort between the Syracuse University Bioengineering Department and MIND (Media Interface and Networking Design) Lab in order to better examine ways in which media interfaces affect human cognition and performance, and and how the mind can be augmented or enhanced by these applications. The lab utilizes functional near infrared spectroscopy (fNIRS), a non-invasive neuroimaging technique similar to fMRI and EEG. It employs optical lasers and sensors combined into channels to illuminate the brain and map various regions. After the brain is stimulated spatially, linked emitters and detectors sense chromophore concentration with near-infrared light attenuations. This gives an indirect measure of the brain activity by making a 3D map of hemoglobin flow patterns, which show activated areas with more quickly changing blood volume. The data collected indirectly reveals the area of the brain that was active, which is monitored while subjects conduct a variety of cognitive computerized tasks (such as the Stroop test).

The current set-up, which focuses all of the optical channels on the prefrontal cortex, is an ill-fitting cap. While it gives direct access to the surface of the scalp, the fNIRS optodes are uncomfortable, with subjects commonly complaining of a headache, and the cap is not adjustable in order to target precise regions. The comfortability and accuracy is paramount to the success of the device, and to that end the project investigates the design of adjustable mounting frames that rest on top of the surface of the subject's head, while also incorporating calibration software to adjust the fit based on head size.

This design involves isolating the fNIRS to those spots when data from functional networks for a particular cognitive resource is sought in order to limit the noise coming from elsewhere in the brain. For example, in a visual attention task, the device takes readings only from specific sections within the occipital, parietal, and prefrontal lobes and give the temporal interplay between the activities of only those three regions as opposed to data of the whole brain. This advancement would be useful in obtaining objective measures of the baseline and typical deviations for different cognitive tasks. Collecting these measures are important in order to better analyze healthy and diseased states and determine where the origin of afflictions may be.

Methods: Materials utilized directly for this project include neoprene fabric, rubber o-rings and matrix material, plastic clips, carbon fiber rods, LED lights and associated wiring, and Arduino Uno. The aforementioned work in tandem with the Hitachi ETG-4100 optical topography system.

The first step in developing the device involved prototyping a hatted framework that allows a number of fNIRS probes to be clipped into a rubber matrix at once. This design was required to work for the full brain as well as for smaller brain regions and thus have flexible node placement options. More specifically, a telescoping mechanism on the horizontal and vertical rods of the structure provide adjustability of the framework in all directions for head size. Due to the light weight carbon fiber rods, weight of the cap is decreased from previous versions. The clipped matrices (Figure 1) enable easy vertical movement of fNIRS probes on the subject's skull.

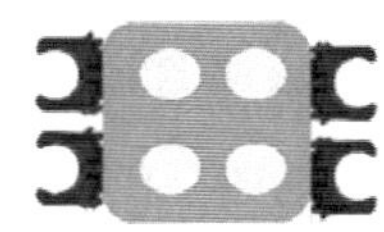

Figure 1

A major design goal is the integration of an interlocking snap-in framework that accomplishes the above without digging into subjects' scalp. This format ensures easy adjustment of node placement in a grid, or spatially linked form, via plastic connected tracks that can be easily pulled apart or clicked into each other depending on the specific areas targeted.

This device is able to accurately map functional network node placement and subsequent data collection for the DMN, a task-positive executive attention network. A secondary functional network being explored is the extended language network, implicated in mind-wandering (Figure 2). The cap configuration developed utilizes an algorithm that first calibrates the fit of the cap by lining up head size measurements with locations of brain area, outputting measurements for telescoping adjustment. The functional network desired is then input into the code and the Arduino Uno, connected to LED lights integrated into the structure of the cap, light up to indicate where probes should be placed.

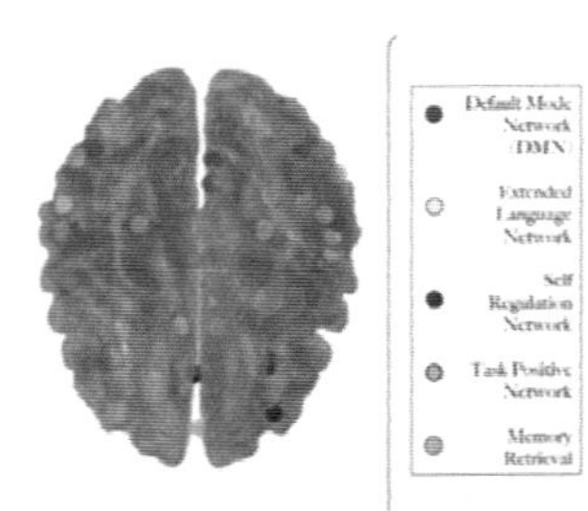

Figure 2

Results: The performance criteria the cap is being tested on include comfort from an end-user standpoint, which is subjective and survey-based with the goal of 75% of users rating the cap as comfortable. The cap has been decreased in weight on the head from 15 lbs to 5-8 lbs depending on configuration. A second point is adjustability, with success rate measured by the unfailing ability to adjust to maximum and minimum adult head size ranges. Third, the stability of the cap is taken into account for accuracy, with time to failure for consistent measurements of a specific region being recorded. Accuracy is also measured by comparing brain output from the new device against the previous results for the same subjects to determine any statistically significant differences. This also involves measuring distance between precise probe placement to solve for statistical difference.

Conclusions: The significance of the success of the new device is substantial in bringing to light the option for low-cost neuroimaging alternatives.

Acknowledgements: Dr. Dessa Bergen-Cico and Trevor Grant for research data in the MIND Lab and Douglas Yung, the capstone coordinator.

Classification For Visual Objects Categorization Using EEG

Hongbo Du, Ismail Jouny.
Department of Electrical and Computer Engineering, Lafayette College, Easton, Pennsylvania.

Introduction: Classification for Electroencephalogram (EEG) from animal/non-animal categorization experiment has been studied. EEG signals were collected when target images and non-target images were presented on screen in front of the participants for categorization [1]. Wavelet transform (WT) and blind source separation (BSS) have been used as feature extraction methods. Convolutional neural networks (CNN), support vector machine (SVM) [2] and extreme learning machine (ELM) [3] have been used as the classifiers. WT with SVM achieved the highest classification rate of 83.06%.

Methods: Fourteen subjects participated in the experiment. Target and non-target images were equal likely flashed for 20ms and participants had to categorize whether there was an animal in the image within 1000ms. Thirty-two-channel EEG data were recorded in 3 seconds and sampled at 1000 Hz [1]. CNN, SVM and ELM are applied as classifiers for raw data as well as the processed data using WT and BSS. Wavelet transformation is a linear transform of signals which can maintain information in both time domain and frequency domain. A wavelet is a wave-like oscillation with an amplitude that begins at zero, increases and then decreases back to zero [4]. Discrete wavelet transform was used, and EEG signal were decomposed into mutually orthogonal set of wavelets. Morse wavelet was used for decomposing the data and the second level of the wavelet transform is subtracted as the feature. SVM is a classification and regression analysis algorithm [2]. A twelve-layer CNN was designed. The input data are 32x3000 matrices where 32 represents the channels and 3000 represents the sampling points and are sent into the input layer of the CNN. 3x3 filters were used in convolutional layers to convolve across the width and height of the input volume, computing the dot product between the filter and input sample. Max pooling layers with a 2x2 pool size and stride 2 are applied for non-linear down-sampling. There are 3 convolutional layers, 2 max pooling layers, 3 batch normalization layers, 1 fully connected layer, 1 SoftMax layer and the input and output layers.

Results:

Twenty runs were conducted to compute the accuracy rate for the all algorithms. Each participant has 600 trials of the experiment. 80% of the data were used for training and rest 20% for testing for the three classifiers. Data were randomized at each run of the classification. Results for all methods are shown on Table 1. The method using wavelet transform for feature extraction and SVM to classify achieved 83.06% classification rate on average. The classification rates of other methods are slightly lower.

Method	Classification Rate (%)
SVM	80.13
ELM	78.25
CNN	75.56
WT+SVM	83.06
WT+ELM	72.16
WT+CNN	75.24
BSS+SVM	77.36
BSS+ELM	74.89
BSS+CNN	71.48

Table 1. Classification Rate for each method

Conclusions: The results showed a decent performance when using wavelet transform for feature extraction and SVM to classify the two classes: animal and non-animals on all electrode channels and all trails. To further improve the classification rate, channel selection algorithm may be needed to keep channels with important information and remove others. In addition, other feature extraction methods and classifiers can be used.

References:

1. Delorme A. et al. Neurocomputing. 2002;44(3):1057-1064.
2. Cortes C. et al. Machine Learning. 1995;20(3):273-297.
3. Huang G. et. al. Neurocomputing. 2006;70(1):489-501.
4. Daubechies. IEEE Trans. Inf. Theory. 1990;36(5):529-531.

Continuous Multi-Limb Monitoring Device for Correlation Between Blood Clot Formation and Bed Rest for Stroke Patients

*Julia Donovan, David Edgar, Sydney Robinson, Eugene Chabot PhD, Brian Silver, MD**
University of Rhode Island, Kingston, RI, USA *University of Massachusetts Medical School, Worcester, MA, USA

Significance: After a moderate to severe stroke, patients are likely to be put on a period of bed rest. While this treatment is necessary for patient recovery, it has some drawbacks. This inactivity has negative health effects, most prominently the formation of blood clots. There are up to 900,000 people affected each year in the United States alone by blood clots [1]. Bed rest increases the risk of blood clots due to slow blood flow in limbs as muscles are not contracting as often [2]. It is critical for patients to move in order to reduce the risk, but it is unknown how often movement is required to effectively eliminate it. There is a need for a device that is capable of tracking and storing triaxial acceleration, triaxial velocity, and movement of limbs continuously for 24 hours in order to correlate blood clot formation with total movement. This device has high commercialization potential on the market as it can serve the large patient population who are at risk of formation of blood clots. The device could also act as a warning device for patients in the future. Devices on the market currently do not store movement data for the duration of time necessary, are not continuous, or are not located on the legs of the patients. Recent studies have utilized an accelerometer as a means to assess patient limb movement, however, devices have not correlated the amount of movement with blood clot formation [3,4]. We seek to develop a device that is wearable on the legs in a clinical setting and that can store the acceleration data for later data analysis on Microsoft Excel.

Design Description: As shown in the block diagram in Fig. 1, the design of the device is such that the accelerometers (LIS302SG, STMicroelectronics, Geneva, Switzerland) are attached just above the ankles of the patient. The accelerometers used in the device weigh 30mg and are 3.0mm x5.0mm x0.92mm in size. The material of the bands used to attach the device to the patient is a sweatband-like material made out of synthetic microfibers. The ribbon wires connect the accelerometers back to an Arduino UNO microcontroller (Arduino LLC, Somerville, MA) for data processing and storage onto a microSD card. The patients will have a bedside apparatus that stores the Arduino and the microSD card. After the data is stored, during either a clinical trial or for the physician's use, the microSD card can be easily removed from the bedside apparatus and plugged into a device with a USB port. The device takes data points ten times a second in order to effectively capture patient movement. There is a 555 timer attached to the Arduino to control the data sampling from the accelerometer at 10 Hz. Data analysis is streamlined through the use of the Excel template designed to handle the data points and compute the total volume of movement.

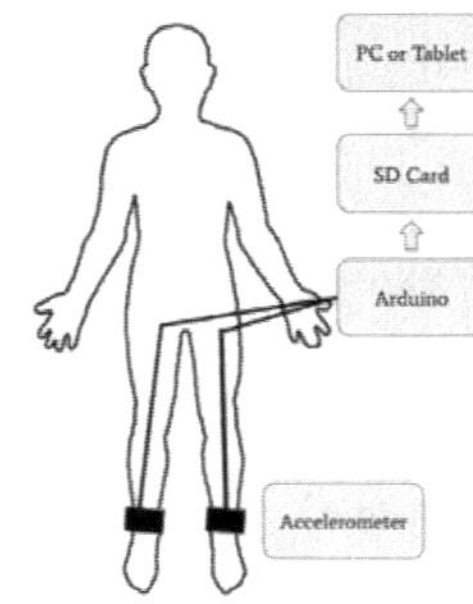

***Figure 1:** Block diagram of the Instrumentation*

Eq. 1: $Digital\ Output = Sensor\ reading\ (V) * \frac{3.3V}{1023}$

Eq. 2: $a = \frac{Digital\ Output - 1.65V}{0.48V} * 9.81 m/s^2$

As shown in Eq. 1, the Arduino reads the data from the accelerometer and multiplies it by the reference voltage (3.3V) divided by the number of analog inputs (1023). Eq 2 converts the solution from Eq 1 to accurate acceleration readings (m/s^2). As shown in Fig 2, the voltage exhibited while the accelerometer is not moving is 1.65V. The voltage equivalent of gravity is 0.48V, and multiplying this fraction by 9.81 m/s^2 converts the solution into the proper units as seen in Eq 2. These equations are inputted into the Arduino code so that the accelerations with the proper units are automatically saved onto the microSD card for ease of use for the physician.

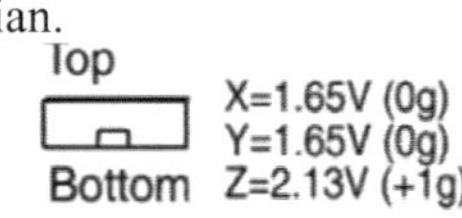

***Figure 2:** The accelerometer measures the acceleration of the limb in Volts/g. The accelerations that the LIS302SG reads at this orientation are given.*

In order to eliminate the effects of acceleration from gravity, the algorithm calculates the difference in magnitude of acceleration between subsequent data points as a movement (Eq 3). If the difference between two data points is zero, then no motion is detected.

Eq 3: $a_{diff} = a_{n+1} - a_n$

Motion artifacts have presented a challenge to collecting accurate data. A threshold of 0.5 (m/s^2) has been implemented to the difference in magnitude of acceleration acting as a high pass filter. The Excel template graphs the acceleration (m/s^2), velocity (m/s), and computes the total amount of movement (m) by computing a double integral (trapezoidal method) of the acceleration data.

Design Criteria: The device has been designed with parameters for use in a clinical setting with patients who have limited mobility. While not a truly wearable system, the device cannot impede on the patient's movements, especially during their physical therapy regimens. It is important that the device is lightweight and small in size to have minimal impacts on patient mobility. Additionally, the ribbon wires connecting the system should allow for a full range of motion by the patient. The way the accelerometers attach to the patient has to be comfortable for long durations of time (24-48 hours) while being rugged enough to protect the accelerometer and its connections from daily use. To accurately record limb movement, an accelerometer must be placed on each limb being monitored. Also, the design must be compliant with ISO/IEC 14971:2007 Medical devices. This standard states that there must be design criteria met to have patient safety paramount as well as have the device be verified. The bedside apparatus protects the system from both water damage and overheating. This eliminates the patient safety risk of burns and shortages. The standard also requires that the device be verified, which ensures that the data is accurate to what it should be reading. Additional constraints were put forth by the institutional review board study (IRB). The device needs to be able to record and store the acceleration data for a physician to analyze after the trial. The device continuously records data to a microSD card for data analysis to be done after the duration of the trial. Microsoft Excel was chosen to be the software for the data analysis because of the ease of use as well as the physician's established knowledge of this software. The sampling rate needs to be chosen based on the storage capacity of the microSD card as well as ensuring that all instances of movements are accurately recorded.

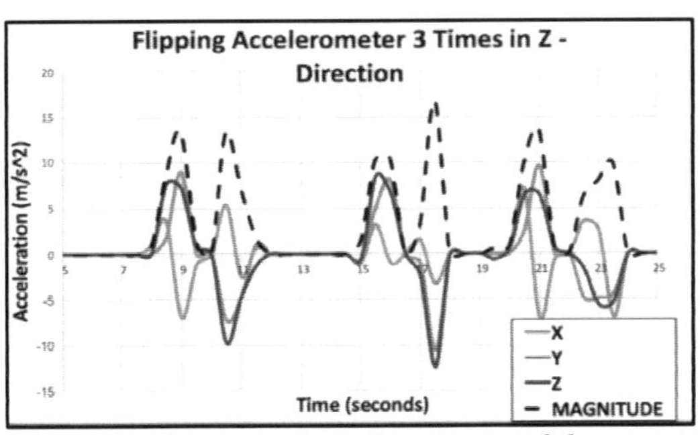

Figure 3: *The triaxial accelerations and the magnitude of acceleration for a series of three controlled movements. During this test, the ankle band was flipped three times about the X-axis in the Z-direction. The first magnitude peak is flipping the accelerometer in one direction and the next peak is returning to the initial position.*

Results: The algorithm graphically shows that there are three instances of movement in Fig 3. These controlled movements are successfully detected by the algorithm and graphically represented by the Excel template. Qualitative analysis of the graphs shows that there are three movements. This proof of concept demonstrates the device's ability to accurately record movement when the ankle band moves. When wearing the system, the accelerometers will record and store the acceleration (m/s^2) of any movements over time. In Fig. 3, the accelerations in three dimensions have been recorded and match the actual movement of the subject's limbs. When the subject was not moving, the acceleration is recorded as zero and a flat line on the graph is shown. In Fig. 3, between 7 and 13 seconds, the magnitude of acceleration spikes up to 14 (m/s^2) twice and then returns to zero. This happened when the subject was wearing the accelerometer on their wrist and asked to turn their forearm over and then back to its original position. The subject did this movement three times and each time, the magnitude of acceleration spiked twice, once for turning their arm and another for turning it back. Although the subject tried to remain as controlled as possible, the movements varied and the magnitudes were not exactly the same each time.

Future Implications: The parameters of the device have been met in order to properly record acceleration, velocity, and the total volume of movement over time. The results that are shown in Fig. 3 proves that the acceleration and magnitude of movement can be recorded in three dimensions. The device has been designed with parameters for an IRB study on patients in collaboration with a physician. The IRB study will monitor bedridden stroke patients' limb movements over 24 hours. The goal of the IRB study will be to find a correlation between the amount of movement a patient conducts to their risk of having a blood clot. With the outcome of the human study, the results will be used to change the device design for home use, acting as an indicator of when movement is required. Relaxations were made because the device is intended to be under constant review of a physician and additional robustness will be needed for an unmanned study. In the future, the device can be commercialized to serve the patient populations on bedrest who are at risk of blood clots.

References:

[1] 'Data and Statistics on Venous Thromboembolism | CDC. (n.d.). Retrieved from https://www.cdc.gov/ncbddd/dvt/data.html

[2] P. Kyrle. Lancet. 2005;365(9465):1163-1171

[3] Sunhwa, S, Hee, K, Jung, J. J Phys Ther Sci. (2014); 26(7): 1009–1011.

[4] Narai, E, Hagino, H & Komatsu, T. J Geriatr Phys Ther. 2016;39(4):171-177

Acknowledgments: Our team would like to acknowledge Dr. Ying Sun for his contributions to the design of the device. We would also like to acknowledge Tanya Wang for her technical assistance.

P3-62

Novel Uterine Manipulator to Reduce Risk of Uterine Perforation and Infection Post-Surgery

Courtney Evans. Kristine Bombardiere, Abigail Emerson, Gabriella Ferrara.
Stevens Institute of Technology.

Introduction: Uterine manipulators are unique tools which allow surgeons to control the uterus externally for improved access and visualization during laparoscopic gynecological surgery such as hysterectomies, fibroidectomies, and endometriosis laparoscopies. While uterine manipulators are vital in uterine surgery, current devices have been known to perforate the uterus during 1.6% of hysteroscopic surgeries.[1] Another common complication post-hysterectomy is infection, which occurs in up to 9% of cases.[2] We hypothesize that uterine perforation and infection may be a result of high forces on the intrauterine tissue from the uterine manipulator. The intrauterine balloon on current devices was developed to maintain the seating of the device in the uterus during manipulation; however, the Stevens Uterine Manipulator (SUM) repurposes the intrauterine balloon. The prototype increases the contact surface area of the intrauterine balloon to reduce pressure on the tissues, thus reducing forces on the tissue.

Methods: A test was conducted which aimed to determine the contact area of the intrauterine balloon of the SUM using a mock intrauterine cavity in order to demonstrate the efficacy of the concept. The governing equation is shown in Equation 1.

$$F = \frac{P}{A}$$

Equation 1. Force, pressure, and area relationship.

The test was conducted using both the SUM and a currently marketed competitor device (CD) in order to ensure the surface area contacted by the SUM is greater than that of the CD. The artificial cavity was coated with 5 ml of 10 mg/ml Bovine Serum Albumin (BSA) (Sigma-Aldrich, Inc.) each time. After insertion, manipulation, and removal, the balloon was soaked in a 1% sodium dodecyl sulfate (SDS) (Thermo Scientific), which removed all the BSA coating the balloon. The BSA and SDS solution was then reacted with BCA Reagent (Thermo Scientific) per ISO Standard 17644:2017 and measured with a spectrophotometer (Thermo Fisher Scientific). The recorded absorbance values were plotted on a standard curve with a correlation of at least 0.99 to calculate the concentration of BSA in each sample. The test was conducted for 10 balloons from each device and results were compared.

Results: The BSA and SDS solution from the SUM intrauterine balloon is expected to have higher absorbance values than those of the CD. When placed on a standard curve, it should be seen that the higher absorbance values correlate to a higher concentration of BSA in the solution. The anticipated concentrations of BSA in mg/ml are shown in Figure 1.

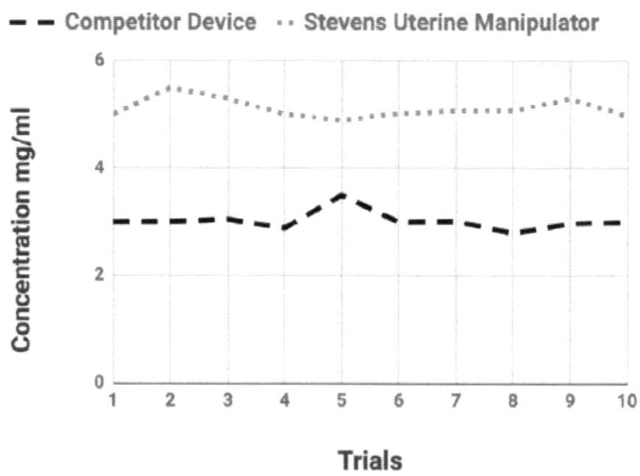

Figure 1. Concentration of Biologics Transfer to Quantify Contact Area of Balloon with Intrauterine Tissue.

As seen by the graph above in Figure 1, the SUM intrauterine balloon is expected to have a consistently and statistically significant concentration of BSA higher than that of the CD.

Conclusions: The concentrations of BSA are proportional to the surface contact of the intrauterine balloon with the intrauterine tissue. The higher concentrations of BSA extracted from the intrauterine balloon on the SUM suggest that the balloon contacted a greater surface area inside the mock intrauterine cavity than that of the CD. The greater the surface area contacted by the balloon, the less pressure is exerted on the endometrial tissue and the lower the forces are on the intrauterine tissue. Therefore, we believe that the risk of perforation is lessened with the SUM, thus demonstrating its safety and efficacy. In future studies, it would be beneficial to manipulate the CD and SUM in cadaver uteri with worst-case sizing to evaluate the safety of the SUM in comparison to the CD pertaining to the risk of uterine perforation.

References:

1. Shakir F, Diab Y. The perforated uterus. *J Obstet Gynaecol*, 2013;15:256-261.

2. Clarke-Pearson D, Geller E. Complications of hysterectomy. *Obstet Gynecol*, 2013;121(3):654-673.

Acknowledgements: The authors thank CooperSurgical, Inc., Emily Noonan, and Dr. Vikki Hazelwood.

Optical Density based Measurement of Cell Count within Polymeric Particles

Dylan Trawinski, Ayesha Aijaz, PhD, Biju Parekkadan, PhD
Department of Biomedical Engineering, Rutgers University

Introduction:. Rapid and accurate determination of cellular growth is critical for many downstream processes such as utilization in cellular therapies, production of biotherapeutics or toxicologic studies. Optical density based measurement of cell density is routinely employed for microbial growth analysis.[2] Suspension adapted culture using spherical microparticles is now being utilized for large scale production of anchorage dependent cells. Herein, we develop a method that can be potentially used for in-line monitoring of cell growth within polymeric capsules without the need for cell recovery prior to analysis. This analytical method overcomes the limitation of widely used MTS assay that requires compromising the sample, yielding a discontinuation for its use in the study.[3] Furthermore, the assay is time consuming, requiring a four hour metabolization time interval before samples may be read.[3] While developing the protocol, this study additionally optimized the process through determining the ideal wavelength for adherent cells, suspension cells, and polymeric particles.

Methods: MSCs and jurkats were cultured in T175 flasks containing Alpha Minimum Essential Medium (Sigma-Aldrich St. Louis, MO) and Roswell Park Memorial Institute Medium (Gibco Gaithersburg, MD), respectively. Empty polyethylene glycol diacrylate (PEGDA) microcapsules, suspended or encapsulated jurkat cells and monolayer or encapsulated mesenchymal stem cells (MSCs) were used as different groups in this study. MSCs and jurkats were used as representatives of adherent and suspension cells respectively. 3D microcapsules were formed by combining 10 kDa PEGDA in 10 mM HEPES buffered saline and photoinitiators 1.5% v/v triethanolamine (TEOA), 37mM 1-Vinyl-2-pyrrolidinone (NVP), and 10 μM eosin Y disodium salt. A hydrophobic photoinitiator solution containing 2,2-dimethoxy-2-phenylacetophenone in 1-vinyl-2-pyrrolidinone (300 mg/mL) was combined in mineral oil (3 μL/mL) to which the cell-prepolymer suspension was added and vortexed for 4 seconds in ambient light, then exposed to white light for 20 seconds with a vortex pulse at 10 seconds. 1X, 2X, and 4X serial dilutions were executed as experimental groups in standard 96 well plates containing phosphorus buffered saline (PBS). The data was repeated in triplicate and absorbance spectra were recorded at 200-800 nm using a Varioskan LUX Multimode Microplate Reader (Thermo Fisher Scientific Waltham, MA). The averaged absorbance values were used to construct standard curves and regression plots were created. Correlation coefficients and equations for each linear regression were displayed on the corresponding figure. Analysis of variance (ANOVA) was conducted on 1X, 2X, and 4X groups. An MTS assay was executed to confirm the cell counts produced by the protocol.

Results: Absorbance peaks were detected at 260 nm with subsequent increase from 275 to 290 nm with an absorbance maximum at 290 nm. Spectra in the region of 300 nm to 800 nm do not demonstrate marked changes and no absorbance peaks were detected in this region. The ratio of OD_{280} and OD_{260} between 0.5 and 1.8 is indicative of nucleic acid and protein content of cells. For 1 x 10^6 MSCs, the optical density values were $OD_{260} = 0.28$, $OD_{280} = 0.29$; $OD_{280}/OD_{260} = 1.042$ (Fig. 1A) and for 1 x 10^6 T cells $OD_{260} = 0.85$, $OD_{280} = 1.159$; $OD_{280}/OD_{260} = 1.37$ (Fig. 1B). Following a range finding experiment, the lowest recorded detectable cell count was 6.25 x 10^5 cells captured at 290 nm for both adherent and suspension cells (Fig. 1C, D). ANOVA yielded no statistical significance between 1X, 2X, and 4X groups ($p>0.05$). Using this analytical method, the measured cell counts were 1.89 ± 0.39 for MSCs and 2.58 ± 0.82 for jurkats.
0.5 x 10^6 and 3.3 ± 1.2 x 10^6 encapsulated MSCs and T cells, respectively.

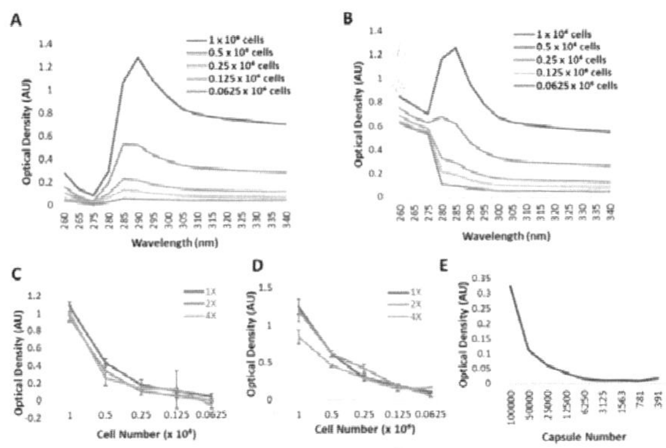

Figure 1: Absorbance spectra of (A) MSCs, (B) Jurkat T cells; optical density at 290 nm for (C) MSCs, (D) Jurkat T cells, (E) polymeric capsules.

Conclusions: The optimal wavelength determined for adherent cells, suspension cells, and polymeric particles was 290 nm since this wavelength allows the largest range of cells/particles to be detected. The statistical insignificance yielded by the ANOVA test shows no advantage to using any particular dilution.

References:

1. Nilapwar SM. Methods Enzymol. 2011;(500):59-75
2. Stevenson K. Scientific Reports. 2016;(6)38828
3. Riss TL. Assay Guidance Manual. 2013;(1)357

Acknowledgements: Matthew Teryek for his assistance in capsule formation. TechAdvance grant as a funding source.

Non-Invasive Arterial Pressure Sensor under Low and High Deflections from Force Signals

Angela Meseha, Dr. Gary Drzewiecki, Ph.D.
Rutgers University.

Introduction:

Arterial tonometry is a method of measuring blood pressure within an artery. A special device, also known as a pressure sensor, can be used in order to gather data and diagnose related conditions. Invasive forms of this device are known to cause discomfort to patients, while non- invasive forms suffer from a decrease in accuracy. However, in order for non-invasive conditions to have a comparable level of accuracy, pressure can be applied to an artery of interest, allowing for a force-reflection-pressure relationship to be formed. This non-invasive method, also known as, continuous non-invasive measurement of arterial blood pressure or CNIBP, requires large vessel deformations and a highly contracted pressure sensor. A device of this sort is able to receive arterial blood pressure readings with less than a 5% error at the radial artery. It should be noted; an evident amount of error may result when there is a slight deviation from the artery wall applanation position.

Methods:

Being able to understand the relationship between pressure, force, and deflection is key in order to accurately receive a pulse pressure reading. A contact force would be used in order to examine this force-deflection-pressure relationship.

Flattening of an artery, also known as, applanation tonometry, is well known for being one the most accurate ways to retrieve arterial pressure. Using this method, we find another non-invasive method, alongside with the generic 'cuff', that can accurately yet precisely measure blood pressure given a certain amount of time. The contact stress, while also incorporating deflections, given over a certain area can be converted into pressure. Without proper positioning of the tonometer, a less accurate reading may be given, which may throw off the entirety of the results. With this being said, this device ensures that at low deflections, a low percent error is possible.

The tonometer will need relatively large deformations in order to accurately receive a reading. This explains how flattening the artery more ensures a more accurate reading. Using the carotid pulse measurement help overcome the problem of high errors in pulse pressure calibration at low deflections.

Results:

A physical model was derived based on previous coding that was done. Using information from past tonometers, an accurate model was able to be determined. A formula was derived in order to ensure that at any given contact area, an accurate reading could be given. This also can be applied to deflections at any given position. An assumption is made in which we assume that the artery is a cylindrical shape. We also assume that since the artery has a larger force pressing against it, it is gradually and symmetrically flattening out. Without these assumption, we cannot assume the amount of area, which would therefore, throw off the results of pulse pressure.

The experiments reveal that the more centered the device is against an artery, the more accurate the results end up being. Based on this, we can still see that the device is user-friendly.

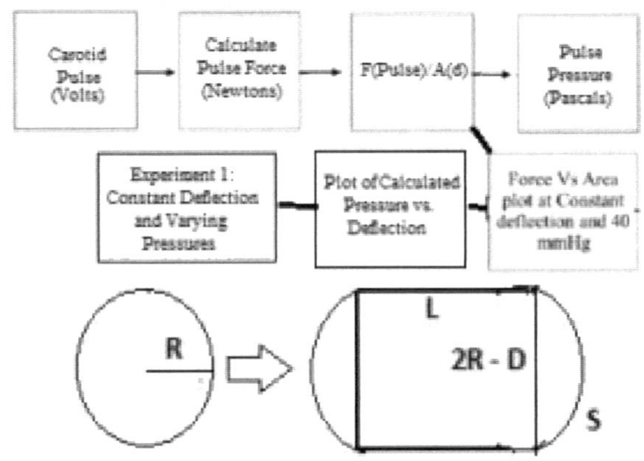

Figure 1 & 2. Conversion of Carotid Pulse to Pulse Pressure, Undeformed artery to deformed artery in order to measure arterial pulse pressure.

Conclusions:

The continuous non-invasive blood pressure measurement, also known as, CNIBP, proves to be a better alternative when compared to the generic 'cuff'. Not only does it take measurements in a timely manner, but also, receives more precise measurements. With the correct use of this device, arterial pressure measurements may be more readily obtained and is accessible for all types of users.

References:

- Drzewiecki, G.M., Melbin, J, Noordergraaf, A. Ann Biomed Eng., 1989;17:325-359.
- Drzewiecki, G.M., Melbin, J, Noordergraaf, A. Elsevier, 1983;16:141-152.
- Matthys, K, Verdonck, P. Technol Health Care, 2002; 10: 65-76.

Acknowledgements:

- Aresty Research Center- Undergraduate Research Fellowship
- New Jersey Space Grant Consortium Summer Fellowship
- Dr. Ronke Olabisi's supervision
- Dr. Gary Drzewiecki's supervision and advising
- Research Scholars Academy

Platform for Developing Beat-to-Beat Heart Rate Detection Algorithms Based on Photoplethysmogram

Jack Cammarata, Jake Matte, Ying Sun
Dept. of Electrical, Computer and Biomedical Engineering, University of Rhode Island, Kingston, RI, USA

Introduction: The photoplethysmogram (PPG) is a light based heart rate detection system that uses light to detect the amount of blood flowing through the circulatory system. Electrocardiograms (ECG) are considered the gold standard for use on patients due to its accuracy. The ECG uses electrodes that are stuck onto the body and utilize conducting gel to enhance the electrical signals from the heart. The electrical signals the heart creates when it pumps are picked up by the electrodes and a heart rate is produced. These signals share the same purpose however behave very differently from one another. The best way to improve the reliability of the PPG is by using the ECG as a model for an ideal system. This platform provides an easy to understand comparison of the two types of signals while featuring a customizable circuit and application for modification of the platform. Figure 1 shows the application in use and the full circuit including the PPG and ECG. This platform was used to produce a simple algorithm to produce a heart rate from the PPG signal by comparing it to the accuracy of the ECG and adapt a similar result.

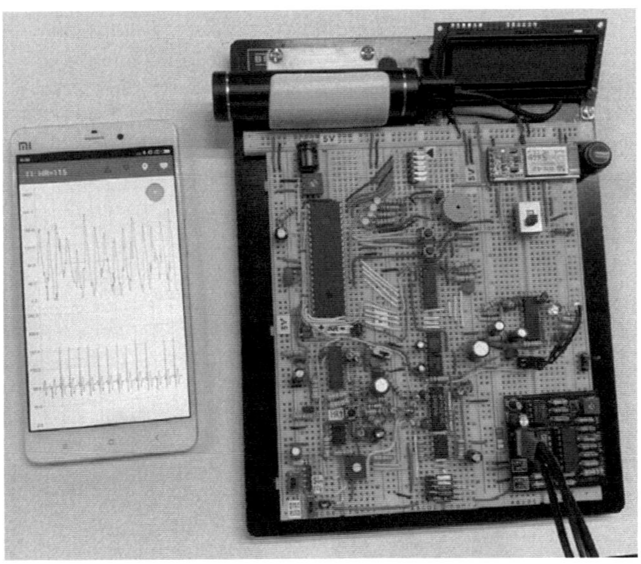

Figure 1: Right: Circuit side of the platform including both the PPG and ECG from a single subject. *Left:* Smartphone application displaying the waveforms; PPG on top and ECG on bottom.

Methods: This platform was created as a tool to compare waveforms for signal observation and algorithm development. It is currently focused on heart rate detection through both PPG and ECG systems. This platform uses both the PPG and ECG signal types from the same subject and displays them side by side for comparison. The circuit features easily accessible components for optimization and troubleshooting. The microchip is programmed in C using the MPLabX IDE and compiler and the application was written in Java using Android Studio. Both softwares are set up to be easy to interpret and modify if needed for improvements or use with other systems.

```
case 11:                    // Function 10: PPG/ECG
    if (do_ECG) {                              // HR meter based on PPG
        do_ECG = 0;
        d2 = d1;                               // Move oldest difference t
        d1 = d0;                               // Move older difference to
        d0 = (int)data0 - data1;               // Store new difference in
        ecg_count++;                           // Increment RR-interval
        mobd = 0;                              // mobd = 0, unless sign co
        if (d0 > 0 && d1 > 0 && d2 > 0){       // (1) If 3 consecu
            mobd = d0 * d1;                    // Multiply first two diffe
            mobd = mobd * d2;                  // Multiply the oldest diff
        }
        if (d0 < 0 && d1 < 0 && d2 < 0){       // (1) If 3 consecu
            d0 = -d0; d1 = -d1; d2 = -d2;
            mobd = d0 * d1;                    // Multiply first two diffe
            mobd = mobd * d2;                  // Multiply the oldest diff
        }
        if (refractory){                       // Avoid detecting extraneo
            refractory++;
            if (refractory == 37){  // Delay for 5.344 x 37 = 1
                refractory = 0;     // Reset refractory flag to
                PORTBbits.RB3 = 0;  // Turn buzzer/LED off (Pin
            }
        }
        else if (mobd > threshold){ // If a peak is detected,
            refractory = 1;                    // Set refractory flag
            PORTBbits.RB3 = 1;                 // Turn buzzer/LED on (Pin
            hr = 11227/ecg_count;              // 842 / 4.5 = 187 Hz,  187
            ecg_count = 0;                     // Reset RRI counter
            if (hr > 40 && hr < 150) hr0 = (unsigned char)hr;
        }
        if (mobd < 255) mobd0 = (unsigned char)mobd;
        else mobd0 = 255;
        if (refractory == 5) {
            if (enableBT) {
                TransmitBT(255);
                TransmitBT(hr0);
                TransmitBT(128);
            }
            else PrintNum(hr0, 77);
        }
```

Figure 2: The microchip code for function 11 including the MOBD algorithm and the bluetooth data transfer to the phone application.

We focused on use with PPG and ECG due to the difference in the waveforms but similarity in the information being conveyed. The ECG is considered the gold standard signal for heart rate detection due to its reliability in detecting proper heart beats. The system uses electrodes stuck to the body around the heart. These electrodes utilize conducting gel to detect the electrical signals sent by the heart every time it contracts. This results in a very clear signal and makes it much more consistent than the PPG. The reason for this is that the PPG has a very noisy signal due to the light based detection system. The PPG uses either green or infrared photodiodes to detect when a large volume of blood of flowing through the vein or artery. Current systems use this method to detect each beat and averages them over a fifteen second period which can lead to averaging errors thus yielding an incorrect heart rate. This platform is meant to find consistencies and differences between the signals to compare them for the aid of the creation of PPG heart rate detection algorithms. The application is mainly used to display both signal types simultaneously for easy comparison of heart rates from the same subject. This is also easily modifiable and has the options to have multiple functions. The accessibility of the circuitry, microchip and application allows the use of different signals and systems and can be easily optimized for ideal use. These features were used to create an algorithm for continuous beat-to-beat heart rate detection from the PPG system. Similarly to the ECG heart rate system used on the platform, the PPG heart rate detection algorithm that has been developed uses multiplication of backwards differences (MOBD) to find the frequency between each spike in the signal. The MOBD algorithm detects peaks in the signal and compares two at a time in order to find the frequency. The difference between the MOBD algorithms is that the ECG uses the upstroke and downstroke while the PPG uses just the upstroke of the larger spike. The platform will be used to further optimize this heart rate detection algorithm and explore new possible methods.

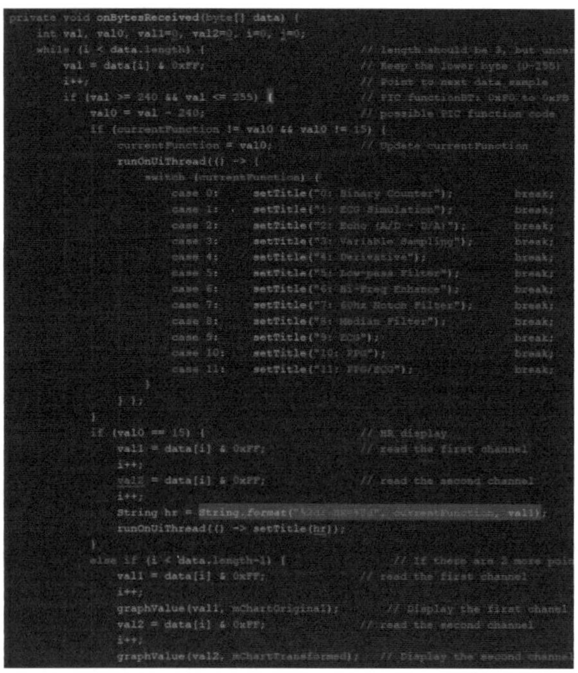

Figure 3: The code for the main activity of the Android application written in C using Android Studio.

Results: This platform allows users to compare signals for data interpretation and modification. It also allows for easy changes in the systems to fit the user's needs. The current application displays the ECG and PPG heart rates simultaneously from a single subject. The waveforms give the same information with very different waveforms. This data was used to develop a continuous beat-to-beat heart rate algorithm for the PPG system. This algorithm uses MOBD on the steep upstroke during the systole period. This system can still be improved by optimizing the software for both the application and microchip and improving the hardware that is used to obtain a less noisy PPG signal.

Conclusions: The current platform is still a prototype but works well with the heart beat detection system that is in place. The main focuses of this project were to create a continuous time PPG and have a platform to show how it compares with the gold standard ECG. The difficulty of the continuous time PPG lead to the development of a system for easy modification allowing it to be used for developmental purposes. This platform is unique in that it allows for the display of both the ECG and PPG waveforms from the same subject. Modifications can be easily made to both the code and hardware to allow this system to have many uses with different signals. The platform is still being used to develop more accurate beat-to-beat detection algorithms as well as optimize data collection of this system.

References:

1. Sun Y, Ph.D. Suppappola S, Wrublewski T A, Microcontroller-Based Real-Time QRS Detection, November/December 1992

2. Andrew Reisner, M.D. Phillip A. Shaltis, Ph.D. Devin McCombie; H Harry Asada, Ph.D. Utility of the Photoplethysmogram in Circulatory Monitoring, Anesthesiology 5 2008, Vol.108, May 2008.

3. Melester, M.t. "The Design of an All Digital Phase-Locked Loop." 38th IEEE Vehicular Technology Conference, doi:10.1109/vetec.1988.195403.

Classification of Tumor in Mammograms with Deep Convolutional Neural Networks

Tianyu Zhu, Ismail Jouny
Lafayette College

Introduction: Breast cancer is the most common cancer diagnosed in women around the world. As stated in a report of American Cancer Society, in 2017, an estimated 252,710 new cases of invasive breast cancer will be diagnosed among women and 2,470 cases will be diagnosed in men. Breast cancer typically produces no symptoms when the tumor is small and most easily treated, thus it is important to detect it in the early stage in order to minimize further exacerbation. Hence, mammogram, X-ray of the breast, is the preferred way to find the tumor and diagnose breast cancer in early detection. Therefore, interpretation and analysis of mammograms could be relatively important and difficult. In recent years, deep learning has brought revolutionary changes to the scientific field including medical field. In medical image processing, CNN is trained on large datasets to recognize and classify the tumor in mammogram automatically. This research mainly focuses on tumor classification by using a specific CNN called AlexNet, which competed in the ImageNet Large Scale Visual Recognition Challenge in 2012.

Methods: The dataset used in this study is from the digital archives of University of South Florida. And a subset of 80 normal, 120 benign and 120 malignant mammograms have been chosen from the large number of images. All these mammograms were from volumes that had been scanned using Lumisys scanner with 12 bits at a resolution of 50 microns. Various image processing techniques were applied on the dataset before testing for the purpose of cleaning, resizing and enhancing. Techniques includes contour segmentation, filtering, hole filling, sharpening and complementing. A specific convolutional neural network called AlexNet is used as our classifier, which designed by the SuperVision group. Two tests have been conducted in the study. First is to classify the dataset by normal (has no tumor) and abnormal (has tumor). Second is to classify the dataset by benign and malignant. During the training process of CNN, both Transfer Learning and Feature Extraction method were applied to the dataset separately.

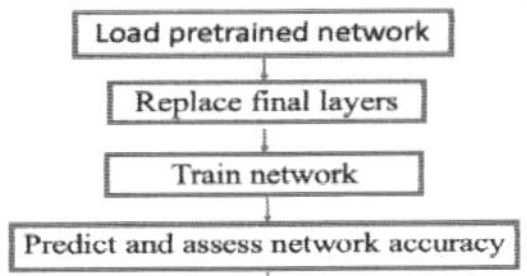

Fig. 1 transfer learning process

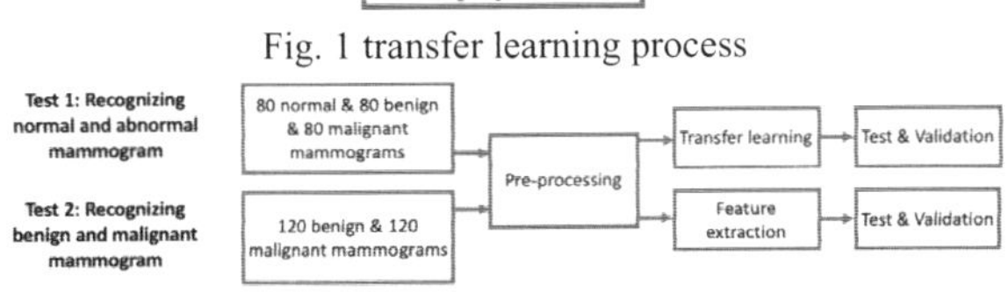

Fig. 2 testing procedure

Results: Result shows both transfer learning and feature extraction methods give nearly 99% accuracy rate and nearly 1% false positive rate for recognizing of normal and abnormal image and approximate 60% accuracy rate and 20% false positive rate for recognizing of benign and malignant image.

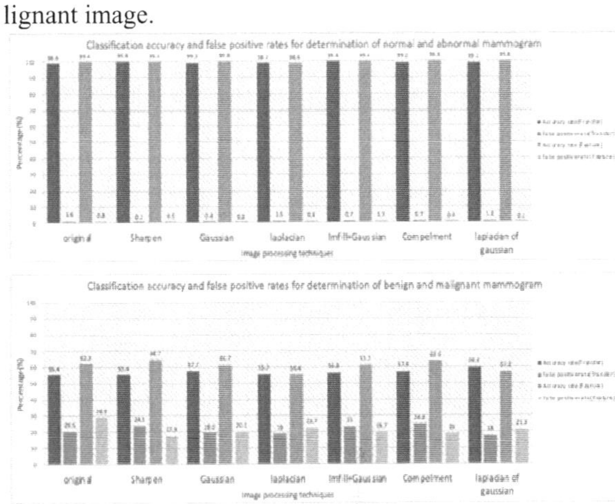

Fig. 3 testing result

Conclusions: The classification accuracy rates of recognizing normal and abnormal mammograms for both transfer learning and feature extraction method adopted with different pre-processing technique reached approximately 99%. And the false positive rate is relatively low. For determining whether a mammogram is benign or malignant, the general classification accuracy rate is approximately 60% for feature extraction method and 56% for transfer learning method. Among five pre-processing techniques, Gaussian filter and image hole filling and image complement give the better performance than other techniques.

References:

Gonzalez, Rafael C., and Richard E. Woods. Digital Image Processing. Pearson, 2018.

1. M. Heath, K.W.Bowyer, D.Kopans et al, "Current status of the Digital Database for Screening Mammography", Digital Mammography, Kluwer Academic Publishers, 1998, pp.457-460
2. Shapiro, L. G. & Stockman, G. C: "Computer Vision", page 137, 150. Prentice Hall, 2001
3. Alex Krizhevsky, Ilya Sutskever, Geoffrey E. Hinton,"ImageNet Classification with Deep Convolutional Neural Networks"

Acknowledgements: I would like to thank Prof. Jouny for his expert advice and extraordinary support in this research. I acknowledge the Electrical and Computer Engineering Department of Lafayette College for financial support.

3D Orientation and Adjustment System for Automated Venipuncture Device

Zachary, Lopez, UG

Venipuncture and blood analysis are plagued with complications associated with puncture accuracy and time-consuming blood analysis. To reduce these complications a robotic venipuncture device is being developed to efficiently and autonomously place a needle accurately into a desired vein, extract a sample of blood, and conduct analysis on the sample. The intention of this research is to develop a 3D orientation and adjustment system that will calculate the deviation of the device from the normal vector of the anterior plane of the forearm, and adjust the needle accordingly to guide it accurately into the desired vein. Analysis will be conducted on the Razor Inertial Measurement System and the Adafruit BNO055 Inertial Measurement System to determine the accuracy of each systems calculation of it's Euler angles. Ultimately, through the application of high pass filters, low pass filters, and calibration, we aim to reduce the drift of the gyroscope to 0.2°/second.

Novel Drug Delivery System using Anti-Angiogenic Peptides for Glioblastoma Multiforme

Anna Mathew, Zain Siddiqui, Peter Nguyen, Biplab Sarkar, Vivek Kumar.
New Jersey Institute of Technology

Glioblastoma, also known as glioblastoma multiforme (GBM), is an extremely aggressive type of nervous system tumor that forms on the tissue of the brain. Nourished by a large and complex network of blood vessels, the tumor cells reproduce quickly, making GBM one of the deadliest cancers. Glioblastoma accounts for roughly half of all brain and central nervous system cancers, with a one-year survival rate of less than forty percent and a five-year survival rate of less than five percent [1]. Current standard of care for patients with newly diagnosed glioblastoma includes surgery, radiation therapy, and chemotherapy. One such treatment involves an in-situ drug delivery system. Following surgery, carmustine wafers are implanted in the tumor cavity. The wafers then slowly dissolve while releasing medication that will kill the remaining cancerous cells. These wafers, however, lack the ability to homogeneously coat and conform to the tissue voids of the resected tumor cavities, amongst other issues. In this project, a space-conforming peptide modified with an anti-angiogenic mimic will be developed in the form of an injectable hydrogel that self assembles into ^{2} -sheet-based nanofibers. This modified peptide is able to conform to the unique shape of the cavity and exhibits anti-angiogenic properties to address the abundant vascularization associated with GBM. Under this project, there are three main objectives. First, the injectable self-assembling ^{2} -sheet peptide modified with an anti-angiogenic mimic will be developed based on an established protocol for peptide synthesis. The completion of the first stage will be based on the successful development of the hydrogel that meets the existing standards for peptide synthesis. Following this advancement, the peptide will be tested for its anti-angiogenic properties through cell culture experiments using human umbilical vein endothelial cells and brain microvascular endothelial cells. It is expected that the hydrogel will allow for a dose-dependent destruction of the capillary networks of these cultures, confirming this in-situ drug delivery system as an effective method against neovascularization and cell proliferation. These culture experiments will also be performed using glioblastoma cells to test for the peptide's innate cytotoxicity, as well as anti-angiogenic properties. The observed response of targeted cell killing could demonstrate the peptide as an immunotherapy, allowing for the body's own immune system to target remaining cancerous cells. After several rounds of testing in cell cultures and evaluation of the preliminary data, a full protocol for the administration of the peptide will be developed by testing the peptide in an animal model. This stage will involve either a subcutaneous or intracranial injection of the peptide in the selected animal to allow for a more thorough understanding of the mechanism by which the peptide delivers the treatment and provide the closest understanding to its performance in a human. Lengthened lifespan of the test subjects will be confirmation of the strong efficacy of the peptide as a drug delivery system.

References

1. Types of Brain Tumors | Types of High Grade Gliomas. (2018). Retrieved October 5, 2018, from https://tocagen.com/patients/brain-cancer/types-of-brain-cancer/

P3-69

Android Based Prosthetic Training Using Augmented Reality

Merci Ujeneza, Anya Duroha, Corvah Akoiwola, Eugene Chabot, Ph.D., Ying Sun, Ph.D.
University of Rhode Island, Kingston, RI, USA.

Significance: Following limb amputation, rehabilitation is critical in order to reliably use the prostheses, incorporate it into their daily lives and eventually become independent [1]. Rehabilitation includes a substantial prosthetic training period to gain the skill to operate the device efficiently. The most common prosthetics used today are myoelectric prosthetics, which uses surface electromyography (EMG) to monitor muscle activity as the user contracts specific muscles to drive the prosthetic. However, despite technological advances to prosthetics and training, many users have noted that they lack intuitive and natural control [2]. One proposed solution is introducing feedback to enhance functionality, control, and provided the amputee a way to monitor their progress. Prosthetic training usually follows the device arrival. The therapist work with the amputee to encourage voluntary activation of target muscle sites, but a recent study indicates that nearly 50% of amputees reject the device, and 30% of amputees self-report that they eventually abandon the device all together [2]. Evidence supports that the outcomes can be improved if the prosthetic training commences after amputation, and prior to arrival. Current training methods and tools are reportedly expensive, limit the user to in-office use, are manufacturer specific, lack training transferability to different tasks and lack personalized training capabilities [2].

Introduced in this paper is a low-cost, accessible hardware and augmented reality (AR) framework, shown in figure 1, that would provide feedback to the user as they are performing the exercises. Specifically, the system would train the user to control the contraction of their muscle, the level of activation needed, and how to isolate specific muscle groups. The system uses AR to allow the amputee to be immersed in a scenario that would be representative of their daily lives. Using AR, which adds digital elements into the real world, provides the amputee with an engaging way to monitor/evaluate performance, it isolates rehabilitation to that specific limb and defines EMG sites and thresholds.

Design Description: In this system, the hardware consists of an EMG circuit that collects EMG muscle effort indications such as EMG amplitude. This is then used as the control input of an AR environment, in which an object is rendered in space. Though this may seem complex the system can be broken down in multiple parts. The hardware can be consolidated into two components: the envelope detector, and Bluetooth. The envelope detector uses the EMG signal as the input. It then recognizes the amplitude variations within the signal and the resulting output is a controlled signal where the peaks are connected giving a smoother non-modulated signal. The EMG signal is obtained from a target muscle group, the demodulated output signal allows the user to clearly see what intended activation looks like, thereby classifying the needed force to move the object. Many different types of amputations were taken into account during the design process, i.e. transtibial (below-knee), transhumeral (above-elbow) and more.

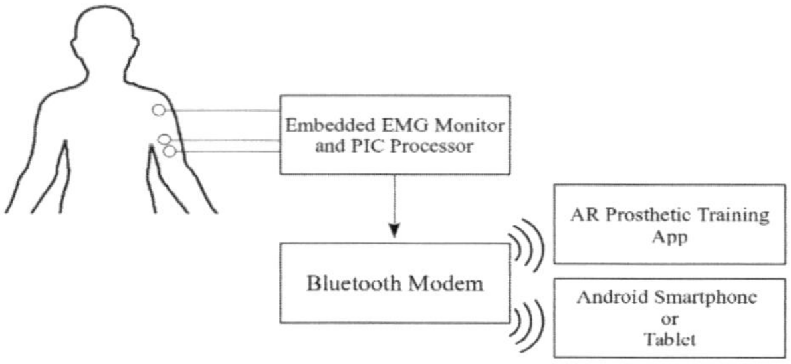

Figure 1. System block diagram.

When the target muscle group was chosen, the type of movement of the object was considered. There is a notable difference between virtual reality and AR. Virtual reality transposes users into a virtual environment, blocking out the real world while enhancing the senses within that space. In comparison, AR, adds elements into the user's reality that "augments" the current state of presence. The overall goal of the application was to have the object move with muscle activation and to accomplish minor tasks. The main innovation of the app is to provide feedback to the user as they are performing the pre-prosthetic exercise. Specifically, the Android-based system would train the user to manipulate an EMG signal to control an object rendered in space.

Design Criteria: As previously mentioned, the design of the system combines several electronic components. The first electronic component is the EMG circuit that detects the impulse signals from the muscle as shown in figure 2. This design opts to use an envelope detector to accurately demodulate the signal and provide a binary way to move the object. The envelope detector consists of a 10μF capacitor, 100kΩ resistor and a 1N914 diode. The cut-off frequency was calculated based on the envelope detector component value which was 0.159 Hertz (Hz), with this cut-off value the power of the signal is still preserved. The complete circuit includes a functioning PIC processor paired with an Android device for configuration and early testing. The second part of the system; the AR-based training app was developed using Unity, an application development software for Android. Unity allowed for the creation of a virtual scene that would present a virtual object to the users. The application has three functions: Bluetooth communication, data processing, and data display.

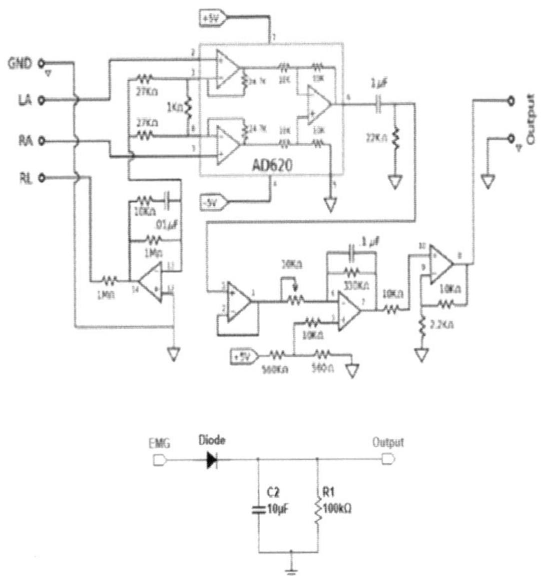

Figure 2. EMG PCB (left) and envelope detector (right).

Results: Using EMG circuit presented in figure 2, we attached leads to the target muscle group; the bicep brachii muscle of the left arm, to model transhumeral amputation. The leads shown in figure 2 as RA and LA, were placed on the muscle and the ground lead (RL), was placed on the deltoid. The current design demonstrates Bluetooth capabilities allowing for plotting of the output of the envelope detector as shown in figure 3.

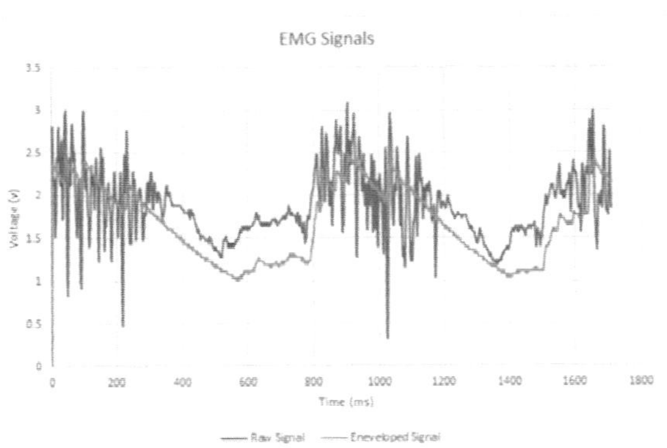

Figure 3: Raw (blue) and enveloped (red) EMG signals

Hardware and software were demonstrated with an android app, which was used to demonstrate in real time how well it was connected to the user. When the image is accepted an object appears and can be moved, rotated, and resized by touch simulation. The android phone connects to the Bluetooth modem. The data was displayed by showing the movement of the object rendered in space. The initial prototype involves using the phone's camera to recognize a designated image target. The image target is used as a way to make the object appear in space shown in Figure 4. The use of Bluetooth allows for training anywhere, from home to the doctor office.

Figure 4: Augment Reality (AR) application

Future Implications: A novel device was proposed to train individuals and as it stands the app works in parts. One part is the augmented world in Unity and the other is the Bluetooth interface in Android Studio. These parts will be integrated once the Bluetooth app is made into a plugin that can be called in Unity. In addition, the results show that the envelope detector can be used to successfully track the intended activation input. The approach is taken only has a degree of freedom (DOF), up or down represented by articulation at the elbow. Future works include adding more target muscles to better model different limb loss and muscle movements. Applying a threshold value to the enveloped signal hardcoded in the software will control the movement of the object in the AR. Therefore, patients can practice EMG signal control of multiple muscles at once. This entire system can be compatible with most low-cost Android devices and will enhance functionality and control of prosthetics.

References:

1. Chadwell, A. The Reality of Myoelectric Prostheses 2016;10, 7.

2. Winslow, B. Game-Based Training for Myoelectric Prosthesis Control 2018; 6: 94.

P3-70

Point of Care Ultrasound for Spine Imaging in Space Explorations

Nina Ninua*, Pragya Hooda*, Daniel Lefever*, Hedy Makris*, and Ilker Hacihaliloglu
Department of Biomedical Engineering, Rutgers University

Introduction: Astronaut's spinal health has been vastly overlooked even though it is known that they experience spinal elongation and back pain. Prolonged space flights have exposed astronauts to low gravity conditions causing the torso to lengthen and puts stress on the intervertebral discs. Microgravity eliminates the need to move in ways that put loads on the lumbar spine therefore causing paraspinal muscle atrophy where 19% of the cross-sectional area is decreased postflight [1]. Point-of-care ultrasound (US) has been investigated as a potential imaging modality for evaluating the changes of the spine and identifying spinal disorder during spaceflight missions [1,2]. One of the disadvantages of US is related to the manual data collection. A single-degree deviation angle by the operator can reduce the signal strength by 50% [3]. Manual data collection is especially challenging while imaging spine due to the complex anatomy and ligaments surrounding the vertebrae [3]. During this work, we propose an automated method for vertebral centerline position estimation. Enhanced US images are cross-correlated to extract the midline location providing a center line location where the transducer should be located during spine imaging. Validation is performed in in vivo US scans.

Methods: The proposed method is based on the enhancement of bone surfaces from US data by extracting local phase-based image features using bandpass quadrature filters [4]. Normalized cross-correlation is calculated from the extracted local phase images in order to automatically detect the midline. The cross-correlation is calculated between the horizontally flipped versions of the images with itself. The absolute maximum correlation coefficient, in the vertical direction, is chosen as the corresponding midline. In order to bias the calculated correlation coefficients to favor good matching between darker pixels over good matching between brighter pixels, we reverse the gray scale values before the correlation calculation. Similar method was used previously in [5] for processing B-mode US data. In order to show the improvements achieved, by using local phase features instead of the B-mode US data, we repeat the same process using B-mode US data. After obtaining approval from the Rutgers University institutional review board (IRB), 33 US spine images were captured from 2 subjects after obtaining written consent. Scans were collected by an expert ultrasound technician by aligning the transducer on the center of the vertebra. The centerline was manually localized from the collected data by the expert technician and compared to the center line extracted using the proposed method by calculating the Euclidean distance.

Results: Investigating Figure 1 we can see that the lowest error results are obtained when using local phase image features. The mean error for centerline extraction when B-mode US data was used was 3.37 mm (SD: 2.76 mm). The mean error for centerline extraction when local-phase image features were used was 1.31 mm (SD: 1.27 mm). Figure 2 represents the GUI interface developed as part of this project.

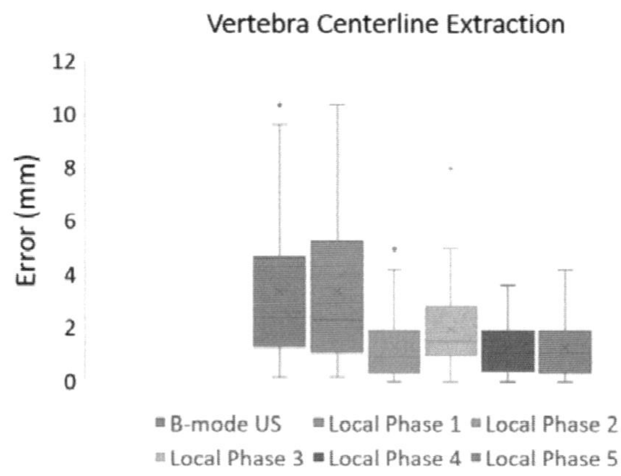

Figure 1. Quantitative results obtained from the proposed vertebra centerline extraction method.

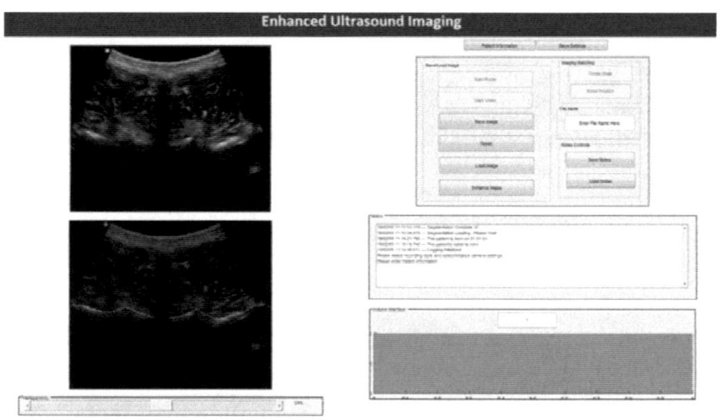

Figure 2. GUI Interface

Conclusions: We have presented an automated method for the extraction of vertebrae center line from in vivo spine US data. Preliminary results show that local-phase feature based centerline extraction outperforms B-mode US image-based centerline extraction ($p<0.05$ using paired t-test). Future work will include improving the accuracy of the methods and validation on extensive data.

References:

1. David A.G., Frontiers in Physiology 8 (2017): 1126.
2. Garcia, K.M, J Ultrasound Med, 37: 987-999.
3. Hacihaliloglu I. Technology. 2017; 5(02):74-80.
4. Hacihaliloglu I., International journal of computer assisted radiology and surgery 12 (6), 2017, 951-960.
5. Toews AR, SPIE Medical Imaging, 2018 Mar 13 (Vol. 10576, p. 1057630).

*Equal contributions.

A Novel Approach to Enhancing Proprioceptive Feedback through the Control of Vibration Motor Frequency

Shreya Anjaria, Brandon Herb, Eric Radbourne, Richard Schermerhorn, Becky Tucci, DPT, Vikki Hazelwood, PhD
Stevens Institute of Technology
Hoboken, NJ 07030, USA

Introduction: Proprioception is the perceptive sense of relative position and movement of joints and neighboring body parts. Impairments to the somatosensory cortex caused by traumatic brain injuries or strokes often result in a lack of proprioceptive feedback, limiting motor capabilities[1]. In order to assist individuals with proprioceptive deficiencies in the lower body, ProprioPRO, a standalone device, has been designed to introduce vibrational stimuli to the legs, providing an external marker for joint position while retraining the proprioceptive pathways. This is accomplished through the stimulation of specific proprioceptors, which can be activated by external vibration within the frequency range of 70-120 Hz[2]. By improving proprioceptive feedback and enhancing sensory maps of the body, ProprioPRO aims to improve the quality of life for patients with proprioceptive deficits, especially those with inclination to participate in sports.

Methods: Nylon/spandex leggings were fitted with multiple pockets to hold the electrical components in place. Vibrational motors were positioned at the knee and ankle in four different locations. A housing enclosure was 3D printed for the various electrical components. An expanded view of the pockets and its components can be found in Figure 1 below.

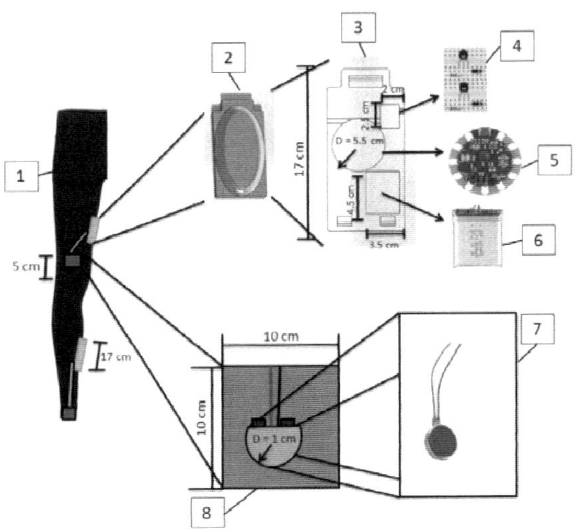

Figure 1: ProprioPRO Leg Sleeve Design
1. Legging, 2. Enclosure Top, 3. Inside of Enclosure, 4. Transistor Circuits, 5. Arduino Lilypad, 6. Battery, 7. Vibration Motor, 8. Motor Pocket

An Arduino Lilypad (Arduino, Somerville, MA) was used to control voltage supply to a vibrational motor (Adafruit, New York City, NY). A Datapower Lithium Ion Battery (SparkFun Electronics, Boulder, CO) supplied power. Attached to the battery was an XZN BMA180 Triple Axis Accelerometer (Bosch Sensortec, Reutlingen, Germany) that was assembled to another Arduino Lilypad, which was powered by and communicated with a laptop computer. An Arduino program was used to record the overall accelerometer output voltage (combined accelerations of x, y, and z directions) and time at a frequency of 1000 Hz was recorded. This time and output voltage was plotted in Excel and a frequency analysis was performed in Matlab. This process was repeated to test voltages between 0 and 5 V at an interval of 0.5 V for a total of 10 trials.

Results:

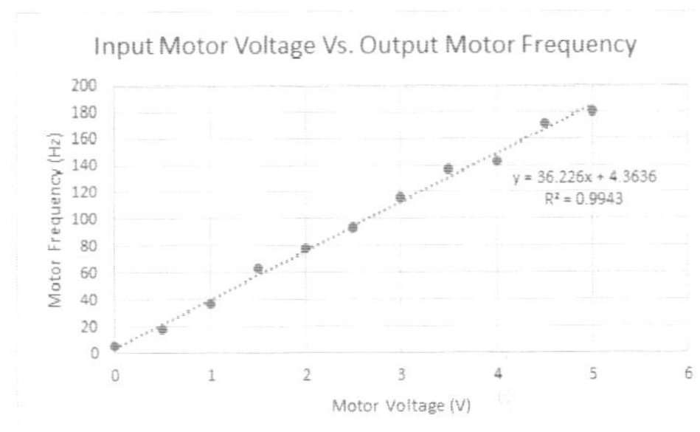

Figure 2: Plot of Input Motor Voltages vs. Output Motor Frequencies

Discussion: The output voltage of the accelerometer followed a sinusoidal pattern. Since the output voltage of the accelerometer is linearly related to acceleration, the Fourier transform of this accelerometer output voltage was taken in order to find the frequency of oscillation of the vibrating motors. This resulted in voltage and frequency values from the 5V input, as well as inputs from 0-5V at 0.5V intervals. When these values were plotted in Figure 2, the relationship between motor voltage and output motor frequency was found to closely follow a linear correlation (coefficient of determination of $R^2 = 0.9943$) of $y = 36.336x + 4.3636$, where y is the motor frequency in Hz, and x is the motor voltage in V. From this equation, the range of input voltages supplied to the motor that fall within the desired motor frequency range of 70-120 Hz was found to be 1.81 - 3.19V. Further testing will focus on the viability of this product in athletic use, such as compression testing and waterproof testing of the 3D printed enclosure. By stimulating the appropriate proprioceptors at the specified frequency range, proprioceptive feedback can be improved, thus improving function and quality of life.

References:

1. Carey, LM. et al. Arch. Phys. Med. Rehab.1993;74:602-611.
2. Montant, R. et al. Human Brain Mapping. 2008.
3. Röijezon, U. et al. Manual Therapy. 2015; 20(3):368-377.

Acknowledgements: We thank advisors Valerie DeAngelo, Deep Parikh and Courtney Mould for technical assistance with this project.

Sound Asleep: A Novel Solution for Active and Passive Noise Control in the Neonatal Intensive Care Unit (NICU)

Rachel Hjerpe, Christian Jensen, Reinout Van Landegem, Paul Moyer, Kristian Quevada, Sally Shady[1], Marwa Khalil[2]
[1]Department of Biomedical Engineering, Stevens Institute of Technology, Hoboken, NJ 07030
[2]Hackensack University Medical Center, Hackensack, NJ 07601

Introduction: Average noise levels in NICU in the United States exceed the American Academy of Pediatrics recommended level by 20 dB[1,2,3]. These excessive noise levels have a detrimental effect on neonatal health and make premature newborns more prone to physiological, behavioral and developmental complications[3]. To address this issue, a noise reduction device was developed to characterize and minimize sound dissipation in NICU incubators. The safety and the efficacy of this device were tested by securing the device to the inside of an incubator analog and generating noise signals at various frequencies. The device then output phase-shifted anti-noise signals, resulting in a net decrease in total noise amplitude via destructive interference. The efficacy of the device was found to be inversely proportional to the frequency of the harmful noise.

Methods: The device is composed of an Arduino Due and two speakers in a 3D-printed ABS plastic housing. The Arduino Due was coded using the compatible and open source Arduino Software Integrated Development Environment (IDE). The device outputs a sinusoidal audio wave at an adjustable frequency to one speaker on the housing. This frequency is used for calculations generating an out-of-phase "anti-noise" signal. The sum of the noise and anti-noise frequencies combine to produce a lower amplitude wave approaching zero via destructive interference. The device and its placement inside of the incubator are displayed below in Figure 1.

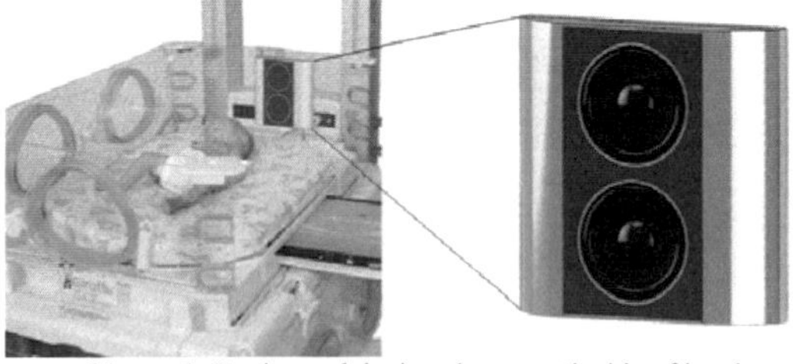

Figure 1. Device and device placement inside of incubator

To quantify the scope of this destructive interference, the Sound Asleep device was secured to the side of an incubator analog. Attenuation capabilities of the device were characterized using a reference microphone at a fixed distance from the device. First, the device was programmed to emit a specific noise frequency. Then, the device was programmed to emit the same noise frequency along with a phase-shifted version of the original frequency. Noise levels both before and after the addition of the phase-shifted signal were recorded by the microphone in terms of signal amplitude. Separate trials were conducted at output frequencies of 25, 45, 65, 250, 500, and 1500 Hz because these frequencies were found to be the most prevalent noise frequencies in the Hackensack NICU.

Results: An amplitude reduction of 20 dB is anticipated for low frequency waveforms. Figure 2 displays the expected amplitudes of noise recorded prior to and following the output of a phase-shifted signal.

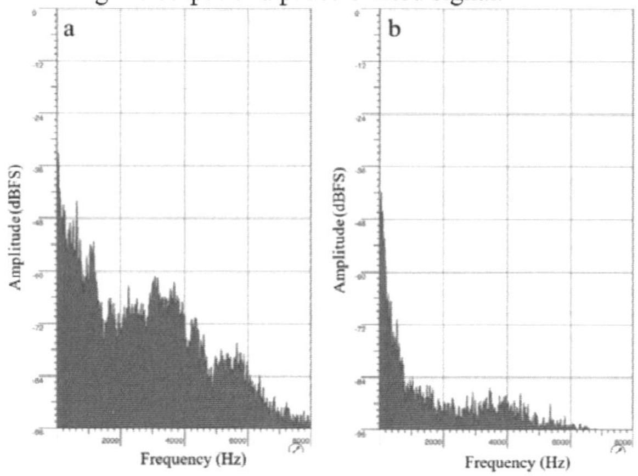

Figure 2. Sound amplitude detected by the reference microphone (a) before and (b) after the Sound Asleep device activation

Conclusions: A reduction in net noise following the emission of a phase-shifted signal is anticipated at various harmful frequency levels commonly heard in the NICU. These results would indicate that active noise cancellation is a safe and effective mode of reducing these loud noises. Future iterations of this technology will implement adaptive filtering technology and a Field Programmable Gate Array to emit signals capable of dynamic, real-time attenuation of the various frequencies found in the NICU. Implementation of this technology in the Neonatal Intensive Care Unit will assuage the short and long term physiological and developmental consequences of infant exposure to excessive noise.

References:

1. Almadhoob A. Cochrane. 2015;1.
2. Valizadeh S. J Caring Sci. 2013;2(1):19-26.
3. Pediatrics. 1997;100(4):724-727.

Acknowledgements:

The Sound Asleep Development Team would like to acknowledge Dr. Vikki Hazelwood, Val DeAngelo and Deep Parikh of the Department of Biomedical Engineering at Stevens Institute of Technology.

Northeast Bioengineering Conference 2019
c/o Ronke Olabisi
3408 Eng Hall / UC Irvine
Irvine, CA, 92697

ISBN 978-1-7138-1407-8

CURRAN ASSOCIATES INC.
proceedings.com